Errata

p. 28, line 1 . . . $\times 10^{-a}$ should read . . . $\times 10^{-az}$.

p. 29, line 27 ‘fern line’ should read ‘firn line’.

p. 53, Table 2.14 1st column heading $\overset{(-)}{\leqslant} 4.3$ should read $\leqslant -4.3$.

p. 61, 2 lines up. For $\dfrac{V^{22} - V^{21}}{2}$ read $\dfrac{V_2^2 - V_1^2}{2}$

p. 65, line 6 up $U^{20}/2N^2$ should read $U_0{}^2/2N^2$.

line 13 up $+U^{20}$ should read $+U_0{}^2$.

p. 79, line 18 $S'_d = \left[\left(\frac{Sb}{S_0}\right) \cdot \frac{\cos i}{\cos Z} S'_d\right]$

should read $S'_d = \left[\left(\frac{S_b}{S_0} \cdot \frac{\cos i}{\cos Z}\right) S_d\right]$

p.131, 11 lines up $= 0.5\bar{U}$ should read $\lambda = 0.5\bar{U}$

p.149, line 6 ‘rohe Föhnwelle’ should read ‘hohe Föhnwelle’.

p.156, line 9. For Seinocco and Pettier read Scinocco and Peltier.

p.164, 12 lines up. $U\mathrm{h} = \int \mathrm{HO}\, udn$ should read $U\mathrm{h} = \int_0^{\mathrm{H}} udn$.

p.209, Table 4.2. R_n column: Quelccaya Ice Cap should read −2 and Mt. Logan −57.

MOUNTAIN WEATHER AND CLIMATE
2nd edition

Mountains and high plateau areas account for 20 per cent of the Earth's land surface. They give rise to a wide range of meteorological phenomena and distinctive climatic characteristics of consequence for ecology, forestry, glaciology and hydrology. *Mountain Weather and Climate* remains the only comprehensive text describing and explaining mountain weather and climate processes. It presents the results of a broad range of studies drawn from across the world.

Following an introductory survey of the historical aspects of mountain meteorology, three chapters deal with the latitudinal, altitudinal and topographic controls of meteorological elements in mountains, circulation systems related to orography, and the climatic characteristics of mountains. The author supplies regional case studies of selected mountain climates from New Guinea to the Yukon, a chapter on bioclimatology that examines human bioclimatology, weather hazards and air pollution, and a concluding chapter on the evidence for and the significance of changes in mountain climates.

Since the first edition of this book appeared a decade ago several important field programmes have been conducted in mountain areas. Notable among these have been the European Alpine Experiment and related investigations of local winds, studies of air drainage in complex terrain in the western United States and field laboratory experiments on air flow over low hills. Results from these investigations and other research are incorporated in this new edition and all relevant new literature is referenced.

Roger G. Barry is Professor of Geography at the University of Colorado and Director, World Data Center-A for Glaciology.

A VOLUME IN THE
ROUTLEDGE PHYSICAL ENVIRONMENT SERIES
Edited by Keith Richards
University of Cambridge

The Routledge Physical Environment series presents authoritative reviews of significant issues in physical geography and the environmental sciences. The series aims to become a complete text library, covering physical themes, specific environments, environmental change, policy and management as well as developments in methodology, techniques and philosophy.

Other titles in the series:

ENVIRONMENTAL HAZARDS: ASSESSING RISK AND REDUCING DISASTER
K. Smith

WATER RESOURCES IN THE ARID REALM
E. Anderson and C. Agnew

ICE AGE EARTH: LATE QUATERNARY GEOLOGY AND CLIMATE
A. Dawson

Forthcoming:

THE GEOMORPHOLOGY OF DESERT DUNES
N. Lancaster

GLACIATED LANDSCAPES
M. Sharp

HUMID TROPICAL ENVIRONMENTS AND LANDSCAPES
R. Walsh

SOILS AND ENVIRONMENT
S. Ellis and A. Mellor

PROCESS, ENVIRONMENT AND LANDFORMS: APPROACHES TO GEOMORPHOLOGY
K. Richards

MOUNTAIN WEATHER AND CLIMATE

2ND EDITION

Roger G. Barry

London and New York

First published 1992
by Routledge
11 New Fetter Lane, London EC4P 4EE

Simultaneously published in the USA and Canada
by Routledge
a division of Routledge, Chapman and Hall, Inc.
29 West 35th Street, New York, NY 10001

Typeset in Scantext September by
Leaper & Gard Ltd, Bristol
Printed and bound in Great Britain by
Biddles Ltd, Guildford and King's Lynn

British Library Cataloguing in Publication Data
Barry, R.G. (Roger Graham) *1935–*
Mountain weather and climate – 2nd ed.
1. Mountains. Climate
I. Title II. Series
551.69143

ISBN 0–415–07112–7
ISBN 0–415–07113–5 pbk

Library of Congress Cataloging in Publication Data
Barry, Roger Graham.
Mountain weather and climate / Roger G. Barry. – 2nd ed.
p. cm. – (Routledge physical environment series)
Includes bibliographical references and index.
ISBN 0–415–07112–7 – ISBN 0–415–07113–5
1. Mountain climate. I. Title. II.Series.
QC993.6.B37 1992
551.6914′3–dc20 91–14579
CIP

CONTENTS

PLATES

FIGURES

TABLES

PREFACE

It is remarkable that, despite almost a century of research into mountain weather and climate, no comprehensive book on the subject has previously been written. After an early upsurge of work at mountain observatories in Britain and North America, interest waned and the field has been mainly left to the European alpine countries. Consequently, a large body of meteorological literature is virtually unknown to most English-speaking scientists. Although W.M. Davis writing on the subject of Mountain Meteorology in 1887 was able to draw on current work published in German by Julius Hann, few researchers in the English-speaking world today have similar linguistic ability.

My interest in the mountain environment can be traced to boyhood hikes in the Pennines of northern England and later in Snowdonia and the Lake District. Serious interest in the subject began with my arrival at the Institute of Arctic and Alpine Research, University of Colorado, in 1968. A good climatic record already existed at several Institute-operated stations in the Front Range, but little meteorological interpretation of the data had been undertaken. While my major research focused on problems in Arctic climate and climatic change, a long-term ambition to develop a better understanding of mountain climates gradually took shape. This received a stimulus in 1975 when I was able to work for two weeks on Mt. Wilhelm in Papua New Guinea and pay a longer visit to Australian National University, Department of Biogeography and Geomorphology, examining the mountain climates of New Guinea.

In North America, meteorological interest in mountains has concentrated heavily on theoretical problems of airflow characteristics and recently on orographic cloud seeding. The foreign literature shows a wider range of concerns. I have not attempted to provide a theoretical treatment, which for many topics would still be rather premature. Instead, I have sought to provide a text useful for specialist courses in climatology as well as for scientists in related disciplines – botany, forestry, glaciology and hydrology – with interest in mountain-related phenomena. Factors of latitude, altitude and topography controlling the meteorological elements are discussed first, then atmospheric circulations related to orography are described. The following chapter deals with climatic characteristics which are affected by circulation systems as well as by the basic controls, or other complex interactions. A selection of case studies representing mountain climates in different latitudinal zones is then given. Bioclimatic topics are considered next, and finally changes in mountain climates are briefly discussed.

The help of many individuals has made it possible for me to complete the undertaking. However, I should especially like to thank my colleague Dr Jack Ives, Director of INSTAAR throughout the time of this work, for interesting me in the topic in the first place; Drs A. Brazell, R.F. Grover, J. Hay, M. Marcus, and U. Radok for valuable comments and suggestions on sections of the text; numerous friends and colleagues who supplied me with copies of their publications; but particularly Dr F. Lauscher in Vienna and Dr F. Fliri in Innsbruck; Marilyn Joel for drafting many of the diagrams and Nancy Hensal and John Adams for additional help in the final stages; Laura Koch and Margaret Strauch for shouldering most of the typing; and Janice Price and Mary Ann Kernan at Methuen for their patience and editorial help. Finally, of course, without the support and understanding of my family for many hours spent on the task it would still be an idea; to them – Valerie, Rachel and Christina – the book is dedicated.

Roger G. Barry
University of Colorado
Boulder
August 1980

PREFACE TO THE SECOND EDITION

In the decade since this book appeared there has been substantial research interest in mountain-related weather phenomena. An extensive literature on lee cyclogenesis and local winds has already resulted from the ALPEX programme in the Alps, and the deployment of new remote-sensing systems has added valuable observational data. Two Sino–American symposia have focused on mountain meteorology and there have been other national or regional conferences in North America and Europe. There is an emerging literature for various mountain ranges of the world – the Andes, the New Zealand Alps, and the mountains of Central Asia, for example – but these papers mostly lack the collective scope needed to add additional regional case studies.

The basic structure of the book is unchanged, but the opportunity has been taken to make a variety of additions, and corrections, in response to reviews of the first edition. The references have also been updated and an author index added to make them easier to locate.

I wish to thank Tristan Palmer of Routledge for his patience, Margaret Strauch for her word-processing expertise, Ann Brennan for editing assistance, Maria Neary for drafting, and Christina Barry for her care in compiling the author index. The library facilities at the Institute of Geography, ETH, and the Swiss Meteorological Agency, in Zurich, provided an invaluable resource during my completion of the revisions. My thanks are due to the Institute of Geography, ETH, for the award of a Visiting Professorship, during part of a sabbatical leave from the University of Colorado, which supported the preparation of this new edition.

Roger G. Barry
Cooperative Institute for
Research in Environmental Sciences,
and Department of Geography
University of Colorado, Boulder
September 1990

ACKNOWLEDGEMENTS

The following individuals and publishers have kindly granted permission for the reproduction of illustrations and other material:

American Meteorological Society – Figures 2.19, 2.20, 3.8, 3.13, 3.16, 3.20, 3.23, 3.26, 3.27, 3.28, 4.5, 4.21, 6.3; Dr A.H. Auer, Jr. New Zealand Meteorological Service – Plate 4; Dr I. Auer – Figure 7.1; Professor D. Boyer – Plates 3, 5, 6; Dr W.A.R. Brinkmann – Figures 3.15, 5.9; Robert Bumpas (NCAR Photographs) – Plate 8; Deutscher Wetterdienst – Plates 13A, B; Professor Dr I. Dirmhirn – Tables 2.6–2.8, 2.10, 2.11; Ellis Horwood Ltd – Figure 2.28; Professor Dr F. Fliri – Figures 2.12, 5.10, 5.11; Professor Dr H. Flohn – Figure 2.21; *Geografiska Annaler* – Figures 5.2, 5.5; Gordon and Breach Science Publishers – Figures 3.1, 3.2; Her Majesty's Stationery Office – Figures 2.14, 2.15, 2.16; Institute of British Geographers – Figure 2.16; *International Journal of Biometeorology* (Swets and Zeitlinger BV) – Figure 6.1; Dr G. Kiladis – Plates 2, 11; Kluwer Academic Publishers – Figure 2.26; Professor Dr W. Lauer – Figure 4.14; Professor Dr F. Lauscher – Figures 2.3, 2.17, 4.13, 4.15, 4.16; Professor H.H. Lettau – Figure 3.30; Dr R.R. Long – Figure 3.10; Metheun & Co. Ltd – Figures 1.1, 1.2, 2.21, 2.29, 4.8, 4.17, 4.20, 4.22, 5.7; MTP Press Ltd – Figure 6.2; National Snow and Ice Data Center – Plate 7; Dr W.D. Neff – Plate 10; Norwegian Defence Research Establishment – Figure 3.21; Oldenbourg Verlag – 4.28; D. Reidel Publishing Co. – Figure 4.6; Royal Meteorological Society – Figures 3.6, 3.9, 4.9, 4.11, 4.17; Royal Society, London – Plate 6; Smithsonian Institution Press – Figure 2.1; Société Météorologique de France – Figure 2.38; Sonnblick Verein – Plate 1; Springer Verlag – Figures 3.18, 3.22, 3.24, 4.1, 4.16, 5.12; Plates 3, 5; Dr K. Steffen – Plates 9, 12, 14, 15; Dr F. Steinhauser – Figures 7.2, 7.3; Swedish Geophysical Society (*Tellus*) – Figure 3.4; Dr A.S. Thom – Figure 5.15; Dr H. Turner – Figures 2.35, 2.39, 2.4; Professor P.D. Tyson – Figure 3.29; University of Colorado Press – Figure 5.16; Dr E. Wahl – Figures 2.23, 2.24; Dr E.R. Walker – Figures 4.19, 4.20; Dr H. Wanner – Figure 5.16; Dr E. Wessely – Figure 2.11; World Meteorological Organization – Figures 2.30, 3.7, 4.23; Professor M.M. Yoshino – Figures 2.34, 3.14, 3.19.

1

MOUNTAINS AND THEIR CLIMATOLOGICAL STUDY

INTRODUCTION

The mountain environment has always been regarded with awe. The Greeks believed Mount Olympus to be the abode of the gods, to the Norse the Jötunheim was the home of the Jotuns, or ice giants, while to the Tibetans, Mount Everest (Chomo Longmu) is the 'goddess of the snows'. Deities have been associated with many conspicuous peaks. Sengem Sama with Fujiyama (3778 m) in Japan and Shiva-Parvati with Kailas (6713 m) in Tibet, although at other times mountains have been identified with malevolent spirits, the Diablerets in the Swiss Valais, for example. This dualism perhaps reflects the opposites of tranquility and danger encountered at different times in the mountain environment. Climatological features of mountains, especially their associated cloud forms, are represented in many names and local expressions. On seeing the distant ranges of New Zealand, the ancestral Maoris named the land Aotearoa, 'the long white cloud'. Table Mountain, South Africa, is well known for the 'tablecloth' cloud which frequently caps it. Wind systems associated with mountains have also given rise to special names now widely applied, such as föhn, chinook and bora, and others still used only locally.

Today, the majestic scenery of mountain regions makes them prime recreation and wilderness country. Such areas provide major gathering grounds for water supplies for consumption and for hydroelectric power generation, they are often major forest reserves, as well as sometimes containing valuable mineral resources. Mountain weather is often severe, even in summer, presenting risks to the unwary visitor and, in high mountains, altitude effects can cause serious physiological conditions. Despite their significance in such respects, and the fact that mountain ranges account for 20 per cent of the earth's land surface, the meteorology of most mountain areas is little known. Weather stations are few and tend to be located at conveniently accessible sites, often in valleys, rather than at points selected with a view to obtaining representative data.

Climatic studies in mountain areas have frequently been carried out by biologists concerned with particular ecological problems, or by hydrologists interested in snow runoff, rather than by meteorologists. Consequently, much of the information that does exist tends to be widely scattered in the scientific literature and it is often viewed only in the context of a particular local problem.

It is the aim of this book to bring together the major strands of our existing knowledge of weather and climate in the mountains. The first part of the book deals with the basic controls of the climatic and meteorological phenomena and the second part with particular applications of mountain climatology and meteorology. By illustrating the general climatic principles, a basis can also be provided for estimating the range of conditions likely to be experienced in mountain areas of sparse observational data.

CHARACTERISTICS OF MOUNTAIN AREAS

Definitions of mountain areas are unavoidably arbitrary. Usually no qualitative, or even quantitative, distinction is made between mountains and hills. Common usage in North America suggests that 600 metres or more of local relief distinguishes mountains from hills (Thompson 1964). Such an altitudinal range is sufficient to cause vertical differentiation of climatic elements and vegetation cover. Finch and Trewartha (1949) propose that a relief of 1800 m can serve as the criterion for mountains of 'Sierran type'. Such a range of relief also implies the presence of steep slopes. In an attempt to provide a rational basis for definition, Troll (1973) delimits *high mountains* by reference to particular landscape features. The most significant ones are the upper timberline, the snow line during the Pleistocene epoch (which gave rise to distinctive glacial

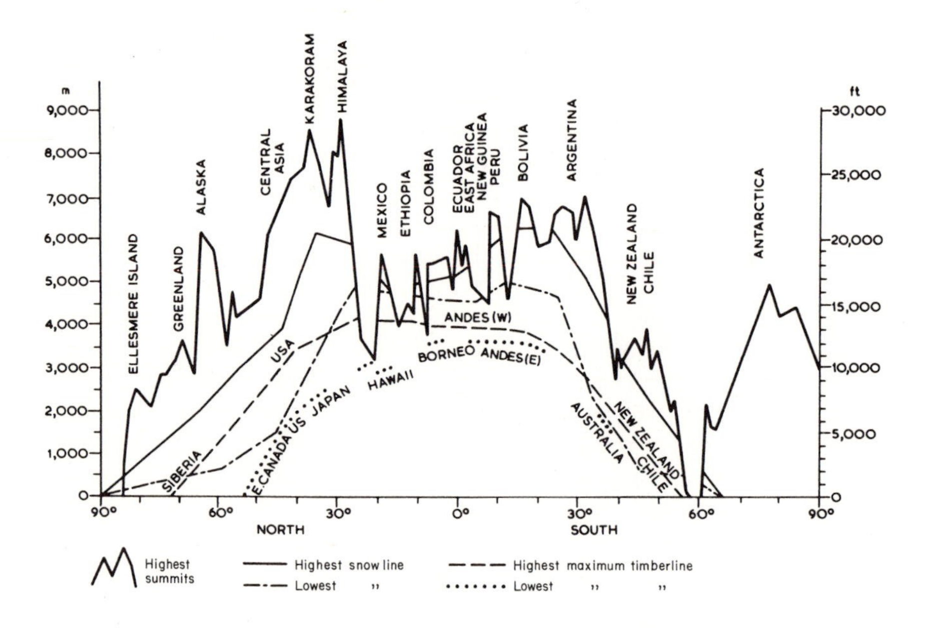

Figure 1.1 Latitudinal cross-section of the highest summits, highest and lowest snow lines, and highest and lowest upper limits of timber line
Source: From Barry and Ives 1974

landforms) and the lower limit of periglacial processes (solifluction, etc.). It is apparent that each of these features is related to the effects of past or present climate and to microclimatic conditions at or near ground level.

On the basis of Troll's criteria, the lower limit of the high mountain belt occurs at elevations of a few hundred metres above sea level in northern Scandinavia, 1600–1700 m in central Europe, about 3300 m in the Rocky Mountains at 40°N, and 4500 m in the equatorial cordillera of South America (see Figure 1.1). In arid central Asia, where trees are absent and the snow line rises to above 5500 m, the only feasible criterion remaining is that of relief.

Troll's approach derives from the German distinction between *Hochgebirge* (high mountains), such as the Alps and Tatra, and the lower and gentler ridges of the *Mittelgebirge* which include the Riesengebirge and Vosges. It is not altogether suitable from the climatological standpoint since, although the altitudinal limit varies with latitude in such a way as to define *alpine* areas and their biota (cf. Barry and Ives 1974), it is the elevational and slope effect which causes many of the special features of mountain climates.

The treatment in this book emphasizes high mountain effects, due to altitude, although since airflow modifications that arise at even modest topographic barriers cause important differences between upland and lowland climates, such effects are also discussed.

The land surface occupied by mountains has recently been recalculated by Louis (1975). He estimates the areas of mountain ranges and plateaux as shown in Table 1.1.

The locations of the major mountain ranges and highland areas of the world and their climatic zonation are shown in Figure 1.2. The most latitudinally extensive mountain chains are the cordilleras of western North and South America. The most extensive east–west ranges are the Himalaya and adjoining ranges of central Asia. Reference should also be made to the vast highland plateau exceeding 3000 m in Tibet and the even larger ice plateaux of Greenland and Antarctica. All of these regions have major significance for weather and

Table 1.1 The global area of mountains and high plateaux

	Mountains	*Plateaux (10^6 Km^2)*	*Mountains/land surface (%)*[a]
3000 m[b]	—6—		4.0
2000–3000 m	4	6	2.7
1000–2000 m	5	19	3.4
0–1000 m	15	92	10.1
Total	30	117	20.2

Notes: [a]The total land surface is about 149 million km^2, oceanic islands covering 2 million km^2 are not included in the listed areas. [b]All land above 3000 m
Source: After Louis, 1975

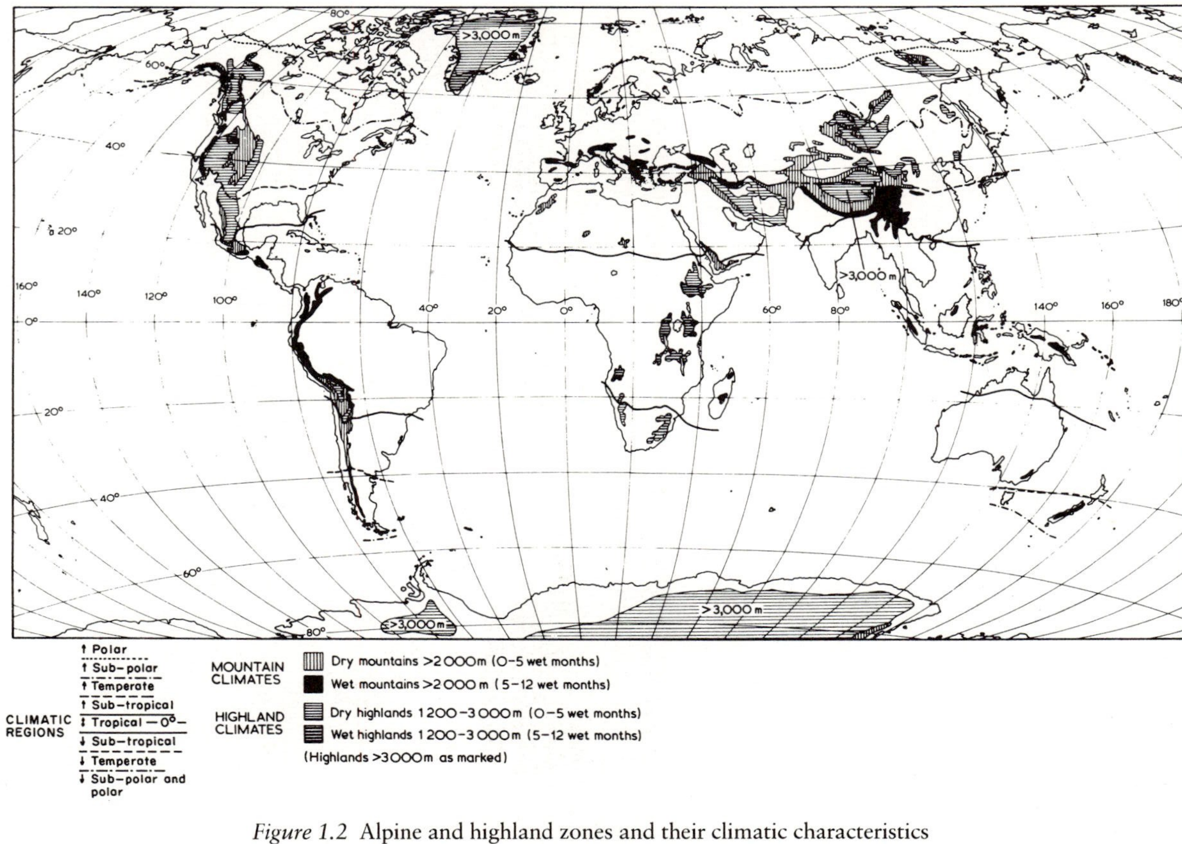

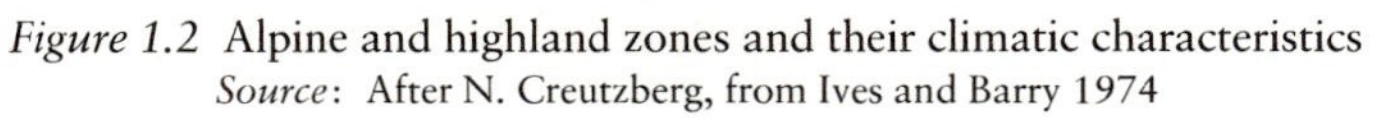

Figure 1.2 Alpine and highland zones and their climatic characteristics
Source: After N. Creutzberg, from Ives and Barry 1974

climate at scales up to that of the general circulation of the atmosphere. In contrast, major, but isolated, volcanic peaks which occur in east Africa and elsewhere, have their own distinctive effects on local weather and climate.

HISTORY OF RESEARCH INTO MOUNTAIN WEATHER AND CLIMATE

Intensive scientific study of mountain weather conditions did not begin until the mid-nineteenth century although awareness of the changes in meteorological elements with altitude came much earlier. The effect of altitude on pressure was proved in September 1648 when Florin Périer, at the request of his brother-in-law Blaise Pascal, operated a simple Torricelian mercury tube at the summit and base of the Puy de Dôme in France. In August 1787, H. B. de Saussure (1796), who was a keen mountaineer, made observations of relative humidity during an ascent of Mont Blanc using the hair hygrometer which he had developed. In July 1788, he and his son maintained two-hourly meteorological observations on the Col du Géant (3360 m) near Mont Blanc while comparative observations were made at Chamonix (1050 m) and Geneva (375 m). From these data he was able to study temperature lapse rate and its diurnal variation, obtaining an estimate

Plate 1 The Sonnblick Observatory (3106 m) in the Austrian Alps viewed from the east. The new facility was opened in 1985 (courtesy of the Sonnblick Verein).

Table 1.2 Principal mountain observatories

Name	*Country/state*	*Location*	*Elevation*[a] *(m)*	*Records*	*References*
Asia					
Mt. Fuji	Japan	35°21′N, 138°44′E	3716	1888–1931 (summers) 1932 →	Fujimara, 1971[b]; Solomon, 1979
O Mei Shan	China	29°28′N, 103°41′E	3383	1932–1933	Lauscher, 1979b
Europe					
Fanaräken	Norway	61°31′N, 7°E	2062	1932→ [d]	Spinnangr and Eide, 1948; Manley, 1949[b]
Ben Nevis	Scotland	56°48′N, 5°0′W	1343	1893–1904	Buchan and Omond, 1890–1910[b] v Hann, 1912; Roy, 1983[b]
Brocken	Germany	51°48′N, 10°37′E	1142	1895→ [d]	
Fichtelberg	Germany	50°26′N, 12°57′E	1213	1891→	Pleiss, 1961[b]
Sniezka (Schneekoppe)	Poland	50°44′N, 15°44′E	1603	1881→ [d]	Hellman, 1916
Hohenpeissenberg	Germany	47°48′N, 11°01′E	989	1781→ [c]	Grunow *et al*., 1957[b]; Lauscher, 1981
Zugspitze	Germany	47°25′N, 10°59′E	2962	1900→ [c]	Hauer, 1950[b]
Sonnblick	Austria	47°03′N, 12°57′E	3106	1886→ [c]	*Jahresbericht des Sonnblick – Vereines* (1892–); Steinhauser, 1938[b]; Böhm, 1986
Hoch Obir	Austria	46°30′N, 14°29′E	2044	1847–1943[c]	Lukesch, 1952
Säntis	Switzerland	47°15′N, 9°20′E	2500	1882→ [c]	
Jungfraujoch	Switzerland	46°33′N, 7°58′E	3577	1923→ [c]	Maurer and Lütschg, 1931[b]
Davos Weissfluhjoch	Switzerland	46°50′N, 9°49′E	2540	1936→ [d]	*Winterberichte*, 1950→; Zingg, 1961
Bjelasnica	Yugoslavia	43°42′N 18°15′E	2067	1895–1915	Forster *et al*., 1919
Mont Blanc	France	45°50′N, 6°52′E	4359	1887–1893 (summers)	Vallot, 1893–98; Hann, 1899 Tutton, 1925[b]
Pic du Midi de Bigorre	France	42°56′N, 0°08′E	2860	1881→ [d]	Klengel, 1894; Bücher and Bücher, 1973
Puy de Dôme	France	45°47′N, 2°57′E	1467	1878→	Woeikof, 1892

Mt. Etna	Italy	37°44′N, 15°0′E	2950	1892–1906	Obermayer, 1908
Izana	Teneriffe	28°18′N, 16°30′W	2367	1915→ [c]	Tzchirner, 1925; Lauscher, 1975
Mount Olympus	Greece	40°03′N, 22°21′E	2817	1963→ (summers)	Livadas, 1963 Kyriazopoulos, 1966
North America					
Mount Washington	New Hampshire	44°16′N, 71°18′W	1915	1870–92; 1932→	Stone, 1934[b]; Smith, 1964, 1982; Leitch, 1978; Mount Washington Observatory, 1959
Pike's Peak	Colorado	38°50′N, 105°02′W	4311	1874–88; 1892–94	US Army, Chief Signal Officer, 1889
Lick Observatory (Mount Hamilton)	California	37°20′N, 121°38′W	1283	1880→	Reed, 1914[b]
Mauna Loa	Hawaii	19°32′N, 155°35′W	3399	1959→	Price and Pales, 1963; Miller, 1978[b]
South America					
El Misti	Peru	16°19′S, 71°23′W	5822	1893–95	Bailey, 1908[b]
Corrido de Cori	Argentina	25°06′S, 68°20′W	5100	1942	Miller, 1976
Cristo Redentor	Argentina	32°50′S, 70°05′W	3800	1935→ [c]	Prohaska, 1957
Collahuasi	Chile	21°0′S, 68°45′W	4810	1914–15	Lauscher, 1979a
Chuquicamata	Chile	21°07′S, 68°31′W	2710	1914–15	Lauscher, 1979a

Notes: [a]The largest published station reference height is given where possible. Elevations may differ by a few metres in different sources; sometimes this is due to reference to the corrected barometer height.
[b]These references generally describe the site and the history of the installation. They also include data tabulations in most cases.
[c]Annual values for these stations are contained in World Weather Records (Clayton 1944, 1947; US Dept of Commerce 1959, 1966, 1968)
[d]Annual values for these stations are contained in Meteorological Office (1973) which also lists sources (in most cases for the period 1931–60).

close to that of Julius von Hann a century later. His writings discussed eighteenth-century theories of the reason for low temperatures in the mountain and he came closer to modern views than most physicists of his day (Barry 1978). De Saussure also attempted to measure altitudinal variations of evaporation and sky colour and was fascinated by numerous other mountain weather phenomena and man's response to high altitude conditions. He can rightfully be regarded as the 'first mountain meteorologist'.

In the 1850s, meteorological measurements begin to be made systematically on high mountains, often in association with astronomical studies, as on the Peak of Tenerife (Canary Islands) (Smyth 1859). In the United States, the earliest extensive observations were those made in the summers of 1853 to 1859 on Mount Washington, New Hampshire (1915 m) (see Stone 1934). This was soon followed by the establishment of observatories by the US Signal Service on Mount Washington in 1870 and on Pike's Peak, Colorado (4311 m) in 1874 (Rotch 1892). Observations were also made on Mount Mitchell, North Carolina (2037 m) in the summer of 1873 (Howgate and Sackett 1873). In Europe, similar developments took place following a suggestion made by J. von Hann at the second International Meteorological Congress in Rome, 1879. Observatories were established by the major European countries (Rotch 1886; Roschkott 1934), particularly in the Alps, where many of these stations are still operating. The impressive location of the Sonnblick Observatory (Böhm 1986) is shown in Plate 1. A summary of the location and operation of the major observatories/ weather stations is given in Table 1.2.

After the initial enthusiasm for mountain weather data in the United States, several problems led to a decline in interest in maintaining mountain observatories (Stone 1934). On the technical side, the telegraph lines were hard to maintain, while a suitable basis for incorporating the data into the synoptic weather-map analyses, then based almost solely on surface weather observations, did not exist. Both Mount Washington and Pike's Peak were closed by the Weather Bureau in the 1890s, and in Scotland the same fate befell the Ben Nevis Observatory in 1904, due to a lack of funds (Roy, 1983). The value of such observatories in connection with upper air studies was raised again in the 1930s when aerological networks were first being established (Bjerknes *et al.* 1934). Mount Evans, Colorado, for example, was used as a site for ozone measurements and new determinations of ultraviolet radiation (Stair and Hand 1939). Mountain stations can operate in any weather conditions and collect data for 24 hours of the day, whereas soundings are made only twice per day and may be restricted by weather conditions. Partly as a result of such concerns, the Mount Washington Observatory was re-established during the International Polar Year and continues in operation (Smith, 1964; 1982). The only recent development is the establishment of Mauna Loa Observatory (Price and Pales 1963), which has assumed major importance as a bench mark monitoring station for solar radiation and atmospheric gases. The Zugspitze Observatory in Germany has served as a base for aerosol, atmospheric electricity and radioactivity studies

(Reiter 1964) and the Weissfluhjoch in Switzerland for snow research (*Winterberichte*, 1950).

In other areas of the world, mountain weather data were primarily collected by survey parties, such as those in the Himalaya (Hill 1881), or by expeditions like those from Harvard University to the Peruvian Andes between 1893 and 1895 (Bailey 1908). There were many such scientific expeditions, some specifically for meteorological purposes. These included early attempts to determine the extraterrestrial solar radiation (see p. 33). In addition, climatic records have been collected from many second-order or auxiliary stations in mountain regions around the world. For long-term records (with published climatic mean data for 1931–60), there are about twenty stations around the world located above 2000 m according to Lauscher (1973). However, some of these are situated on high plateaux, in mountain passes, or in high valleys. There are also more than 200 high stations with shorter periods of record.

The best-known mountain ranges, in meteorological terms, are undoubtedly the European Alps (e.g. Bénévent 1926; Fliri 1974, 1975), while the least known are the mountain systems of central Asia and the Andes. In recent years, meteorological research has been carried out in areas as diverse as the Caucasus, the Mt St Elias Range (Yukon) and Mt Wilhelm (Papua New Guinea). Meetings on alpine meteorology have been held biennially in Europe since 1950 (Lauscher 1963; Obrebska-Starkel 1983, 1990) and other symposia have focused on special topics of mountain weather and climate (Reiter and Rasmussen 1967; World Meteorological Organization 1972; Reiter *et al.* 1984). The intense interest in airflow phenomena in and around the Alps – lee cyclogenesis, local winds, mountain drag, and differential heating effects – led to the design of an Alpine Experiment (ALPEX) as part of the Global Atmospheric Research Programme (Kuettner 1982; Smith 1986) and similar interest in the influence of the Tibetan Plateau has developed (Xu 1986).

THE STUDY OF MOUNTAIN WEATHER AND CLIMATE

The study of mountain weather and climate is hampered in three respects. First, many mountain areas are remote from major centres of human activity and tend, therefore, to be neglected by scientists. This problem is compounded by the difficulty of physical access, inhibiting the installation and maintenance of weather stations. Second, the nature of mountain terrain sets up such variety of local weather conditions that any station is likely to be representative of only a limited range of sites. Third, there are serious difficulties to be faced in making standard weather observations at mountain stations. Some aspects of the last two problems are worth elaborating.

The conventional approach to climatic description involves long-term observation records spanning 30 years or more at a site selected to represent the 'regional' environment. Stations are usually located in open terrain, away from the influence of buildings and other obstacles to airflow. In the mountains, however, we are dealing with at least three types of situation – summit, slope, and valley bottom – apart from considerations of slope orientation, slope angle, topographic screening, and irregularities of small-scale relief. These factors necessitate either a very dense network of stations or some other approach to determining mountain climate. In the future, the use of ground-based and satellite remote sensing combined with intensive case studies of particular phenomena may provide the best solution. The measurement problem is due to the generally severe nature of mountain weather with, in many localities, frequent strong winds and a high proportion of the precipitation occurring as snow. In this respect the problems are analogous to those encountered in polar regions, but special problems may arise, for example, due to the frequent occurrence of cloud at station level. Local topography may also cause observations to be unrepresentative, at least for certain elements and for particular wind directions. This question has been debated in the case of the Zugspitze Observatory by Küttner (1949) and Reichel (1949).

In most climate classifications, mountain areas are ignored, or regarded as undifferentiated 'highland'. Vegetation and climate zones on the east slope of the Colorado Front Range are analysed by Greenland *et al.* (1985). They show that standard climate classifications are of limited applicability. For example, Niwot Ridge (3749 m) is variously classed as polar tundra (Köppen), perhumid tundra (Thornthwaite), or arid subpolar/alpine (Holdridge). Greenland *et al.* suggest that more appropriate bioclimatic variables are summer mean temperature, growing season soil moisture deficit, and the ratio of growing season thawing degree days to growing season precipitation. All three indices show consistent altitudinal gradients. For New Zealand mountains, Coulter (1973) proposes a threefold classification – (i) high annual precipitation (>250 cm) with long periods of cloud, fog and precipitation as on the west coast of South Island or the North Island volcanoes; (ii) low annual precipitation (100–150 cm), high wind speeds, and low temperatures with short frost-free periods, as at higher altitudes of central Otago; (iii) moderate annual precipitation, higher temperatures than category (ii) and periods of strong drying conditions, as on the eastern ranges of both islands (see also McCracken 1980). On a global scale, additional categories are undoubtedly necessary. It seems that key variables to be considered should include incoming solar radiation, mean temperatures of the warmest and coldest months, duration of the frost-free season and growing season, annual precipitation, snow cover duration, and wind speed.

A description of mountain climate rests on two considerations. What is the purpose of any particular characterization? What can feasibly be investigated in terms of the available technology and resources, including existing knowledge? In the present context, the characterization of mountain climates is in terms of the special meso- and micro-scale meteorological phenomena that occur in mountain areas. In other words, the concern is with the ways in which mountains give rise to distinctive weather and climatic regimes and their nature.

Mountains have three types of effect on weather in their vicinity. First, there is

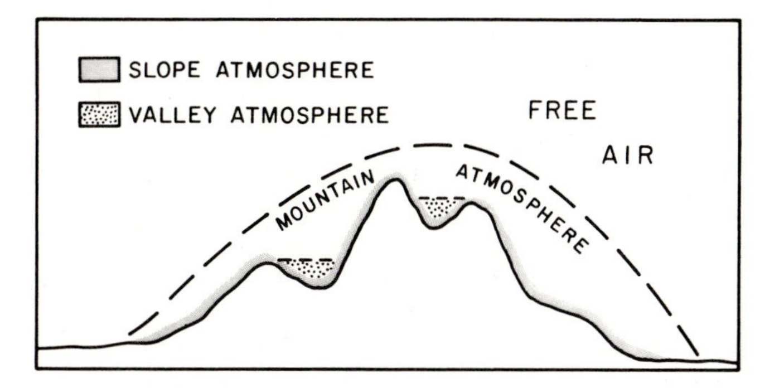

Figure 1.3 The mountain atmosphere
Source: After Ekhart 1948

the substantial modification of synoptic weather systems or airflows, by dynamic and thermodynamic processes, through a considerable depth of the atmosphere. Second, there is the recurrent generation of distinctive regional weather conditions, involving dynamically and thermally induced wind systems, cloudiness and precipitation regimes and so on. Both of these major effects require that the mountain ranges be extensive in width and height and uninterrupted by deep transverse valleys and passes in their long dimension. Both also contribute to shaping the year-round character of the mountain climate, although each occurs with particular types of synoptic situation. For example, thermal wind regimes are best developed with light pressure gradients and clear skies. The third type of mountain effect is a result of slope and aspect variations. It operates primarily at the local scale of tens to hundreds of metres to form a mosaic of topoclimates. In the case of slope and valley wind systems in particular, however, it can operate on a regional scale causing recurrent patterns in the climatic distributions in major mountain systems such as the Himalaya.

The question as to what can feasibly be studied has both a practical and a conceptual aspect. The usual networks of reporting stations are designed to describe synoptic weather systems and regional climatic conditions (on a scale of 100 km or more) but, in the mountains, synoptic systems are greatly modified by the topography. The recommended spacing of weather stations (Brooks 1947) on the plains (and mountains), respectively, is: 1 station per 26 000 km^2 (1300 km^2) for temperature, wind velocity and radiation; 1 station per 13 000 km^2 (1300 km^2) for precipitation; and 1 station per 6500 km^2 (500 km^2) for snow data. These station densities exist in only a few mountain areas such as the Alps. Also, local contrasts of slope angle and orientation give rise to such large variation in local climatic conditions that it seems doubtful whether the concept of a 'regional mountain climate' has much validity or value. It is more meaningful to describe the typical range of climatic elements produced in particular topographic situations, according to the type of airflows that occur, bearing in mind the major controls of altitude, latitude, and continental location.

Ekhart (1936, 1948) suggests that the atmosphere over mountains can be separated into the slope atmosphere – a few hundred metres thick, a valley atmosphere dominated by thermally included circulations and, in extensive mountain ranges, an enveloping mountain atmosphere where the airflow and weather systems are subject to major modification (Figure 1.3). Outside the latter is the 'free air'. In the case of isolated peaks, however, there is considerable mixing of the slope air with the free atmosphere, and the broader 'mountain atmosphere' may be non-existent. Moreover, the degree to which Ekhart's divisions are identifiable changes with season and, more especially, with the large-scale synoptic pressure field. In light pressure gradients, with clear weather, the valley and slope atmospheres may be decoupled from the surrounding atmosphere, whereas in strong airflows and cloudy conditions the only distinctive features tend to be those associated with mechanical sheltering effects.

Yoshino (1975) has attempted to develop a similar *climatic* regionalization for areas with hilly terrain. His scheme interposes a mesoclimatic scale between local (topoclimate) and macroclimate and the uplands are distinguished from the lowlands at this scale in the hierarchy. However, the gradation of altitudinal effects suggests that this is probably an unworkable division. Also, the use of mesoclimate in this context is potentially misleading. [It seems preferable to restrict its use to the climatology of mesometeorological-scale weather systems.] An attempt to summarize the identifiable levels of climatic divisions in major mountain areas, based partly on Yoshino's ideas, is shown in Figure 1.4. The mountain climate of isolated peaks or smaller features is perhaps best regarded as a topoclimatic variant of the particular regional (macro-) climate.

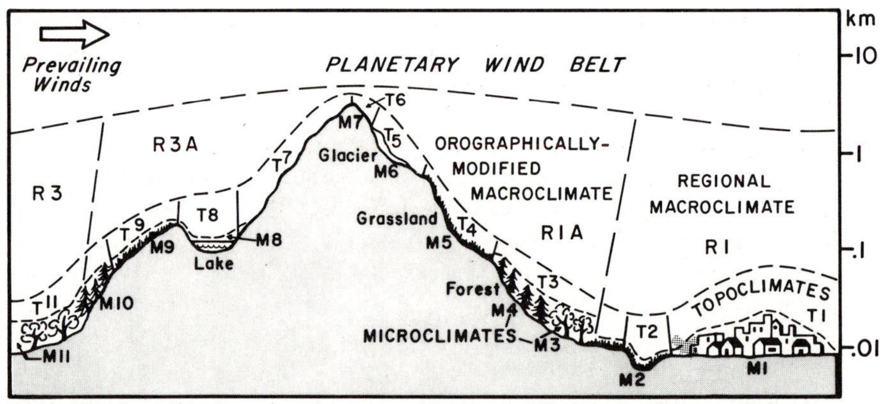

Note: The boundary between the orographically-modified macroclimates R1A and R3A will vary seasonally and synoptically.

Figure 1.4 Scales of climatic zonation in mountainous terrain
Notes: R = regional macroclimate, T = topoclimate, M = microclimate
Source: After Yoshino 1975

The following chapters discuss the controls on mountain weather and climate from the standpoint of the geographical determinants of the meteorological elements, and then in terms of the dynamic and thermodynamic effects of mountain barriers on weather systems and airflow.

REFERENCES

Bailey, S.I. (1908) 'Peruvian meteorology: Observations made at auxiliary stations, 1892–1895,' *Ann Astron. Obs. Harvard Coll.*, 49(2), 104–232.

Barry, R.G. (1978) 'H.B. de Saussure: the first mountain meteorologist', *Bull. Am. Met. Soc.*, 59, 702–5.

Barry, R.G. and Ives, J.D. (1974) 'Introduction', in J.D. Ives and R.G. Barry (eds) *Arctic and Alpine Environments*, pp. 1–13, London, Methuen.

Bénévent, E. (1926) 'Le Climat des Alpes Françaises', *Mémorial de l'Office National Météorologique de France.* 14.

Bjerknes, J. *et al.* (1934) 'For mountain observatories', *Bull. Am. Met. Soc.*, 15, 993–4.

Böhm, R. (1986) *Der Sonnblick. Die 100 jährige Geschichte des Observatoriums und seiner Forschungstätigkeit*, Vienna, Osterreichischer Bundesverlag.

Brooks, C.F. (1947) 'Recommended climatological networks based on the representativeness of climatic stations for different elements', *Trans. Amer. Geophys. Union.* 28: 845–6.

Buchan, A. (1890) 'The meteorology of Ben Nevis', *Trans. R. Soc. Edinb.*, 34.

Buchan, A. and Omond, R. (1902) 'The meteorology of the Ben Nevis observatories, Part II', *Trans. R. Soc. Edinb.*, 43.

Buchan, A. and Omond, R. (1905) 'Part III', *Trans R. Soc. Edinb.*, 43.

Buchan, A. and Omond, R. (1910) 'Parts IV and V', *Trans. R. Soc. Edinb.*, 44.

Bücher, A. and Bücher, N. (1973) 'La température au sommet du Pic du Midi de Bigorre', *La Météorologie*, 5(28), 19–50.

Clayton, H.H. (1944a) 'World weather records', *Smithson. Misc. Coll.*, 79.

Clayton, H.H. (1944b) 'World weather records, 1921–1930', *Smithson. Misc. Coll.*, 90.

Clayton, H.H. (1947) 'World weather records, 1931–40,' *Smithson. Misc. Coll.*, 105.

Coulter, J.D. (1973) 'Ecological aspects of climate', in G.R. Williams (ed.) *The Natural History of New Zealand*, Wellington, A.H. and A.W. Reed.

de Saussure, H.B. (1779–96) *Voyages dans les Alpes, précédés d'un essai sur l'histoire naturelle des environs de Geneve*, vol. 2, chapters XXXV and LIII (1786) and vol. 4, Chapters V–IX (1796), Neuchatel, L. Fauche-Borel.

Ekhart, E. (1936) 'La classification des Alpes au point de vue météorologique,' *Geofis. pura appl.*, 17, 136–41.

Ekhart, E. (1948) 'De la structure thermique et de l'atmosphère dans la montagne', *La Météorologie* (Ser. 4), 9, 3–26. (Translated as 'On the thermal structure of the mountain atmosphere' in Whiteman, C.D. and Dreiseitl, E. (eds) (1984) *Alpine Meteorology*, PNL–5141, ASCOT–84–3; pp. 73–93, Richland, WA, Battelle, Pacific Northwest Laboratories.)

Finch, V.C. and Trewartha, G.T. (1949) *Elements of Geography*, New York, McGraw-Hill.

Fliri, R. (1974) 'Niederschlag und Lufttemperatur in Alpenraum', *Wiss. Alpenvereinshefte*, 24.

Fliri, F. (1975) *Das Klima der Alpen im Raume von Tirol*, Innsbruck, Universitätsverlag Wagner.

Forster, A., Hann, J., and Harisch, O. (1919) 'Das meteorologische Observatorium auf der

Bjelasnica (2067 m) bei Sarajevo', *26–28 Jahresbericht des Sonnblick-Vereines für die Jahre 1917–18*. pp. 12–18, Vienna.

Fujimara, I. (1971) 'The climate and weather of Mt Fuji', in *Fujisan-sôgô-gakujutsuchôsa-hôkoku (Mt Fuji, Scientific Report)*, 215–304. Tokyo, Fujikyuko Co. Ltd. (In Japanese; English summaries: 293–6, 343–5, also figure and table captions.)

Greenland, D.E., Burbank, J., Key, J., Klinger, L., Moorhouse, J., Oaks, S. and Shankman, D. (1985) 'The bioclimates of the Colorado Front Range', *Mountain Res. Devel.*, 5, 251–62.

Grunow, J., Greve, K. and Heigel, K. (1957) 'Das Observatorium Hohenpeissenberg 1781–1955', *Berichte d. Deutschen Wetterdienstes*, 5(36).

Hann, J. (1899) Einige Ergebnisse der meteorologischen Beobachtungen am Observatorium Vallot auf dem Montblanc (4359 m)', *Met Zeit.* 16, 198–204.

Hauer, H. (1950) 'Klima und Wetter der Zugspitze', *Berichte d. Deutschen Wetterdienstes in der US-Zone*, 16.

Hellman, G. (1916) 'Das meteorologische Observatorium auf der Schneekoppe', *24 Jahresbericht des Sonnblick-Vereines für das Jahr 1915*, pp. 6–9, Vienna.

Hill, S.A. (1881) 'The meteorology of the North-West Himalaya', *Indian Met. Mem.*, 1 (vi), 377–429.

Howgate, H.W. and Sackett, D.H. (1873) 'Report of meteorological observations made at Mount Mitchell during the months of May, June, July, August and part of September, 1873', *Annual Rep., U.S. Army Signal Corps*, Washington, D.C., Paper no. 16, 770–947.

Ives, J.D. and Barry, R.G. (eds) (1974) *Arctic and Alpine Environments*, London, Methuen.

Jahresbericht des Sonnblick-Vereines, Vienna, 1892 onwards.

Klengel, F. (1894) 'Die Resultate der meterologischen Beobachtungen am Pic du Midi, 2860 meter', *Met. Zeit.*, 11, 53–64.

Küttner, J.P. (1949) 'Messprobleme auf Bergstation', *Met Rdsch.*, 2, 37–41.

Küttner, J.P. (1982) 'An overview of ALPEX', *Ann Met.*, 19, 3–12.

Kyriazopoulos, B.D. (1966) 'The meteorological observatory of Ayios Antonios Peak of Mount Olympus', *Meteorologika*, 8.

Lauscher, F. (1963) 'Wissenschaftliche Ergebnisse der Alpin-Meteorologischen Tagungen', *Geofis, Met.*, 11, 295–303.

Lauscher, F. (1973) 'Neues von Höhenstation in vier Kontinenten', *68–69 Jahresbericht des Sonnblick-Vereines für die Jahre 1970–71*, pp. 65–7, Vienna.

Lauscher, F. (1975) 'Naturforschung auf den Kanarischen Inseln von Humboldt bis zur Gegenwart', *72–73 Jahresbericht des Sonnblick-Vereines für die Jahre 1974–75*, pp. 61–75, Vienna.

Lauscher, F. (1979a) 'Ergebnisse der Beobachtungen an den nordchilenischen Hochgebirgsstationen Collahuasi und Chuquicamata', *74–75 Jahresbericht des Sonnblick-Vereines für die Jahre 1976–77*, pp. 43–66, Vienna.

Lauscher, F. (1979b) 'Schnee in China', *Wetter u. Leben*, 30, 148–64.

Lauscher, F. (1981) 'Hohenpeissenberg 1781–80', *Wetter u. Leben*, 33, 239–48.

Leitch, D.C. (1978) 'Air temperatures on Mount Washington, New Hampshire, and Niwot Ridge, Colorado', *Mt. Washington Observatory Bull.*, 19, 50–4.

Livadas, G.C. (1963) 'The new Mt.Olympus Research Center', *Geofis. Met.*, 11, 234.

Louis, H. (1975) 'Neugefasstes Höhendiagramm der Erde', *Bayer. Akad. Wiss.* (Math.-Naturwiss. Klasse), 305–26.

Lukesch, J. (1952) 'Die Geschichte des meteorologischen Observatoriums auf dem Hochobir, 2041 m', *48 Jahresbericht des Sonnblick-Vereines für das Jahr 1950*, pp. 25–30, Vienna.

McCracken, I.J. (1980) 'Mountain climate in the Craigieburn Range, New Zealand', in U.

Benecke and M.R. David (eds) *Mountain Environments and Subalpine Growth.* pp. 41–59, Wellington: Tech. Paper, No. 70, Forest Res. Inst., New Zealand Forest Service.

Manley, G. (1949) 'Fanaråken: the mountain station in Norway', *Weather*, 4, 352–4.

Maurer, J. and Lütschg O. (1931) 'Zur Meteorologie und Hydrologie des Jungfraugebietes', *Jungfraujoch Hochalpine Forschungstation*, 33–44.

Meteorological Office (1973) *Tables of Temperature, Relative Humidity, Precipitation and Sunshine for the World*, Pt. III, 'Europe and the Azores' (MO 856c) London, HMSO.

Miller, A. (1976) 'The climate of Chile', in W. Schwerdtfeger (ed.) *Climates of Central and South America*, pp. 113–45, Amsterdam, Elsevier.

Miller, J. (ed.) (1978) *Mauna Loa Observatory: a 20th Anniversary Report*, NOAA Environmental Research Laboratories. US Dept. of Commerce, Boulder, Co.

Mount Washington Observatory (1959 -), News Bulletin, Gorham, NH v. 1- .

Obermayer, A. von (1908) 'Das Observatorium auf dem Atna', *16 Jahresbericht des Sonnblick Vereines für das Jahr 1907*, pp. 3–11, Vienna.

Obrebska-Starkel, B. (1983) 'Progress in research on Carpathian climate', *J. Climatol.*, 3, 199–205.

Obrebska-Starkel, B. (1990) 'Recent studies on Carpathian meteorology and climatology', *Internat. J. Climatol.*, 10, 79–88.

Pleiss, H. (1961) 'Wetter und Klima des Fichtelberges', *Abhand. Met. Hydrol. Dienst. der DDR*, 8(62).

Price, S. and Pales, J.C. (1963) 'Mauna Loa Observatory: the first five years', *Mon. Weather Rev.*, 91, 665–80.

Prohaska, F. (1957) 'Uber die meteorologischen Station der Hohen Kordillere Argentiniens', *51–53 Jahresbericht des Sonnblick-Vereines für die Jahre 1953–55*, pp. 45–55, Vienna.

Reed, W.G. (1914) 'Meteorology at the Lick Observatory', *Mon. Weather Rev.*, 42, 339–45.

Reichel, E. (1949) 'Zur der Messungen auf Gipfelstationen', *Met. Rdsch.*, 2, 41–2.

Reiter, E.R. and Rasmussen, J.L. (eds) (1967) *Proceedings of the Symposium on Mountain Meteorology*, Atmos, Sci. Paper No. 122, Fort Collins, Colorado State University.

Reiter, E.R., Zhu, B-Z., and Qian, Y-F. (eds) (1984), *Proceedings of the First Sino–American Workshop on Mountain Meteorology*, Amer. Met. Soc., Boston, MA, 699 pp.

Reiter, R. (1964) *Felder, Ströme und Aerosole in der unteren Troposphäre nach Untersuchungen in Hochgebirge bis 3000m NN* (Wissenschaftliche Forschungsberichte 71, Darmstadt, D Steinkopf.

Roschkott, A. (1934) 'Die Höhenobservatorien in internationalen Wetterdienst', *42 Jahresbericht des Sonnblick-Vereines für das Jahr 1933*, pp. 51–3, Vienna.

Rotch, A.A. (1886) 'The mountain meterological stations of Europe', *Am. Met. J.*, 3, 1524.

Rotch, A.L. (1892) 'The mountain meteorological stations of the United States', *Am. Met. J.*, 8, 396–405.

Roy, M.G. (1983) 'The Ben Nevis meteorological observatory, 1883–1904', Met. Mag., 112, 118–29.

Smith, A.A. (1964) 'The Mount Washington Observatory', *Weather*, 19, 374–9, 384–6.

Smith, A.A. (1982) 'The Mount Washington Observatory - 50 years old', *Bull. Amer. Met. Soc.*, 623, 986–95.

Smith, R.B. (1986) 'Current status of ALPEX research in the United States', *Bull. Amer. Met. Soc.*, 67, 310–18.

Smyth, C.P. (1859) 'Astronomical experiments on the Peak of Teneriffe', *Phil Trans. R. Soc.*, London, 148, 465–533.

Solomon, H. (1979) 'Air temperatures on Mount Fuji, 1951–1960', *Mt. Washington*

Observatory Bull., 20, 55–60.

Spinnangr, F. and Eide, O. (1948) 'On the climate of the lofty mountain region of Southern Norway', *Met. Ann*, 2(13), 403–65.

Stair, R. and Hand, I.F. (1939) 'Methods and results of ozone measurements over Mount Evans, Colorado', *Mon. Weather Rev.*, 67, 331–8.

Steinhauser, F. (1938) *Die Meteorologie des Sonnblicks. 1. Teil. Beiträge zu Hochgebirgs-meteorologie nach Ergebnissen 50-jähriger Beobachtungen des Sonnblick-Observatoriums, 3106 m*, Vienna, Springer.

Stone, R.G. (1934) 'The history of mountain meteorology in the United States and the Mount Washington Observatory', *Trans. Am. Geophys. Union*, 15, 124–33.

Thompson, W.F. (1964) 'How and why to distinguish between mountains and hills', *Prof. Geogr.*, 16, 6–8

Troll, C. (1973) 'High mountain belts between the polar caps and the equator: their definition and lower limit', *Arct. Alp. Res.*, 5(3, Part 2), A19–A27.

Tutton, A.E.H. (1925) 'The story of the Mont Blanc observatories', *Nature*, 115, 803–5.

Tzchirner, B. (1925) 'Ergebnisse der Temperaturregistrierungen in drei Höhenstationen auf Tenerifa', *33 Jahresbericht des Sonnblick-Vereines für das Jahr 1924*, pp. 19–22, Vienna.

US Army, Chief Signal Officer (1889) 'Meteorological observations made on the summit of Pike's Peak, Colorado. January 1874 to June 1888', *Ann. Astron. Obs. Harvard Coll.*, 22.

US Dept. of Commerce (1959) *World Weather Records 1941–50*, Washington, D.C., Weather Bureau.

US Dept. of Commerce (1966) *World Weather Records, 1951–60*, vol. 2, Europe, Washington, D.C., ESSA.

US Dept. of Commerce (1968) *World Weather Records, 1951–60*, vol. 6, Africa, Washington, D.C., ESSA.

Vallot, J. (1893–98) *Annales de l'Observatoire de Mont Blanc*, vol. 1 (1893), 1–45; vol. 2 (1896), 5–67; vol. 3 (1898), 1–43; Paris, Steinheil.

von Hann, J. (1912) 'The meteorology of the Ben Nevis observatories', *Q.J.R. Met. Soc.* 38, 51–62.

Winterberichte, Eidgenössische Institut für Schnee und Lawinenforschung, Nos. 1–25, Davos Platz, Switzerland; Nos. 26–39, Bern; No. 40- , Weissfluhjoch/Davos. (No. 1 contains meteorological data for 1936/37 to 1945/46; subsequent numbers give an annual tabulation.)

Woeikof, A. (1892) 'Klima des Puy de Dôme in Centralfrankreich', *Met. Zeit.*, 9, 361–80.

World Meteorological Organization (1972) *Distribution of precipitation in mountainous areas*, WMO Tech Note no. 326, vol. 1 and vol. 2, Geneva, World Meteorological Organization.

Xu, Y-G. (ed.) (1986), *Proceedings of the International Symposium on the Qinghai-Xizang Plateau and Mountain Meteorology*, Science Press, Beijing; Amer. Met. Soc., Boston, MA: 1036 pp.

Yoshino, M.M. (1975) *Climate in a Small Area: An introduction to local meteorology*, Tokyo, University of Tokyo Press.

Zingg, Th. (1961) 'Beitrag zum Klima von Weissfluhjoch', *Schnee und Lawinen in den Schweizeralpen*, 24 (Winter 1959/60), 102–27.

Proceedings of alpine meteorology conferences

I	1950	(Milan and Turin) *Geofis. pura appl.* 17(3–4), 81–245 (1950).
II	1952	(Obergurgl) *Wetter u. Leben*, 5, 1–54 (1953).
III	1954	(Davos-Platz) *Wetter u. Leben*. 6, 187–211 (1954).

IV 1956 (Chamonix) *La Météorologie*, 4(45–6), 111–377 (1957).

V 1958 (Garmisch-Partenkirchen) *Ber. Dtsch, Wetterdienstes*, 8(54) (1959).

VI 1960 (Bled) VI *Internationale Tagung für alpine Meteorologie, Bled, Jugoslawien, 14–16 September 1960*, Institut Hydro-Meteorologique Federal de la Republique Populaire Federative de Yougoslavie, Beograd (1962).

VII 1962 (Sauze d'Oulx-Sestrière) *Geofis. Met.*, 11 (1963).

VIII 1964 (Viliach) *Carinthia 11*, Sonderheft 24 (1965).

IX 1966 (Brig and Zermatt) *Veröff. Schweizer. Meteorol. Zentralanstalt*, No. 4, 366 (1967).

X 1968 (Grenoble-St. Martin d'Hères) *X Congres International de Météorologie Alpine, La Météorologie*, Numéro Spécial (1969).

XI 1970 (Oberstdorf) *Ann. Met.* NF 5 (1971).

XII 1972 (Sarajevo) *Zbornik Met. Hydrol. Radova*, 5 (1974).

XIII 1974 (Saint Vincent, Valle d'Aosta) *Alti del Treolidesimo Congresso Internationale di Meteorologia Alpina, Riv. Geofis. Sci. Alpini*, 1 (Speziale) (1975).

XIV 1976 (Rauris, Salzburg) *Veröff. Met. Geophys. Zentralanstalt*, Publ. No. 227, 1 Teil; Publ. No. 228, 2 Teil. (*Arbeiten, Zentralanst. f. Met. Geodynan.*, Vienna, L. 31, 32), 1978.

XV 1978 (Grindelwald) *Veroff. Schweiz. Met. Zentr.*, 40, 1 Teil (1978); 41, 2 Teil (1979).

XV1 1980 (Aix-les-Bains). *Contributions, XVIème Congrès International de Météorologie Alpine.* Société Météorologique de France, Boulogne (1980).

XVII 1982 (Berchtesgaden). *Annal. Met.* No. 19. Deutscher Wetterdienst, Offenbach (1982).

XVIII 1984 (Opatija). *Zbornik Met. Hidrol., Radova* 10. Beograd (1985).

XIX 1986 (Rauris) *19th International Tagung für Alpine Meteorologie. Tagungsbericht.* Österreich. Ges. f. Meteorologie, Vienna (1987).

XX 1988 (Sestola) *Cima '88. 20° Congresso Internazionale di Meteorologia Alpina.* (3 vols.). Servizio Meteorologico Italiano (1990).

XXI 1990 (Engelberg) *Veröff. Schweizer Meteorol. Anstalt.* No. 48 (1990), No. 49 (1991).

Carpathian Meteorology Conference

1980 Tagung Gebirgsmeterologie (8 Internat. Conf. Karpatenneteorologie, Freiburg 1977). *Abhand. Met. Dienst.* DDR, XVI(124), Akad. Verlag, Berlin.

2

GEOGRAPHICAL CONTROLS OF MOUNTAIN METEOROLOGICAL ELEMENTS

The geographical factors which most strongly influence mountain climates are latitude, continentality, altitude, and the topography. The fundamental ways in which these affect the basic meteorological elements are examined below; their effects on climatic distributions in a regional context are discussed in Chapters 4 and 5.

LATITUDE

The influence of latitude on the climate of different mountain systems shows up in a variety of ways. First, solar and net radiation and temperature broadly decrease with increasing latitude and, as a result, the elevations of the tree line and of the snow line decrease polewards. This means that the belt of alpine vegetation (above tree line) and the nival belt of permanent snow and ice are represented on much lower mountains in high latitudes than in the tropics (see Figure 1.1, p. 2). Second, the latitude factor is apparent in the relative importance of seasonal and diurnal climatic rhythms. This is determined by the seasonal trend in the daily sun path at different latitudes (Figure 2.1). Seasonal changes of solar radiation, daylength and temperature are basically small in low latitudes, whereas the diurnal amplitude of temperature, for example, is relatively large. Thus, the equatorial mountains of East Africa are characterized by Hedberg (1964) as experiencing 'summer every day and winter every night'. Insufficient data exist to determine the diurnal range on these mountains reliably, but at 3480 m on Mt. Wilhelm (5°40′S), Papua New Guinea, it is about 7–8°C throughout the year, compared with a seasonal range of 0.8°C for mean monthly temperatures (Hnatiuk *et al.*, 1976). In middle and higher latitudes seasonal effects greatly exceed diurnal ones. At 3750 m on Niwot Ridge, Colorado, for example, the seasonal amplitude of mean temperature range is 21°C compared with a daily range of 6–8°C (Barry 1973). If hourly data are available, the diurnal and seasonal regimes of temperature can be conveniently illustrated by a *thermoisopleth* diagram (Troll, 1964). This is demonstrated in Figure 2.2 for Pangrango, Java, and for the Zugspitze, Germany. Latitudinal differences in diurnal temperature range have been examined in detail by Lauscher (1966).

Table 2.1 Measures of continentality at selected mountain stations

Station	*Latitude*	*Altitude (m)*	*Warmest month (°C)*	*Coldest month (°C)*	*Mean annual temperature range (°C)*	*Hygric continentality (deg)*
Mt. Wilhelm/Pindaunde Papua New Guinea	6°S	3480	12.0	10.9	1.1	46
Mucubaji (Sierra Nevada de Merida) Venezuela	5°S	3550	5.9	4.7	1.2	74
Ski Basin (Craigieburn Range) New Zealand	43°S	1550	9.7	−1.4	11.1	44
Mt. Fuji, Japan: Summit	35°N	3776	5.9	−19.5	25.4	–
SW Slope, Taroba	35°N	1300				15
Ben Nevis, Scotland	57°N	1343	5.0	−4.5	9.5	18
Niwot Ridge, Colorado	40°N	3749	8.2	−13.2	21.4	76
Sonnblick, Austria	47°N	3105	1.2	−12.9	14.1	64

Figure 2.3 shows the linear decrease of mean diurnal temperature range versus latitude for exposed mountain sites. In mountain valleys or on high plateaux, the range at low and middle latitudes is considerably larger than on mountain summits due to less mixing of the air with that of the free atmosphere.

These latitudinal differences in temperature regime also have an effect on the precipitation characteristics. On equatorial high mountains, above about 4000 m, snow may fall on any day of the year, particularly over-night. In middle and higher latitudes there is a well-marked and prolonged winter season. In the European Alps, for example, there is typically an average of 350 days per year with snow cover at 3000 m (Geiger 1965, Table 91). Here, snowfall accounts for 80 per cent or more of total annual precipitation (Lauscher, 1976), a higher proportion than at many arctic stations where much of the precipitation occurs as summer rains.

Other latitudinal differences in mountain climates arise due to characteristics of the global atmospheric circulation. Tropical mountains are within the easterly trade wind regime, mid-latitude mountains are within the westerlies. The tropical easterlies decrease with height, whereas the westerlies increase steadily. Associated with these differences in the global wind belts, precipitation systems are primarily convective and small-scale in low latitudes, but cyclonic and generally large-scale in middle and higher latitudes.

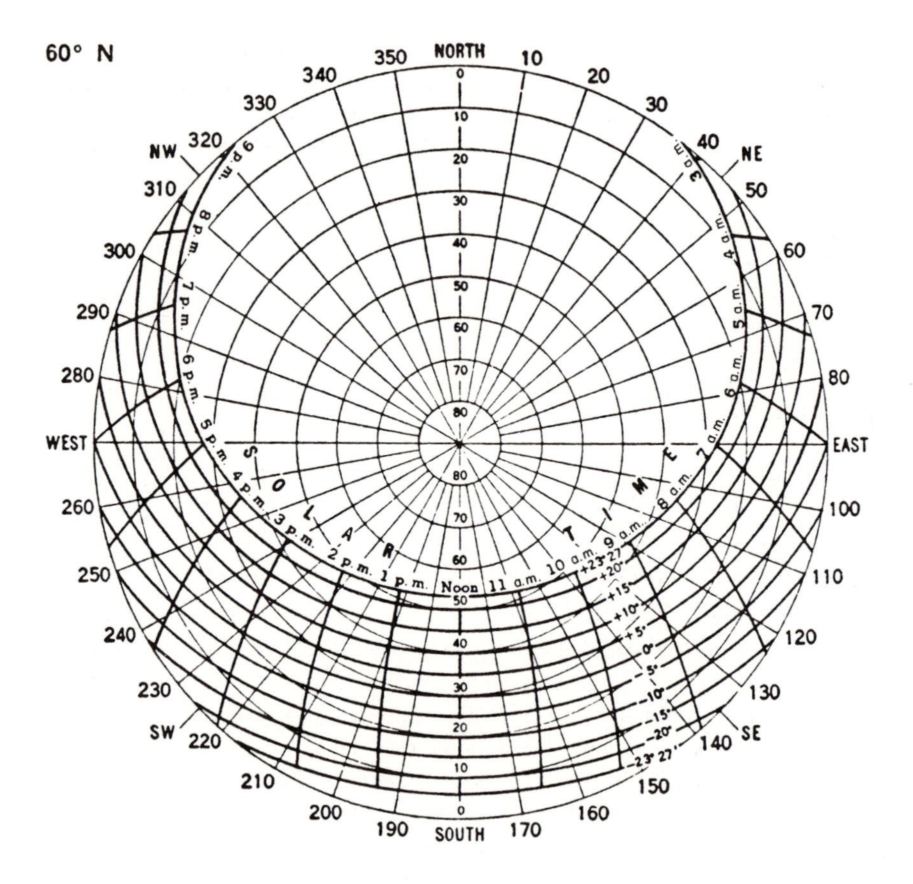

Figure 2.1(a) Daily sun paths at latitudes 60°N (2.1a; above) and 30°N (2.1b; right)

Notes: Approximate dates of the indicated declination angles are as follows:

Declination	*Approx. dates*
+23°27′	June 22
+20°	May 21, July 24
+15°	May 1, Aug. 12
+10°	Apr. 16, Aug. 28
+5°	Apr. 3, Sept. 10
0°	Mar. 21, Sept. 23
−5°	Mar. 8, Oct. 6
−10°	Feb. 23, Oct. 20
−15°	Feb. 9, Nov. 3
−20°	Jan. 21, Nov. 22
−23°27′	Dec. 22

Source: From Smithsonian Meteorological Tables, 6th edn

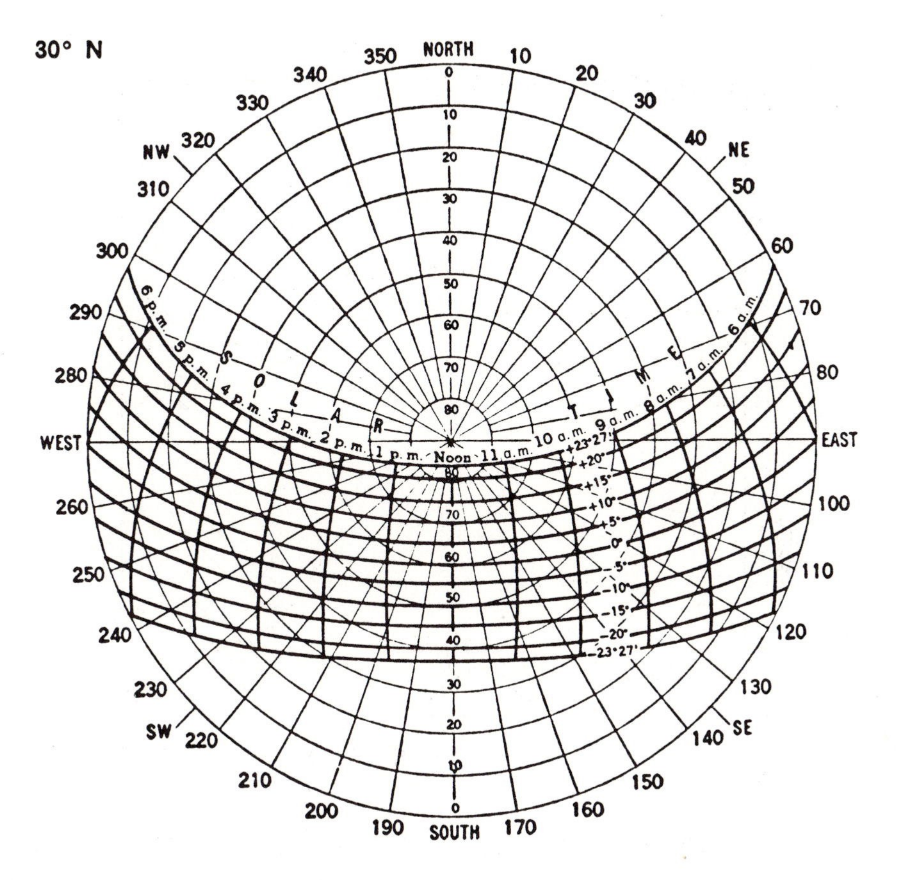

Figure 2.1(b)

CONTINENTALITY

The degree of continentality/oceanicity of a climatic region is most commonly expressed by the annual range of mean monthly temperatures. Various indices of continentality have been devised (W. Gorczynski, N. Ivanov); most of these incorporate a weighting of the annual temperature amplitude (by the inverse of the sine of the latitude angle) to standardize for the latitudinal effects on seasonality discussed above. Such adjustments are not essential, however, since it is only a relative measure.

The basic cause of continental effects is the well known difference in the heat capacity of land and water. Typically, the heat capacity of a sandy soil is × 2 (wet) to × 3 (dry) that of water, for unit volumes. As a result of this factor, and the deeper vertical transfer of heat by turbulent mixing in the upper ocean compared with the limited heat conduction of soils, the annual and diurnal

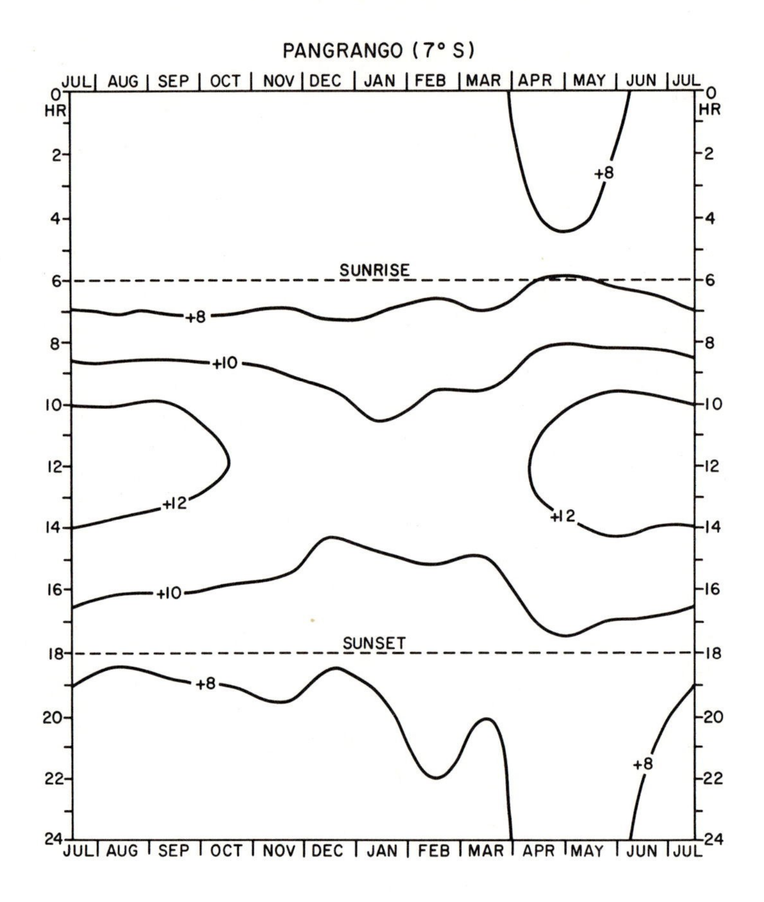

Figure 2.2(a) Thermoisopleth diagram of mean hourly temperatures (°C) Pangrango, Java, 7°S, 3022 m. After Troll 1964

ranges of surface and air temperature are much larger in continental climates than over the oceans. The predominance of maritime air masses are masses in west coast mountain ranges extends the oceanic influence inland.

The annual temperature amplitudes for similar latitudes and elevations at maritime and continental mountain stations are illustrated in Table 2.1. In addition to the smaller temperature range in maritime mountain ranges, the months of maximum and minimum values tend to lag the solar radiation maximum and minimum by up to two months, compared with only one month in continental interiors. Overall, continentality effects are more pronounced in the northern hemisphere than the southern, due to the 2:1 ratio of their respective land areas.

Oceanic versus continental conditions associated with these basic physical processes of surface heating and air mass transfer are also apparent in terms of

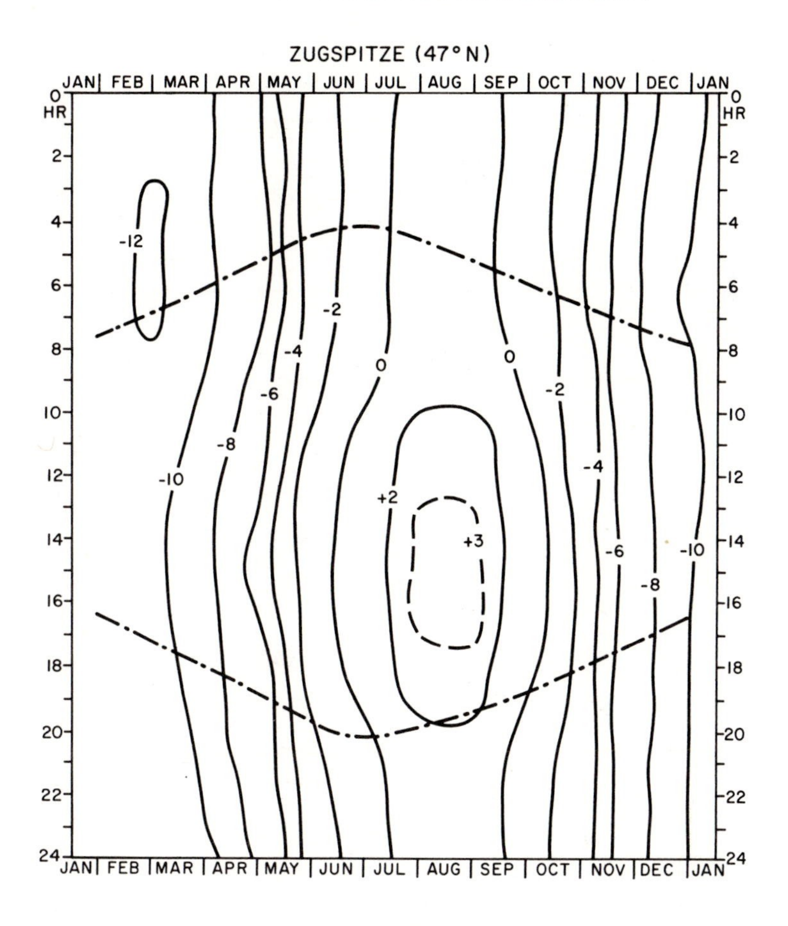

Figure 2.2(b) Themoisopleth diagram of mean hourly temperatures (°C) for Zugspitze, Germany, 47°N, 2962 m. After Hauer 1950

cloud and precipitation regimes. Mid- and high-latitude west coast ranges have a high frequency of cyclonic activity, stratiform cloud and overcast skies, as in British Columbia and Alaska. Precipitation is usually greater in the winter half year, but there are no dry months. Conversely, in the interiors of the northern continents, the winter season is dominated by cold anticyclones and the summers by heat lows and convective storms. There are generally low total amounts of cloud in winter and predominantly convective cloud regimes in summer. These contrasts apply in some degree to coastal versus continental mountain ranges.

Hygric continentality is a term introduced by Gams (1931) in his studies of vegetation in the Alps. He demonstrated that the distribution of forest types is related to the ratio of a station's altitude (m) to its annual precipitation (mm), Z/P. The index may be expressed as $\tan^{-1}$ (Z/P), which is 45° when Z/P = 1. Aulitsky *et al.* (1982) show values ranging from about 65° (Z/P = 2.1) at the

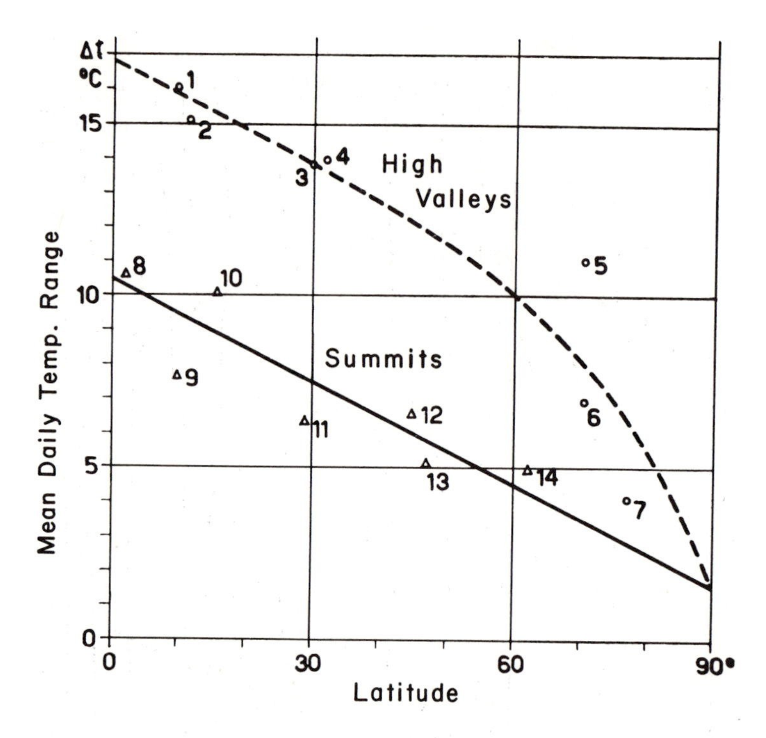

Figure 2.3 Mean daily temperature range versus latitude for a number of high valley and summit stations
Source: After Lauscher 1966

treeline in Poschach Obergurgl, Austria (2072 m) and Stillberg, Switzerland, in the central Alps to only 31° (Z/P = 0.6) at Lunz-Dürnstein (1860 m). Figure 2.4 illustrates how forest types are related to hygric continentality, snow cover duration and annual temperature range. For example, larch and stone pine (*Pinus cembra*) woods occur on sites with relatively low annual precipitation (high hygric continentality), but modest annual temperature range and long-lasting snow cover; beech–oak woods occur at the opposite extremes, with larch–spruce woods on intermediate sites. On a global scale, values of hygric continentality range up to 75° on mountains in Colorado and Venezuela and as low as 15°–20° in maritime locations such as western Scotland and Japan (Table 2.1).

Extensive mountain massifs and high plateaux set up their own large-scale and local scale circulations. Such large-scale effects on diurnal and seasonal circulations ('plateau monsoons') have recently been demonstrated for the Tibetan Plateau and the plateaux and mountain ranges of the southwestern United States (Tang and Reiter 1984; Reiter and Tang 1984). For example, the summer-time plateau wind circulation in the Great Basin area of the southwestern United States reverses diurnally over a depth of 2 km. Gao and Li (1981) show that the Tibetan Plateau (30°–38°N, 70°–100°E) creates a lateral boundary

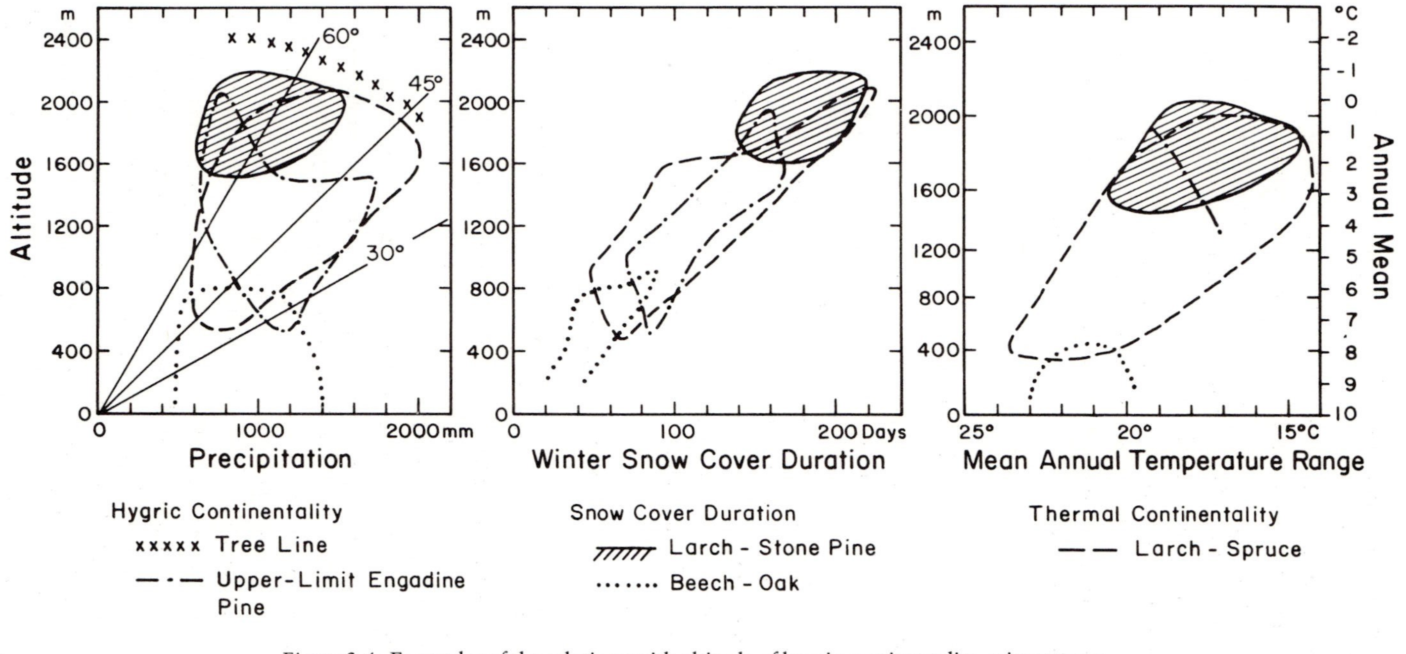

Figure 2.4 Examples of the relations with altitude of hygric continentality, winter snow cover duration and thermal continentality, and tree species in Austria
Source: After Aulitsky *et al*. 1982

layer that enlarges its effective dimensions to 22°–43°N, 62°–115°E in summer, and 27°–40°N, 70°–105°E in winter.

High mountains also protrude into the middle troposphere where the atmospheric circulation may differ considerably from that at sea level. For these reasons, mountain ranges located in a semi-arid macroclimatic zone, for example, may have distinctly different climatic characteristics and vegetational assemblages from the adjacent lowlands.

In view of the complexity of the effects of continentality in mountain areas, generalizations are of little value. Some particular aspects of continentality are discussed in the regional case studies (Chapter 5), in the following sections on meteorological elements, and in Chapter 4 on climatic characteristics of mountains.

ALTITUDE

The effect of altitude on climatic elements is of such primary importance that it is considered under several subheadings. First, its general effects on the atmospheric state variables of pressure, density, and vapour pressure are examined. Then, processes that determine the altitudinal variations of the radiation components, air temperature, and wind velocity are reviewed. Components of the energy and moisture budgets are discussed in Chapter 4 since their distribution in mountain areas involves the interaction of airflow with the topography on several scales, as well as direct effects of altitude.

Air pressure and density

The relationship between altitude and pressure was first demonstrated more than three centuries ago. It is the most precisely documented aspect of altitudinal influence on meteorological elements, although the mean condition is of little direct significance for weather phenomena. It is short-term departures from the average state that are important in this respect.

In an ideal incompressible fluid, the pressure at any depth (h) can be expressed as $p = g\rho h$, where p = pressure, g = gravity and ρ = air density. Where the air density also changes with height (z) we have $dp/p = -gdz/RT$ where R = the specific gas constant for dry air, T = temperature. The relationship between p, ρ, and z is often defined for a *Standard Atmosphere* where mean sea-level pressure = 1013.25 mb, surface temperature = 288 K and dT/dz = 6.5 K km^{-1} to an altitude of 11 km (see Table 2.2). The Standard Atmosphere represents a hypothetical state approximating annual conditions in middle latitudes. The concept was first developed in France during the mid-nineteenth century in connection with ballooning and mountaineering (Minzer 1962).

In the tropical atmosphere, pressure is about 15 mb greater at 3000 m, and 20 mb greater at 5000 m, than at the same levels in middle latitudes (Prohaska 1970). This is due to higher virtual temperatures in the tropical atmosphere.

Table 2.2 The Standard Atmosphere

Z′(m)	*P(mb)*	*T(°C)*	$\rho(kg\ m^{-3})$
0	1013.25	15.0	1.2250
1000	898.8	8.5	1.1117
2000	795.0	2.0	1.0581
3000	701.2	−4.5	0.90925
4000	616.4	−11.0	0.81935
5000	540.5	−17.5	0.78643
6000	472.2	−24.0	0.66011

Notes: Z' = geopotential altitude. The difference between geopotential and geometric (Z) altitude is within 0.3% to 6000 m. $Z' < Z$ in low and middle latitudes; $Z' > Z$ in polar latitudes
Source: COESA 1962

(Virtual temperature is the temperature for which dry air at the same pressure would have the same density as the air sample $T_v = T/(1-0.375\ e/p)$, where e = vapour pressure.)

Thus, pressure levels are some 200–300 m higher in the tropics than the levels given by the Standard Atmosphere in Table 2.2

The effects of reduced pressure and density with height are of particular importance in connection with radiation conditions, as discussed, on p. 34 and for human bioclimatology (see pp. 344–52).

Vapour pressure

This is the partial pressure exerted by water vapour as one of the atmospheric gases. It is typically about 1 per cent of the sea level pressure. The limiting pressure – the saturation value – is determined only by the air temperature. Since temperatures at high altitudes are low, vapour pressures in mountain areas are also low, and the decrease is proportionately greater in the lower layers. For Mt. Fuji, Japan, the vapour pressure averages 3.3 mb at 3776 m compared with 11 mb at 1000 m and 14.5 mb at sea level (Fujimara 1971, cited by Yoshino 1975: 203).

Few specific studies of vapour pressure in mountain regions have been made, although most mountain stations record it, and there is extensive information for the free air from radisonde soundings. Various empirical formulae have been developed to express the general exponential form of vapour pressure decrease with altitude. For example,

$$e_z = e_0 exp(-\beta z),$$

where e is in units kg m^{-3}, e_0 denotes vapour pressure at the surface, z = height (km). According to Reitan (1963), β is approximately 0.44 km^{-1}. Kuz'min (1972) gives:

$$e_z = e_0 \times 10^{-a}$$

where $a = 0.20$ for the free air, $= 0.159$ for mountain areas of central Asia. These empirical coefficients denote the vertical diffusion of water vapour, which must vary in efficiency both regionally and seasonably. On a daily basis the moisture profile is strongly determined by synpotic- and meso-scale vertical motion.

For Sonnblick, Austria (3106 m) the July mean is about 6.5 mb, compared with 13.5 mb at Salzburg (430 m). Corresponding January averages are 1.8 mb and 4.9 mb, implying a slight increase here in the seasonal range with altitude (Steinhauser 1936). On Sonnblick there is a moderate diurnal variation in summer of 1.2 mb, with a maximum at about 1400 hours and a minimum at about 0500 (Steinhauser 1938). In winter, the variation is of similar phase, but negligible amplitude (0.1 mb), and this is also apparent from relative humidity data (Rathschuler 1949).

Interesting data for the tropical Andes are provided by Prohaska (1970) in units of absolute humidity. Seasonal ranges in moisture content are large, but

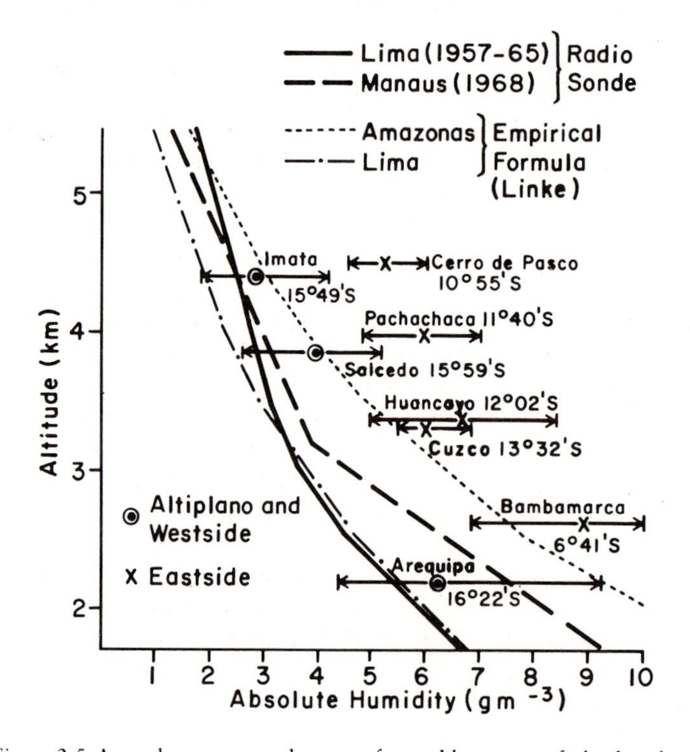

Figure 2.5 Annual averages and range of monthly means of absolute humidity ($g\ m^{-3}$) as a function of altitude in tropical South America

Notes: Station data, free-air radiosonde values, and gradients calculated from Linke's equation are shown

Source: After Prohaska 1970

diminish with altitude (Figure 2.5). Even so, dry season values in these tropical high deserts (e.g. Imata) considerably exceed those in winter on summits in the Alps. Figure 2.5 also demonstrates that moisture contents in the Andean highlands are larger than those at the same altitudes in the free air. Storr (1970) notes that this is generally the case up to at least 4 km, based on studies by Khrgian (1965) in the Caucasus and his own investigations at Marmot Creek in the Rocky Mountains near Calgary. At 1200 m, the excess in summer averages 2 mb at Marmot Creek and 2.5 mb in the Caucasus. Its cause is attributed to evapotranspiration by Khrgian, but Storr notes that the seasonal and diurnal occurrence of the excess does not fully support this view. He suggests that day-time upvalley and upslope air circulation is likely to carry moister valley air to higher elevations, with local variations in the strength and duration of these winds causing additional variability. However, this hypothesis is not fully in accord with Storr's finding that the excess was greater at 0500 LST than at 1700 LST at half of the sites in June and July 1967. Further studies of this problem are still needed. In mountains with permanent snow and ice cover, the slope-free air difference in vapour content will decrease above the zone of glacier ablation – the fern line – since there is no longer a moisture source (Kuz'min 1972: 122).

Vapour pressure is important climatically in three ways. First, it reduces the transmission of infra-red radiation and, to a lesser extent, solar radiation. Second, it influences the saturation deficit (the difference between the saturation and actual vapour pressure) which is an index of bioclimatic significance sometimes termed the 'drying power' of the air. Third, it affects the total air density inversely, and this may also be important biologically in terms of the hypoxic effects of oxygen deficiency at high altitudes.

The fact of a vapour pressure excess in the mountain atmosphere should act to lower the condensation level, other factors remaining constant. It will also tend to reduce the transmission of infra-red radiation, by comparison with the free air, thereby leading to higher atmospheric temperatures (Storr 1970).

Solar radiation

Mountain observatories were of special importance in early studies of solar radiation and the *solar constant* – the average flux of solar radiation received on a surface perpendicular to the solar beam outside of the earth's atmosphere at the mean distance of the earth from the sun. The earliest mountaintop measurements were made on Mt. Blanc by J. Violle in 1875. Langley (1882, 1884) also made actinometer observations on a special expedition to Mt. Whitney, California, in 1881, but their estimates of the solar constant were considerably higher than the currently assumed value of 1368 W m^{-2} (1.96 cal cm^{-2} min^{-1}). (Violle obtained a value of 2.8 cal cm^{-2} min^{-1} and Langley (1884) 3.0–3.5 cal cm^{-2} min^{-1}.) Long-term spectral measurements were initiated at Davos (1560 m) during 1908–1910 by Dorno (1911).

Before examining observational results, the nature of atmospheric effects on solar radiation will be briefly reviewed. Detailed theoretical treatments may be found in the texts of Sivkov (1971), Kondratyev (1969) and others. We first consider a pure, dry atmosphere. In this case, solar radiation is affected by molecular (Rayleigh) scattering and absorption by the atmospheric gases.

The path length of the solar beam through the atmosphere is expressed in terms of *optical air mass, m*

$$m = \frac{1}{\sin \theta}, \text{ where } \theta = \text{solar altitude}$$

For most practical purposes, this formulation is sufficiently accurate when $\theta > 10°$. At sea level, the relationship between optical air mass and solar altitude is: for $m = 1$, $\theta = 90°$, $m = 2$, $\theta = 30°$, and $m = 4$, $\theta = 14°$. For comparative radiation calculations at different altitudes, the absolute optical air mass $M = m(p/p_0)$, where p = station pressure and $p_0 = 1000$ mb, is used to allow for the effects of air density on transmission. Thus at 500 mb a value of 2 for M corresponds to $m = 4$ and $\theta = 14°$. For an ideal (pure, dry) atmosphere, the direct solar radiation received at the 500 mb level (approximately 5.5 km) is 5–12 per cent greater, according to solar altitude, than at sea level (Table 2.3). This corresponds to an average increase of 1–2 per cent km^{-1}.

Table 2.3 Altitude effect on (direct) solar radiation in an ideal atmosphere ($W\ m^{-2}$)

Level	Optical mass (*m*) 1	2	3	6
	Corresponding solar altitude (θ) 90	30	19.3	9.3
Extra-Terrestrial	1370	1359	1350	1335
500 mb	1299	1238	1183	1962
750 mb	1269	1188	1123	985
1000 mb	1244	1146	1073	922

Source: Modified from Kastrov 1956

The depletion of the direct solar beam irradiance by atmospheric absorption and backscatter is referred to as the relative opacity of the atmosphere or its *turbidity*. Some indicies of turbidity are defined for all wavelength radiation, others are for monochromatic visible radiation, used for studies of particulate aerosol loading. There are various indices of turbidity due to A. Ångstrøm, Schüepp (1949) and others (see Lowry 1980b; Valko 1961). Valko (1980) shows that the turbidity at Swiss mountain stations is typically 4 to 5 times less than that in the lowlands. Beer's Law for the direct flux through an atmospheric layer can be expressed:

$$S = S_0 exp(-\tau M)$$

where S_0 and S are the fluxes at the top and bottom of the layer, respectively, τ is a transmissivity coefficient (< 1.0), and M is the absolute optical air mass to the level of S. An analogous all wavelength expression is

$$S = S_0(\tau)^M$$

A useful index, K, developed by Lowry (1980b) for zenith path transmissivity (τ_p) can be defined as

$$K = -d(ln\ \tau_p)/dp$$
$$\text{where } (ln\ S - ln\ S_p) = ln(\tau_p)$$

For a clean, dry atmosphere, at $p = 1$ (where p = mb/1000)

$$\ln(\tau_p) = Kp(ln\ 0.906)$$
$$\text{or } \tau_p = (0.906)^{Kp}$$

The lines indicating values of K in Figure 2.6 represent the amounts of depletion of direct beam zenith flux through an atmosphere with homogeneously-distributed absorption and scattering, relative to clean dry air ($K = 1$). Figure 2.6 illustrates schematic soundings for a clean dry atmosphere with ozone (2.5 per

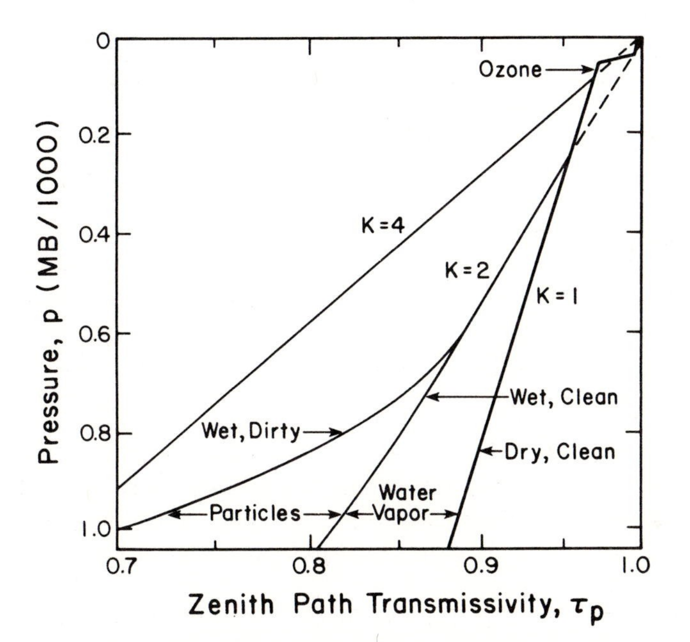

Figure 2.6 Profiles of zenith-path transmissivity for a clean, dry atmosphere with ozone, a clean, wet atmosphere and a dirty, wet atmosphere; profiles of the theoretical transmissivity index (K) of W P Lowry are also shown for $K = 1$, 2 and 4
Source: After Lowry 1980b

Table 2.4 Direct radiation on a perpendicular surface at 47°N, as a percentage of the extraterrestrial total

	Altitude (m)				
	200	*1000*	*2000*	*3000*	*Extra-terrestrial total* (W m^{-2})
15 December	37	48	58	61%	488
15 June	51	58	67	72%	865

Source: After Steinhauser, 1939

cent absorption), a clean moist atmosphere, and a moist dirty atmosphere. Note that the zenith path transmissivity at the surface ($p = 1$) is approximately 0.7.

For a real atmosphere, the effects of absorption of radiation by water vapour and of attenuation by particulate matter (Mie scattering) must be taken into account, in addition to absorption by the atmospheric gases and molecular scattering. Ozone absorption is strongly wavelength dependent, occurring primarily below 0.34 μm. Molecular scattering is also inversely dependent on wavelength, giving rise to the blue colour of the sky (more scattering at shorter wavelengths), whereas aerosol scattering and absorption of solar radiation change only slightly in the visible and UV wavelengths. Generally, in mountain areas, however, the aerosol content tends to be low and most of the atmospheric water vapour is below about 700 mb, diminishing these effects.

Aerosol nuclei concentrations decrease by an order of magnitude between the lowlands (10^{10} m^{-3}) and summits at about 3 km in Europe (10^{9} m^{-3}), for example. This is a prime reason for the generally good visibility in mountain regions.

Water vapour influences atmospheric transmissivity through refraction effects but, since it is concentrated in the lowest 2–3 km of the atmosphere, the effect on the optical mass is negligible according to Sivkov, except where $m > 6$ (Sivkov 1971: 28). Drummond and Ångstrøm (1967) show that under cloudless skies at 3380 m on Mauna Loa, Hawaii, the water vapour absorption accounts for about 90 W m^{-2} (0.13 cal cm^{-2} min^{-1}) at noon. The same is true for aerosols, except with high turbidity at very low solar altitudes. Ångstrøm and Drummond (1966) show turbidity effects to be negligible, at high altitude stations between 16° and 36°N, for wavelengths $> 0.7\mu$m; these wavelengths account for 50 per cent of all solar radiation.

Early estimates of the solar constant were made by extrapolating actinometer (later pyrheliometer) measurements of integrated solar radiation, made at different path lengths, to a theoretical zero value of m (Maurer 1912) Abbott and Fowle (1908) noted that this method would underestimate extra-terrestrial radiation if the extrapolations were for relatively large values of m. Instead, the tech-

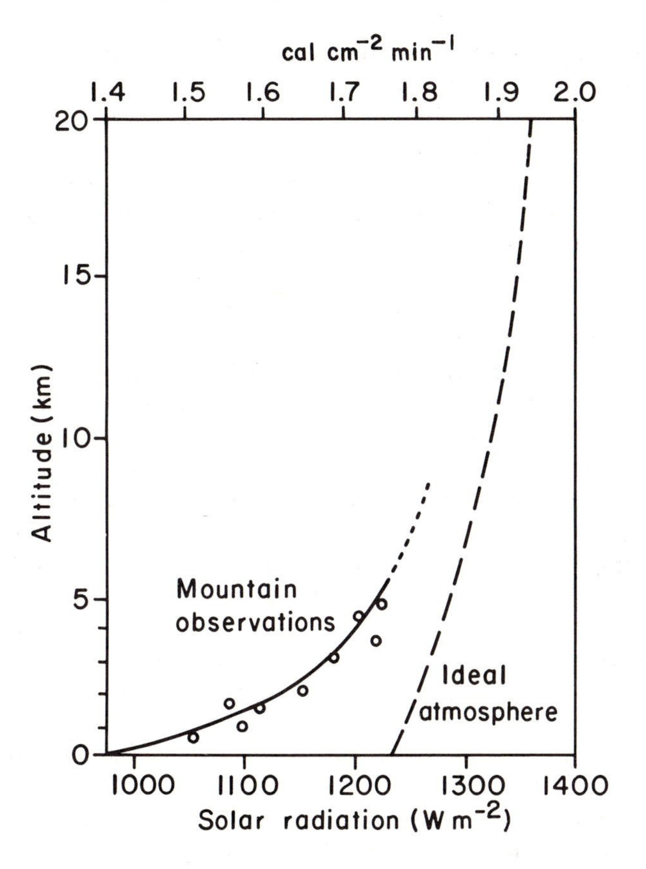

Figure 2.7 Direct solar radiation versus altitude in an ideal atmosphere for m = 1 (after Kastrov, in Kondratyev 1969: 262) and as observed at mountain stations
Source: Based on Abetti 1957, Kimball 1927, and Pope 1977

nique of spectrobolometry, pioneered by Langley on Mt. Whitney in 1881, was being perfected by Abbott and his associates. This involves observations of *relative* intensity in narrow spectral bands at different solar angles. Transmission coefficients are thereby determined for each ray. These results are then scaled by absolute pyrheliometric measurements of total solar radiation. Thus, Abbott and Fowle (1908) were able to calculate a value of 1470 W m^{-2} from observations at Washington and Mt. Wilson (1750 m), California. It is important to note that nearly all values cited during the preceding 25 years were some 50–100 per cent *greater*, although Radau (1877) indicated general agreement on a figure of about 1220 W m^{-2}.

The uncertainty at this time is illustrated by K. Ångstrøm's changing views. After estimating a solar constant of 2800 W m^{-2} (4.0 cal cm^{-2} min^{-1}) in 1890, he subsequently withdrew this value. Although he had carried out careful studies of radiation on the Pico de Teyde, Tenerife (3683 m) in 1895 and 1896, he made no attempt to re-calculate the solar constant because of doubts about the actinometric data (Ångstrøm, 1900).

Following new spectrobolometer measurements with ultraviolet filters on Mt. Whitney (4420 m), California, in 1909–10, Abbott and Fowle (1911) revised their estimate of mean solar constant to 1343 W m^{-2}. Pyrheliometric data obtained 30 years later by the US Weather Bureau on Mt. Evans, Colorado, gave 1349 W m^{-2} (Hand *et al.* 1943), both very close to the modern value derived from satellite measurements (see note 1, p. 96). High mountain sites continue to be useful for radiation studies. Drummond and Ångstrøm (1967) repeated earlier measurements of direct beam solar radiation normal to the surface at solar noon and found that, for cloudless skies, there is an almost constant maximum value of 1174 W m^{-2} (1.68 cal cm^{-2} min^{-1}) when corrected to mean solar distance. This was in close agreement with results of K. Ångstrøm (1900). Abbott and Fowle (1908) and Bishop *et al.* (1966). The difference between this and the solar constant is accounted for by upper atmospheric absorption.

Some of the most extensive studies of altitudinal effects on solar radiation have been made in the European Alps. Steinhauser (1939) analysed the direct radiation, on a surface normal to the beam, and showed that, as a percentage of the extra-terrestrial radiation, there is a rapid increase up to about 2000 m, after which the rate of increase declines (Table 2.4). In fact, based on observations from many high mountain stations, the increase is broadly exponential (Figure 2.7), due to the concentration of water vapour in the lower troposphere. This effect is pronounced by comparison with the theoretical distribution for an ideal (pure, dry) atmosphere.

Empirical expressions for direct beam radiation under cloudless skies versus altitude were originally developed by Klein (1948) and Becker and Boyd (1957). Lowry (1980a) proposes a more physically realistic approach relating zenith transmissivity to 'standard atmosphere' pressures and a turbidity index. He shows an improvement in the explained variance over the expression of Klein for the same data.

Table 2.5 Relationships between clear-sky direct beam transmissivity of solar radiation (τ_p) with pressure level (p) and turbidity (K)

Pressure level	*Moist atmosphere*[1]		*Dry atmosphere*[2]	
p (mb/1000)	τ_p	*K*	τ_p	*K*
1	0.678	3.937	0.757	2.820
0.9	0.758	3.119	0.806	2.418
0.8	0.817	2.559	0.844	2.148
0.7	0.860	2.183	0.873	1.966
0.5	0.914	1.822	0.914	1.822

Notes: 1 $\tau_p = 0.975 - 0.0127 \exp(3.1529p)$; 2 $\tau_p = 0.975 - 0.0171 \exp(2.5468p)$
Source: After Lowry, 1980B

Table 2.6 Altitude and cloudiness effects on global solar radiation in the Austrian Alps (W m^{-2})

	Cloud cover					
		December			*June*	
Altitude (m)	*0/10*	*5/10*	*10/10*	*0/10*	*5/10*	*10/10*
200	63	43	14	335	241	75
1000	73	51	19	362	262	100
2000	80	59	26	387	290	142
3000	83	63	36	404	314	196
		March			*September*	
200	187	126	41	216	146	43
1000	208	150	58	237	161	55
2000	226	172	87	256	179	75
3000	234	187	122	270	193	96

Source: After Sauberer and Dirmhirn 1958

Lowry's (1980) physically based models for atmospheric depletion show an improvement in the explained variance over the regression formulation of Klein for the same data. The best fit is obtained with:

$$\ln (0.975 - \tau_p) = \ln M + NP$$

M and N are constants for the moist and dry atmospheres of the data. Lowry (1980b) proposes standard relationships between clear sky direct beam transmissivity with pressure level and turbidity which are summarized in Table 2.5.

Measurements made at the Jungfraujoch, Sonnblick and Zugspitze observatories have provided a wealth of material for analysis (Sauberer and Dirmhirn 1958). Table 2.6 shows that for cloudless conditions the global solar radiation is 32 per cent greater at 3000 m than at 200 m in December, 25 per cent greater in March and September and 22 per cent greater in June. This implies an increase of 7–10 per cent km^{-1} which, as noted by Lauscher (1937), is far greater than the rate of increase for direct radiation in an ideal atmosphere. For overcast conditions the global solar radiation increased by 9–11 per cent km^{-1} in all months. The ratio of global radiation for overcast to cloudless skies increases considerably with altitude, from about 0.22 at 200 m, in all seasons, to about 0.40 at 3000 m in winter and fall and about 0.50 in summer and spring (Sauberer and Dirmhirn 1958). The higher ratios indicate that cloud cover has a smaller influence on the radiation conditions.

The effect of altitude on diffuse, or sky, radiation has been studied in detail by Dirmhirn (1951). Under cloudless skies, the sky radiation decreases with altitude owing to the reduction in air density and therefore in scattering, but multiple reflections from adjacent peaks may obscure this to some extent, especially when there is snow cover. Table 2.7 summarizes her results for cloudless and overcast

Table 2.7 Altitude effects on sky radiation in the Austrian Alps (W m^{-2})

	December		*June*	
Altitude (m)	*Cloudless*	*Overcast*	*Cloudless*	*Overcast*
200	13.9	14.5	48	75
1000	12.2	18.5	41	100
2000	10.4	26.0	34	142
3000	9.3	36.5	30	196

Notes: All radiation data in this and other tables have been uniformly converted to W m^{-2}, for convenience of comparisons, using the conversions appropriate for the particular time unit (see Appendix).
Source: After Sauberer and Dirmhirn 1958

skies. With overcast conditions, sky radiation is much more intensive on the mountains owing to the generally shallower cloud layers overhead. Table 2.8 compares diffuse radiation for stations at several elevations in the Alps and shows a 280 per cent increase in December at the Sonnblick, and a 175 per cent increase in June compared with Vienna. However, it should be noted that an analysis of the actual mean values of diffuse radiation at these stations shows increases of only 119 per cent in December and 128 per cent in June at the Sonnblick, compared with Vienna (Neuwirth 1979). The diffuse component represents 50–55 per cent of the global radiation on an annual basis in Austria. For the Sonnblick, this figure ranges from 70 per cent in May to 32 per cent in October, which is usually a sunny month.

The effect of cloud cover on solar radiation as a function of altitude is complex and, again, the most detailed results are available from the Alps (Sauberer and Dirmhirn 1958; Thams, 1961a,b). In June and December there is a nearly linear relationship between global solar radiation and cloud amount in the mountains at 3000 m, whereas at lower elevations thicker clouds cause a sharper decline for conditions of overcast (Figure 2.8). A generalized relationship between diffuse radiation and cloud amount at four elevations is shown in Figure

Table 2.8 Mean daily totals of diffuse radiation in the Alps with overcast conditions (W m^{-2})

	December	*March*	*June*	*September*
Vienna (202m)	10	42	73	45
Rauris (950m)	20	57	110	73
Davos (1600m)	40	93	136	75
Sonnblick (3106m)	38	124	202	98

Source: After Dirmhirn 1951

2.9. The diffuse radiation increases up to a limiting value of cloud amount that varies according to altitude. This limit is about 6/10 cover over the lowlands, but increases to 9/10 at about 2000 m. This effect represents the predominance of thinner cloud layers at the higher stations (Thams 1961b; Bener 1963) as well as multiple reflections between snow cover and cloud layers. Observations by Thams (1961a) at Locarno (380 m) confirm the general shape of the curve for lowland stations in Figure 2.9, although his graph peaks at about 5/10 cloud cover where diffuse radiation contributes 260 (280) per cent of the amount for cloudless skies in winter (summer), respectively.

The altitudinal dependence of the ratio of diffuse to global solar radiation has been estimated for cloudless skies by Klein (1948). For a solar elevation of 65°, the ratio is around 0.16 near sea level and 0.08 at 4400 m (Mount Whitney,

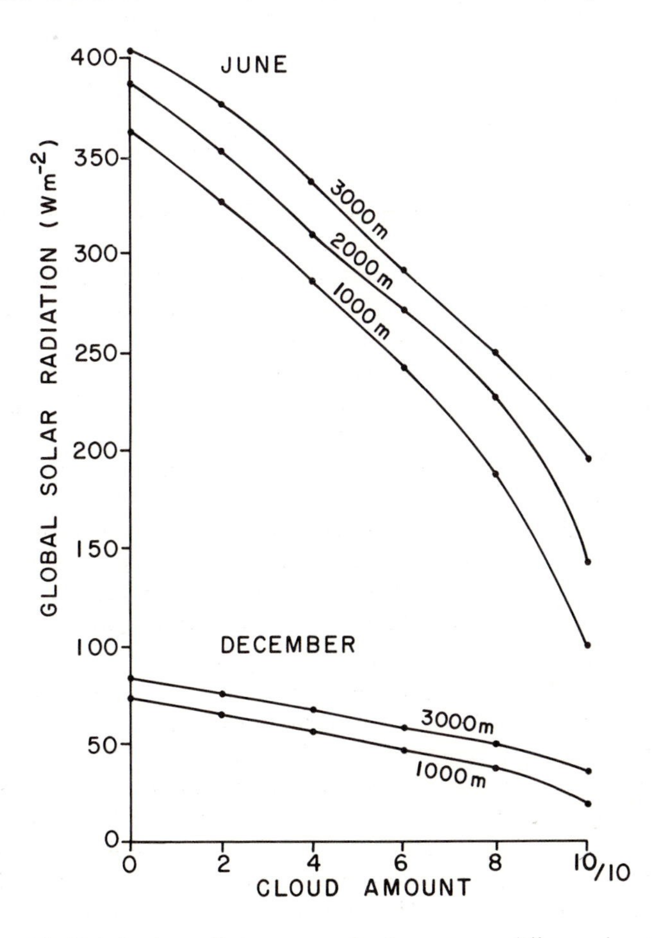

Figure 2.8 Global solar radiation versus cloud amount at different elevations in the Austrian Alps in June and December

Source: Based on Sauberer and Dirmhirn 1958

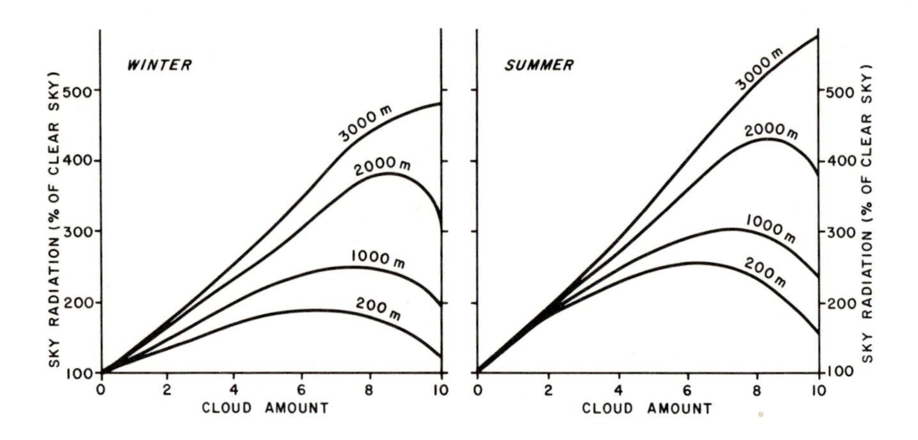

Figure 2.9 Diffuse (sky) radiation versus cloud amount in winter and summer at different elevations in the Alps
Source: From Sauberer and Dirmhirn, 1958

California). The ratio increases at lower solar altitudes. A preliminary estimate for generalized mean annual values in middle latitudes has been prepared by Flach (1966). Figure 2.10 shows the decrease in diffuse radiation with increased atmospheric transparency; the minimum of global radiation at 1.0–1.5 km represents the effect of water vapour absorption.

The effects of decreasing air density are not constant with wavelength and there is a wide range of estimates of the magnitude of the effects at short wavelengths. Free-air data indicate that the increase in ultraviolet radiation between sea level and 4 km altitude for $\theta = 90°$ decreases with increasing wavelength, although the absolute intensities are of course greater at the longer UV wavelengths (Elterman 1964). The increases are $\times$ 2.5 at 0.30 μm, $\times$ 2.0 at 0.32 μm and $\times$1.8 at 0.34 μm. The respective intensities in these three wavelengths at 4 km altitude are: 0.06 J m^{-2} min^{-1} $Å^{-1}$, 21.2 J m^{-2} min^{-1} $Å^{-1}$ and 41.6 J m^{-2} min^{-1} $Å^{-1}$, (1 Ångstrøm = $10^{-4}\mu$m). Gates and Janke (1966) estimated that alpine areas (3650 m) at 40°N receive 1.5 times more total ultraviolet radiation ($<$ 0.32 μm) for m = 1.05, and 2.2 times more for m = 2, than at sea level. Caldwell (1968), however, measured increases from sea level to 3650 m in Colorado (August–September 1966) that were only 4 per cent and 50 per cent for m = 1.05 and m = 2 respectively, for the biologically effective irradiance (UV-B) between 0.28 and 0.315 μm (which is weighted towards the shortest wavelength). He also reported an absolute decrease in sky ultraviolet radiation for these wavelengths with increasing elevation above 1500 m due to reduced atmospheric scattering. This corresponds with earlier findings in the Alps (Bckel 1936).

Limited measurements of global UV-B in Utah (Caldwell 1980), show total irradiance increasing 27 per cent from 2.77 W m^{-2} at Logan (1463 m) to 3.51 W

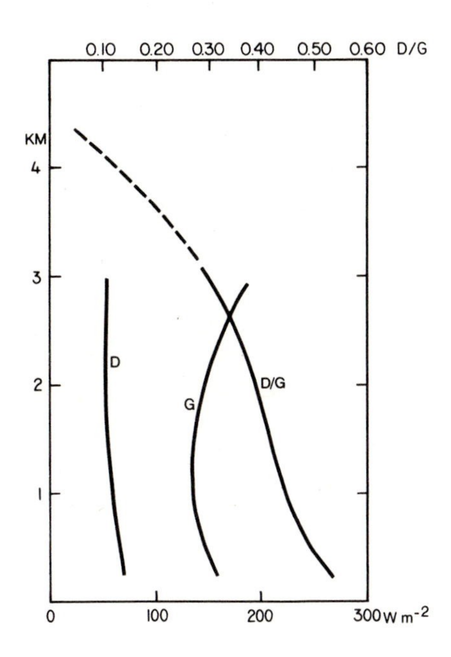

Figure 2.10 Average profiles of global (G) and diffuse (D) solar radiation and their ratio in middle latitudes
Source: After Flach, 1966

m^{-2} at Snowbird (3352 m) for a solar altitude of 63.5° compared with a corresponding total shortwave radiation increase of only 5 per cent. Using the action spectrum for DNA-effective irradiance, the increase over the same altitudinal range amounts to 35 per cent.

A summary of research carried out in the Alps, particularly by O. Eckel, indicates that direct UV-B radiation increases between 200 m and 3500 m by 100 per cent in summer and 280 per cent in winter, whereas the corresponding increases for global UV-B radiation are only 34 per cent and 72 per cent, respectively (Sauberer and Dirmhirn 1958:99–100). Air mass values for these data are not given, although the increases are generally in line with Caldwell. More recently, using an interference filter and photo-elements in the range 0.32–0.34 μm, Wessely (1969) found that in late April, 1964, the relative direct-beam UV radiation was reduced to about 90 per cent of that at the Sonnblick (3106 m) at 2700 m and to about 73 per cent at 1600 m (Figure 2.11).

Some of the most detailed continuous measurements of UV-B radiation have been made in the Bavarian Alps (Reiter *et al.* 1972). Absolute UV for the 0.305–0.335 μm band (centred on 0.314 μm) has a daily mean value for 1975–9 on the Zugspitze (2964 m) of 2.6 W m^{-2}. The corresponding extraterrestrial UV radiation is 28 W m^{-2}. Under cloudless skies, mean daily values in June average 6.3 W m^{-2}. The UV/global radiation ratio on clear days is 1.0 per cent in winter,

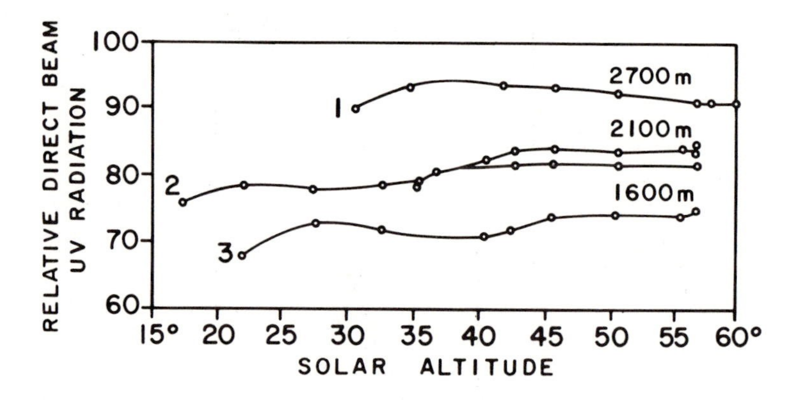

Figure 2.11 Relative values of direct-beam ultraviolet radiation versus solar altitude at three elevations in Austria, 27–28 April 1964; comparison is made with the Sonnblick Observatory = 100 per cent
Source: After Wessely, 1969

1.3 per cent in summer, and on cloudy days is 1.5 per cent and 2.2 per cent, respectively. A summary of the annual, mid-winter, and mid-summer daily values is given in Table 2.9. Reiter *et al.* (1982) also provide information on altitudinal gradients based on data for individual days. The mean increases in UV radiation between Garmisch (740 m) and the Zugspitze for these days are 27 per cent km^{-1} for cloudless days and 13 per cent km^{-1} on all days in summer; corresponding rates in winter are 17 and 14 per cent km^{-1}, respectively. These seasonal differences are attributable to variations in cloud amount and cloud vertical structure.

Cloud cover should in general attenuate UV radiation to a lesser degree than total shortwave radiation, because a larger proportion (40 to 75 per cent for global UV-B irradiance) is in the form of diffuse radiation (Caldwell 1980). This

Table 2.9 Mean values, 1975–9, of global radiation and ultraviolet radiation in the Bavarian Alps ($W m^{-2}$)

		Global Radiation			*Ultraviolet Radiation*		
Altitude (km)		0.7	1.8	3.0	0.7	1.8	3.0
June	All days	220	214	257	2.9	3.5	4.3
	Clear days	350	368	401	4.3	5.4	6.3
December	All days	28	55	59	0.5	0.5	0.6
	Clear days	44	80	91	0.5	0.8	0.8
Annual	All days	128	142	167	1.8	2.0	2.6
	Clear days	197	224	246	2.4	3.1	3.6

Source: After Reiter *et al*. 1982

is apparent in Table 2.9, although the differences are generally modest. At 3 km, the mean annual UV radiation for all days is 72 per cent of that on clear days, compared with 68 per cent for global radiation.

Maximum UV intensities are recorded just below the upper boundary of stratiform cloud layers, rather than in cloudless conditions, as a result of the scattering effect.

Snow cover can have an important effect on downward UV flux in spring. The reflected radiation undergoes atmospheric scattering causing an increase in downward diffuse UV irradiance. The effect increases at shorter wavelengths due to increased molecular scattering. Caldwell (1980) cites a 20 to 60 per cent increase (depending on solar altitude) of downward global UV flux (0.33 μm) over snow cover at Davos (1560 m), as measured by P. Bener.

For the infra-red end of the solar radiation spectrum (> 0.65 μm) there is also an altitudinal dependence. For example, Kondratyev (1969: 234), based on studies by S.P. Popov in the USSR, shows that the fraction of solar infra-red radiation to the total incoming increases from about 64 per cent near sea level to 83 per cent at the 2000 m level, for a constant optical mass of 3. There is a corresponding increase in the infra-red components in polar latitudes which, like the altitudinal effect, is a result of lower vapour contents and therefore reduced attenuation.

Table 2.10 Altitude and cloudiness effects on infra-red radiation in the Austrian Alps (for thick, low cloud) (W m^{-2})

A *Atmospheric back radiation*[a]

Altitude (m)	*December* 0/10	*December* 5/10	*December* 10/10	*June* 0/10	*June* 5/10	*June* 10/10
	Cloud cover					
200	227	255	304	323	345	390
1000	210	237	287	295	308	370
2000	193	225	275	260	292	342
3000	176	206	255	228	255	302

B *Outgoing radiation (mean values)*

Altitude (m)	*December* Bare ground	*December* Snow cover	*June* Bare ground	*June* Snow cover
200	289	301	385	–
1000	270	287	366	–
2000	255	274	355	320
3000	240	255	304	302

Note: January values are slightly lower than those for December and July values are slightly higher than those for June, in response to air temperatures primarily. The same months are used to allow comparison with Tables 2.6–2.8

Source: After Sauberer and Dirmhirn 1958

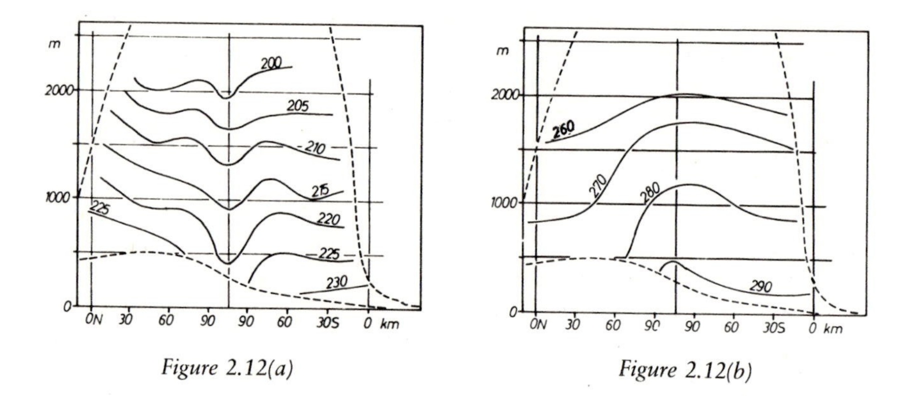

Figure 2.12(a) *Figure 2.12(b)*

Calculated infra-red radiation (a) incoming (b) outgoing, for a north–south section across the Alps
Source: From Fliri, 1971

Infra-red radiation

While the effect of reduced atmospheric density with altitude is important for solar radiation, the maximum absorptance by an atmospheric column under clear skies is only about 15 per cent of the incoming extra-terrestrial solar radiation. Infra-red radiation fluxes are significantly affected by the increased atmospheric transparency at high elevation and by the lower air temperatures.

There have been few measurements of infra-red radiation at mountain stations, except in the Alps, although there have been profile studies by balloons and aircraft which help to provide additional information. LeDrew (1975) demonstrates, from observations and model estimates on Niwot Ridge, Colorado, that estimates of atmospheric emittance under clear skies cannot be made using semi-theoretical models of the Brunt-type (1932) without special adjustments. Such models use screen-level climatological values and are successful because a large contribution to the atmospheric back radiation originates in the lowest 100 m or so. At 3500 m (640 mb), however, the corresponding optical length, in terms of its vapour content, is of the order of 1750 m. The coefficients of the Brunt expression (and other similar ones) lead to overestimates of the downward flux. Calculations of this type have been made for Austria by Sauberer and Dirmhirn (1958), but constants for the equations were based on measurements at Vienna and the Sonnblick.

In general, both the infra-red radiation emitted from the surface and the atmospheric back radiation decrease with altitude (see Table 2.10). This arises due to the lower effective temperature and, in the case of the atmospheric emittance, as a result of the smaller vapour content in the overlying air column. Calculations by Fliri (1971) over the Alps also illustrate these trends (Figure 2.12). There is a small increase in the (negative) net infra-red radiation with altitude according to Fliri, whereas the results of Sauberer and Dirmhirn (1958: 78) indicate that, for a bare soil surface under 5/10 cloud cover, the net infra-red radiation ($I\downarrow - I\uparrow$) decreases slightly with altitude:

	200 m	*1000 m*	*2000 m*	*3000 m*	$W\ m^{-2}$
January	−58.1	−57.6	57.2	−56.7	
July	−81.4	−74.2	−74.2	−74.6	

They consider that realistic regional values, taking into account surface conditions and sky cover, cannot yet be estimated.

Net radiation

Net radiation (Rn) is usually most strongly affected by the absorbed solar radiation, $S(1 - \alpha)$, since: $Rn = S(1 - \alpha) + I\downarrow - I\uparrow$. On an annual basis, the increased duration of snow cover at higher elevations causes the absorbed short-wave radiation to be reduced and, consequently, net radiation tends to decrease with elevation. The small increase in net infra-red radiation adds to this effect. Budyko (1974: 192) notes that, below the snow line, the net radiation changes

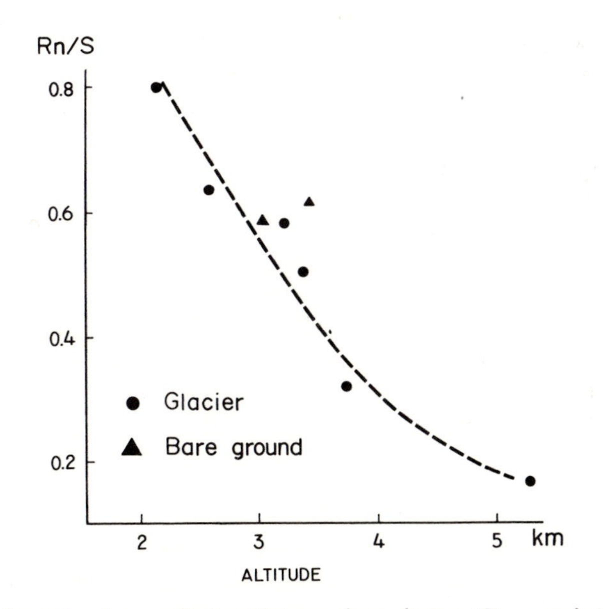

Figure 2.13 The ratio of net radiation (R_n) to solar radiation (S) versus height in the Caucasus in summer
Source: After Voloshina 1966

Table 2.11 Mean net radiation at different altitudes in Austria (W m^{-2})

Altitude (m)	*March*	*June*	*December*
500	32.0	141.5	6.8
1000	24.2	151.6	15.0
1500	25.2	164.8	15.0

Source: After Sauberer and Dirmhirn, 1958

little with altitude, since an increase in absorbed short-wave radiation (resulting from the higher totals of global solar radiation) is offset to some extent by the increase in net infra-red radiation. This altitudinal pattern will occur, at least during the snow-free season.

In the humid equatorial Andes, net radiation declines by about 20 per cent from 1000 to 3100 in elevation, and in Peru and Bolivia it decreases by 30 per cent from about 105 W m^{-2} (for daily values) at 1000 m to 75 W m^{-2} around 4000 m in association with increased effective infra-red radiation loss (Frère *et al.* 1975). The first region has a bimodal regime of annual precipitation and of incoming solar radiation, whereas in the southern tropical Andes the annual regimes are unimodal. Net radiation as a percentage of incoming solar radiation decreases from 55–60 per cent below 1000 m in the equatorial Andes to 43 per cent at 4000 m; corresponding figures for the southern tropical Andes are 50–55 per cent and 34 per cent, respectively. Where there is snow cover, the sharp increase in albedo causes a large reduction in absorbed shortwave radiation and, therefore, in net radiation. Over glacier surfaces in the Caucasus, there is, in summer months, a progressive decrease in the ratio of net to solar radiation with height (Figure 2.13), from about 0.8 at 2000 m to below 0.2 at 5000 m (Voloshina 1966). However, this is primarily related to an albedo gradient ranging from 0.28 at 2130 m to 0.74 at 5300 m.

For Austria, mean monthly values of R_n have been calculated by Sauberer and Dirmhirm (1958: 83) between 500 and 1500 m (Table 2.11). These are based on sunshine duration at 43 stations and the long-wave fluxes illustrated in Table 2.10 (p. 41). The figures show an altitudinal decrease of R_n only to about 1000 m in March and December and altitudinal increase in June when there is no snow cover effect up to at least 1500 m elevation.

Temperature

During the eighteenth century there was still considerable controversy as to the cause of the general temperature decrease with height. De Saussure, working in the Mont Blanc massif, was one of the first physical scientists to approach a realistic explanation of the cause of cold in mountains (see Barry 1978). Since the atmosphere is relatively transparent to solar radiation, 45 per cent of the incoming total is absorbed at the earth's surface as a global average. The atmosphere is heated primarily through absorption of terrestrial infra-red radiation (although the *net* effect is still one of cooling, especially above cloud layers) and by turbulent heat transfer from the ground.

The average temperature decrease with height, or the *environmental lapse rate*, approximates 6°C km^{-1} in the free atmosphere. At night and in winter, the gradient may be temporarily reversed, over limited vertical distances, in a layer of *temperature inversion.* This may occur due to nocturnal radiative cooling at the surface, large-scale subsidence in an anticyclone, or advection of a warm air mass over a colder surface.

There is an upper limit to the absolute rate of temperature decrease with height, due to the hydrostatic stability of the atmosphere. This limit – the dry adiabatic lapse rate (DALR) of 9.8°C km^{-1} – is the rate at which an unsaturated air parcel cools when it is displaced upward. The environmental lapse rate may exceed the DALR, especially as a result of surface heating. In this situation, the density difference between the surface air and that overlying it causes overturning. When air is saturated, the cooling rate of an air parcel displaced upward depends on its initial temperature, but is always less than the DALR due to the release of latent heat by the condensation process. Above 20°C, this saturated adiabatic lapse rate (SALR) is less than 5°C km^{-1}, whereas at sub-zero temperatures, the available moisture content of the air is so small that only a very limited amount of latent heat can be released. At −40°C, the SALR is almost identical with the DALR (see also p. 220).

Average lapse rates show considerable variability in relation to climatic zone, as well as to season (Lautensach and Bogel 1956; Hastenrath 1968). Highest values tend to be reached in summer over tropical deserts, whereas the strongest negative rates, due to temperature inversions, occur in eastern Siberia, north-west Canada and the polar regions in winter. Such differences render the practice of adjusting average station temperatures, or pressures, to sea level likely to produce non-comparable and therefore misleading results.

Another factor which affects the altitudinal decrease of temperature is the type of airmass. Yoshino (1966) shows that lapse rates are generally greater for northerly than southerly winds (in the northern hemisphere). For example, continental polar air is frequent in February in Japan which tends to make lapse rates a maximum in this month. Mean values for 1939–48 between Mt. Fuji and Kofu, with an altitudinal range of 3500 m, are 6.1°C km^{-1} in February compared with 5.4°C km^{-1} in November.

By comparison with the surrounding atmosphere, the slope air over a mountain is affected by radiative and turbulent heat exchanges. These processes modify the temperature structure over the massif so that lapse rates on a mountain slope may differ from those in the free atmosphere according to the time of day. In the Austrian Alps, von Hann (1906: 102) found the following values between Kolm Saigurn, 1600 m, and Sonnblick, 3106 m.

Winter 4.9°C km^{-1} at 2 a.m., 6.6°C km^{-1} at midday;
Summer 6.0°C km^{-1} at 2 a.m., 8.9°C km^{-1} at midday.

The influence of nocturnal inversions at night and of near-adiabatic conditions by day, together with the effects of föhn winds and katabatic drainage, led von Ficker (1926) to the view that 'true' lapse rates cannot be determined in mountain regions. However, a unique approach to direct measurement was performed in the Austrian Alps in a little noted study by Brocks (1940). A goniometer was used to determine density differences in air layers between five locations near Salzburg on two clear autumn days in 1938. Brocks found that the diurnal amplitude of lapse rate decreases with altitude more rapidly in the free air over the plains than over the mountains, and also that the mountain atmosphere extends above mean ridge height.

It is important to distinguish between the effects of local topography which cause diurnal changes in lapse rate, as illustrated above, and large-scale topographic effects that modify the atmospheric structure. Tabony (1985) outlines three idealized topographic situations – an isolated mountain, a plateau of limited extent, and an extensive plateau (Figure 2.14). The temperature curve is representative of a surface inversion on a clear winter night. In the first case, the variations in mountain temperature are similar to those in the free atmosphere, diurnally and seasonally. In the third case, the entire profile is displaced upward creating a seasonally-modified atmosphere and lower surface temperatures over an extensive plateau. In the case of the limited plateau, lapse rates are like those in the seasonally-modified atmosphere, but the terrain does not displace the profile upward and temperatures are independent of altitude. Tabony notes that, in the Austrian Alps, winter temperature profiles show a 6°C km^{-1} lapse rate near the surface and an isothermal layer between 700 m and 1400 m. The main valleys enclosed by mountains are at about 700 m; above 1400 m the mountain slopes project into the free atmosphere, while locations between 700–1400 m are still enclosed. Radiosonde soundings near the Alps show slightly warmer conditions in the lower layers in winter, with a surface inversion. Thus, mean winter temperatures up to 500–600 m above the surface are modified, comparable to the depth of diurnal variations, and this layer has been displaced upward.

Thermal contrasts between low-level snow-free ground and snow-covered mountain slopes can also modify lapse rates. The period of snow melt when air temperatures are held close to 0°C is especially important. The date of snow melt is retarded considerably at higher altitudes; it occurs in the Austrian Alps in late March at 2000 m, in late April at 2500 m, and in June at 3000 m according to Geiger (1965). Moreover, snow depths are greater at higher altitudes and the altitudinal gradient of snow depth is increased in spring as the snow boundary rises. Thus, a steep lapse rate is to be expected across the zone of the melting snow.

British data show that altitudinal effects are dominant for height gradients of

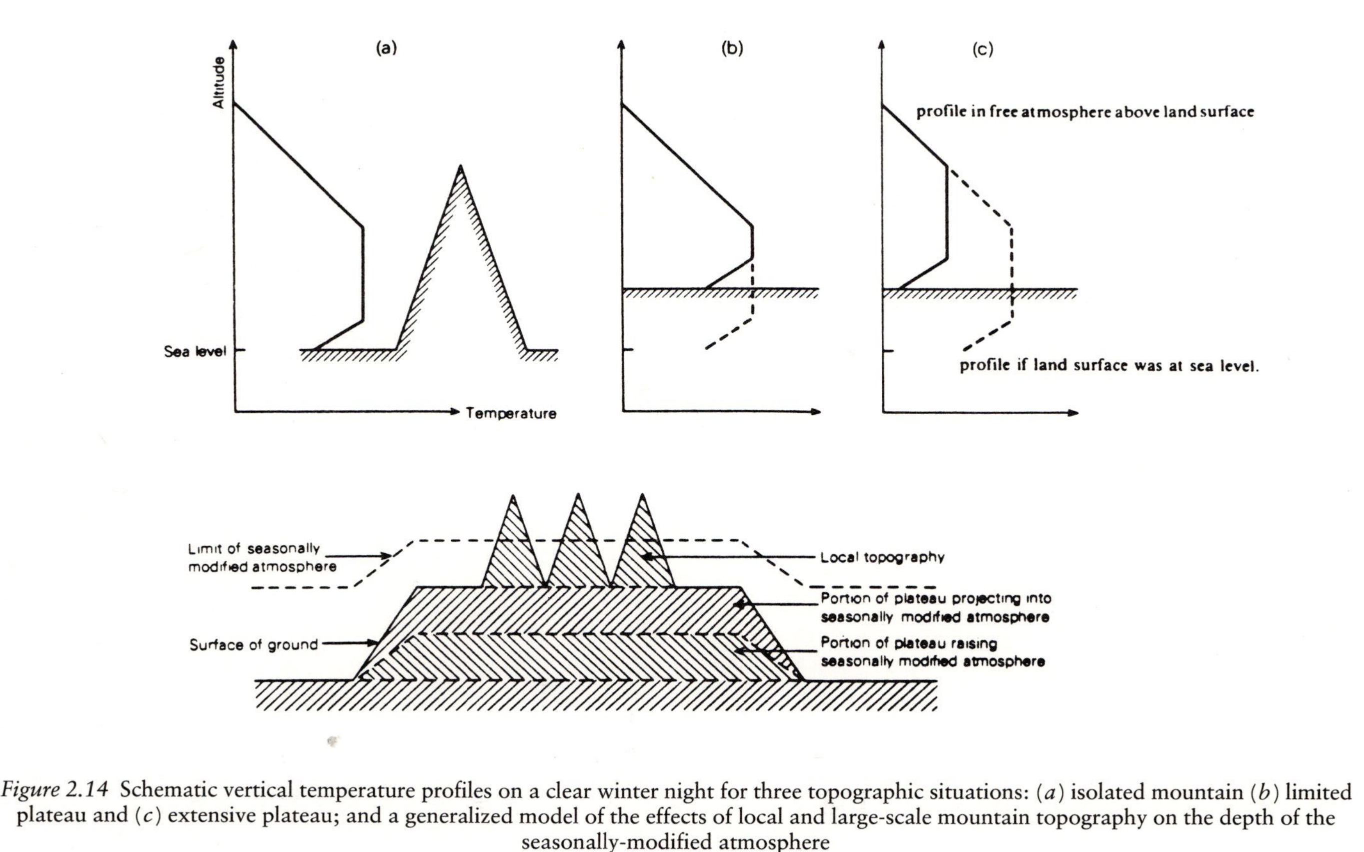

Figure 2.14 Schematic vertical temperature profiles on a clear winter night for three topographic situations: (*a*) isolated mountain (*b*) limited plateau and (*c*) extensive plateau; and a generalized model of the effects of local and large-scale mountain topography on the depth of the seasonally-modified atmosphere

Source: After Tabony 1985 (reproduced with permission of HMSO)

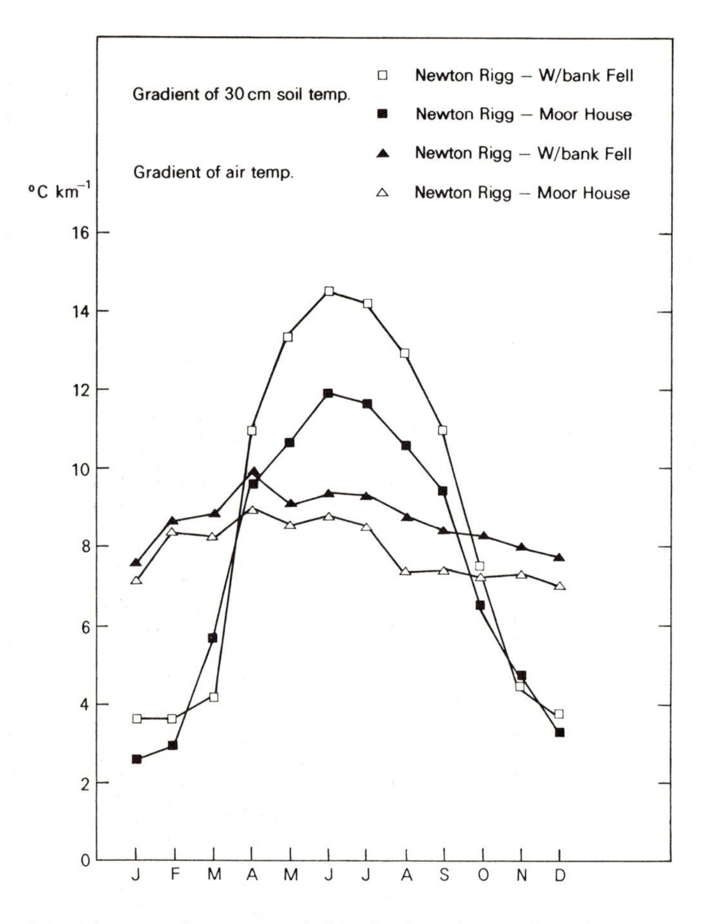

Figure 2.15 The annual variation of altitudinal gradients of air temperature and 30 cm soil temperature between two upland stations in the Pennines of northern England and the lowland station of Newton Rigg (171 m)
Source: From Green and Harding, 1979

mean maximum temperatures, whereas local topography is at least as important for minimum temperatures (Harding 1978). The mean gradient of maximum temperatures in Great Britain is about 8–9°C km^{-1} but with a general winter minimum (6–7°C km^{-1}) and spring maximum (9–10°C km^{-1}) (see Figure 2.15). The spring maximum is apparently *not* due to airflow direction, but may be related to the seasonal increase of instability; the atmosphere is generally more stable in winter. Both the form and magnitude of this pattern are also present in gradients of mean temperature in southern Norway and central France (Harding 1978: Figure 3). Using data from the Pennines of northern England, Harding (1979) shows that a linear relationship exists between the gradient of daily *maximum* temperature and the difference in sunshine hours between lowland

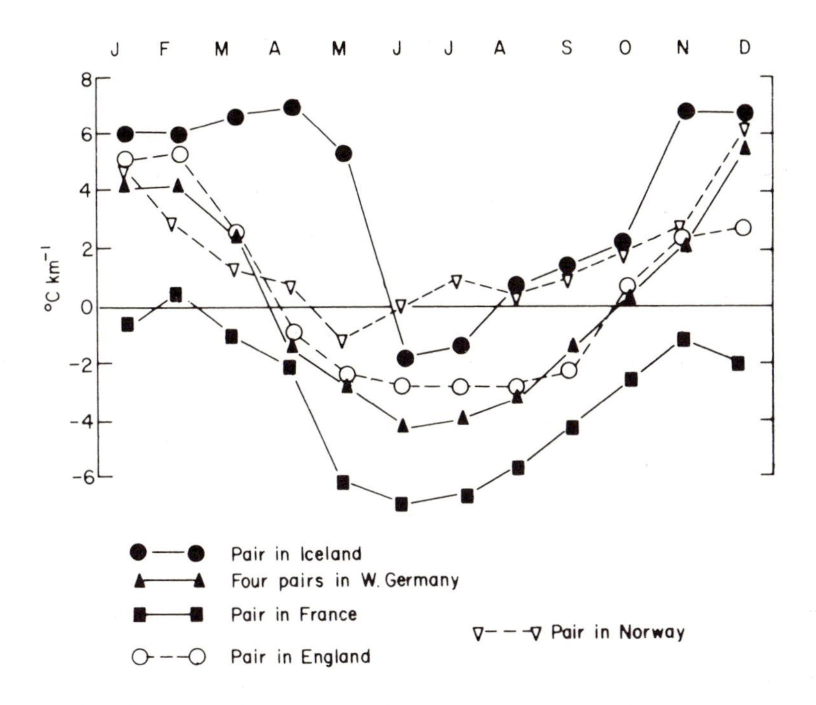

Figure 2.16 Differences between altitudinal gradients of soil and air temperature at pairs of stations in Europe (positive values = temperature gradient in air exceeds soil)
Source: From Green and Harding 1980

and upland sites. (The upland stations are between 500–800 m a.s.l.) Where there is a positive excess of two hours of sunshine at lowland stations, the altitudinal gradient of maximum temperatures averages about 9.5°C km^{-1}. The reduction of solar radiation by cloud cover in the uplands will reduce the sensible heat flux into the boundary layer, thereby potentially lowering screen level maximum temperatures. For a two-hour reduction of the upland sunshine duration, maxima may be lowered about 0.5°C according to Harding's estimates. The observed gradient of maximum temperature increases approximately 0.45°C km^{-1} per hour of increase in lowland minus upland sunshine duration.

In the case of altitudinal gradients of soil temperature, recent work in Great Britain shows clearly that the effects of soil properties are subordinate to meteorological controls (Gloyne 1971; Green and Harding 1979). Gloyne found the annual mean altitudinal gradients of air and soil temperature to be similar, based on 0900 GMT observations which approximate the daily minimum temperature in the soil. Records of daily maximum and minimum soil temperatures for two upland stations in Wales at just over 300 m a.s.l. and a lowland station (30 m) show strong seasonal variations in gradient, ranging from 1°C km^{-1} in winter to 6°C km^{-1} in summer for maxima and from 0°C to 4°C km^{-1} for minima (Green and Harding 1979). This matches the earlier finding of

Harrison (1975) for maximum soil temperatures in the same locale. Confirmation that this seasonal pattern is not a result of coastal influences or of soil properties, is provided by Green and Harding's analysis of 30 cm soil temperatures between three station pairs in and around the Pennines (with the upland stations at 400–550 m) and ten other station pairs with an altitude difference of 200–300 m in northern England and Scotland. Subsequent work by the same authors for stations in Europe shows it to be a general result, with the seasonal range of soil temperature being of similar magnitude down to at least 1 m depth (Green and Harding 1980).

Whereas the seasonal amplitude of the altitudinal gradient of air temperature is only 2°C km^{-1} in Britain, that for soil temperatures is 10–11°C km^{-1} (Figure 2.15). Similar amplitudes in these respective gradients are recorded in most other European countries (Green and Harding 1980). The seasonal amplitude shows little spatial variation in the case of soil temperature gradient (although absolute values vary), but the magnitude of the altitudinal gradient of air temperature decreases with increasing continentality in Europe and the seasonal amplitude of the gradient shows a corresponding increase. Figure 2.16 illustrates *differences* in altitudinal gradients of air and soil temperatures for several locations. The seasonal patterns are remarkably similar, with steeper altitudinal gradients of soil temperature in summer and of air temperature in winter. Absolute values of the gradient differences are somewhat affected by horizontal climatic gradients between the station pairs and also perhaps by latitudinal influences. The excess of soil temperature lapse rates over the dry adiabatic rate in summer implies that the upland soils are relatively cool. This is probably due to the greater use of available energy for evaporation from the moist upland peats (Oliver 1962; Green and Harding 1979). The cause of the excess of upland soil temperatures over air temperature in winter is more problematic. Transient snow cover should only diminish the diurnal cycle without raising the mean temperature and it does not appear that the soil could maintain the required upward flux of sensible heat (approximately 25 W m^{-2}) according to Green and Harding (1979). The intermittent freezing of soil water is a further possible heat source, but again this effect does not match the observed increase of soil maximum and minimum temperatures in Wales, nor the occurrence of this positive soil–air difference with frozen ground and snow cover in the mountains of East Germany in January–February 1963 (Green and Harding 1980).

While the altitudinal gradients of soil and air temperatures may be broadly similar, the absolute values differ. In the Colorado Rocky Mountains, the mean annual 0°C isotherm of air temperature occurs at about 3225 m whereas that for the soil (15–30 cm depth) is almost 400 m higher (Thompson 1990). Although the insulating effect of snow cover on winter ground temperatures must play a role, a multiple regression analysis of weekly soil temperatures with air temperature, precipitation, wind speed and snow depth shows that only a few per cent of the variance in soil temperatures is accounted for by the inclusion of variables additional to air temperatures (Barry 1972). Absorption of solar radiation by

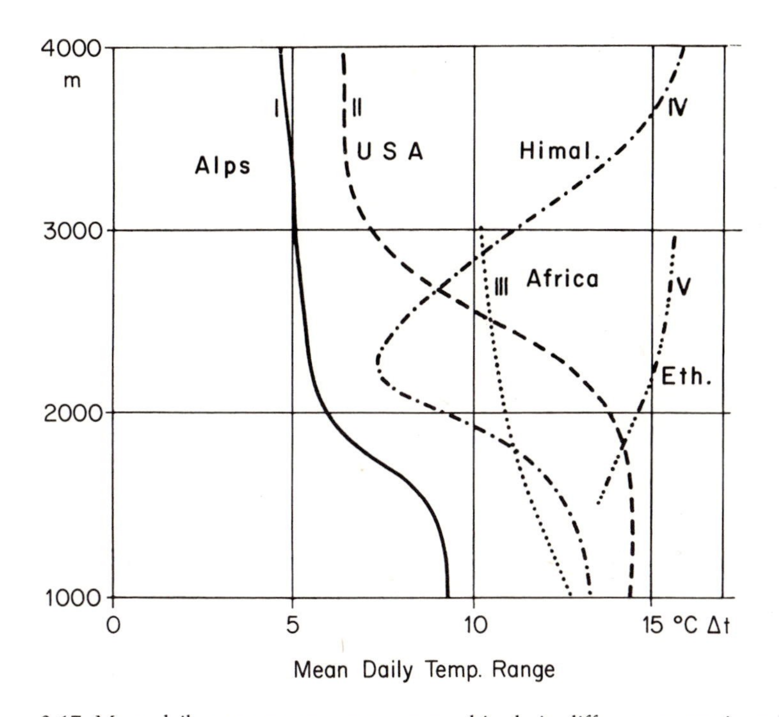

Figure 2.17 Mean daily temperature range versus altitude in different mountain and highland areas: I, Alps; II, western USA; III, East Africa; IV, Himalaya; V, Ethiopian highlands

Source: After Lauscher, 1966

the ground in summer is the likely cause.

The diurnal range of temperature in the free air decreases with increasing altitude. Typically, in the middle latitudes, it is of the order of 1.5°C at 850 mb, 1.0°C at 700 mb and 0.7°C at 500 mb. There is a phase lag with increasing height; observations in the Alps in summer 1982 showed that the temperature maximum at 850 mb occurs about half an hour later than at the ground. Surprisingly, however, the phase was reversed at higher levels, probably as a result of local advection effects (Richner and Phillips 1984). Mountain slope and summit data reflect the same decreasing amplitude of diurnal air temperature range with increasing altitude, reflecting the mixing of slope air with the free atmosphere. This is clearly apparent in middle latitudes where the westerlies increase with height. Figure 2.17 illustrates this for the Alps and North America. The greater dryness of the air over western North America may account for the larger range compared with the Alps. Lauscher (1966) suggests that the high moisture content of the air is the cause of the small altitudinal change on the mountains of equatorial Africa. The contrasting curve for the Tibetan-Himalayan region in Figure 2.17 appears to result from cloud and precipitation effects and site location. There is a minimum diurnal range of temperature in the zone of

Table 2.12 Components of the daily temperature fluctuation (°C), Austrian Alps

	All days (January 1931–June 1934)		*Clear days (< 3/10 cloud)*	
Periodic	*Winter*	*Summer*	*Winter*	*Summer*
Sonnblick, 3106 m	1.0	2.0	1.6	4.2
Rauris, 943 m	7.6	10.7	12.4	17.3
Aperiodic				
Sonnblick, 3106 m	4.4	4.4	4.0	5.4
Rauris, 943 m	10.2	13.1	14.4	17.9

Source: After Steinhauser 1937

maximum precipitation on the southern slopes and then a substantial increase of the range at stations located in high-lying valleys such as Leh (3496 m) and Lhasa (3685 m). This site factor is probably likewise responsible for the profile of the curve for Ethiopia in Figure 2.17. A further study of worldwide data by Linacre (1982) shows evidence of an increasing daily range between sea level and 200 m, apparently related to coastal sea breezes. Between 750 m and 3400 m, the diurnal range generally decreases in response to stronger winds or greater cloudiness, as noted above.

The controls of the daily temperature fluctuation have been demonstrated by Steinhauser (1937) for stations in the Sonnblick area, Austria, by a comparison of their periodic and aperiodic components. The periodic component is determined from the mean daily range of a long record; the aperiodic one from the average of the daily extreme temperatures (Conrad, 1944: 202). Thus, in Table 2.12, the periodic component is part of the aperiodic fluctuation. The periodic component was found to be a smaller proportion of the daily fluctuation at the mountain stations in both seasons indicating the role of synoptic variability, whereas at valley stations, such as Rauris, the large amplitude depends primarily on regular local changes, especially in clear weather.

In 1913, von Hann noted that temperatures observed at summit stations (Sonnblick and Obir, Austria), on average, are lower than those in the free air at

Table 2.13 Mean temperature differences between the Brocken (1134 m) and the free atmosphere at Wernigerode, April 1957–March 1962

	J	*F*	*M*	*A*	*M*	*J*	*J*	*A*	*S*	*O*	*N*	*D*	*Year*
00 GMT	−2.2	−1.9	−1.8	−1.4	−1.5	−1.7	−1.9	−1.9	−2.0	−2.3	−1.8	−1.9	−1.8
12 GMT	−1.2	−0.5	−0.6	−0.4	−0.7	−1.0	0.5	0.3	0.7	0.3	−0.8	−0.4	0.0

Note: Local time is one hour ahead of GMT
Source: After Hänsel 1962

Table 2.14 Frequency (%) of temperature differences (°C), Zugspitze minus free air, 1910–28

		(–) *⩽ 4.3*	*−4.2 to −1.3*	*−1.2 to 1.7*	*⩾ 1.8 (°C)*	*No. of obs.*
October–March	a.m.	15.8	35.1	41.8	7.3	876
	p.m.	9.7	27.3	45.1	17.9	421
April–September	a.m.	6.5	32.7	51.5	9.3	1442
	p.m.	2.6	18.9	50.5	28.0	503

Source: After Peppler 1931

the same level. He referred to this as an apparent paradox, in view of the expected effect of the mountain acting as a heat source. Numerous comparisons of mountain temperatures and comparable data from aircraft or balloon soundings have been made (Kleinschmidt 1913; von Ficker 1913; Ferguson 1934; Samson 1965). The most thorough analyses of this question, however, have been carried out by Peppler (1931) in the northern Alps, Eide (1948) in Norway and von Hänsel (1962) for the Harz Mountains, Germany (see Tables 2.13–2.15; Figure 2.18).

Peppler found a strong dependence of the temperature difference on cloud amounts over the mountain. The Zugspitze is always colder on clear mornings in any season (Table 2.15), but the difference is small with overcast skies and in the afternoon, with > 8/10 cover on the mountain, temperatures are on average a little higher than in the free air, particularly in summer.

Overheating of the thermometers in an instrument shelter during temporary calm conditions is unlikely to be a frequent event on most mountain summits and can, therefore, probably be ruled out as the cause of positive departures in summer. The negative differences increase with increasing wind speed, up to about 6 m s^{-1}, and then decrease, except on summer afternoons. Maximum negative values occur with northerly winds on the Zugspitze, but in the case of

Table 2.15 Average temperature differences (°C), Zugspitze minus free air, according to cloud conditions

		Cloud amount (tenths)			
		0–1	*4–5*	*8–9*	*10*
October–March	a.m.	−2.8	−2.2	−1.6	−0.3
	p.m.	−2.2	−0.8	0.1	0.5
April–September	a.m.	−2.0	−0.9	−0.5	−0.2
	p.m.	−1.2	−0.4	1.0	0.8

Source: After Peppler 1931

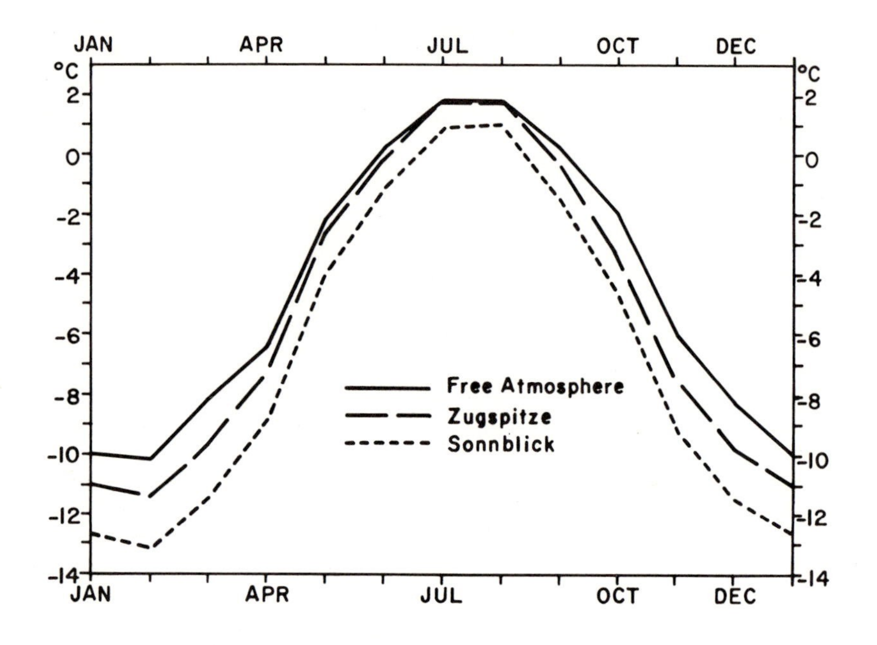

Figure 2.18 Mean daily temperatures in the free air and at mountain stations in the Alps
Source: After Hauer 1950

Säntis peak with southerly winds. Variations according to wind direction arise as a result of lee effects, including föhn occurrences (see also Peppler 1935). In Norway, Eide (1948) also found that the negative temperature difference on mountain summits, which has an annual mean value of −2.5°C in the case of Gaustatoppen (60°N, 9°E; 1792 m) compared with the free air over Kjeller 140 km to the east, increases almost linearly with increasing wind speed. The correlation between summit winds and summit-free air temperature difference is +0.61 in the case of Fanaråken (61.5°N, 8°E; 2061 m), and +0.46 for Gaustatoppen where the wind data seem subject to local site effects. The negative temperature difference is larger for south-easterly winds than for north-westerly winds at Gaustatoppen, whereas on Fanaråken it is greater with westerly than with easterly winds. In each case, smaller negative departures on the summits occur when the wind direction is such that the station is on the leeward side of the mountains. Eide also points out that some of the temperature difference between northerly and southerly winds at Gaustatoppen could be a result of the thickness relationship with temperatures being lowest in an upper low pressure centre. The pressure differences between Gaustatoppen and Kjeller are greatest for isobars oriented north–south and a 4-mb pressure gradient would account for the 0.7°C difference between northerly and southerly airflows. However, this pressure gradient seems to be too large for the distance involved.

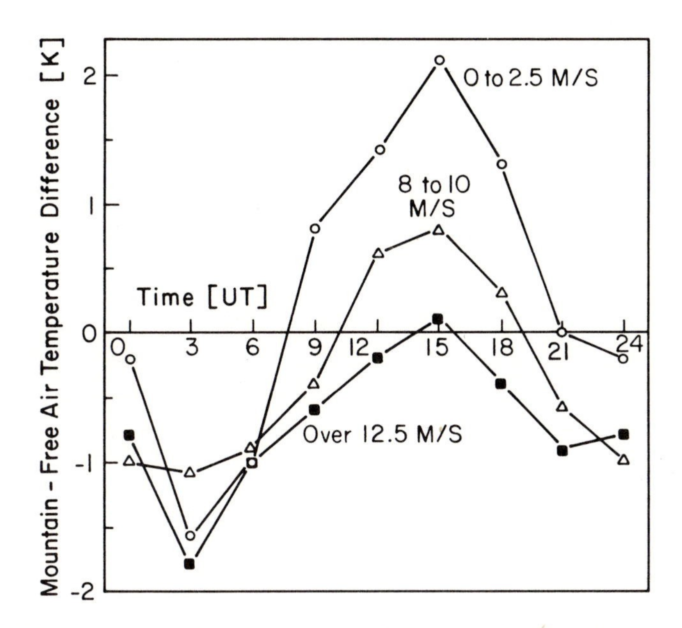

Figure 2.19 Mean summit-free air temperature differences (*K*) in the Alps as a function of time (*UT*) and wind speed

Source: After Richner and Phillips 1984

The question has been re-examined in the Alps by Richner and Phillips (1984). Frequent rawinsonde flights were made in summer 1982 and the data compared with mountain-top station values, after corrections for diurnal variations. The mountain stations were slightly cooler, but the twenty-four mean differences, although statistically significant, were less than 1°C. However, in light winds the difference can reverse with the mountain summits 0.3°C warmer than the free air (averaged over 24 hours) and up to 2°C warmer at the summit in early afternoon (Figure 2.19). Cloud effects are only apparent for cloudfree and overcast conditions (Figure 2.20), in contrast to the dependence noted by Peppler. Undoubtedly, the pre–1940s measurements in the free air are likely to have been less accurate or reliable. Nevertheless, Richner and Phillip's results are only for a single summer (94 ascents) and for separations of about 100 km between station pairs. Moreover, even modern instruments can show differences of 1°C between sondes. The conclusion is that the mountain top–free air difference is still not conclusively determined. Another recent study (McCutchan 1983) comparing eight stations on the south-facing slope of the San Bernadino Mountains, California, with rawinsonde data collected only 2 km to the west on 19 days in summer–autumn 1975 shows the mountain slope (590–1621 m) to be warmer than the free air at 10 a.m. and 4 p.m. and colder at 4 a.m. (PST). The humidity mixing ratio was higher on the mountain than in the free air at 10 a.m. and 4 p.m. and lower at 4 a.m. The differences were larger than those of Richner and Phillips in the Alps.

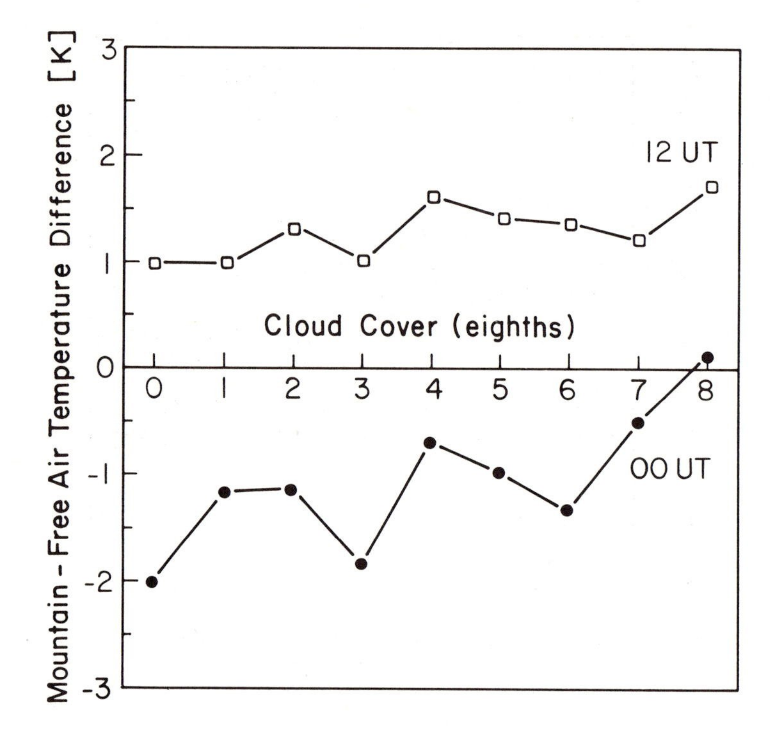

Figure 2.20 Mean summit-free air temperature differences (*K*) in the Alps as a function of cloud cover (eights) for 00 and 12 UT
Source: After Richner and Phillips 1984

From several studies, the primary control of free air–summit temperature differences seems to be the atmospheric temperature structure. Peppler found that, in the Alps, mountain temperatures are closest to those in the free air when the lapse rate is nearly adiabatic between one and three kilometres whereas, with isothermal or inversion conditions, temperatures in both summer and winter are considerably lower on mountain summits. Eide reported a negative correlation of −0.36 between the summit–free air temperature difference between Gaustatoppen and Kjeller and the lapse rate of Kjeller. This question has been elaborated by Ekhart (1939). Based on data for the period 1930–7 at stations in the northern Alps, and in the free air at the same levels over Munich, he showed that the mean diurnal variation of temperature (between 0700 and 1400) was greater at the mountain stations. Following Peppler, he attributed this to adiabatic cooling, due to forced ascent, up to about 1 km, but suggested that above 2 km, the main effect is due to thermal upcurrents. Eide proposes that air mass lifting takes place in proportion to the wind velocity. The average lifting at both Fanaråken and Gaustatoppen is estimated to be of the order of 500 m, assuming dry adiabatic ascent but since the process will be at least partly a saturated-adiabatic one, this figure is probably an underestimate. In addition, where the mountains are isolated peaks, it is clear that 'mountain air' is subject to greater mixing with the free air than would be the case over an extensive mountain massif or elevated plateau. This mixing limits the degree of day-time, and mainly summer-time, heating of the mountain surface. At night radiative cooling from the mountain will probably cause a mean negative temperature departure even in the absence of lifting effects on the air mass. The Brocken data, however, indicate little apparent effect due to the presence or absence of snow cover on the nocturnal cooling rate. The peak is colder than the free air at midnight by an average of 1.4°C to 2.2°C in all months of the year (Table 2.13, p. 52).

The existence of generally negative temperature differences between mountain summits and the free air is important to the question of the so-called mass-elevation (*Massenerhebung*) effect. This concept was introduced by A. de Quervain (1904) to account for the observed tendency for temperature-related parameters such as treeline and snowline to occur at higher elevations in the central Alps than on their outer margins. In studies using this concept, terrain elevation is spatially averaged. The idea has been widely applied in ecological studies and the particular case of the Alps is examined further in Chapter 5 (p. 299). Here, we are concerned with general meteorological considerations.

Flohn (1953) first proposed that elevated plateau surfaces, such as those of Tibet and the Altiplano in South America, are warmer in summer than the adjacent free air as a result of the altitudinal increase in solar radiation and the relative constancy of the effective infra-red radiation with height. The data in the previous sections (p. 42–4) (see also Table 2.10, p. 41), confirm this view. Two factors contribute to the heating effect in the mountain atmosphere. One is sensible heat transfer from the surface, the other is the latent heat of condensation due to precipitation from orographically induced cumulus development. Flohn demonstrated that the flux of sensible heat into the atmosphere over the Alai-Pamir Mountains in summer is of the same order of magnitude as that over sea level desert surfaces in the same latitudes (Flohn 1968, 1974).

Recent calculations indicate a total daily energy transfer from the plateau to the atmosphere of 230 W m^{-2} in June (Yeh 1982). Sensible heat flux is important over the drier western part of the Tibetan plateau, where it reaches 220 W m^{-2}. This is indicated by afternoon lapse rates which reach up to 9.0°C km^{-1} between 4 and 7 km in the Soviet Pamir (*c.* 38°N, 74°E). Over southeastern Tibet and the eastern Himalaya-Assam, however, convective activity, due to forced ascent of air against the mountains ('Stau'), provides a major heat input from latent heat of condensation. East of 85°E, the latent and sensible heat terms are almost equal (90 and 100 W m^{-2}, respectively). The maximum heating rates in June for the layer between 600 and 150 mb amount to +1.8°C day^{-1} from sensible heat, +1.4°C day^{-1} from latent heat, and radiative cooling of −1.5° day^{-1} giving a net heating of +1.7°C day^{-1} (Yeh 1982). Various estimates suggest that the heating is about 2°C day^{-1} over the eastern half of the plateau (Chen *et al.* 1985). Advective

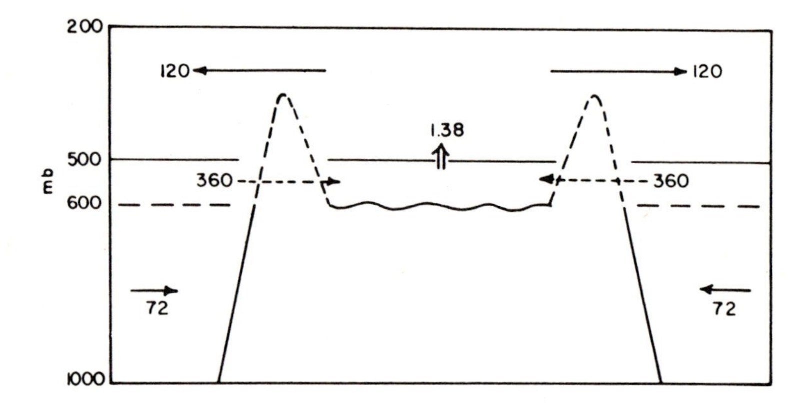

Figure 2.21 Components of the mean daily thermal circulation above Tibet, cm s^{-1}. The main inflow is between 600–500 mb across the mountains
Source: From Flohn, 1974

transports export the net heat and there is mean ascent of about 1 cm s^{-1} as illustrated in Figure 2.21. The heating effect is of great importance in the development of an upper tropospheric anticyclone centred above 30°N, 85°E. Yeh notes that the summertime surface low and upper high vary diurnally, being much stronger in late afternoon than in the early morning hours. Similar atmospheric heating estimates have been made by Gutman and Schwerdtfeger (1965) and Rao and Erdogan (1989) for the Andes where an upper anticyclone forms in summer over the Altiplano (Kreuels *et al.* 1975). Rao and Erdogan suggest that the main source of heating in January is latent heat released over the eastern margins of the Bolivian Plateau.

Flohn (1953) notes that the windward 'Stau' effect is operative over mountain ranges in all latitudes, even low ones, and adds to the dynamic effect of mountain barriers in the mid-latitude westerlies by causing a ridge of higher pressure over the range (see Chapter 3, p. 110). However, the sensible heat contribution is generally only effective in summer and is absent where the mountains are snow-covered. Hence, the temperatures observed on 15 days in summer 1956, between 4700 and 7000 m on Pobeda Peak in the Tien Shan Mountains, averaged 1.8°C less than free-air temperatures over Alma Ata according to Borisov *et al.* (1958). It is evident that the *Massenerhebung* concept can only be applied after careful consideration of the region and the meteorological factors that are involved.

Despite the differences between slope conditions and those in the free-air, radiosonde data can sometimes be used as a guide to mountain temperatures. A study in coastal British Columbia (Peterson 1969) shows that the frequency of below-freezing temperatures in winter can be estimated from 04.00 h radiosonde soundings, although periods of temperature inversion were excluded from the analysis.

In general, patterns of mountain temperature distribution can be described

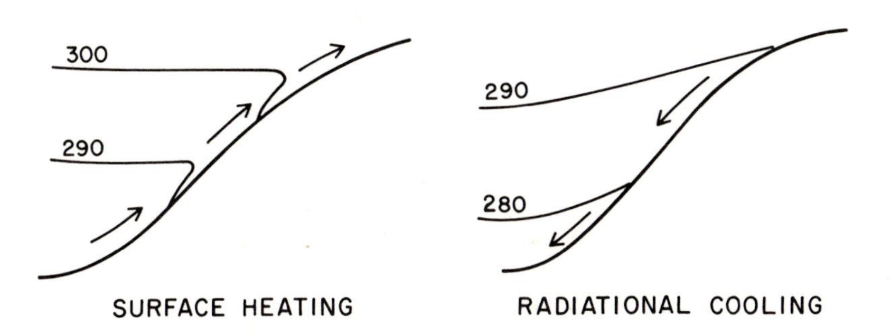

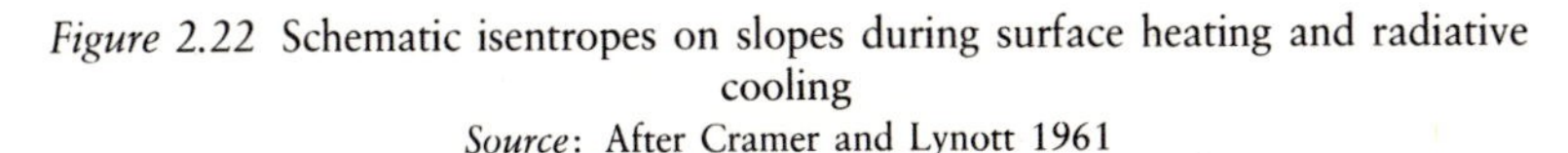

Figure 2.22 Schematic isentropes on slopes during surface heating and radiative cooling
Source: After Cramer and Lynott 1961

conveniently by means of potential temperature analysis. The potential temperature, θ, represents the temperature of an air parcel brought adiabatically to 1000 mb pressure, thus $\theta = T(1000/p) \exp (R/c_p)$, where p = pressure (mb), T = temperature (°K), R = gas constant for dry air and c_p = specific heat of dry air at constant pressure; $R/c_p = 0.288$. This conversion is readily carried out graphically on a thermodynamic chart if station elevation is converted to pressure. Little error is generally introduced if Standard Atmosphere equivalents are used. Surfaces of potential temperature, or *isentropic surfaces* represent the path of air moving adiabatically (unaffected by diabatic processes such as radiative warming or cooling). Cramer (1972) shows that isentropes of surface air temperature generally parallel the terrain contours in the morning hours when the air is stably stratified. As slopes are warmed and a mixed layer forms, the isentropes tend to intersect the topography. This is especially evident on cross-sections (Figure 2.22). Sloping isentropes denote atmospheric discontinuities such as fronts, sea breezes in coastal areas, and slope breezes or mountain/valley winds. Thus, such potential temperature analyses are a useful basis for inferring likely airflow patterns in mountain areas, especially during summer situations with weak pressure gradients.

If the objective of a temperature analysis is to examine air mass characteristics or the location of fronts, in a mountain area, it is preferable to analyse fields of equivalent potential temperature,

$$\theta_E = T_z + \left\{\frac{L_v r}{c_p}\right\} + \left\{\frac{g \Delta z}{c_p}\right\}$$

where L_v = latent heat of condensation ($2.5 \times 10^6 \mathrm{J\ kg^{-1}}$)
r = humidity mixing ratio ($\mathrm{g\ kg^{-1}}$)
T_z = temperature (°K) at level z
Δz = height difference (km) between z and the reference height (msl or 1000 mb)

$g \;= 9.81 \text{ m s}^{-2}$

$c_p = $ specific heat of dry air (1004 J kg^{-1} K^{-1})

For reduction to sea level,

$\theta_E = T + 2.5r + 9.81z$

θ_E represents the temperature attained by an air parcel following adiabatic condensation of all water vapour in the air at constant pressure. It is conservative for unsaturated and saturated adiabatic processes.

Wind

The most important characteristics of wind velocity over mountains are related to their topographic, rather than their altitudinal, effects. Nevertheless, it is appropriate to make brief note of the latter here. In middle and high latitudes it is normal to expect that, on average, there will be an increase of wind speed with height, due to the characteristics of the global westerly wind belts (e.g. Reiter 1963). Isolated peaks and exposed ridges experience high average and extreme speeds as a result of the limited frictional effect of the terrain on the motion of the free air. In some locations, terrain configuration may even increase wind speeds near the surface above those in the adjacent free air. Thus, Mount Washington, New Hampshire (1915 m) has a mean speed of 23 m s^{-1} in winter and 12 m s^{-1} in summer (Eustis 1942) and has recorded a peak gust of 103 m s^{-1}. More typical is the Sonnblick, Austria (3106 m) with a mean speed of 7 m s^{-1} (Steinhauser 1938). In contrast, winds on equatorial high mountains appear to be much lighter. In New Guinea at 4250 m on Mt. Jaya (Mt. Carstenz), a mean value of only 2 m s^{-1} is reported during December–February (Allison and Bennett 1976) and on El Misti, Peru (4760 m) there is an estimated mean speed of about 5 m s^{-1} with a recorded maximum of 16 m s^{-1} (Bailey 1908). Generally, in the tropics, the easterly trade winds weaken with height. In the winter season, on their poleward margins, they may give way to westerly winds associated with the extra-tropical westerly air circulation. Synoptically, this is most likely when polar troughs in the upper air penetrate into tropical latitudes. In southern Asia there is a marked seasonal change-over, from strong westerly flow over the Himalaya between about October and May, on average, to moderate easterly winds from late June through September. This topic is dealt with more fully in Chapter 5, p. 302.

The effect of mountains on the wind flow over them aroused early interest through the manned-balloon flights of von Ficker (1913) especially. On the basis of these observations, pilot balloons and kite surroundings, Georgii (1922, 1923) argued that wind speeds generally increase above mountain summits up to a level corresponding to about 30 per cent of their absolute altitude, which he termed the 'influence height'. Most of his observations were from the 820 m Feldberg (Taunus), however, and A. Wagner questioned the generality of the results, also pointing out that the *relative* relief would more likely determine an influence height. This topic is discussed more fully on p. 70.

The two basic factors which affect wind speeds on mountain summits operate in opposition to each other. The vertical compression of airflow over a mountain causes acceleration of the air, while frictional effects cause retardation. Frictional drag in the lower layers of the atmosphere is caused partly by 'skin friction' (shear stress), due to small-scale roughness elements ($<$ 10 m dimensions) and by 'form drag' caused by topographic features 0.1–1 km in size that set up dynamic pressure perturbations. In mountain areas, the latter contributes the largest proportion of the total friction. Over simple two-dimensional terrain, drag increases in proportion to $(\text{slope})^2$, up to the point where flow separation from the surface takes place according to Taylor *et al.* (1989). Balloon measurements in the central Alps indicate that drag influences extend up to about 1 km above the local mean ridge altitude of 3 km (Müller *et al.* 1980). Special soundings made during ALPEX show that the airflow over the central Swiss Alps is decelerated up to about 4 km (600 mb). Ohmura (1990) suggests that momentum transfer between the atmosphere and the mountains takes place from about 4 km down to 500 m below the ridge tops (3000 m in the study area near the Rhône Glacier, Gletsch). Many of the wind profiles indicated a wind maximum at 1.5 km above the ridges, with speeds greater than over the adjacent lowlands. Such acceleration occurs during warm air advection which sets up an inversion above the valley atmosphere.

Microbarograph measurements along a cross-section from Zurich to Lugano during ALPEX indicate a mean pressure drag of +0.8 Pa ($\pm$5 Pa) according to Davies and Phillips (1985). They note that a 5 Pa drag would be sufficient to deplete the momentum of the entire troposphere in traversing the Alps. However, due to its largely east–west orientation, the Alpine region exerts only a slight total mean drag as a result of alternating synoptic regimes with northerly and southerly flows.

By considering friction and compression, Schell (1936) attempted to explain contrasting observations with tethered balloons on three summits in the Caucasus at about 1300 m. He concluded that in the case of an isolated peak, or an exposed ridge, the compressional effect outweighs frictional retardation, giving stronger winds up to about 50–100 m over the summit than in the overlying free air. The acceleration due to 'compression' is attributable to a 1–2 mb pressure reduction as a result of the streamline curvature over the crestline – the so-called 'Bernoulli effect' (Davidson *et al.* 1964). The Bernoulli effect may cause mountain-top pressure observations to be unrepresentative of the large-scale flow at that level.

For steady incompressible, frictionless flow along a streamline, Bernoulli's equation states:

$$\frac{V^{22} - V^{21}}{2} + \frac{P_2 - P_1}{\rho} + g(Z_2 - Z_1) = 0.$$

The first term denotes kinetic energy, the second the work done by the pressure

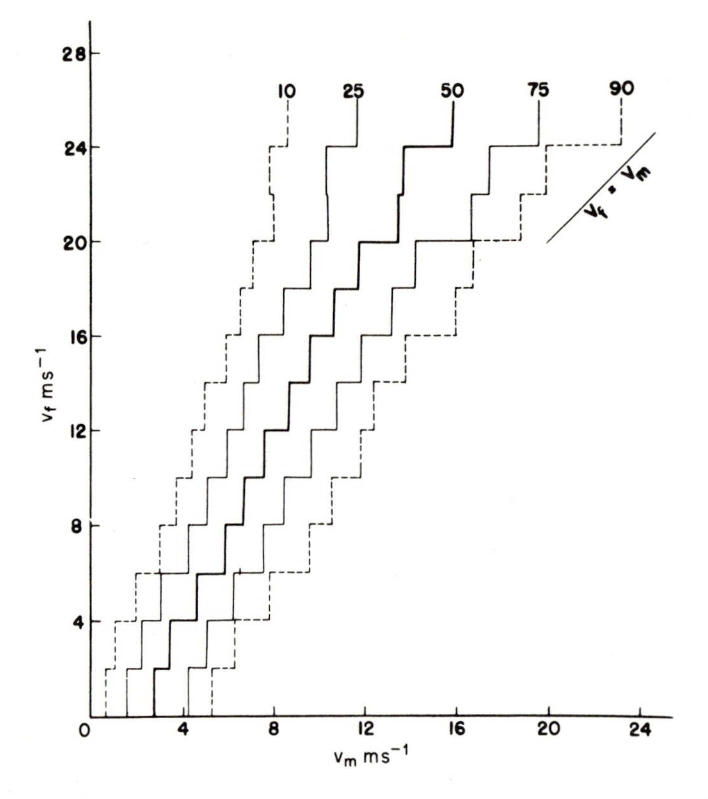

Figure 2.23 Wind speeds observed on mountain summits (V_m) in Europe and at the same level in the free air (V_f). Percentile lines for the frequency distribution of values are shown

Source: From Wahl 1966

force, and the third potential energy. Typically a 1-mb pressure drop could cause an increase of 4–5 m s^{-1} in wind speed.

Georgii recognized that mountain effects must depend considerably on wind direction and wind speed, as well as on lapse rate, but there are still few data on mountain and free-air winds to determine the general nature of these relationships. Schumacher (1923) summarized his studies of data for three Alpine stations. For Säntis, Sonnblick and Zugspitze, the mean annual speeds average about 0.8 times those in the free air although the ratio usually exceeds unity for wind speeds below 4 m s^{-1} in the free air. It also depends heavily on the terrain in relation to the general wind direction and on the precise anemometer location on the summit. Thus, on the Sonnblick in the central Alps, southerly winds are stronger than those in the free air, whereas westerly and north-east to easterly winds are weaker. On the Zugspitze, on the northern margin of the Alps, summit speeds exceed the free-air value only for winds from the southerly quadrant.

An extensive survey of wind observations on mountain summits and in the

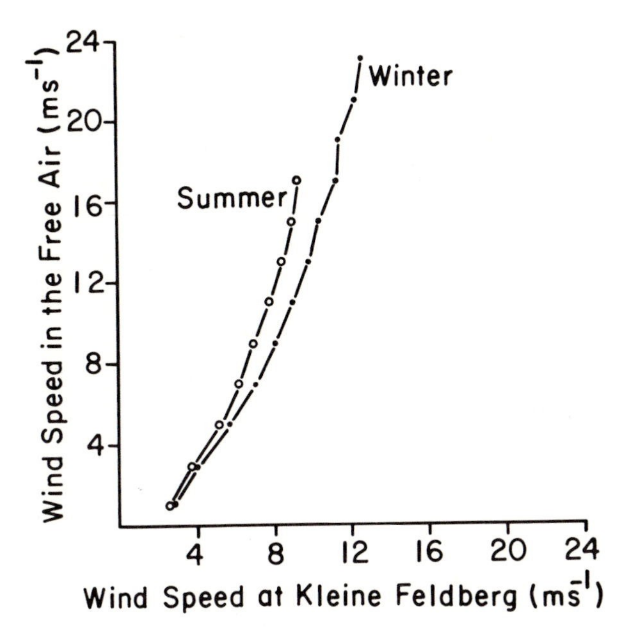

Figure 2.24 Mean wind speeds at the Kleine Feldberg (805 m) and in the free air over Wiesbaden in summer and winter
Source: After Wahl 1966

free air has been carried out by Wahl (1966). From data for European stations, he found that, in general, speeds on summits average approximately half of the corresponding free-air values. For the median values, a generalized regression equation is:

$$V_m = 2.1 + 0.5\ V_a$$

where V_m = summit wind speed (m s^{-1}) and
V_a = corresponding free-air wind speed

The relationship for different frequency ranges is shown in Figure 2.23. Where frictional effects are large, the ratio may fall as low as 0.3. Hence, summit speeds are both relatively and absolutely higher in winter when the general increase in atmospheric stability reduces frictional effects. This is illustrated in Figure 2.24 for measurements on the Kleine Feldberg (805 m) compared with sounding data over Wiesbaden, Germany. Wahl also notes that the summit free-air ratio sometimes exceeds 1.01. Such occurrences are determined mainly by the nature of the site and by the wind direction in relation to terrain configuration. The exceptional speeds on Mt. Washington, for example, are recorded there because it is the highest summit in the centre of an extensive area of mountain ridges. Eustis (1942) reported summit free-air ratios of 1.4 in summer and 1.8 in winter for this location.

TOPOGRAPHY

The interaction between topography and meteorological elements involves several basic characteristics of any relief feature. The overall dimensions and the orientation of a mountain range with respect to prevailing winds are important for large-scale processes, relative relief and terrain shape are particularly important on a regional scale, while slope angle and aspect cause striking local differentiation of climates.

Dimensional effects

The effects that an orographic barrier produces on air motion depend first on the dimensional characteristics of the barrier – its height, length, width, and the spacing between successive ridges – and, second, on the properties of the airflow itself – the wind direction relative to the barrier, the vertical profiles of wind and of stability. Each of the three dimensions of a mountain barrier interacts with a particular atmospheric scale parameter (Smith 1979). Hence, the vertical dimension of the mountain should be compared with the atmospheric depth, as measured by the 'density scale height' (~ 8.5 km); this is the thickness of a hypothetical incompressible atmosphere of constant density. Modelling studies indicate that the large-scale flow responds to an 'envelope' topography that intersects the mountain peaks, rather than to an average height of the peaks and valleys. The length of the range, in relation to the wind component perpendicular to it, greatly influences the degree of airflow perturbation. An airstream will separate around an isolated mountain whereas a range several hundred kilometres long may cause blocking of the flow, forced uplift, flow deflection, or some combination of these.

Barrier width interacts with five different atmospheric length scales according to Smith (1979). These are:

1 The thickness of the boundary layer (~ 300 m).
2 The distance the air travels downwind during a buoyancy oscillation (~ 1 km).
3 The downwind travel during the condensation and precipitation processes (~ 1 h; 10–100 km).
4 The downwind travel during one rotation of the earth (~ 10^3 km).
5 The earth's radius (6000 km), which determines the magnitude of the effect of the earth's curvature on the large-scale flow.

Scales 1 and 2 are involved in boundary layer turbulence and buoyancy effects such as gravity waves (see pp. 69 and 125), scale 3 involves precipitation processes (section on p. 226), and scales 4 and 5 involve planetary waves and synoptic-scale processes such as orographic cyclogenesis (see pp. 108 and 119). These atmospheric length scales interact with topographic dimensions. For example, average slopes in the Alps are around 1/2 for horizontal scales of

100 m to 10 km, whereas for a horizontal scale of 300 km the relief amplitude is of the order of 4 km giving slope gradients of only 1/40 (Green 1984). Air will flow more readily over gentle slopes than locally steep ones.

It is useful to consider when and why air crosses a topographic barrier rather than flowing around it. From an energy standpoint, the air arriving at a barrier must have sufficient kinetic energy in order to rise over it against the force of gravity (Stringer 1972). The level of exhaustion of kinetic energy for an air parcel rising from the surface (and affected by friction) is approximately $0.64U/\sqrt{S}$, where U = surface speed upwind (m s^{-1}) and $S = g(\Gamma - \gamma)/T$ is the static stability, representing the net balance of buoyancy forces and gravity; Γ = the adiabatic lapse rate and γ = the environmental lapse rate ($-\partial T/\partial z$) (Sheppard 1956; Wilson 1974) (see note 2, p. 96). For an isothermal atmosphere, with a temperature of 270°K, $1/\sqrt{S} = 53$ s and, if $U = 10$ m s^{-1}, the air can surmount a barrier of approximately 320 m. Hence, in an isothermal atmosphere, a surface speed upwind of 19 m s^{-1} is required for the air just to reach the crest of a 1000 m barrier.

Sheppard's criterion is unrealistic according to Smith (1980; 1990), since it implies that minimum wind speeds should be observed at the top of the barrier, which is not the case. Snyder *et al.* (1985) suggest that it provides a lower limit for the height of the layer which is decelerated and diverted around a mountain barrier. Trombetti and Tampieri (1987) note that it neglects accelerations due to the pressure field and momentum losses due to viscosity. This argument is extended by Smith (1990) who contends that pressure differences at a constant level rather than potential energy are the main control on wind speed. From a Lagrangian form of the hydrostatic equation, Smith develops the following expression for the kinetic energy:

$$U^2 = -2N^2\left(\int_{z_0}^{\infty} \Delta z dz_0\right) + U^{20}$$

$$\text{where } N^2 = -\frac{g}{\rho_0}\left(\frac{\partial\theta}{\partial z}\right) \text{ is the stratification,}$$

Δz = the vertical displacement of a streamline ($z - z_0$)
z_0 = the initial upstream level of the streamline
and U_0 = the upstream velocity of z_0.

The equation implies that the flow speed is determined by the *integral* of the displacement Δz above a point, not by the local displacement. It follows that stagnation ($U = 0$) develops where $\int^{\infty} \Delta z dz = U^{20}/2N^2$. Weak winds occur where the integral displacement is large. Smith concludes that increasing pressure along a streamline (due to lifting and positive density anomalies aloft) leads to flow stagnation and, at the lower boundary on the obstacle, to flow splitting.

The magnitude of $1/\sqrt{S}$ increases with decreasing stability. For an inversion with $\gamma = +6.5$ K km^{-1}, $1\sqrt{S} = 41$ s, while for lapse conditions with

$\gamma = +\ 6.5$ K km^{-1}, $1/\sqrt{S} = 91$ s. Thus with $U = 10$ m s^{-1}, the airstream can just surmount a barrier of 545 m for these lapse conditions, compared with one of only 245 m elevation for the corresponding inversion condition. If the airflow lacks the necessary kinetic energy to rise over the barrier it is deflected across the isobars towards lower pressure, thereby acquiring kinetic energy (Wilson 1974). After some time this deflection may have extended sufficiently far upwind to provide the airstream with the energy necessary to rise over the barrier. This implies that the isentropic (potential temperature) surfaces rise over the barrier so that the air can flow parallel to them. On the lee side of a ridge, surplus energy may appear as waves in the airflow (kinetic energy), or it goes into potential energy by deflection of the air towards higher pressure.

Wilson (1974) also shows that the width of the barrier that will cause blockage of the airflow is given approximately by 0.36 U/f, where f = the Coriolis parameter. For $f = 1.0 \times 10^{-4}$ s^{-1} and $U = 10$ m s^{-1} this critical width is 36 km.

The general classes of flow for an ideal fluid encountering an obstacle can be described with reference to the Froude number (F) which is the ratio of internal viscous forces to gravitational forces (Nicholls 1973).

$$F = \frac{U}{\sqrt{(hS)}} = \frac{1}{l\sqrt{h}}$$

where U = the undisturbed flow of velocity,
h = the mountain height,
S = the static stability

The Froude number can also be interpreted as the ratio of kinetic energy of the air encountering a barrier to the potential energy necessary to surmount the barrier. For flow over a barrier, F describes the ratio of the natural wavelength associated with vertical oscillations in the airflow to the wavelength of the barrier's cross-section.

The effect of three-dimensional hills in deflecting stable or neutrally-stratified airflows has been extensively modelled using laboratory towing tanks and wind tunnels. Hunt and Snyder (1980) introduced the concept of a 'dividing streamline' to identify the height, H_s, at which a strongly stratified airstream crosses a hill (height, h) rather than being deflected (Figure 2.25). They suggest $H_s = h(1 - F)$ as the criterion for flow impacting a hill or surmounting it, where the Froude number, F, is between 0 and 1. The application of this criterion to wind studies on San Antonia Mountain, an isolated peak in New Mexico, is shown by McCutchan and Fox (1986) and Wooldridge *et al.* (1987). H_s lowers as F increases; for example, on San Antonia Mountain (h = 670 m), $H \sim 400$ m for $F = 0.35$ and ~ 200 m for $F \sim 0.65$–0.70. Snyder *et al.* (1985) demonstrate that H_s can be understood in terms of Sheppard's (1956) formula (discussed on p. 65), where the kinetic energy of the parcel at H_s is equated to the potential energy gained by lifting the parcel from H_s to the top of the hill. They confirm

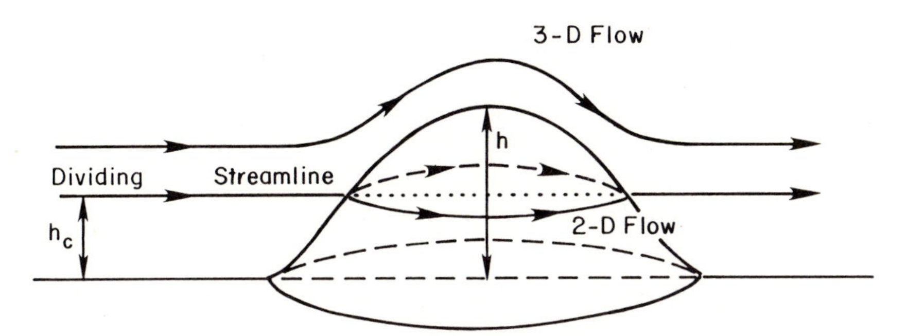

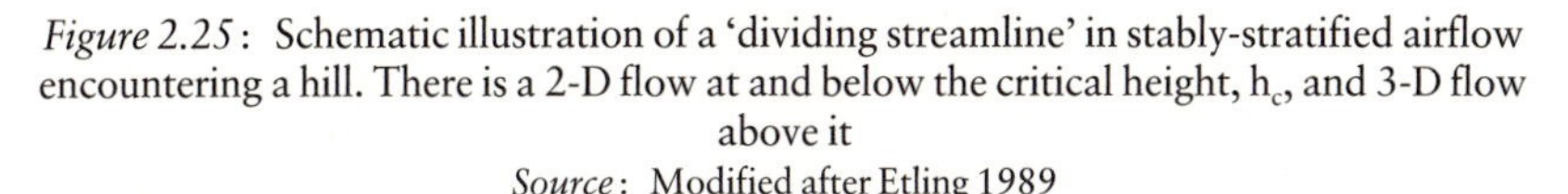

Figure 2.25: Schematic illustration of a 'dividing streamline' in stably-stratified airflow encountering a hill. There is a 2-D flow at and below the critical height, h_c, and 3-D flow above it

Source: Modified after Etling 1989

that the Sheppard relationship is valid for axisymmetric hills of varying shapes (bell, cone, hemisphere) with stable atmospheric density profiles.

The nature of stratified flow over or around an obstacle is determined by the degree of stratification, but also by the obstacle dimensions and the upwind velocity profile (Hunt and Richards 1984). The degree of stratification can be expressed in terms of both obstacle height (h), $F_h = U_0/Nh$, and obstacle length, $F_L = U_0/NL$, where $N = \sqrt{S}/2\pi$ (the Brunt-Väisälä frequency see p. 128) describes the natural frequency of vertical oscillations in the airstream.

Generalized flow behaviour over a single hill (Hunt and Richards 1984) is illustrated in Figure 2.26. For a neutral atmosphere, $F_h \gg 1$ and $F_L \gg 1$, the flow has a logarithmic profile of velocity and potential flow is maintained with a largely smooth disturbance over the barrier. The flow depends on the upwind velocity profile and hill shape. There is an extensive 'influence zone' and a turbulent wake downstream. The depth of atmosphere perturbed by the terrain is of a scale comparable to half the width of the barrier (L_0). This is because the air tends to move in circular paths with a vertical scale equivalent to the horizontal dimension of the sources and sinks that set up perturbations in the flow (Hunt and Simpson 1982). At the other extreme, with strong stratification, F_L and $F_h \rightarrow 0$ ($U_0/Nh < 1$), the flow is subject to upstream blocking and most of it goes around the hill. For moderate stratification, $F_L < 1$ but $F_h > 1$, the negative buoyancy is sufficient to prevent flow from crossing the barrier and modifies the mean flow above the hill; lee waves and rotors form downstream (see p. 127). The flow is sensitive to the upwind temperature gradient. Typical flow parameter values are illustrated in note 3 (see p. 96).

It is apparent, therefore, why 4–6 km-high ranges are, in most cases, substantial obstacles to airflow and weather systems encountering them and thus form major climatic divides. This is especially so where they are oriented more or less perpendicular to the flow direction as in the case of the Cordillera and Rocky Mountains of North America, or the Andes, and to a lesser degree in the case of

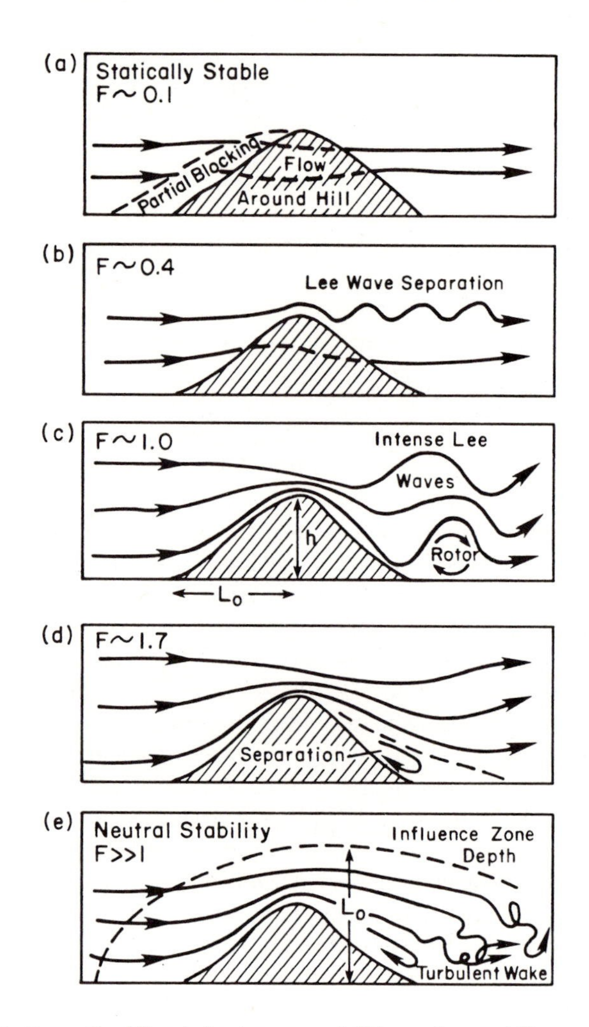

Figure 2.26 Generalized flow behaviour over a hill for various stability conditions. The Froude number (F) is expressed in terms of obstacle height
Source: After Stull 1988

the Southern Alps of New Zealand. The effects on airflow characteristics are much less pronounced if the barrier is more nearly parallel to the predominant airflow direction for at least part of the year, as in the case of the Pyrenees, the Himalaya and the ranges of New Guinea. However, while these broad generalizations may be appropriate in a climatic sense, the instantaneous airflow can be directed across any range by dynamic and thermodynamic changes in the pressure field inducing local flow accelerations.

The extent to which a mountain range is dissected by deep valleys also determines its effectiveness as a climatic barrier. Major through-valleys afford a direct passage for air motion. For example, the 30 km wide Palghat Gap through the Western Ghats at 11°N in India shows no topographic effects on rainfall, although wind speeds at the exit show a two- to three-fold increase over those in the gap (Ramachandran 1972; Ramachandran *et al.* 1980). This is similar to the dynamic effect observed when a jet of air passes through a nozzle into a large chamber. If the jet is tangential to a convex surface it remains attached to this surface for some distance downstream, despite the increasing deflection of the flow. This is termed the *Coanda effect.* Giles (1976) suggests that the winds observed to the east of the Carpathian Mountains and Transylvanian Alps in Romania appear to reflect this principle since they are deflected south-westward towards the adjacent convex surface of the range. A local low pressure is caused by the entrainment of air into the flow on the south side of the Transylvanian Alps. The deflection of the mistral south-eastward over the Rhône delta may indicate a similar effect around the Maritime Alps.

Apart from its effect on the airflow, the orientation of mountain ranges also modifies the regimes of temperature, evaporation, convection, and thermally induced wind circulations, as a result of the augmentation (reduction) of solar radiation receipts on equatorward- (poleward) facing slopes. This topic is discussed below. In many cases, the dissection of mountain ranges by valley systems creates a complex pattern of slope facets and therefore of topoclimates. Nevertheless, the orientation of the major ranges in the European Alps, for example, causes snow lines to be about 200 m higher on southern slopes than northern ones.

Relief effects

The role of relief in modifying the wind velocity over mountains and hills was recognized in the studies of 'influence height' in the 1920s (see p. 60), but only recently have more adequate observations and theories become available to describe such effects. The influence zone can be described in terms of the horizontal and vertical deflection of streamlines in the airflow, both upstream and downstream of the barrier. However, its effects are expressed through many other meteorological variables, including the drag effect on wind velocity (see p. 60) and changes in atmospheric heating forced by the surface (p.57).

Studies on atmospheric flow modification due to relief have focused largely on low hills of the order of 1 km long and 100 m high. For such hills, stability and geostrophic forces have negligible effects on the air motion, whereas for hills a few kilometres wide, stability becomes important. The effects of such topography on wind velocity are important in terms of understanding the dispersal of pollutants and assessing potential wind energy availability (Hunt and Richards 1984; Taylor *et al.* 1987; World Meteorological Organization, Secretariat 1981).

Local flow perturbations over hills are related to two basic processes that modify the pressure gradient on a local scale (Taylor 1977). For two-dimensional flow in a neutral atmosphere (ignoring shear stress and Coriolis effects), the dynamic pressure gradient (P_D) on the upwind side of the hill is approximately

$P_D = -pU\Delta U/L$, and if $L\Delta U \simeq hU$

$P_D \simeq -pU^2h/L^2$

where L = width and h = height of the hill.

p = air density

U = upwind velocity and ΔU = the increase over the hill

The hydrostatic pressure gradient induced by heating (P_H) is approximately

$P_H = -pg\Delta\theta h^*/\theta L$

where θ = upwind potential temperature,

$\Delta\theta$ = increase over the hill,

h^* = the smaller of h and h_s, the height of the superadiabatic layer near the surface,

g = acceleration due to gravity.

By a dimensional analysis of the ratio of P_D/P_H, for selected values of L, h, and h^*, assuming that $U = 10$ m s^{-1}, $g \sim 10$ m s^{-2}, and $\Delta\theta/\theta = 0.1$, Taylor shows that dynamic pressure effects are important within the boundary layer ($L \sim 100$ m, $h = 10$ m, $h^* > h$). For large-scale hills ($L \sim 10$ km, $h = h^* = 100$ m) thermal effects dominate, while for steeper slopes of 1:10, the dynamic and thermal effects are of similar magnitude when $h^* \leqslant h$.

Most theoretical and observational work has so far concentrated on low hills with slopes of about 1 in 5. In the work of Jackson and Hunt (1975) and Bradley (1980) this is expressed by the 'aspect ratio' (L^*/h, where L^* the half-width is the horizontal distance from the summit to the point at which the height is half of its maximum value, h). The flow is treated in a linear analysis of the equations of motion in terms of a shallow inner layer, where the dynamics are modified by turbulent transfers with the surface, and an outer layer of inviscid potential flow *i.e.* viscosity is unimportant for the motion and the flow moves as a set of vertical columns with relative height above the surface conserved following each fluid element (Hunt *et al* 1988; Carruthers and Hunt 1990) (see Figure 2.27). Accelerations and buoyancy effects set up within this outer region extend to heights equivalent to L^*, or higher if internal waves are generated (Hunt and Richards 1984). It can be assumed that at some level above the hill, the horizontal wind velocity will be the same as in the undisturbed upwind flow (U). Near the surface, however, the boundary layer is displaced upward by the terrain, setting up a local pressure gradient which modifies the horizontal velocity. Pressures are reduced over the hill and increased slightly upwind and downwind, accelerating the flow over the hill top. This 'fractional speed up ratio' $\Delta S = (\Delta U/U)_z$, or the excess of wind speed, ΔU, at height z over the hill top compared with the speed U at the same absolute elevation upwind (Note 3 p. 96, see also Figure 2.27), varies according to the upwind velocity profile, the airflow stratification, the aspect ratio of the hill, and surface roughness length

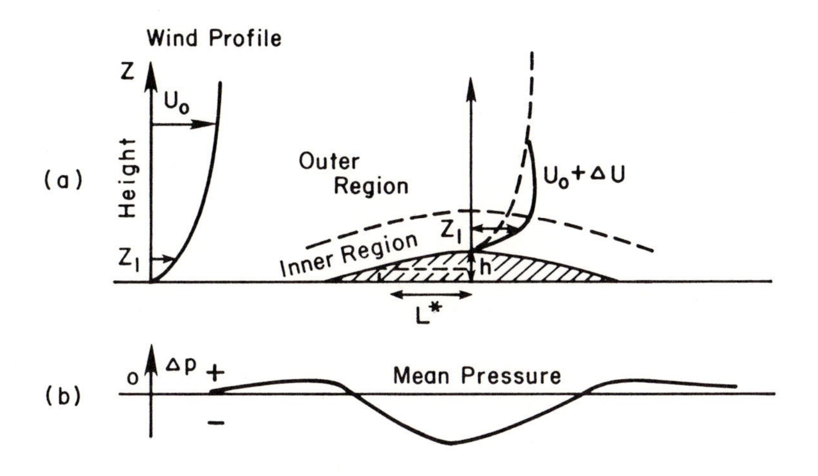

Figure 2.27 Schematic illustration of the speed-up of boundary layer winds (ΔU) over a low hill (height, h; half-width, L*) and the corresponding pattern of pressure anomalies (Δp). The flow characteristics in the inner and outer regions are described in the text
Source: Modified after Taylor *et al.* 1987; Hunt and Simpson 1982

(Hunt and Richards 1984). Taylor and Lee (1984) suggest that $\Delta S = (Bh/L^*)$ exp $(-A(\Delta z/L^*))$, where A and B are constants, $A \sim 3$ to 4 and $B \sim 2.0$–1.6, varying with hill shape. The speed-up increases by a factor of about 2 between unstable and stable boundary layers and ΔS reaches a maximum typically within 3–5 m of the summit surface. Taylor *et al.* (1987) consider this to mark the inner layer depth, whereas Jackson and Hunt (1975) indicate that the boundary corresponds to $\sim 0.05\ L^*$ (of the order of 25 m). At Black Mountain, Canberra, for example, a local wind maximum is observed at $h/5 = 35$ m, for cases of neutral stability (Bradley 1980). At Brent Knoll, Somerset ($h = 130$ m, $L^* \sim 300$ m), the speed at 2 m for cases of neutral westerly flow increased from 10 m s^{-1} upwind to 22–23 m s^{-1} over the crest, with a minimum of 4–5 m s^{-1} in the lee (Mason and Sykes 1979). The observed and predicted ΔS are in close agreement.

Field and modelling studies have been combined in the Askervein Hill Project in the Outer Hebrides, Scotland (Salmon *et al.* 1988; Teunissen *et al.* 1987). The hill has 116 m relative relief; it is 2 km long perpendicular to the prevailing southwesterly winds and 1 km wide. Fractional speed-up over the summit exceeds 0.7 at 10 m height, with velocities almost constant down to 3 m above the surface, where the largest increases in speed are observed ($\Delta S \geqslant 1.0$). Effects of the hill in retarding the flow at the surface are detectable 800 m upwind.

The typical maximum speed-up ratios for different types of small-scale topography can be summarized as follows (Hunt 1980; Taylor and Lee 1984):

$\Delta S_{max} \simeq 2h/L^*$ for two-dimensional ridges (negative for valleys);

$\simeq 1.6h/L^*$ for three-dimensional axisymmetric hills;
$\simeq 0.8h/L^*$ for two-dimensional scarps.

These estimates apply for winds $\geqslant 6$ m s^{-1}, near-neutral stability, horizontal terrain scales of $\leqslant 1$ km, and $h/L^* \leqslant 0.5$. Taylor and Lee (1984) and Walmsley *et al.* (1989) provide a convenient step-by-step procedure to estimate wind-speed for specific sites. Such information is important for siting wind turbines since the energy is proportional to U^3. In addition to summit speed-up estimates, Taylor and Lee (1984) provide formulae for surface roughness corrections. An application to the terrain of Grindstone Island in the Gulf of St. Lawrence, Canada, indicates a slow-down due to roughness of 20 per cent compared with a 30 per cent speed-up effect. Average winds are only 10 per cent greater than those observed at 10 m over the Gulf (Taylor *et al.* 1987).

The speed-up ratio of large hills also appears to fit the general Jackson–Hunt model, as extended by Carruthers and Choularton (1982). Gallagher *et al.* (1988) report that flow speed-up over Great Dun Fell (847 m) in northern England is dominated by the elevated air stratification, except when the flow is blocked. Extension of this approach for an inversion-capped boundary layer (Carruthers and Choularton 1982) shows that the speed-up over a hill depends on the height of the inversion (I_z) relative to the obstacle. When $I_z \geqslant 2h$ there is increased speed-up as the flow in the inversion layer tends to be subcritical; whereas if $I_z \leqslant 2h$ the speed-up is reduced. When the stability of the air above the inversion increases, it can lead to asymmetric flow with an upwind lull and a rapid increase in speed at the summit and downwind.

Linear models of airflow over hills give considerable errors in the lee of the obstacle. Non-linear models are impractical for detailed studies of complex terrain, but new approaches using a Fourier transform of the terrain and velocity field have enabled useful models to be developed for neutral and stratified flows (Carruthers and Hunt 1990; see also note 4 p. 96). To obtain flow variables at a point the transform is inverted numerically. The model is suitable for hill slopes $< 1/4$ and $F_h \gg 1$ (i.e. flow over the hill).

The shape or profile of obstacles is also important. Sharp breaks of slope set up more turbulence in the air passing over them than if the slope is gradual. Breaks of slope greatly increase the tendency for the airflow to separate from the ground and form vertical eddies or rotors. This may occur as a lee eddy downwind of a salient edge, or as a 'bolster' at the foot of a steep windward-facing slope (Scorer 1955). Where the general airflow is transverse to a large valley, an eddy may occupy the entire valley without disturbance to the flow above, or there may be downward motion following the valley side. These effects are illustrated in Figure 2.28. Separation of the flow from the surface is favoured by thermal instability and inhibited by radiative cooling at the surface.

Flow separation is characterized by large reductions in mean wind speed and increased variance in the flow components (Taylor *et al.* 1987), but recirculating flow is not a necessary accompaniment. Separation develops at a critical height

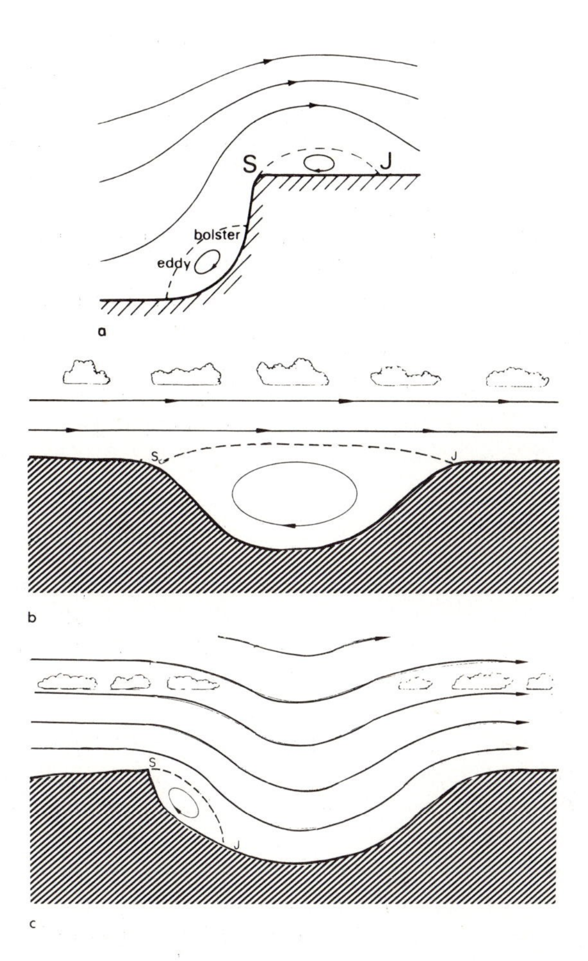

Figure 2.28 Examples of flow separation
(*a*) Separation at a cliff top (S), joining at J. A 'bolster' eddy resulting from flow divergence is shown at the base of the steep windward slope; (*b*) Separation on a lee slope with a valley eddy. The upper flow is unaffected; (*c*) Separation with a small lee slope eddy. A deep broad valley may cause the air to sink resulting in cloud dissipation above it.
Source From Scorer 1978

above the surface where the surface stress becomes zero and it should depend on L^*/z_0 (of the order of $10^4 - 10^5$), where z_0 = the roughness length, the height where the extrapolated neutral wind profile = 0. (Taylor *et al.* 1987).

Studies on San Antonia Mountain, New Mexico, by Wooldridge *et al.* (1987) indicate that for $F_h \ll 1$ separation is controlled by the pressure distribution relative to the lee wave pattern, whereas for $F_h \gg 1$ it is determined by the boundary-layer flow. A recirculation occurs near the base of the lee slope for

$0.4 < F_h < 0.7$ and this develops higher up the lee slope as F increases. Separation may occur at a hill summit for $F \sim 1.7$ according to Hunt and Richards (1984). It may also occur on upwind slopes under stratified flow, when $F_L \geqslant 1$, while being suppressed on lee slopes. On San Antonia Mountain, upwind separation is reported with neutral flow; the deflected flow may rise to about 2 h (see Figure 2.26, p. 68).

Tampieri (1987) provides estimates of the height and location of the separation point for different stability conditions and valley shapes. In general, it occurs lower down a valley slope than on a single hill, and is lower for sinusoidal topography than irregularly-spaced ridges. It is characteristic on lee slopes > 20° ($h/L^* \geqslant 0.35$); the separation distance from the hill top down the lee slope is of the order of $\lambda/2$, where λ is the lee wavelength (Hunt and Snyder 1980). The region of separation or reversed flow can extend up to 10 h downwind of a hill, with a vertical extent between a fraction of h and up to 2 h (Hunt and Simpson 1982).

The effects of air flowing over a succession of ridges and valleys is also of interest. For neutral static stability airflow over the valleys of South Wales, UK, with slopes of 32°, marked separation occurs when the flow is cross-valley (Mason and King 1984). In a further study, both unstable and neutral flows show separation, whereas around dawn and dusk the flow remains attached (Mason 1987). This may be of considerable significance for pollution dispersal since it ensures that the valley is well ventilated.

Valley sweeping requires $F \geqslant 1.5$ according to Kimura and Manins (1988), but depends also on the topography up- and down-wind, as well as on terrain shape. They suggest that complete sweeping of periodic valleys occurs for $F \sim 2.8$ and blocked flow with $F \sim 1$.

The threshold between ventilation and stagnation in a valley depends on the ratio between the separation distance (D) between the ridges and the lee wavelength (λ), and also on the ratio D/L^* according to Hunt and Richards (1984). With strong periodic waves, beneath an upper inversion layer, the criterion is $D \sim \lambda$. In stable conditions, significant flow interactions may be set up by successive ridges. Thus, higher speeds in the lee of one ridge may exceed the slowing down caused by the upstream influence of a second ridge downwind. In general, cross-valley flows are associated with stagnation and re-circulation in the valley, except for a narrow range of F_h, F_L and D/L^* conditions (Tampieri and Hunt 1985; Tampieri 1987). For example, separation is reduced for $0.5 < F_L < 1$; for $F_L \sim 1$, the ratio D/L^* must be about 4 to 6 for ventilation. Similar results are illustrated by Lee *et al.* (1987) in a towing tank experiment with a pair of ridges. Typically, stagnation begins near the valley bottom and slowly deepens. A detailed review of the laboratory modelling of flow over hills is given by Meroney (1990).

The relative relief of a topographic obstacle need not be great in order for it to influence the airflow or the occurrence of climatic elements such as precipitation. Even 'microrelief' of 50 m or less can affect precipitation distributions according to studies by Bergeron (1960) near Uppsala, Sweden. There is little or no effect

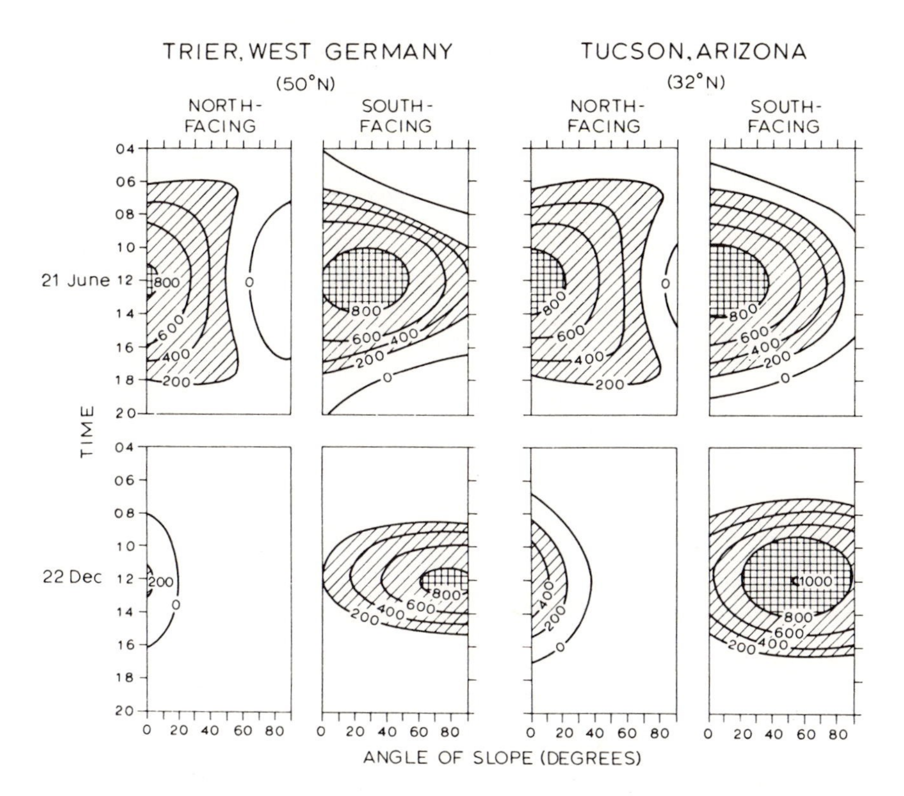

Figure 2.29 Average direct beam solar radiation (Wm^{-2}) incident at the surface under cloudless skies at Trier, West Germany, and Tucson, Arizona, as a function of slope, aspect, time of day and season of year

Source: From Barry and Chorley 1987, after Geiger 1965 and Sellers 1965

during convective conditions, but precipitation amounts increase over the hills from stratiform cloud. In these situations the flow pattern is more nearly stationary. Since the process of rain droplet growth cannot be accomplished during the short time interval whilst the air passes over a small hill, Bergeron suggested that washout of cloud droplets from low-level *scud* (or *pannus*) cloud must occur due to rain from an upper cloud layer falling through it (see Chapter 4, p. 227). The boundary layer must be close to saturation for scud to form over the hills. Frictional influences are probably important in such situations in initiating mechanical turbulence (forced convection). Over small hummocks of about 10 m, size differences in precipitation are attributable to wind effects on gauge catch, rather than orographic factors as such.

In the case of large barriers to airflow, where the mountain height exceeds that of the planetary boundary layer, many other factors need to be taken into account as noted in the preceding section. Mountain-produced effects include lee waves and flow blockage on the meso scale, frontal modification and lee cyclogenesis on a regional-to-large scale, and planetary wave effects for airflows perpendicular to high, semi-infinite barriers. These features are examined in detail below (see chapter 3).

Slope and aspect

Slope angle and aspect (or orientation) have fundamental effects on radiation income and temperature conditions – which have been the subject of many observational and analytical studies. The radiation falling on slopes of given angle and orientation has been calculated for direct or global radiation at various latitudes by many authors. Geiger (1965: 373–3), for example, tabulates a number of these sources, some of which are based on actual measurements. Other studies are cited by Kondratyev (1969: 342 and 485; also Kondratyev and Manolova 1960; Kondratyev and Federova 1977) and Hay (1977).

The basic effects of season, slope angle and orientation are demonstrated in Figure 2.29 for two mid-latitude stations at the solstices. Except on north-facing slopes, there is a displacement of the maximum intensity of direct radiation from steeper slopes in winter, to gentler ones in summer when the solar altitude is higher. South-facing slopes at the equinoxes show a symmetrical diurnal pattern; since the sun rises due east, it cannot shine immediately on a south-facing slope and the delay is greater the steeper the slope. This effect is even more marked on west-facing slopes, whereas on east-facing ones it is the time of apparent sunset which is affected by slope angle. East- (and west-) facing slopes have a maximum

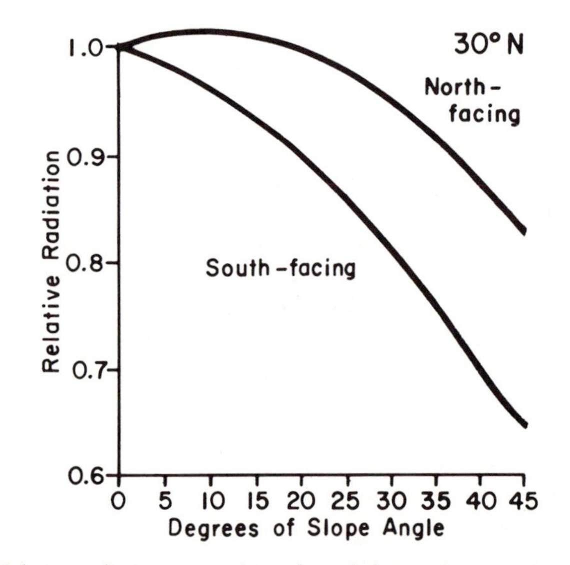

Figure 2.30 Relative radiation on north- and south-facing slopes, at latitude 30°N for daily totals of extra-terrestrial direct beam radiation on 21 June
Source: After Lee 1978

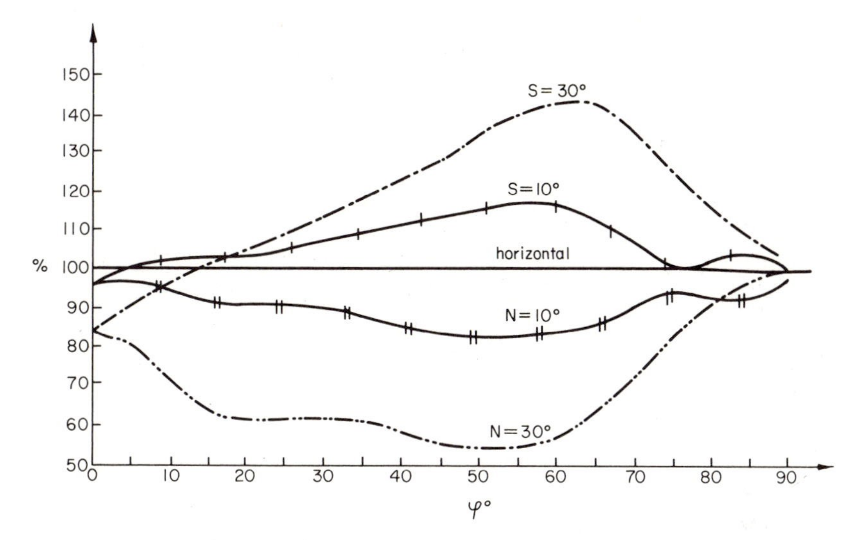

Figure 2.31 Annual totals of possible direct solar radiation according to latitude for 10° and 30° north- and south-facing slopes, in percentages relative to those for a horizontal surface

Source: From Kondratyev and Federova 1977

intensity that shifts according to both solar declination angle, and therefore season, and slope angle. On north-facing slopes in middle latitudes, direct radiation is only received around midday in winter if the solar altitude exceeds the slope angle (see Figure 2.1, p. 20, for example); the maximum is always centred on an inclination of zero.

It is important to recognize that not only the magnitude, but even the sign, of slope effect changes with latitude (Lee 1978: 171). Equatorward of 40° latitude, north-facing slopes in the northern hemisphere at the summer solstice receive *more* direct radiation over the course of the day than south-facing ones, with the difference increasing with slope angle (Figure 2.30). Nevertheless, south-facing slopes receive more in the few hours around noon. This difference in daily totals of direct radiation arises from the day length and the fact that the sun rises and sets to the north of local east and west. On an annual basis, theoretically possible direct radiation totals show the greater relative differences between north- and south-facing slopes in middle latitudes (Figure 2.31) (Kondratyev and Federova 1977). It is interesting to note that a 30°N slope causes a greater deficit in the tropics than near the pole.

We can now examine some of the geometrical considerations involved in slope radiation. The solar radiation received on a sloping surface comprises three components: the direct beam S'_b, the diffuse radiation S'_d, and that reflected from surrounding surfaces (S_r). The instantaneous direct beam solar radiation incident on a slope can be expressed:

$$S'_b = S_b \cos i,$$

where i = the angle between the solar beam and the normal to the slope,

and S_b = direct beam radiation on a plane normal to the beam at the earth's surface.

Various approaches for evaluating this expression in terms of measurable parameters have been devised (see Heywood 1964; Loudon and Petheridge 1965; Kondratyev 1969: 344 *et seq.*; Revfeim 1976; Hay 1979). The method of determining the geographical coordinates of an 'equivalent' horizontal site, proposed by Lee (1962, 1978) has the drawback that it applies only to instantaneous values since the times of sunrise and sunset on slopes differ from those on a horizontal surface. Hence, complex calculations are required to obtain daily or monthly sums.

The direct radiation is determined as follows (Kondratyev and Federova 1977; Hay 1979):

$$\cos i = \cos s \cos Z + \sin s \sin Z \cos (A - A_s)$$

where s = slope angle,

Z = zenith angle (= 90° − solar elevation),

$\cos Z = \sin \Phi \sin S + \cos \Phi \cos \delta \cos \omega t$,

where Φ = latitude,

δ = declination angle of the sun (north positive),

ωt = hour angle of the sun from the apparent noon (clockwise positive),

t = time from solar noon,

ω = earth's angular velocity ($\pi/12$ radians h^{-1}),

A = solar azimuth,

A_s = azimuth of the slope.

$$\cos A = \frac{\sin \Phi \cos Z - \sin \delta}{\cos \Phi \sin Z}$$

and

$$\sin A = \frac{\cos \delta \sin \omega t}{\sin Z}$$

Thus,

$$S'_b = S_b[\cos s \cos Z + \sin s \sin Z \cos(A - A_s)]$$

$$= S_b\Bigg\{\cos s[\sin \Phi \sin \delta + \cos \Phi \cos \delta \cos \omega t] + \sin s\left[\cos A_s\left(\frac{\sin \Phi \cos Z - \sin \delta}{\cos \Phi}\right) + \sin A_s\left(\cos \delta \sin \omega t\right)\right]\Bigg\}$$

The diffuse radiation on a slope, S'_d, differs from that on a horizontal surface, since only part of the sky is visible. Assuming istropic sky radiation (S_d),

$$S'_d = S_d(\cos)^2 \left(\frac{s}{2}\right) = 0.5S_d(1 + \cos s)$$

For gentle slopes up to 20°, the effect is negligible, especially for daily totals, since $\cos^2 (20°/2) = 0.97$, whereas for a slope of 45°, $S'_d = 0.85\ S_d$.

An isotropic model underestimates the clear-sky diffuse radiation on equator-facing slopes since the incident flux is anisotropic. According to Kondratyev and Federova (1977), the diffuse radiation flux from the circumsolar half of the sky (i.e. divided by a plane normal to the solar vertical) is about 75 per cent of the total diffuse radiation falling on a horizontal surface. This azimuthal dependence causes differences according to slope orientation for all slope angles. Horizon screening of a point from the side towards the sun is important for slopes facing either towards or away from the sun (Kondratyev 1969: 489–92). Hay (1979) has developed a model for anisotropic diffuse radiation on slopes and shows that it gives lower systematic and random errors than either the isotropic model or one using a 50:50 combination of circumsolar and isotropic components. Hay's anisotropic formulation is:

$$S'_d = \left[\left(\frac{S_b}{S_0}\right) \cdot \frac{\cos i}{\cos Z} S'_d\right] + \left\{0.5S_d(1.0 + \cos s) \left[1.0 - \left(\frac{S_b}{S_0}\right)\right]\right\}$$

where S_b = the solar constant,

and $\frac{S_b}{S_0} \cos i$ = an 'antistropy index' for the slope.

The first term in the right side of the equation is the circumsolar component and the second accounts for the isotropically distributed radiation flux.

In an alternative approach, the diffuse irradiance on slopes has been calculated by integrating the mean radiance distribution under cloudless skies for the sector seen by the slope (Steven 1977; Steven and Unsworth 1979). Their tabulations show the pronounced effect of slope azimuth for steep slopes and low solar elevations.

The reflection from adjacent slopes (S_r) is not in general readily estimated. However, if we assume a uniform landscape (in terms of albedo) which is diffusively reflective, then the reflective component on the slope is

$$S_r = (S_b + S_d)\ \alpha \sin^2 \left(\frac{s}{2}\right) = 0.5\ (S_b + S_d)\ \alpha\ (1.0 - \cos s),$$

where α = albedo of the landscape.

As for the diffuse component, anisotropy gives rise to important discrepancies from the theoretical S_r, calculated as above, especially in the presence of a snow cover. Relative values of $S_r/(S_b + S_d)$ determined from measurements and from isotropic approximation for S_r have been compared by Kondratyev (1969: 492). He shows that the isotropic estimates are too small for slopes facing the sun. Differences, expressed by the ratio (measured – isotropic estimate)/measured, for 30° slopes with a melting snow cover for a solar elevation of 30° were +49 per cent for slopes facing toward the sun and −46 per cent for slopes facing away from the sun.

The net effect of reflected and diffuse radiation is small compared with direct radiation when the sun is not obscured by cloud and the solar elevation is large ($\geqslant$ 60°). However, the reflected component can provide a larger contribution than the diffuse flux on steep slopes facing away from the sun in the presence of a winter snow cover. Table 2.16 provides an illustration of calculated mean daily values of S'_b, S'_d and S_r for Churchill, Manitoba, in December and June from Hay's (1977) analysis (using a slightly different formulation for S'_d from that presented above). Realistic cloud and surface albedo estimates are used and the starting point for the calculations is *measured* global radiation on a horizontal surface.

Beginning with the work of Garnier and Ohmura (1968, 1970), a variety of computer programs have been developed to calculate solar radiation (under cloudless skies) at any location. Map output can be generated directly from such routines using digital terrain data in gridded form (Williams *et al.* 1972).

The steps involved in such computations are: (1) determination of the extraterrestrial radiation for the required data, time and latitude; (2) calculation of the direct beam radiation for cloudless skies, assuming a specific transmission coefficient for the appropriate atmospheric turbidity conditions: (3) calculation of the direct radiation on a slope of given azimuth and inclination; (4) calculation

Table 2.16 Calculated slope radiation components for Churchill, Manitoba, in December and June (MJ m^{-2} d^{-1}).

Orientation	*Slope angle* 0°		20°			40°			90°		
	S_b	S_d	S_b	S_d	S_r	S_b	S_d	S_r	S_b	S_d	S_r
December: Surface albedo = 0.60, cloud amount = 51%											
North			0.0	0.65	0.03	0.0	0.35	0.11	0.0	0.01	0.47
West[a]	0.63	0.92	0.80	0.89	0.03	1.02	0.81	0.11	1.14	0.42	0.47
South			2.60	1.13	0.03	4.26	1.26	0.11	5.87	1.13	0.47
June: Surface albedo = 0.25, cloud amount = 63%											
North			10.85	8.37	0.17	7.25	7.04	0.65	2.10	2.17	2.78
West	13.14	9.08	12.64	8.72	0.17	11.51	7.70	0.65	7.03	3.38	2.78
South			14.08	9.07	0.17	13.55	8.35	0.65	6.49	4.21	2.78

Note: Values for east-facing slopes differ by only a small amount
Source: From Hay 1977

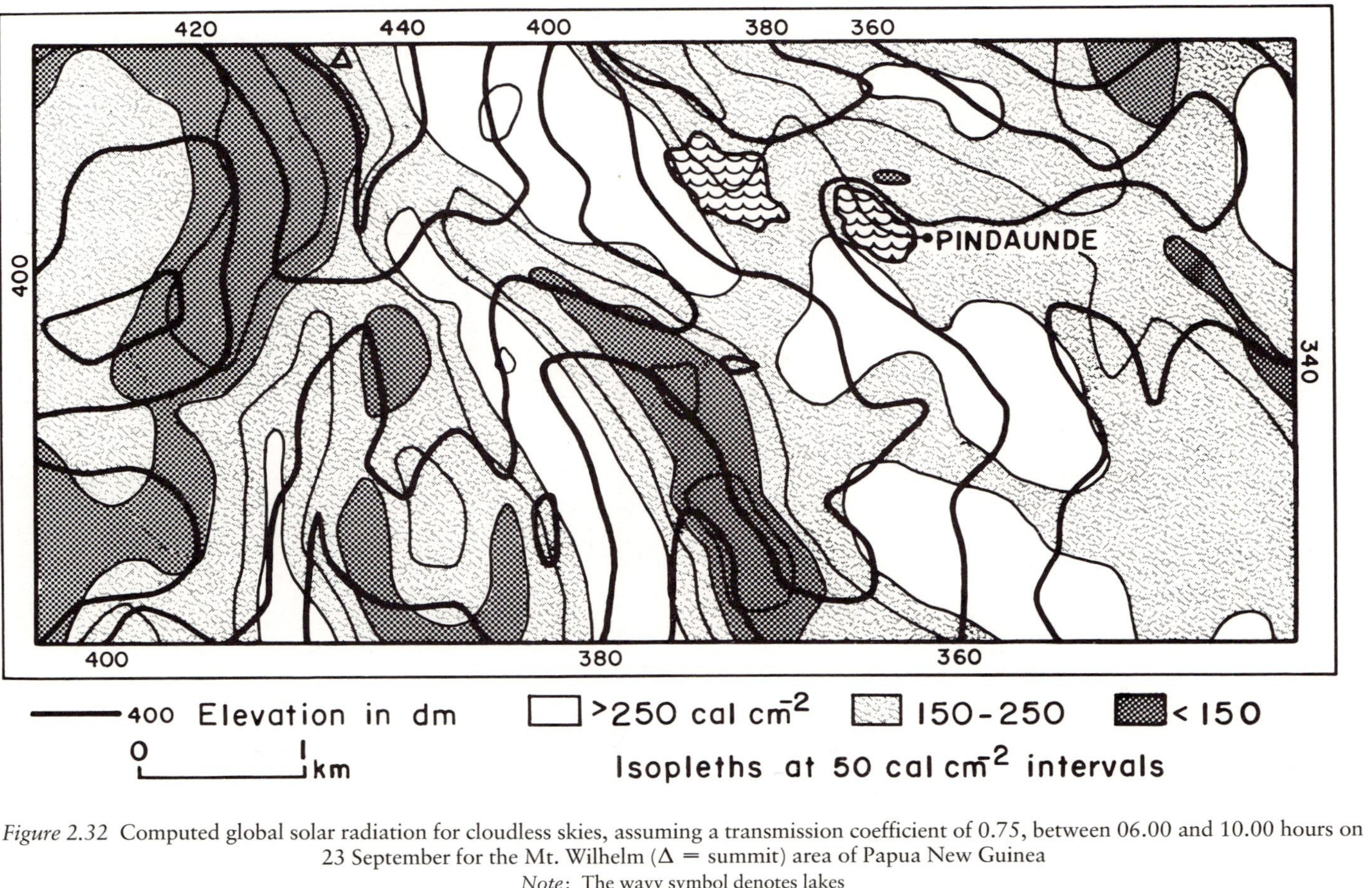

Figure 2.32 Computed global solar radiation for cloudless skies, assuming a transmission coefficient of 0.75, between 06.00 and 10.00 hours on 23 September for the Mt. Wilhelm (Δ = summit) area of Papua New Guinea

Note: The wavy symbol denotes lakes

Source: From Barry, 1978

of the sky radiation according to a dependence on solar altitude and slope azimuth and inclination; and (5) calculation of the effect of horizon screening by neighbouring mountains. The details of these calculations are beyond the scope of our discussion here, but they are readily available in the cited references. An example of a map generated by such a procedure is given in Figure 2.32 – for the Mount Wilhelm area, Papua New Guinea. It shows theoretical global solar radiation totals between sunrise and 10 a.m., illustrating the pronounced effect of the terrain. The slope contrasts are clearly evident in observed daily soil temperatures down at least 7 cm (Barry 1978). Cloud cover eliminates these slope contrasts later in the day. The improvements for anisotropic effects, discussed above, and extension of the methods to deal with spatially variable albedo and varying cloud cover is already well advanced (Hay 1977, 1979; Dozier 1980).

The effect of slopes in terms of net radiation is especially complex in view of the added role of infra-red fluxes. Observations are very limited, although the theoretical formulation has been treated (Kondratyev 1969: 680–5, for example). By analogy with the above discussion for solar radiation we can write:

$$R'_a = S'_b + S'_d + S_r - r' + \varepsilon' L'_i + \varepsilon' L_{ir} + \varepsilon' L_b - \varepsilon' L'_0,$$

where $(S'_b + S'_d)$ = the global solar radiation on the slope,

S_r = $(S_b + S_d)\,\alpha \sin^2(s/2)$; short-wave reflection onto the slope,

r' = $(S'_b + S'_d + S_r)\,\alpha'$; short-wave radiation reflected by the slope,

L'_i = $L_i \cos^2(s/2)$; atmospheric emission to the slope,

L_{ir} = $(1 - \varepsilon)\,L_i \sin^2(s/2)$; atmospheric emission to the adjacent surface reflected onto the slope,

L_b = $\varepsilon \sigma T^4 \sin^2(s/2)$; infra-red flux from the adjacent surface received by the slope,

L'_0 = $\sigma(T')^4$; infra-red flux from the slope at temperature T',

α = albedo of adjacent surface; α' = slope albedo,

ε = emissivity (absorptivity) of surface; ε' = slope value,

σ = the Stefan–Boltzmann constant.

The various components are illustrated in Figure 2.33. In general, L_{ir} and S_r are of minor importance although as already noted S_r becomes important in the presence of a snow cover. The surface albedo for infra-red radiation is typically 0.05–0.10. A model for calculating atmospheric radiation (L_i and L_{ir}) in mountain areas has recently been developed (Marks and Dozier 1979). It requires estimates of air temperature and vapour pressure, as well as topographic data in order to calculate the 'thermal view factor', the unobscured sky. This is obtained from $\cos^2(\bar{Z}_H)$ where $\bar{Z}_H$ = mean horizon angle from the zenith.

For slopes less than 30°, Kondratyev and Federova (1977) show that the net infra-red radiation on a slope (L'_n) is approximated by $L'_n = L_n \cos s$.

Likewise, for net radiation, Wilson and Garnier (1975) show that the slope radiation can be approximated within 4 per cent for daily totals by:

$$R'_n = S'_b + S'_d - r' + L_i - L_0$$

for slopes less than 20°.

Apart from the direct effects of slope angle and orientation on sunshine hours, radiation totals and therefore soil temperature, slope 'exposure' also affects precipitation input, snow cover, evaporation rates, and windiness. Since air motion is critical for these other elements, however, that topic is treated first in Chapter 3.

Topo- and microclimates

In the high mountains, especially above tree line, plant and animal life is strongly controlled by the climate at and near the ground surface – the *microclimate*. Additionally, the spatial pattern of microclimate forms a mosaic due to the effects of topography which cause distinctive *topoclimates*. Topoclimates show spatial differences in standard climatic elements over distances of 100 m to 1–10 km; microclimates are represented within the vegetation canopy and soil layer and show spatial differences over distances of a few cm (plant surfaces) to 100 m (clearings) (Turner 1980).

The combination of topo- and microclimatic effects largely determines local

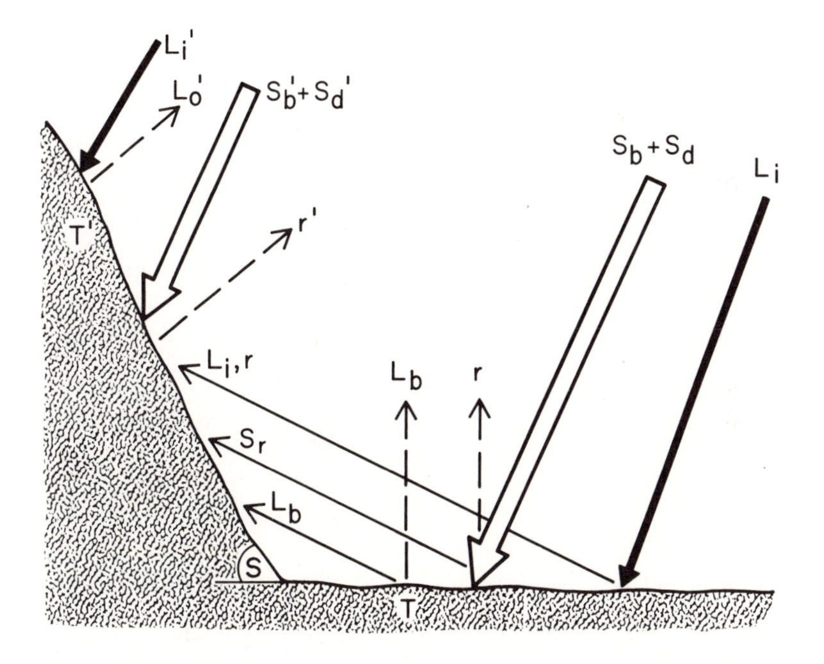

Figure 2.33 Components of solar and infra-red radiation incident on slopes (see text)

variability in plant cover. Small topographic irregularities and differences of slope angle and aspect produce marked contrasts in vegetation due to the combined effects of radiation, evaporation, wind speed and snow accumulation.

Turner (1980) suggests that alpine microclimates can be categorized in terms of (i) slope irradiation (ii) wind exposure (iii) depth of winter snow cover (iv) density and height of ground cover. The first two are independent variables, whereas the latter two are partially dependent variables (see Figure 2.35, p. 87). On this basis the following types of microclimate can be differentiated:

1 sunny windward slope – irradiation and windspeeds high;
2 sunny lee slope – irradiation high, windspeeds low;
3 shaded windward slope – irradiation low, windspeeds high;
4 shaded lee slope – irradiation and windspeeds low.

Appropriate local threshold values can be set for these, or intermediate, categories. The degree of slope and its shape may also be considered. Turner also notes that a key variable for characterizing alpine microclimate is soil heat flow, because it is highly conservative. It is governed by the four variables listed above and by the heat capacity and thermal conductivity of the soil.

Some basic influences of altitude on microclimates are considered first. These arise from the typical altitudinal profiles of the meteorological elements discussed in the section on p. 26.

Microclimatic gradients

The great intensity of solar radiation at high altitudes results in high absolute surface temperatures. During July 1957, Turner (1958) reliably measured extreme temperatures of 80°C on dark humus at 2070 m in the Otztal, Austria, for example. The site had a south-west exposure and 35° slope and the air temperature (at 2 m) was 30°C. Using a radiation thermometer, which registers infra-red emittance from the surface, the author recorded a bare soil temperature of 60°C on a level site at 3480 m on Mt. Wilhelm, Papua New Guinea, during September 1975 (Barry 1978). On this occasion the air temperature was only 15°C. This effect of altitude is apparent from climatic data. Mean annual ground temperatures in the Alps at 1.2 m depth exceeded air temperatures by 0.5°C at 600 m, 2.0°C at 1800 m and 2.9°C at 3000 m (Maurer 1916).

Since air temperatures decrease with altitude, surface heating effects usually produce stronger near-surface temperature gradients at higher elevations (Schwind 1952; Aulitsky 1962), although the soil–air difference depends on weather conditions. From detailed measurements at tree line (2072 m) near Obergürgl, Austria, Aulitsky (1962) shows that there is a linear relationship between mean and extreme values of soil and air temperature when the soil is unfrozen (Figure 2.34). Soil temperatures are lower than air temperature in the transition seasons, due to radiative cooling of the surface in autumn and to lag effects of the snow cover in spring. In the Alps, soils generally reach their lowest

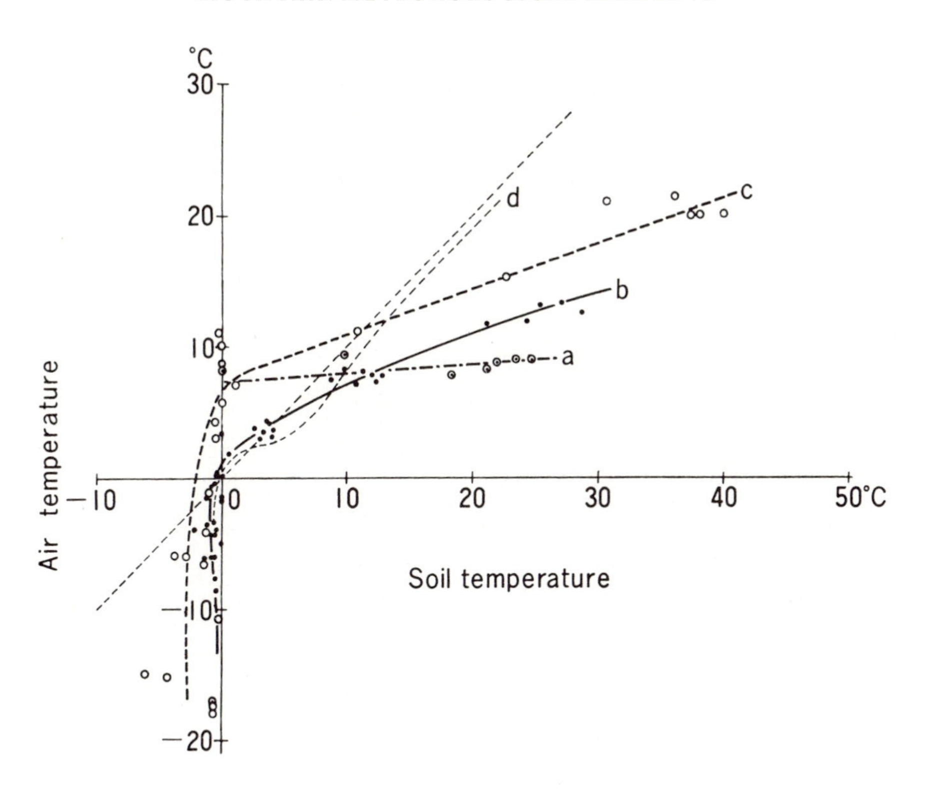

Figure 2.34 Relations between monthly soil temperatures (0–1 cm and 10 cm) and air temperature (2 m) near the forest limit, Obergurgl, Austria (2072 m), June 1954–July 1955

Notes: a Mean amplitudes (0–1 cm); b Mean and mean extremes (0–1 cm); c Absolute extremes (0–1 cm); d Mean and mean extremes (10 cm)
Sources: Aulitsky 1962, from Yoshino 1975

temperatures during autumn through frost penetration, whereas winter snow cover tends to insulate the ground. In shaded lee situations, however, persistently low snow temperatures may enable frost to penetrate through even deep snow covers into the soil.

Similar differences in other vertical micoclimatic gradients are apparent at high altitude. Summit winds generally maintain high average speeds in middle latitudes due to the vertical increase in wind velocity. Consequently, the accentuated vertical wind profile near the ground enhances the vertical fluxes of heat and moisture. This is significant in the Rocky Mountains, for example, in terms of strong advection of rather dry air causing high rates of evapotranspiration from alpine tundra surfaces (LeDrew 1975; Isard and Belding 1986).

Small-scale topography and vegetation cover play a major role in modifying microclimates in mountains, especially near timber line (Aulitsky 1984, 1985).

Plate 2 A 'flagged' Engelmann spruce tree in the alpine forest-tundra ecotone, Niwot Ridge Colorado. Growth occurs only on the downwind side of the stem (courtesy Dr G. Kiladis, CIRES, University of Colorado).

Studies in the subalpine in Switzerland, for example, show that ridges and gullies with a relief of 5–12 m can modify the wind speed by ± 60 per cent when the direction is perpendicular to the ridges (Nägeli 1971). More important is the formation of vertical eddies in the form of a rotor, in the lee of obstacles. Gloyne (1955) shows that these extend horizontally downwind 10–15 times the height of the obstacle. For a vegetation barrier of 50 per cent density, the wind speed is reduced by 80 per cent up to 3–5 times the height of the vegetation downwind. In association with the wind regime, therefore, trees and shrubs generate distinctive and recurrent patterns of snow accumulation in their lee. Drifts form

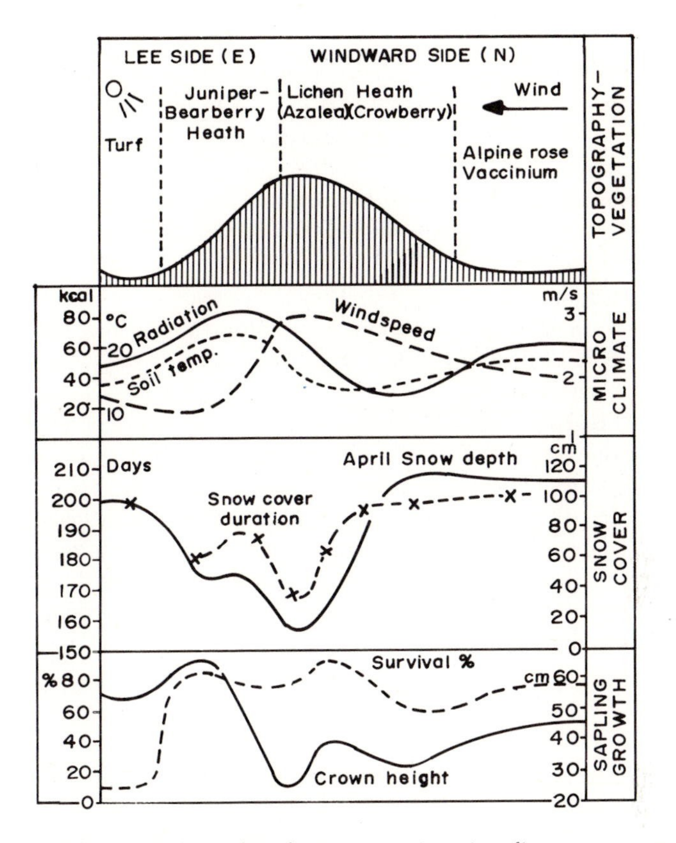

Figure 2.35 Schematic relationships between terrain, microclimate, snow cover and tree growth at the tree line (2170 m), near Davos, Switzerland

Notes: The site is a 40–45° north-east facing slope through an avalanche track and an adjacent spur. Climatic averages are for the growing season: global solar radiation (k cal cm^{-2}) on the slope under cloud-free conditions, surface soil temperature, 1 m wind speed for 06.00–18.00 hours; also mean April snowdepth and duration of snow cover. The bottom graph shows crown growth and percentage survival of 12-year-old larch plantings in pots

Source: After Turner *et al.* 1975

in the lee of each tree cluster (or 'island') and these protect the downwind limbs from winter desiccation, whereas exposed shoots in the crown or on the upwind side gradually go brown and die. The desiccation apparently causes evaporative stress at a time when moisture in the roots and stem, or in the soil, is frozen. Wardle (1974) considers that inadequate pre-winter hardening is also responsible. The desiccation stresses give rise to anatomical and morphological changes in the trees. The latter are evident in the deformation of the tree crown which shows 'flagging' with most of the growth on the lee side of the trunk (Plate 2). In extreme conditions within the upper Krummholz (crooked wood) zone, the tree assumes a prostrate cushion form (Yoshino 1973; Wardle 1974; Grace 1977).

Since the wind effect on trees is well established although not entirely understood, attempts have been made to use tree deformation as an index of local wind conditions in areas of extreme data paucity. Thus, small-scale maps of wind direction have been constructed for an area in the Whiteface Mountains, New York, by Holroyd (1970) and in the Indian Peaks, Colorado, by Holtmeier (1978). Such maps can be used for infer local areas of wind convergence/ divergence which may be important in siting buildings, power lines, ski-lifts and wind power installations, or in predicting pollutant dispersal. More recently, Wade and Hewson (1979) have tried to quantify the relationship between indices of tree deformation and wind speed. They developed regression equations for deformation with mean annual wind speed, although the duration of speeds above a threshold of the order of 3–5 m s^{-1} seems to be a more likely parameter of wind effects on the trees.

In terms of vegetation growth on mountains, ecologists interpret the dominant influence of altitude to be that of *exposure* (Ingram 1958). This concept seems not to have been defined quantitatively although it obviously involves wind velocity and its effect on snow cover – both in the Cairgorms, where Ingram worked, and in the Colorado Rockies, where Wardle and Holtmeier worked. However, wind velocity also affects transpiration stress which involves factors other than wind (Gale 1972). Studies near tree line in the Alps (Davos) show that relatively high wind speeds and early snow melt favour the survival of tree plantings (larch, spruce and pine) on ridges and warm slopes (Schönenberger 1975). As in the Rocky Mountains, deep late-lying snow encourages snow-blight fungus. In contrast, vertical tree growth is favoured by sunny, but wind-sheltered sites. Figure 2.35 schematically illustrates these results (Turner *et al.* 1975). For plants in the alpine belts, Larcher (1980) considers that the climatic stresses imposed by low and strongly fluctuating temperatures have induced various adaptation processes to ensure cold resistance in both summer and winter conditions. At present, however, no definitive interpretation of 'exposure' seems to be feasible.

Topoclimatic effects

Topoclimates are primarily manifestations of slope angle, aspect and horizon effects. These are relatively obvious in the case of radiation and temperature, but second-order effects may come into play through the influence of terrain on wind velocity and in setting up slope and mountain valley winds.

The most obvious topoclimatic effects are those between (1) north-facing slopes, on the one hand, and south-facing on the other; and (2) between valley bottoms and ridgetops. Other effects may also arise according to valley orientation with respect to the mountain range, and valley cross-profile. The primary control of conditions relating to slope orientation is solar radiation, whereas effects of airflow and air drainage are important supplementary factors for ridgetop and valley bottom locations, especially for night-time and winter conditions.

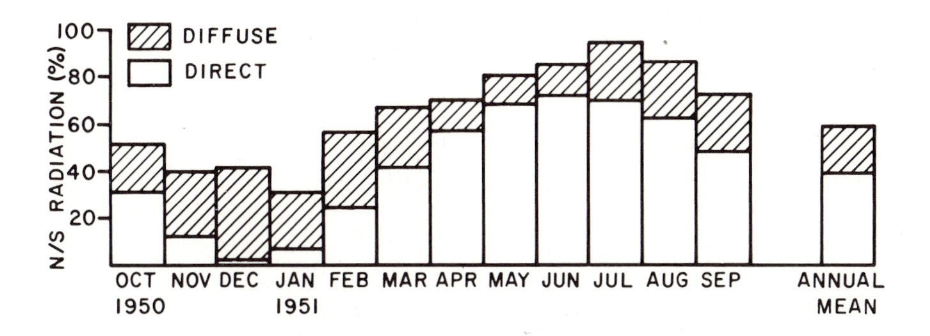

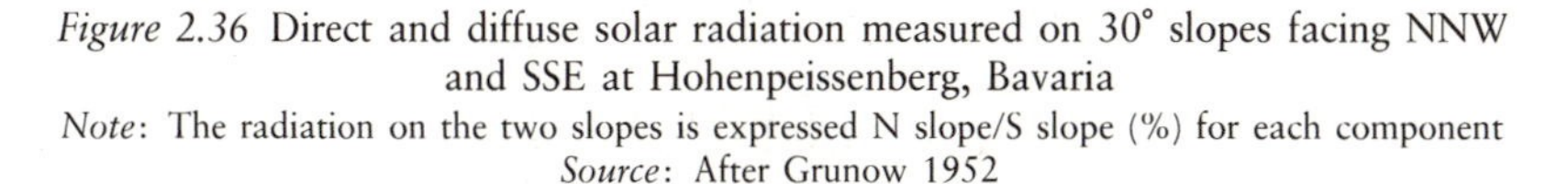
Figure 2.36 Direct and diffuse solar radiation measured on 30° slopes facing NNW and SSE at Hohenpeissenberg, Bavaria

Note: The radiation on the two slopes is expressed N slope/S slope (%) for each component

Source: After Grunow 1952

An extensive series of observations on slope, aspect and elevation effects has been obtained for San Antonia Mountain, in northern New Mexico. It is an isolated conical mountain rising from 2740 m to 3325 m. Hourly data have been collected at three elevations and the four cardinal directions for 3–5 years (McCutchan *et al.* 1982). McCutchan and Fox (1986) show that the effect of aspect (acting to generate slope winds) exceeds that of elevation on wind velocity and, to a lesser degree, on temperature. Also, wind speeds in excess of 5 m s^{-1} eliminate any differences associated with slope, aspect or elevation.

Minimum temperatures at rural stations in Britain show that topographic influences operate at two scales. Tabony (1985) distinguishes the role of 'local shelter' – the height above the valley bottom (represented by the drop in altitude within 3 km of the site), and 'large-scale shelter' associated with cold air drainage (represented by the average height of the terrain above the valley over a 10 km radius). Tabony observes that nocturnal radiational cooling rapidly establishes a cooled layer that is deepest in places that have significant local shelter. During the night, air drainage causes minima in places where the large-scale shelter is pronounced. The temperature drop is limited by the infra-red back radiation from the atmosphere. The large-scale shelter is more important in the winter months, and particularly for events with a two-year return period. The effect of local shelter (height above valley) has a maximum effect on daily temperature range in autumn when there are large altitudinal differences in soil moisture deficit (Tabony 1985).

Slope effects on radiation and ground temperature have been extensively analysed by Grunow (1952) at Hohenpeissenberg, Bavaria. Figure 2.36 illustrates differences in direct and diffuse radiation receipts on NNW- and SSE-facing slopes of about 30°. Global totals differ most in winter when the solar altitude is low; the north-facing slope receives only 32 per cent of that facing south, and almost all of the radiation on the former is diffuse. The associated differences in ground temperatures are shown in Figure 2.37 for daily means and means at

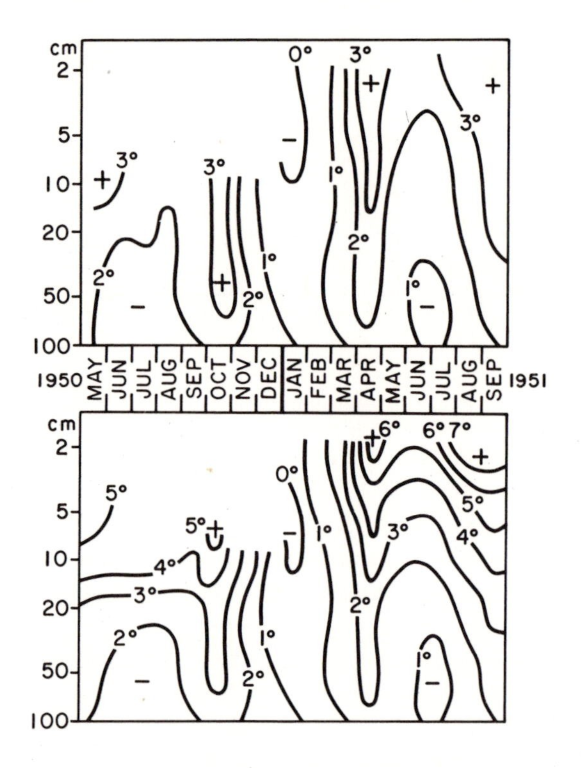

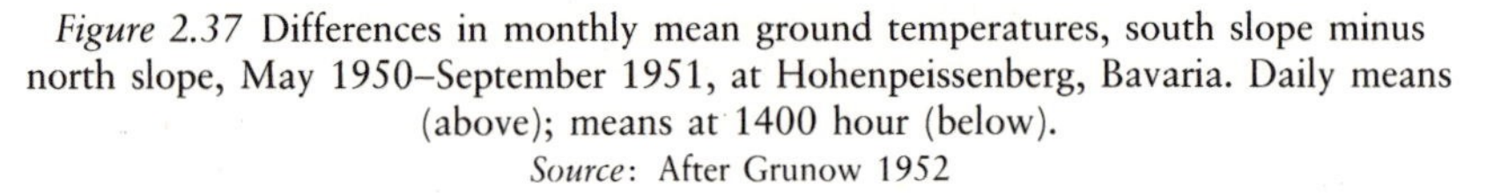

Figure 2.37 Differences in monthly mean ground temperatures, south slope minus north slope, May 1950–September 1951, at Hohenpeissenberg, Bavaria. Daily means (above); means at 1400 hour (below).
Source: After Grunow 1952

1400 hours. The difference in ground temperatures (50–100 cm) shows minima in winter and summer and maxima in the transition seasons. In winter, the snow cover insulates the ground leading to almost no difference between the slopes. The slopes are snow-covered from November through March (April on the north slope) and the north slope also is generally more moist. At 1400 hours, the effect of diurnal heating on the upper soil layer is apparent in summer.

There are three general approaches to topoclimatic differentiation in mountainous terrain. One is by direct measurement of climatic elements at a network of stations; another is from field studies of the distribution of climatic effects, such as frost damage or wind shaping, on vegetation. The third method is based on calculation of radiative fluxes from topographic map analysis. Topoclimatic investigations have been performed in a number of mountain areas, including south-eastern Alaska, the Polish Carpathians, and the European Alps. Some general results from these studies are worthy of note.

The characterization of topoclimates in Poland by Hess *et al.* (1975), based on temperature and relative humidity indices, relates to three broad terrain

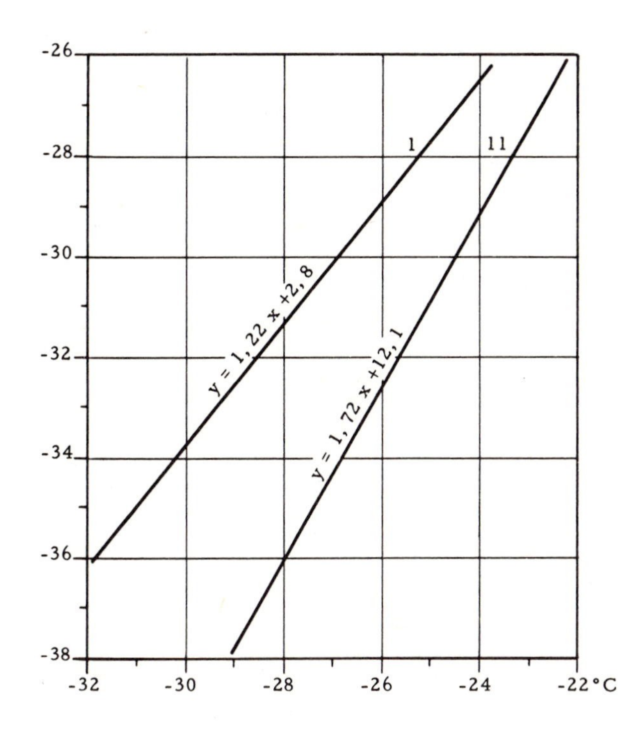

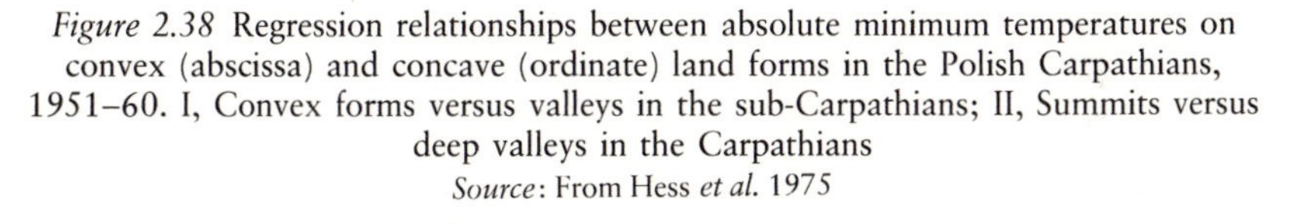
Figure 2.38 Regression relationships between absolute minimum temperatures on convex (abscissa) and concave (ordinate) land forms in the Polish Carpathians, 1951–60. I, Convex forms versus valleys in the sub-Carpathians; II, Summits versus deep valleys in the Carpathians
Source: From Hess *et al.* 1975

categories: (1) valleys, subdivided according to valley bottom, terraces, and valley sides; (2) slopes, including the 'thermal belt' (p. 217), and lower ridges; and (3) ridges, summits and slopes in low–moderate sized mountains. Valleys typically have the greatest diurnal temperature range, and the greatest frequency of frosts and radiation fogs, while the slopes have the smallest temperature range and longest frost-free season. The upper zone has lower mean temperatures and small diurnal temperature variability. Within each category, microclimatic differences are identified according to small-scale terrain features and vegetation cover.

For the West Carpathians of Poland, Hess *et al.* (1976) relate various thermal parameters to mean annual temperature via regression equations; mean annual temperature itself is linearly related to elevation. Separate equations are developed for convex and concave landforms and for north/south slopes. Figure 2.38 shows that extreme minima are considerably lower on concave slopes, for example. Similar sets of equations are also developed for parameters such as the duration of the frost-free season in relation to mean annual minimum

Table 2.17 Global solar radiation in valley location in Austria as a percentage of that on a horizontal surface for different cloud conditions

		North–south valley			*East–west valley*		
Horizon angle		*10°*	*20°*	*30°*	*10°*	*20°*	*30°*
Summer							
Cloud	0	97	92	85	99	98	96
	5/10	98	93.5	88.5	99	97	93.5
	10/10	99	96	93	99	97	93
Winter							
Cloud	0	95	86	74	95	17	17
	5/10	96	88.5	80.5	96.5	47	46
	10/10	99	98	96	99	98	96

Source: From Sauberer and Dirmhirn 1958

temperatures. These are applicable only to sites above the level of valley inversions. Such equations have to be based on local data since the regression coefficients may vary widely in different regions. Consequently, this type of approach requires extensive measurements over a range of weather conditions.

The spatial coherence of some climatic elements is obviously less than others in heterogeneous terrain. For example, correlations of air temperature between stations in different topoclimatic settings within a small area in the Chitistone Pass (1774 m), south-east Alaska, are much weaker for daily minima than for maximum temperatures (Brazel 1974). This is evidently due to the effects of glacier winds and cold air ponding on minimum readings at valley stations. Further studies in the same area (Brazel and Outcalt 1973a) demonstrate that the surface thermal regime is primarily determined by slope angle, aspect, and horizon elevation angle. It is less sensitive to radiative properties of the surface (albedo and emissivity), except where snow or ice is present, and to thermal properties such as volumetric heat capacity, thermal diffusivity and depth to the freezing plane in the soil.

Topographic analysis, involving the calculation of potential radiation on uniform slope facets, was first carried out laboriously by a combination of field or map measurements of slope angle, orientation and horizon screening and field measurements or calculation of direct or global solar radiation for cloudless skies (Turner 1966; Wendler and Ishikawa 1974). The effect of valley orientation and slope angle of the valley sides in Austria on global solar radiation is summarized in Table 2.17. The effects of altitude and slope orientation on radiation have been calculated by Borzenkova (1967) based on theoretical methods of Kondratyev and Manolova (1960) and observations of Aizenshtat (1962). Her results are shown in Table 2.18. For cloudless skies, north-facing slopes of 30° receive only 49 per cent of the annual direct solar radiation falling on a horizontal surface at 3600 m compared with 57 per cent at 400 m. The

Table 2.18 Calculated direct beam solar radiation and net radiation for horizontal and sloping surfaces in the Caucasus

	Horizontal surface			*30°N slope*			*30°S slope*		
	Direct Solar	*Net Radiation*	R_n/S_b	*Direct Solar*	*Net Radiation*	R_n/S_b	*Direct Solar*	*Net Radiation*	R_n/S_b
Annual averages (kJ cm^{-2} yr^{-1}) and ratios									
Cloudless sky									
400 m	683	347	0.51	389	108	0.28	857	490	0.57
3600 m	917	144	0.16	453	−5	–	1125	288	0.26
Overcast									
400 m	142	79	0.56	132	79	0.60	132	79	0.60
3600 m	185	40	0.22	170	40	0.24	170	40	0.24
June averages (kJ cm^{-2} month^{-1}) and ratios									
Cloudless sky									
400 m	89	54	0.61	70	41	0.59	85	54	0.64
3600 m	116	42	0.36	85	28	0.33	107	68	0.64
Overcast									
400 m	18	11	0.61	16	11	0.69	16	11	0.69
3600 m	23	13	0.56	21	13	0.62	21	13	0.62

Source: After Borzenkova 1967; personal communication 1979

corresponding figures for 30° south-facing slopes are 122 per cent of horizontal at 3600 m and 126 per cent at 400 m. At lower elevations there is also more tendency for direct radiation to be reflected onto slopes from adjacent mountains than at higher elevations, although this factor is not calculated. With overcast conditions (low cloud), unit surface areas on all 30° slopes receive about 92 per cent of the global solar radiation on a horizontal surface. Figure 2.29 illustrates the striking differences between 30° north-facing and south-facing slopes in winter months in middle latitudes (p. 75).

For net radiation (Table 2.18), the clear sky values also show an important annual difference between north- and south-facing slopes, especially at lower elevations. At high elevations the difference is most marked in summer months. Aizenshtat (1962) measured net radiation at 3150 m on north and south slopes of the Kumbel Pass (near Shakhristan), on the northern slope of the Turkestan Mountains (40°N; 31–33° slopes). He found that on five clear days in September the day-time total on the south-facing slope (1784 MJ m^{-2}) was 3.4 times greater than on the north-facing one, while the 24-h total (1298 MJ m^{-2}) was respectively 7.5 times greater. (The albedo difference was small; 0.15 on the south-facing and 0.20 on the north-facing slope.)

Slope comparisons of turbulent heat fluxes are rare. In south-eastern Alaska, data for 11 summer days at a valley floor site, a south-east-facing slope, and a north-west-facing slope show large differences in magnitude and phase of the latent heat flux (Table 2.19) (Brazel and Outcalt, 1973a and b). Daily totals of

Table 2.19 Latent heat fluxes ($W\ m^{-2}$) and Bowen ratio at three sites in Chitistone Pass, Alaska for 11 days in July–August

		Latent heat flux ($W\ m^{-2}$)									
Site	*Hour*	*00*	*03*	*06*	*09*	*12*	*15*	*18*	*21*	*Daily value*	*Evaporation (mm)*
Horizontal		10.5	7.0	−66.3	−181.4	−233.8	−177.2	−48.8	10.5	−126.9	4.4
SE Slope (34°)		34.9	34.9	−167.5	−361.5	−321.0	−101.2	34.9	34.9	−97.4	3.4
NW Slope (28°)		−20.9	−20.9	−20.9	−41.9	−80.2	−97.7	−80.2	−17.4	−45.5	1.6
Bowen Ratio											
Horizontal		2.86	4.70	−0.01	0.33	0.39	0.42	−0.21	2.67		
SE Slope		0.80	0.86	0.35	0.36	0.39	0.48	0.48	0.70		
NW Slope		−0.60	−0.93	−1.10	−0.17	0.53	0.86	1.09	0.36		

Note: *Bowen ratio* = sensible heat flux/latent heat flux. Both fluxes are positive towards the surface. Negative values of Bowen ratio occur when the fluxes are of opposite sign.
Source: After Brazel and Outcalt, 1973b

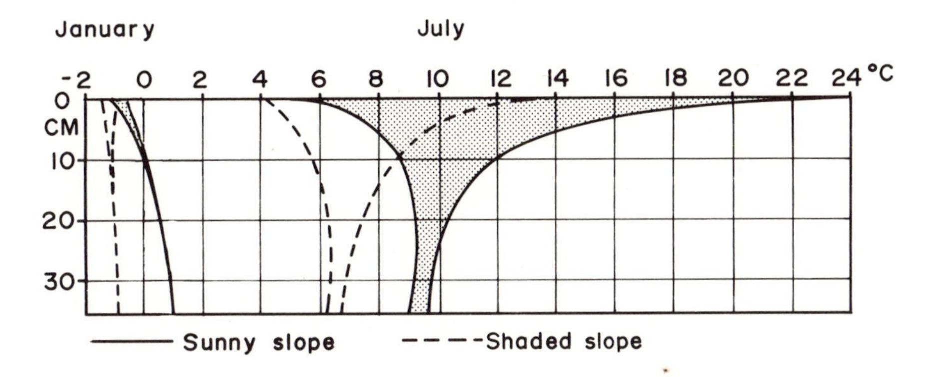

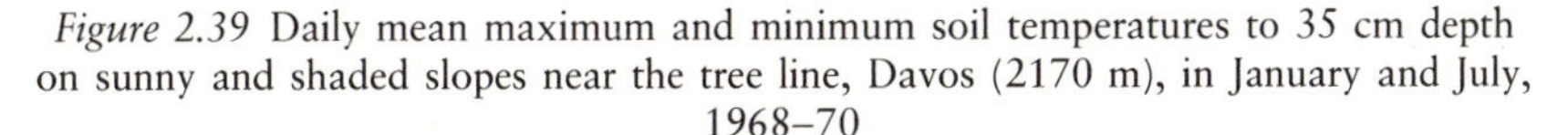

Figure 2.39 Daily mean maximum and minimum soil temperatures to 35 cm depth on sunny and shaded slopes near the tree line, Davos (2170 m), in January and July, 1968–70

Source: After Turner *et al.* 1975

evaporation on the north-east-facing slope are only a third of those on the valley floor, but vegetation cover is a major determinant here. The valley floor and south-east slope were predominantly moss-lichen tundra, with willow on the slope, whereas the site on the north-west slope was fine scree and rubble.

The large contrasts in radiation on sunny and shaded slopes at tree line in the Alps are apparent in mean soil temperature values to about 10 cm depth in January and 25 cm or more in July (Figure 2.38). But, in spite of the calculated radiation differences at high altitudes (Table 2.18), several studies show decreasing differences in soil temperatures between slopes of north and south aspect with altitude. This is true in the Colorado Front Range on 7–10° north- and south-facing slopes between 2200 m and 3750 m, and in the Santa Catalina Mountains in Arizona between 2150 m and 2750 m (Barry and Van Wie 1974). In the Northwest Himalaya, Mani (1962: 7) notes that soil temperatures in summer are almost the same on north- and south-facing slopes above 4500 m. More comparisons of calculated and measured energy fluxes and of observed soil temperatures seem to be required to check these results.

Direct calculation and mapping of incident radiation is now possible from digitized topographic data, as illustrated in Figure 2.32 (p. 81). Olyphant (1984, 1986) uses terrain data and radiance estimates to assess potential ablation on cirques glaciers in the Front Range, Colorado. He finds that the rockwalls enhance net radiation by reducing net longwave loss compared with a level surface. Diurnal solar radiation receipts are very asymmetric as a result of summer afternoon cloudiness. Thus east-facing basins receive large totals before the typical cloud build up. Recently, a computer model has been developed to simulate energy balance components over mountainous topographic surfaces (Dozier and Outcalt 1979). Incoming solar and infra-red radiation are calculated for every point, taking account of slope, exposure and horizon data. Values of air temperature, humidity and wind speed must be specified over the grid and these are used to determine vertical profiles for the calculation of heat fluxes. Roughness length, albedo and soil thermal properties are also specified. Surface temperature, soil heat flow, sensible and latent heat fluxes, and net radiation are simulated and can be mapped for specific times of day. The underlying basis of the model is the equilibrium temperature theory which states that a unique surface temperature satisfies the energy flux equations for a given set of boundary conditions. At present the model is applicable only to cloudless sky conditions and has numerous limitations, but it represents a significantly new approach to studies of topoclimate.

NOTES

1 *These data need reducing by 2 per cent to correct them to the 1956 International Pyrheliometric Scale. Values from early European measurements should likewise be increased by 1.5 per cent.

2 Wilson (1968) illustrates the derivation of this relationship, neglecting frictional effects:

$$\frac{dw}{dt} = -S(z_1 - z_0),$$

where w = vertical motion, $(z_1 - z_0)$ = vertical displacement from the base level. By integration,

$$\Delta\frac{w^2}{2} = \frac{S(z_1 - z_0)^2}{2}$$

If the kinetic energy of the upwind flow (= $U_2/2$) is the source for $\Delta(w^2/2)$, then $(z_1 - z_0) = U/\sqrt{S}$.

3 In these papers, and some other courses, normalized wind speeds, expressed as $(1 + \Delta S)$, or $(\Delta U + U)/U$, are reported.

4 Typical values of the flow parameters cited by Carruthers and Hunt (1990) are as follows:

Neutral flow $F_L \gg 1$; $F_h \gg 1$.
  e.g. $N = 0.005\ s^{-1}$, $H = 20$ m, $L = 100$ m.

Moderately stable stratification $F_L \leqslant 1$, $F_h > 1$.
  e.g. $N = 0.01\ m\ s^{-1}$, $H = 200$ m, $L = 1000$ m, and $U_0 = 5\ m\ s^{-1}$.

Strongly stable stratification $F_h \leqslant 1$, $\Delta u \sim U_0$.
  e.g. $N = 0.04\ m\ s^{-1}$, $H = 200$ m, $U_0 = 5\ m\ s^{-1}$.

REFERENCES

Latitude

Barry, R.G. (1973) 'A climatological transect on the east slope of the Front Range, Colorado', *Arct. Alp. Res.*, 5, 89–110.

Geiger, R. (1965) *The Climate near the Ground*, Cambridge, Mass., Harvard University Press.

Hauer, H. (1950) 'Klima und Wetter der Zugspitze', *Berichte d. Deutschen Wetterdienstes in der US-Zone*, 16.

Hedberg, O. (1964) 'Features of afroalpine plant ecology', *Acta phytogeographica suecica*, 49.
Hnatiuk, R.J., Smith, J.M.B. and McVean, D.N. (1976) *Mt. Wilhelm studies. II. The Climate of Mt. Wilhelm, Australian National University*, Canberra, Biogeography Studies 4.
Lauscher, F. (1966) 'Die Tagesschwankung der Lufttemperatur auf Höhenstation in allen Erdteilen', *60–62 Jahresberichte des Sonnblick-Vereines für die Jahre 1962–64*, pp. 3–17, Vienna.
Lauscher, F. (1976) 'Methoden zur Weltklimatologie der Hydrometeore. Der Anteil des festen Niederschlags am Gesamtniederschlag', *Arch. Met. Geophys. Biokl.*, B, 24, 129–76.
Troll, C. (1964) 'Karte der Jahreszeitenklimate der Erde', *Erdkunde* 18, 5–28.

Continentality

Aulitsky, H., Turner, H. and Mayer, H. (1982) 'Bioklimatische Grundlagen einer standortsgemässen Bewirtschaftung des subalpinen Lärchen-Arvenwaldes', *Eidg. Anst. forstl. Versuchswes., Mitt.*, 58(4), 325–580.
Gams, H. (1931) 'Die klimatische Begrenzung von Pflanzenarealen und die Verteilung der hygrischen Kontinentalität in den Alpen. I. Teil', *Zeitschr. Ges. f. Erdkunde, Berlin*, Nr. 9/10, 321–46.
Gao, Y.-X. and Li, C. (1981) 'Influences of Qinghai-Xizang Plateau on seasonal variation of atmospheric general circulation', in *Geoecological and Ecological Studies of Qingai-Xizang Plateau*, vol. 2, pp. 1477–84, Beijing, Science Press.
Reiter, E.R. and Tang, M. (1984) 'Plateau effects on diurnal circulation patterns', *Mon. Wea. Rev.*, 112, 638–51.
Tang, M. and Reiter, E.R. (1984) 'Plateau monsoons of the northern hemisphere: a comparison between North America and Tibet', *Mon. Wea. Rev.*, 112, 617–37.

Altitude

Air pressure, density and vapour pressure

COESA (1962) 'US Standard Atmosphere, 1962', U.S. Comm., Extension Standard Atmosphere, Washington D.C.
Fujimara, I. (1971) 'The climate and weather of Mt. Fuji', *Fujisan-sogogakujutsuchosa-hokoku* (*Mt. Fuji, Scientific Report*), pp. 215–304, Tokyo, Fujikyuko Co. Ltd.
Khrgian, A. Kh. (1965) 'Atmospheric moisture distribution over mountain country', *Atmos. Ocean. Phys. Izvestiya Acad. Sci. USSR*, 1(4), 233–6.
Kuz'min, P.P. (1972) *Melting of Snow Cover* (Russian original 1961), Jerusalem, Israel Prog. Sci. Transl.
Minzer, R.A. (1962) *A History of Standard and Model Atmospheres, 1847 to 1962*, Bedford, Mass., TR 62-6-N, Geophysics Corp. America.
Prohaska, F. (1970) 'Distinctive bioclimatic parameters of the subtropical-tropical Andes', *Int. J. Biomet.*, 14, 1–12.
Rathschuler, E., (1949) 'Uber die Anderung des Tagesganges der Luftfeuchtigkeit mit der Hohe', *Arch. Met. Geophys. Biokl.*, B1, 17–31.
Reitan, C.H. (1963) 'Surface dew point and water vapor aloft', *J. appl. Met.*, 2, 776–9.
Steinhauser, F. (1936) 'Uber die Haufigkeitsverteilungen des Dampfdruckes in Gebirge und in der Niederung und ihre Beziehungen zueinander', *Met. Zeit.*, 53, 415–19.
Steinhauser, F. (1938) *Die Meteorologie des Sonnblicks. I. Beiträge zur Hochgebirgsmeteorologie*, Vienna, J. Springer.

Storr, D. (1970) 'A comparison of vapour pressure at mountain stations with that in the free atmosphere', *Can Met. Res. Rep.*, 1/70, Dept. of Transport, Canada.

Yoshino, M.M. (1975) 'Climate in a small area', *An Introduction to Local Meteorology*, Tokyo, University of Tokyo Press.

Radiation

Abbott, C.G. and Fowle, F.E., Jr. (1908) 'Determination of the intensity of solar radiation outside the earth's atmosphere, otherwise termed the "solar constant of radiation"', *Ann. Astrophys. Obs. Smithson. Inst., Washington*, 2(1), 11–126.

Abbott, C.G. and Fowle, F.E., Jr. (1911) 'The value of the solar constant of radiation', *Astrophys. J.*, 33, 191–6.

Abetti, G. (1957) *The Sun* (transl. J.B. Sidgwick), London, Macmillan.

Ångstrøm, A.K. and Drummond, A.J. (1966) 'Note on solar radiation in mountainous regions at high altitude', *Tellus*, 18, 801–5.

Ångstrøm, K. (1900) 'Intensité de la radiation à differentes altitudes. Recherches faites à Tenériffe. 1895 et 1896', *Nova Acta Reg. Soc. Sci. Upsal.*, 20 (3).

Becker, C.F. and Boyd, J.S. (1957) 'Solar variability on surfaces in the United States as affected by season, orientation, latitude, altitude and cloudiness', *J. Solar Energy*, 1, 13–21.

Bener, P. (1963) 'Der Einfluss der Bewölkung auf die Himmelstrahlung', *Arch. Met. Geophys. Biokl.*, B, 12, 442–57.

Bishop, B.C., Ångstrøm, A.K., Drummond, A.J. and Roche, J.J. (1966) 'Solar radiation measurements in the high Himalayas (Everest region)', *J. appl. Met.*, 5, 94–104.

Brunt, D. (1932) 'Notes on radiation in the atmosphere', *Q.J.R. Met. Soc.*, 58, 389–420.

Budyko, M.I. (1974) *Climate and Life*, pp. 189–92, New York, Academic Press.

Caldwell, M.M. (1968) 'Solar ultraviolet radiation as an ecological factor for alpine plants', *Ecol. Monogr.*, 38, 243–68.

Caldwell, M.M. (1980) 'Light quality with special reference to UV at high altitudes', in U. Benecke and M.R. Davis (eds) *Mountain Environments and Subalpine Tree Growth*, pp. 61–79, Wellington, Tech. Pap. No. 70, Forest Res. Inst., New Zealand Forest Service.

Dirmhirn, I. (1951) 'Untersuchungen der Himmelstrahlung in den Ostalpen mit besonder Berücksichtigung ihrer Höhenabhängigkeit', *Arch. Met. Geophys. Biokl.* B, 2, 301–46.

Dorno, C. (1911) *Studie über Licht und Luft des Hochgebirges*, Braunschweig, Vieweg und Sohn.

Drummond, A.J. and Ångstrøm, A.K. (1967) 'Solar radiation measurements on Mauna Loa (Hawaii) and their bearing on atmospheric transmission', *Solar Energy*, 11, 133–44.

Eckel, O. (1936) 'Uber einige Eigenschaften der ultravioletten Himmelstrahlung in verschiedenen Meereshöhen und bei Föhnlage', *Met. Zeit.*, 53, 90–4.

Elterman, L. (1964) 'Atmospheric attenuation model, 1964, in the ultraviolet, visible and infra-red regions for altitudes to 50 km', Env. Res. Pap. No. 46, AFCRL-64-740, Cambridge, Massachusetts, U.S. Air Force Cambridge Research Laboratories.

Flach, E. (1966) 'Geographische Verteilung der Globalstrahlung und Himmelstrahlung', *Arch. Met. Geophys. Biokl.*, B, 14, 161–83.

Fliri, F. (1971) 'Neue klimatologische Querprofile der Alpen-ein Energiehaushalt', *Ann. Met.*, N.F. 5, 93–7.

Frère, M., Rijks, J.Q. and Rea, J. (1975) '*Estudios Agroclimatologico de la Zona Andina*', Informe Technico. Rome: Food and Agricultural Organization of the United Nations.

Gates, D.M. and Janke, R. (1966) 'The energy environment of the alpine tundra', *Oecol. Planta*, 1, 39–62.

Geiger, R. (1965) *The Climate near the Ground*, Cambridge, Mass., Harvard University Press.

Hand, I.F., Conover, J.H. and Boland, W.A. (1943) 'Simultaneous pyrheliometric measurements at different heights on Mount Washington, N.H.', *Mon. Weather Rev.*, 71, 65–9.

Kastrov, V.G. (1956) 'Solnechnaia radiatsiia v troposfere v sluchae absoliutno chistnogo i sukhogo vozdukha', *Trudy Tsentr. Aerol. Obs.*, 16, 23–30.

Kimball, H.H. (1927) 'Measurements of solar radiation intensity and determinations of its depletion by the atmosphere with bibliography of pyrheliometric measurements', *Mon. Weather Rev.*, 55, 155–69.

Klein, W.H. (1948) 'Calculation of solar radiation and the solar heat load on man', *J. Met.*, 5, 119–29.

Kondratyev, K. Ya. (1969) *Radiation in the Atmosphere*, New York, Academic Press.

Langley, S.P. (1882) 'The Mount Whitney expedition', *Nature*, 26, 314–17.

Langley, S.P. (1884) 'Researches on solar heat and its absorption by the earth's atmosphere. A report on the Mount Whitney expedition', U.S. War Dept., Prof. Pap. Signal Service XV, Washington, D.C.

Lauscher, F. (1937) 'Die Zunahme der Intensität der Sonnenstrahlung mit der Hohe', *Gerlands Beitr. Geophys*, 50, 202–15.

LeDrew, E.F. (1975) 'The estimation of clear sky atmospheric emittance at high altitudes', *Arct. Alp. Res.*, 7, 227–36.

Lowry, W.P. (1980a) 'Clear-sky direct-beam solar radiation versus altitude: a proposal for standard soundings', *J. appl. Met.*, 19, 1323–17.

Lowry, W.P. (1980b) *Direct and diffuse solar radiation: Variations with atmospheric turbidity and altitude*, Urbana-Champaign, Institute for Environmental Studies, University of Illinois. Research Report No. 6.

Maurer, J. (1912) 'Aus älteren und neuen Messungen der Sonnenstrahlung auf hochalpinen Stationen', *Met, Zeit.*, 29, 561–9.

Neuwirth, F. (1979) 'Beziehungen zwischen Globalstrahlung, Himmelstrahlung und extraterrestrischer Strahlung in Osterreich', *Arch. Met. Geophys. Biokl.*, B, 27, 1–13.

Pope, J.H. (1977) 'Computations of solar insolation at Boulder, Colorado', NOAA Tec. Mem., NESS 93, Washington, D.C.

Radau, R. (1877) *Actinometrie*, Paris, Gauthier-Villars.

Reiter, R., Carnuth, W. and Sladkovic, R. (1972) 'Ultraviolettstrahlung in alpinen Höhenlagen', *Wetter u. Leben*, 24, 231–47.

Reiter, R., Mungert, K. and Sladovic, R. (1982) 'Results of 5–year concurrent recordings of global, diffuse and UV radiation at three levels (700, 1800, and 3000 m a.s.l.) in the northern Alps', *Arch. Met. Geoph. Biokl.*, B30, 1–28.

Sauberer, F. and Dirmhirn, I. (1958) 'Das Strahlungsklima', in F. Steinhauser, O. Eckel, and F. Lauscher (eds) *Klimatographie von Osterreich*, pp. 13–102, Vienna, Springer.

Schüepp, W. (1949) 'Die Bestimmung der Komponenten der atmosphärischen Trübung aus Aktinometermessungern', Arch. Met. Geophys. Biokl., B, 1, 257–346.

Sivkov, S.I. (1971) *Computation of Solar Radiation Characteristics*, Leningrad, Gidromet, Izdat. Jerusalem, Israel Prog. Sci. Translations.

Steinhauser, F. (1939) 'Die Zunahme der Intensität der direkten Sonnenstrahlung mit der Höhe im Alpengebiet und die Verteilung der "Trubung" in den unterer Luftschichten', *Met. Zeit.*, 56, 173–81.

Thams, J.C. (1961a) 'The influence of the alps on the radiation climate', in *Recent Progress in Photobiology*, Proc. 3rd Int. Congress. pp. 76–91, New York, Elsevier.

Thams, J. C. (1961b) 'Der Einfluss der Bewölkungsmenge und art auf die Grösse der diffusen Himmelstrahlung', *Geof. pura appl.*, 48, 181–92.

Valko, P. (1961) 'Untersuchung über die vertikal Trübungsschlichtung der Atmosphäre',

Arch. Met. Geophys. Biokl., 11, 148–210.

Valko, P. (1980) 'Some empirical properties of solar radiation and related parameters', in *An Introduction to Meteorological Measurements and Data Handling for Solar Energy Applications*, Chapter 8, DOE/ER-0084, Washington, DC, US Dept. of Energy.

Voloshina, A.P. (1966) *Teplovoi balans poverkhnosti vysokogornykh lednikov v letnii period*, Moscow, Izdat. Nauka.

Wessely, E. (1969) 'Messung der UV-Strahlung mit Interferenzfilter und Photoelementen bei 332 μm', Ph. D. dissertation, University of Vienna.

Temperature

Barry, R.G. (1972) *Climatic environment of the east slope of the Colorado Front Range*, Occas, Pap. no. 3, Inst. Arctic Alp. Res., Boulder, CO, University of Colorado, 206 pp.

Barry, R.G. (1978) 'H-B. de Saussure: the first mountain meteorologist', *Bull. Am. Met. Soc.*, 59, 702–5.

Borisov, A.M., Grudzinski, M.E. and Khrgian, A. Kh. (1958) 'O meteorologicheskikh usloviakh vysokogornogo Tian-Shanya (Meteorological conditions of the high Tien-Shan)', *Trudy Tsentr. Aerol. Obs.*, 21, 175–99.

Brocks, K. (1940) 'Lokale Unterschiede und zeitliche Anderungen der Dichteschichtung in der Gebirgsatmosphare', *Met. Zeit.*, 57, 62–73.

Chen, L., Reiter, E.R. and Feng, Z. (1985) 'The atmospheric heat source over the Tibetan Plateau; May–August 1979', *Mon. Wea. Rev*, 113, 1771–90.

Conrad, V. (1944) *Methods of Climatology*, Cambridge, Mass., Harvard University Press.

Cramer, O.P. and Lynott, R.E. (1961) 'Cross-section analysis in the study of windflow over mountainous terrain', *Bull. Am. Met. Soc.*, 42, 693–702.

Cramer, O.P. (1972) 'Potential temperature analysis for mountainous terrain', *J. Appl. Met.*, 11, 44–50.

de Quervain, A. (1904) 'Die Hebung der atmosphärischen Isothermen in der Schweizer Alpen und ihre Beziehung zu deren Höhengrenzen', *Gerlands Beitr. Geophys.*, 6, 481–533.

Eide, O. (1948) 'On the temperature difference between mountain peak and free atmosphere at the same level. II. Gaustatoppen-Kjeller', *Met. Annal.*, 2(3), 183–206.

Ekhart, E. (1939) 'Mittlere Temperaturverhältnisse der Alpen und der freien Atmosphäre über dem Alpenvorland. Ein Beitrag zur dreidimensionalen Klimatologie. 1. Die Temperaturverhältnisse der Alpen', *Met. Zeit.*, 56, 12–26; 2. Die Temperaturverhältnisse der freien Atmosphare über München und Vergleich mit den Alpen', *Met. Zeit.*, 56, 49–57.

Ferguson, S.P. (1934) 'Aerological studies on Mt. Washington', *Trans. Am. Geophys. Union*, 15, 114–17.

Ficker, H. von (1913) 'Temperaturdifferenz zwischen freier Atmosphäre und Berggipfeln', *Met. Zeit.*, 30, 278–304

Ficker, H. von (1926) 'Vertikale Temperaturgradienten im Gebirge', *Veroff. Preuss. Met. Inst.*, 335, 45–62.

Flohn, H. (1953) 'Hochgebirge und allgemeine Zirkulation. II. Die Gebirge als Wärmequellen', *Arch. Met. Geophys. Biokl.*, A, 5, 265–79.

Flohn, H. (1968) 'Contributions to a meteorology of the Tibetan Highlands'. Atmos. Sci. Pap. no. 130, Ft. Collins, Colorado State University.

Flohn, H. (1974) 'Contribution to a comparative meteorology of mountain areas'. in J.D. Ives and R.G. Barry (eds) *Arctic and Alpine Environments*, pp. 55–71, London, Methuen.

Geiger, R. (1965) *The Climate near the Ground*, Cambridge, Mass., Harvard University Press.

Gloyne R.W. (1971) 'A note on the average annual mean of daily earth temperature in the United Kingdom', *Met. Mag.*, 100, 1–6.

Green F.H.W. and Harding, R.J. (1979) 'The effect of altitude on soil temperature', *Met. Mag*, 108, 81–91.

Green, F.H.W. and Harding, R.J. (1980) 'Altitudinal gradients of soil temperatures in Europe', *Trans. Inst. Brit. Geog.*, 5, 243–54.

Gutman, G.J. and Schwerdtfeger, W. (1965) 'The role of latent and sensible heat for the development of a high pressure system over the subtropical Andes in the summer', *Met. Rdsch.*, 18, 69–75.

Hann, J. von (1906) *Lehrbuch der Meteorologie*, Leipzig, C.H. Tauchinitz.

Hann, J. von (1913) 'Die Berge kälter als die Atmosphäre, ein meteorologisches Paradoxon', *Met. Zeit.*, 30, 304–6.

Hänsel, C. von (1962) 'Die Unterschiede von Temperatur und relativer Feuchtigkeit zwischen Brocken und umgebender freier Atmosphäre', *Zeit. Met.* 16, 248–52.

Harding, R.J. (1978) 'The variation of the altitudinal gradient of temperature within the British Isles', *Geogr. Ann.* A, 60, 43–9.

Harding, R.J. (1979) 'Altitudinal gradients of temperature in the northern Pennines', *Weather*, 34, 190–201.

Harrison, S.J. (1975) 'The elevation component of soil temperature variation', *Weather*, 30, 397–409.

Hastenrath, S.L. (1968) 'Der regionale und jahrzeitliche Wandel des vertikalen Temperaturgradienten und seine Behandlung als Wärmhaushaltsproblem', *Met. Rdsch.*, 21, 46–51.

Hauer, H. (1950) 'Klima und Wetter der Zugspitze', *Berichte d. Deutschen Wetterdienstes in der US-Zone*, 16.

Kleinschmidt, E. (1913) 'Die Temperaturverhältnisse in der freien Atmosphäre und auf Berggipfeln nach den Massungen der Drachenstation am Bodensee und der Observatorien auf der Säntis und der Zugspitze', *Beitr. Phys. frei. Atmos.*, 6, 1–18.

Kreuels, R., Fraedrich, K. and Ruprecht, E. (1975) 'An aerological climatology of South America, *Met. Rdsch.*, 28, 17–26.

Lauscher, F. (1966) 'Die Tagesschwankung der Lufttemperatur auf Höhenstationer in allen Erdteilen, *60–62 Jahresbericht des Sonnblick-Vereines für die Jahre 1962–64*, pp. 3–17, Vienna.

Lautensach, H. and Bogel, R. (1956) Der Jahrsgang des mittleren geographischen Höhengradienten der Luftemperatur in den verschiedenen Klimagebieten der Erde, *Erdkunde*, 10, 270–82.

Linacre, E. (1982) 'The effect of altitude on the daily range of temperature', *J. Climatol.* 2, 375–82.

McCutchan, M.H. (1983) 'Comparing temperature and humidity on a mountain slope and in the free air nearby', *Mon. Wea. Rev.*, 111, 836–45.

Oliver, J. (1962) 'The thermal regime of upland peat soils in a maritime temperate climate, *Geogr. Ann.*, 44, 293–302.

Peppler, W. (1931) 'Zur Frage des Temperaturunterschiedes zwischen den Berggipfeln und der freien Atmosphäre', *Beitr. Phys. frei. Atmos.*, 17, 247–63.

Peppler, W. (1935) 'Ergänzung zu meiner Arbeit: "Zur Frage des Temperaturunterschiedes zwischen den Berggipfeln und der freien Atmosphäre"', *Beitr. Phys. frei. Atmos.*, 21, 172–76.

Peterson, E.B. (1969) 'Radiosonde data for characterization oı a mountain environment in British Columbia', *Ecology*, 50, 200–5.

Rao, G.V. and Erdogan, S. (1989) 'The atmospheric heat source over the Bolivian Plateau

for a mean January', *Boundary-Layer Met.*, 46, 13–33.
Richner, H. and Phillips, P.D. (1984) 'A comparison of temperature from mountaintops and the free atmosphere – their diurnal variation and mean difference', *Mon. Wea. Rev.*, 112, 1328–40.
Samson, C.A. (1965) 'A comparison of mountain slope and radiosonde observations', *Month. Weather Rev.*, 95, 327–30.
Steinhauser, F. (1937) 'Uber die täglichen Temperaturschwankungen im Gebirge', *Gerlands Beitr. Geophys.*, 50, 360–7.
Tabony, R.C. (1985) 'The variation of surface temperature with altitude; *Met. Mag.* 114, 37–48.
Thompson, W.F. (1990) 'Climate related landscapes in world mountains: Criteria and map', *Zeit. f. Geomorph., Suppl.* 78, 92 pp.
Troll, C. (1964) 'Karte der Jahreszeiten-Klimate der Erde', *Erdkunde*, 18, 5–28.
Yeh, D.-Z. (1982) 'Some aspects of the thermal influences of Qinghai-Tibetan plateau on the atmospheric circulation', *Arch. Met. Geophys. Biocl.*, A31, 205–20.
Yoshino, M.M. (1966) 'Some aspects of air temperature climate of the high mountains in Japan', *Jap. Progr. Climat.*, November, 21–7.

Wind

Allison, I. and Bennett, J. (1976) 'Climate and microclimate', in G.S. Hope, *et al.* (eds) *The Equatorial Glaciers of New Guinea*, pp. 61–80, Rotterdam, A.A. Balkema.
Bailey, S.I. (1908) 'Peruvian meteorology: observations made at the auxiliary stations. 1892–1895', *Ann. Astron. Obs. Harvard Coll.*, 49(2), 104–232.
Davidson, B., Gerbier, S.D., Papagiankis, S.D. and Rijkoort, P.J. (1964) 'Sites for wind-power installations', *W.M.O. Tech. Note* 63, Geneva, World Meteorological Organization.
Davies, H.C. and Phillips, P.D. (1985) 'Mountain drag along the Gotthard section during ALPEX', *J. Atmos. Sci.* 42, 2093–109.
Eustis, R.S. (1942) 'The winds over New England in relation to topography', *Bull. Am. Met. Soc.*, 28, 383–7.
Ficker, H. von (1913) 'Die Wirkung der Berge auf Luftströmungen', *Met. Zeit.*, 30, 608–10.
Georgii, W. (1922) 'Die Windbeeinflussung durch Gebirge', *Beitr. Phys. frei Atmos.*, 10, 178–84.
Georgii, W. (1923) 'Die Luftströmung über Gebirge', *Met. Zeit.*, 40, 108–12, 309–11.
Müller, F., Ohmura, A., Schroff, K., Funk, M., Pfirter, K., Bernath, A. and Steffen, K. (1980) 'Combined ice, water and energy balances of a glacierized basin of the Swiss Alps – the Rhonegletscher Project', in *Geography in Switzerland, Geogr. Helvet.*, 35(5), 57–69.
Ohmura, A. (1990) 'On the wind profile over the Alps', *Proc. xxi Internat. Tagung f. Alpine Meteorologie, Veröff, Schweiz. Met. Austalt* 48, 102–5.
Reiter, E.R. (1963) *Jet-stream Meteorology*, Chicago, University of Chicago Press.
Schell, I.I. (1936) 'On the vertical distribution of wind velocity over mountain summits', *Bul. Am. Met. Soc.*, 17, 295–300.
Schumacher, C. (1923) 'Der Wind in der freien Atmosphäre und auf Säntis, Zupgspitze und Sonnblick', *Beitre. Phys. frei Atmos.* 11, 20–42.
Steinhauser, F. (1938) *Die Meteorologie des Sonnblicks, I. Beiträge zur Hochgebirgs-meteorologie*, Vienna, J. Springer.
Taylor, P.A. Sykes, R.I. and Mason, P.J. (1989) 'On the parameterization of drag over small-scale topography in neutrally-stratified boundary layer flow', *Boundary-Layer Met.*, 48, 409–22.

Wahl, E. (1966) 'Windspeed on mountains', Milwaukee, Meteorology Dept., University of Wisconsin, Final Report AFCRL–66–280.

Topography

Dimensional and relief effects

Bergeron, T. (1960) *Preliminary results of Project Pluvius*, Helsinki, Int. Ass. Sci. Hydrol. Pub. No. 53, 226–37.

Bradley, E.F. (1980) 'An experimental story of the profiles of wind speed, shearing stress and turbulence at the crest of a large hill', *Q.J.R. Met. Soc.*, 106, 101–24.

Carruthers, D.J. and Choularton, T.W. (1982) 'Airflow over hills of moderate slope', *Q.J.R. Met. Soc.*, 108, 603–24.

Carruthers, D.J. and Hunt, J.C.R. (1990) 'Fluid mechanisms of airflow over hills: Turbulence, fluxes, and waves in the boundary layer', in W. Blumen (ed.) *Atmospheric Processes over Complex Terrain*, Met. Monogr. 23(45), pp. 83–103, Boston, Amer. Met. Soc.

Etling, D. (1989) 'On atmospheric vortex streets in the wake of large islands'. *Met. Atmos. Phys.*, 41, 157–64.

Gallagher, M.W., Choularton, T.W. and Hill, M.K. (1988) 'Some observations of airflow over a large hill of moderate slope', *Boundary–Layer Met.* 42, 22–50.

Giles, B.D. (1976) 'Fluidics, the Coanda effect and some orographic winds', *Arch. Met. Geophys. Biokl.*, A, 25, 273–80.

Green, J.S.A. (1984) 'Describing the Alps', *Riv. Met. Aeronaut*, 44: 23–30.

Hunt, J.R. (1980) 'Winds over hills', in J.C. Wyngaard (ed.) *Workshop on the Planetary Boundary Layer*, Boston, American Meteorological Society, pp. 107–44.

Hunt, J.C.R. and Richards, K.J. (1984) 'Stratified airflow over one or two hills', *Boundary-Layer Met.*, 30, 223–259.

Hunt, J.C.R. and Simpson, J.E. (1982) 'Atmospheric boundary layers over non-homogeneous terrain', in E.J. Plate (ed.) *Engineering Meteorology*, Amsterdam, Elsevier, 269–318.

Hunt, J.C.R. and Snyder, W.H. (1980) 'Experiments on stably and neutrally stratified flow over a model three-dimensional hill', *J. Fluid Mech.*, 96, 671–704.

Hunt, J.C.R., Richards, K.J. and Brighton, P.W.M. (1988) 'Stably stratified flow over low hills, *Q.J.R. Met. Soc.*, 114, 859–86.

Jackson, P.S. and Hunt, J.C.R. (1975) 'Turbulent wind flow over a low hill', *Q.J.R. Met. Soc.*, 101, 929–55.

Kimura, F. and Manins, P. (1988) 'Blocking in periodic valleys', *Boundary-Layer Met.*, 44, 137–169.

Lee, J.T., Lawson, R.E., Jr., and March, G.L. (1987) 'Flow visualization experiments on stably stratified flow over ridges and valleys', *Met. Atmos. Phys.*, 37, 183–94.

McCutchan, M.H. and Fox, D.G. (1986) 'Effect of elevation and aspect on wind, temperature and humidity', *J. Clim. appl. Met.*, 25: 1996–2013.

Mason, P.J. (1987) 'Diurnal variations in flow over a succession of ridges and valleys' *Q.J.R. Met Soc.*, 113, 1117–40.

Mason, P.J. and King, J.C. (1984) 'Atmospheric flow over a succession of nearly two-dimensional ridges and valleys', *Q.J.R. Met. Soc.*, 110, 821–45.

Mason, P.J. and Sykes, R.I. (1979) 'Flow over an isolated hill of moderate slope', *Q.J.R. Met. Soc.*, 105, 383–95.

Meroney, R.N. (1990) 'Fluid dynamics of flow over hills/mountains – insights through physical modeling', in W. Blumen (ed.) *Atmospheric Processes over Complex Terrain*, Met. Monogr. 23(45), pp. 145–71, Boston, Amer. Met. Soc.

Nicholls, J.M. (1973) *The Airflow over Mountains. Research, 1958–1972*, WMO Tech. Note no. 127, Geneva, World Meteorological Organization.

Ramachandran, G. (1972) 'The role of orography on wind and rainfall distribution in and around a mountain gap: an observational study', *Ind. J. Met. Geophys.*, 23, 41–4.

Ramachandran, G., Rao, K.V., and Krishna, K. (1980) 'An observational study of the boundary-layer winds in the exit region of the mountain gap', *J. appl. Met.*, 19, 881–8.

Salmon, J.H., Bowen, A.J., Hoff, A.M., Johnson, R., Mickle, R.E., Taylor, P.A., Tetzlaff, G. and Walmsley, J.L. (1988) 'The Askervein Hill Project: Mean wind variations at fixed heights above ground', *Boundary-Layer Met.*, 43, 247–71.

Scorer, R.S. (1955) 'Theory of airflow over mountains. IV. Separation of flow from the surface', *Q.J.R. Met. Soc.*, 81, 340–50.

Scorer, R.S. (1978) *Environmental Aerodynamics*, Chichester, Ellis Horwood.

Sheppard, P.A. (1956) 'Airflow over mountains', *Q.J.R. Met. Soc.*, 82, 528–9.

Smith, R.B. (1979) 'The influence of mountains on the atmosphere', *Adv. Geophys.*, 21, 87–230.

Smith, R.B. (1980) 'Linear theory of stratified hydrostatic flow past an isolated mountain', *Tellus* 32, 348–64.

Smith, R.B. (1990) 'Why can't stably stratified air rise over high ground?' in W. Blumen (ed.) *Atmospheric Processes over Complex Terrain*, Met. Monogr. 23(45), pp. 105–7, Boston, Amer. Met. Soc.

Snyder, W.H., Thompson, R.S., Eskridge, R.E. Lawson, R.E, Jr., Castro, I.P., Lee, J.T., Hunt, J.C.R. and Ogawa, Y. (1985) 'The structure of strongly stratified flow over hills: Dividing streamline concept', *J. Fluid Mech.*, 52, 249–88.

Stringer, E.T. (1972) *Foundations of Climatology*, pp. 141–67 San Francisco, W.H. Freeman and Co.

Stull, R.B. (1988) *An Introduction to Boundary Layer Meteorology*, Dordrecht, Kluwer Academic Publishers.

Tampieri, F. (1987) 'Separation features of boundary-layer flow over valleys', *Boundary-Layer Met.*, 40, 295–307.

Tampieri, F. and Hunt, J.C.R. (1985) 'Two-dimensional stratified fluid flow over valleys: linear theory and a laboratory investigation', *Boundary-Layer Met.*, 32, 257–79.

Taylor, P.A. (1977) 'Numerical studies of neutrally stratified planetary boundary-layer flow above gentle topography. I. Two-dimensional cases', *Boundary-Layer Met.*, 12, 37–60.

Taylor, P.A. and Lee, R.A. (1984) 'Simple guidelines for estimating wind speed variations due to small-scale topographic features', *Climatol. Bull.* Ottawa, 18, 3–32.

Taylor, P.A., Mason, P.J. and Bradley, E.F. (1987) 'Boundary-layer flow over low hills', *Boundary-Layer Met.*, 39, 107–32.

Teunissen, H.W., Shokr, M.E., Bowen, A.J., Wood, C.J. and Green, D.W.R. (1987) 'The Askervein Hill Project: Wind-tunnel simulations at three length scales', *Boundary-Layer Met.*, 40, 1–29.

Trombetti, F. and Tampieri, F. (1987) 'An application of the dividing-streamline concept to the stable airflow over mesoscale mountains', *Mon. Wea. Rev.*, 115, 1802–6.

Walmsley, J.L., Taylor, P.A. and Salmon, J.R. (1989) 'Simple guidelines for estimating wind speed variations due to small-scale topographic features – an update', *Climatol. Bull.*, 23, 3–14.

Wilson, H.P. (1968) *Stability Waves.* Toronto, Dept. of Transport, Met. Branch, TEC703.

Wilson, H.P. (1974) 'A note on mesoscale barriers to surface airflow', *Atmosphere.*, 12, 118–20.

Woodridge, G.L., Fox, D.G. and Furman, R.W. (1987) 'Airflow patterns over and around

a large three-dimensional hill', *Met. Atmos. Phys.*, 37, 259–70.
World Meteorological Organization, Secretariat (1981) '*Meterological Aspects of the Utilization of Wind as an Energy Source*', Tech. Note No. 75, WMO no. 575. Geneva.

Slope and aspect

Barry, R.G. (1978) 'Diurnal effects on topoclimate on an equatorial mountain', *Arbeiten. Zentralanst. Met. Geodynam.*, 32, 72, 1–8.
Barry, R.G. and Chorley, R.J. (1987) *Atmosphere, Weather and Climate*, 5th edn, London, Methuen.
Dozier, J. (1980) 'A clear-sky spectral solar radiation model for snow-covered mountainous terrain', *Water Resources Res.* 16, 709–18.
Garnier, B.J. and Ohmura, A. (1968) 'A method of calculating the direct short wave radiation income of slopes', *J. appl. Met.*, 7, 796–800.
Garnier, B.J. and Ohmura, A. (1970) 'The evaluation of surface variations in solar radiation income', *Solar Energy*, 13, 21–34.
Geiger, R. (1965) *The Climate Near the Ground*, Cambridge, Mass., Harvard University Press.
Hay, J.E. (1977) 'An analysis of solar radiation data for selected locations in Canada', *Climat. Studies*, 32, Atmos. Env. Service, Downsview, Ontario.
Hay, J.E. (1979) *Study of shortwave radiation on non-horizontal surfaces.* Can. Climate Centre, Atmos. Env. Serv. Rep. no. 79–12, Downsview, Ontario.
Heywood, H. (1964) 'Solar radiation on inclined surfaces', *Nature*, 204 (4959), 669–70.
Kondratyev, K. Ya. (1969) *Radiation in the Atmosphere*, pp. 485–502, New York, Academic Press.
Kondratyev, K. Ya. and Federova, M.P. (1977) 'Radiation regime of inclined slopes', W.M.O. Tech. Note no. 152, Geneva, World Meteorological Organization.
Kondratyev, K. Ya. and Manolova, M.P. (1960) 'The radiation balance of slopes', *Solar Energy.*, 4, 14–19.
Lee, R. (1962) 'Theory of the "equivalent slope"', *Mon. Weather Rev.*, 90, 165–6.
Lee, R. (1978) *Forest Microclimatology*, New York, Columbia University Press.
Loudon, A.G. and Petheridge, P. (1965) 'Solar radiation on inclined surfaces', *Nature*, 206 (4984), 603–4.
Marks, D. and Dozier, J. (1979) 'A clear-sky longwave radiation model for remote alpine areas', *Arch. Met. Geophys. Biokl.*, B. 27, 159–87.
Revfeim, K.J.A. (1976) 'Solar radiation at a site on known orientation on the earth's surface', *J. appl. Met.*, 15, 651–6.
Sellers, W.D. (1965) *Physical Climatology*, Chicago, University of Chicago Press.
Steven, M.D. (1977) 'Standard distributions of clear sky radiance', *Q.J.R. Met. Soc.*, 103, 457–65.
Steven, M.D. and Unsworth, M.H. (1979) 'The diffuse solar irradiance of slopes under cloudless skies', *Q.J.R. Met Soc.*, 105, 593–602.
Williams, L.D., Barry, R.G. and Andrews, J.T. (1972) 'Application of computed global radiation for areas of high relief', *J. appl. Met.*, 11, 526–33.
Wilson, R.G. and Garnier, B.J. (1975) 'Calculated and measured net radiation for a slope', *Climat. Bull.* Montreal, no. 17, 1–14.

Topo- and microclimates

Aizenshtat, B.A. (1962) 'Nekotorie cherti radiatsionnogo rezhima teplogo balansa i mikroklimata gornogo perrevala (Some characteristics of the heat balance regime and microclimate of a mountain pass)', *Met. i Gidrol*, 3, 27–32.

Aulitsky, H. (1962) 'Die Bodentemperaturverhältnisse einer zentralalpinen Hochgebirgs-Hangstation II' *Archiv. Met. Geophys. Biokl.*, B, 11, 301–62.

Aulitsky, H. (1984) 'Die Windverhältnisse an einer zentralalpinen Hangstation und ihre ökologische Bedeutung. 1 Teil', *Cbl. ges. Forstw.*, 101, 193–232.

Aulitsky, H. (1985) '*Ibid. 2 Teil*', *Cbl. ges. Forst.*, 102, 55–72.

Barry, R.G. (1978) 'Diurnal effects on topoclimate on an equatorial mountain', *Arbeiten, Zentralanst. Met. Geodynam.*, Vienna, 32, 72, 1–8.

Barry, R.G. and Van Wie, C.C. (1974) 'Topo- and microclimatology in alpine areas', in J.D. Ives and R.G. Barry (eds) *Arctic and Alpine Environments*, pp. 73–83. London, Methuen.

Borzenkova, I. (1967) 'K voprosy o vliyanii mestnykh faktorov na prikhod radiatsii v gornoi mestnosti (Concerning the influence of local factors on the course of radiation in mountain locations)', *Trudy Glav. Geofiz. Obs.*, 209, 70–7.

Brazel, A. (1974) 'A note on topoclimatic variation of air temperature, Chitistone Pass region, Alaska', in *Icefield Ranges Research Project, Scientific Results*, vol. 4, pp. 81–7, New York, American Geographical Society.

Brazel. A. and Outcalt, S.I. (1973a) 'The observation and simulation of diurnal surface thermal contrast in an Alaskan alpine pass', *Arch. Met. Geophys. Biokl.*, B, 21, 157–74.

Brazel, A. and Outcalt, S.I. (1973b) 'The observation and simulation of diurnal evaporation contrasts in an Alaskan alpine pass', *J. appl. Met.*, 12, 1134–43.

Dozier, J. and Outcalt, S.I. (1979) 'An approach toward energy balance simulation over rugged terrain', *Geog. Anal.* 11, 65–85.

Gale, J. (1972) 'Elevation and transpiration: some theoretical considerations with special reference to Mediterranean-type climate', *J. appl. Ecol.*, 9, 691–702.

Gloyne, R.W. (1955) 'Some effects of shelter-belts and windbreaks', *Met. Mag.*, 84, 272–81.

Grace, J. (1977) *Plant Response to Wind*, London, Academic Press.

Grunow, J. (1952) 'Beiträge zum Hangklima', *Berichte dt. Wetterdienstes in der US-Zone*, 8(35), 293–8.

Hess, M., Niedzwiedz, T. and Obrebska-Starkel, B. (1975) 'The methods of constructing climatic maps of various scales for mountains and upland territories exemplified by the maps prepared for southern Poland', *Geogr. Polon.*, 31, 163–87.

Hess, M., Niedzweidz, T. and Obrebska-Starkel, B. (1976) 'The method of characterizing the climate of the mountains and uplands in the macro-, meso- and microscale (exemplified by Southern Poland)', *Zesz. Nauk. Univ. Jagiellon., Prace Geog.*, 43, 83–102.

Holroyd, E.W., III, (1970) 'Prevailing winds on Whiteface Mountain as indicated by flag trees', *For. Sci.*, 16, 222–9.

Holtmeier, F.K. (1978) 'Die bodennahen Winde in den Hochlagen der Indian Peaks Section (Colorado Front Range)', *Münster Geogr. Arbeit.*, 3, 3–47.

Ingram, M. (1958) 'The ecology of the Cairngorms. IV. The *Juncus* zone: *Juncus trifidis* communities', *J. Ecol.*, 46, 707–37.

Isard, S.A. and Belding, M.J. (1986) 'Evapotranspiration from the alpine tundra of Colorado, USA', *Arct. Alp. Res.*, 21, 71–82.

Kondratyev, K. Ya, and Manolova, M.P. (1960) 'The radiation balance of slopes', *Solar Energy*, 4, 14–19.

Larcher, W. (1980) 'Klimastress im Gebirge-Adaptionstraining und Selektions-filter für Pflanzen', *Rheinisch-Westfalien Akad. Wiss.* no. 291, 49–80.

LeDrew, E.F. (1975) 'The energy balance of a mid-latitude alpine site during the growing season, 1973', *Arct. Alp. Res.*, 7, 301–14.

McCutchan, M.H. and Fox, D.G. (1986) 'Effect of elevation and aspect on wind, temperature and humidity', *J. Clim. appl. Met.*, 25: 1996–2013.

McCutchan, M.H., Fox, D.G. and Furman, R.W. (1982) 'San Antonia Mountain Experiment (SAMEX)', *Bull. Amer. Met. Soc.* 63, 1123–31.

Mani, M.S. (1962) *Introduction to High Altitude Entomology*, pp. 1–73, London, Methuen.

Maurer, J. (1916) 'Bodentemperatur und Sonnestrahlung in den Schweizer Alpen', *Met. Zeit.*, 33, 193–9.

Nägeli, W. (1971) 'Der Wind als Standortsfaktor bei Aufforstungen in der subalpinen Stufe', *Mitt. Schweiz. Anst. Förstl. Versuch*, 47, 33–147.

Olyphant, G.A. (1984) 'Insolation topoclimates and potential ablation in alpine snow accumulation basins: Front Range, Colorado', *Water Resources Res.*, 20, 491–8.

Olyphant, G.A. (1986) 'Longwave radiation in mountainous areas and its influence on the energy balance of alpine snowfields', *Water Resources Res.*, 22, 66–6.

Sauberer, F. and Dirmhirn, I. (1958) 'Das Stahlungsklima', *Klimatographie von Osterreich*, 3(1), 13–102.

Schönenberger, W, (1975) 'Standortseinflüsse auf Versuchsaufforstungen an des alpinen Waldgrenze', *Mit Schweiz Anst. forstl. Versuch.* 51(4), 358–428.

Schwind, M. (1952) 'Mikroklimatische Beobachtungen am Wutaischan', *Erdkunde*, 6, 44–5.

Tabony, R.C. (1985) 'Relations between minimum temperature and topography in Great Britain', *J. Climatol.*, 5, 503–20.

Turner, H. (1958) 'Maximal Temperaturen oberflächennaher Bodenschichten an der alpinen Waldgrenze', *Wetter u. Leben*, 10, 1–12.

Turner, H. (1966) 'Die globale Hangbestrahlung als Standortsfaktor bei Aufforstungen in der subalpinen Stufe', *Mitt. Schweiz. Anst. forstl. Versuch.* 42(3), 109–68.

Turner, H. (1980) 'Types of microclimate at high elevations', in U. Benecke and M.R. Davis (eds) *Mountain Environments and Subalpine Tree Growth*, pp. 21–6, Wellington, Tech. Pap. no. 70, Forest Res. Inst., New Zealand Forest Service.

Turner, H., Rochat, P. and Streule, A. (1975) 'Thermische Charakteristik von Hauptstandortstypen im Bereich der oberen Waldgrenze (Stillberg, Dischmatal bei Davos)', *Mitt. Eidgenöss. Anst. forstl. Versuch.*, 51, 95–119.

Wade, J.E. and Hewson, E.W. (1979) 'Trees as a local climatic wind indicator', *J. appl. Met.*, 18, 1182–7.

Wardle, P. (1974) 'Alpine timberlines', in J.D. Ives and R.G. Barry (eds) *Arctic and Alpine Environments*, pp. 371–402, London, Methuen.

Wendler, G. and Ishikawa, N. (1974) 'The effect of slope exposure and mountain screening on the solar radiation of McCall Glacier, Alaska: a contribution to the International Hydrological Decade', *J. Glaciol.*, 13(68), 213–26.

Yoshino, M.M. (1973) 'Studies on wind-shaped trees: their classification, distribution and significance as a climatic indicator', *Climat. Notes* 12, 1–52.

Yoshino, M.M. (1975) *Climate in a Small Area*, Tokyo, pp. 445–459, University of Tokyo Press.

3

CIRCULATION SYSTEMS RELATED TO OROGRAPHY

DYNAMIC MODIFICATION

The effects of topography on air motion operate over a wide range of scales and produce a hierarchy of circulation systems through the mechanism of dynamic and thermal factors. Here, we concentrate on three major types of dynamic process. First, extensive mountain ranges set up planetary-scale wave motion through large-scale rotational effects. Second, mountains give rise to modifications of synoptic-scale weather systems, especially fronts. Third, topography on all scales introduces wave motion through local gravitational effects. While these categories are not always sharply differentiated from one another, they provide a convenient basis for discussion. Detailed accounts of orographic effects on airflow are given in Alaka (1960), Nicholls (1973), Smith (1979a) Hide and White (1980) and a convenient summary has been written by Beer (1976).

Planetary-scale effects

The influence of mountain barriers on the planetary-scale atmospheric circulation involves three principal processes: the transfer of angular momentum to the surface through friction and form drag; the blocking and deflection of airflow; and the modification of energy fluxes, particularly as a result of the airflow effects on cloud cover and precipitation. Various attempts have been made to distinguish the relative importance of these factors in generating standing planetary waves, through diagnostic, theoretical and modelling studies (Kasahara 1980). Orography and diabatic heating (latent heat release, absorption of solar radiation, infra-red cooling and surface sensible heat) each contribute to the forcing of the planetary waves, but their effects are poorly quantified according to Dickinson (1980). As far as the mountain belts are concerned, they tend to be located beneath an upper ridge of high pressure, but this feature is only well-developed over high-middle latitudes in western North America. The significance of these planetary waves is that they influence the formation and movement of pressure systems. Mid-latitude depressions commonly tend to develop or intensify beneath the eastern limb of an upper wave trough, over the eastern seaboard of North America and off the east coast of Asia, for example.

The effects of orography on global climate were discussed by meteorologists in the 1950s in terms of the structure of the planetary standing waves. The high mountains and plateaux of eastern Asia and western North America, for example, exert a strong control on the winter wave pattern, as confirmed by GCM experiments with and without mountains (Manabe and Terpstra 1974). This topic has received new attention in light of the suggested rapid uplift of these plateaux and mountain ranges during the Late Cenozoic, particularly the last 10-million years (Ruddiman *et al.* 1989; Ruddiman and Kutzbach, 1989). Simulations with the NCAR Community Climate Model were performed for January and July with present-day mountains, no mountains, and an intermediate 'half-mountain' case (Kutzbach *et al.* 1989). The three cases show that in January the planetary waves increase in amplitude with terrain uplift and the low-level flow is progressively blocked or diverted around the barriers. In July, monsoon-like circulations form near the Tibetan and Colorado plateaux in response to higher barriers. Patterns of vertical motion and precipitation show corresponding adjustments to increasing elevations. Similar experiments with the Geophysical Fluid Dynamics Laboratory model by Manabe and Broccoli (1990) imply that in the absence of orography, relatively moist climates would be found over central Asia and western interior North America. This could explain the paleobotanical evidence for reduced aridity in those regions during late Tertiary time.

The large-scale effects of an orographic barrier on an airflow crossing it are usually explained as a consequence of the relationship between divergence and vorticity. This is illustrated by the equation for the conservation of potential vorticity

$$\frac{(\zeta + f)}{\Delta p} = \text{constant},$$

where ζ = relative vorticity about a vertical axis (cyclonic = positive in the northern hemisphere)

f = Coriolis parameter (expressing the component of the earth's rotation in a horizontal plane; $f = 2\omega \sin \phi$, where ω = the angular velocity of the earth, ϕ = latitude angle),

Δp = thickness of the air column in pressure units.

It is assumed that the atmosphere is incompressible and that the air motion is adiabatic. This is equivalent to stating that the flow is *isentropic*, i.e. it is along surfaces of constant potential temperature (the temperature of air brought dry-adiabatically to a pressure of 1000 mb).

The equation shows that, if the expression on the left-hand side is to remain constant as an air column approaches a mountain range from the west and Δp decreases, then there must be a corresponding decrease in $(\zeta + f)$. In other words, vertical shrinking of the column must be matched by lateral expansion, implying horizontal divergence. For $(\zeta + f)$ to decrease, either the airstream undergoes anticyclonic curvature or the air must be displaced equatorward where

f is smaller. Conversely, on the downward side of the barrier, Δp increases again with the opposite effects, producing a mountain ridge and lee trough pattern. For flows approaching a mountain ridge from the east, the vertical shrinking of the column leads to a decrease of absolute vorticity. The flow is deflected to higher latitudes and experiences increased earth's vorticity, f. This leads ζ to decrease causing greater anticyclonic curvature and looping the flow. No wave pattern is created by easterly flows (Kasahara, 1980; Queney 1948; Colson 1949; Bolin 1950).

The flow patterns observed over major mountain ranges indicate that the effect on the curvature (or relative vorticity) of the airflow is predominant. Over the Rocky Mountains, for example, there tends to be anticyclonic curvature of the upper westerlies, despite the associated slight poleward displacement (f changes very little by comparison with the reversed sign of ζ). However, Smith (1979b) shows the correct explanation does *not* involve vorticity changes due to vertical stretching. Rather, air parcels crossing a range in a quasi-geostrophic flow undergo volumetric expansion as they rise. This creates anticyclonic vorticity and the effect of the expansion does not cancel out, but causes a non-vanishing circulation.

Laboratory experiments can be used to investigate topographic effects on rotating fluids (Boyer and Davies 1982; Boyer and Chen 1987). A tow-tank containing saltwater and freshwater solutions, to establish a vertical density gradient, is rotated on a turntable. When the system reaches solid body rotation, a model mountain is towed steadily through the fluid. This system can generate eddies, but neglects the β-effect of the latitudinal variation of the Coriolis parameter. Boyer and Chen find qualitative agreement between observations of westerly flow over the Rocky Mountains and simulated features in such a system. These include a ridge over the mountains, a lee trough, and the characteristic slope of the ridge and trough lines; maximum relative cyclonic vorticity is located to the south-east of the central part of the mountains.

The amplitude of the wave disturbance depends strongly on the latitudinal extent of the barrier. Mountain ranges such as the Rocky Mountains, the Andes, and the Himalaya affect the large-scale planetary airflow whereas the extent of the European Alps is insufficient for such effects. The Southern Alps of New Zealand, spanning 5° of latitude, are also probably below the critical threshold for such effects although data for the South Pacific are sparse. Reiter (1963: 382–3) shows two other interesting results. First, a narrow jet stream flow may be deflected anticyclonically, so as not to cross a mountain range, if the wind speed is below a particular threshold. For zonal flow encountering a 2 km high, 1000 km wide, barrier, the critical speed is 20 m s^{-1}. If the curvature of the flow is cyclonic (anticyclonic), this critical value will be correspondingly less (greater), respectively. Second, if we consider the cyclonic and anticyclonic sides of a jetstream axis, the vorticity relationships can lead to airstream diffluence. South of the jet axis, where the absolute vorticity ($\zeta + f$) approaches zero, the mountain barrier effect on the flow leads to a reduction of f, deflecting the current equatorward. On the north side of the jet axis, the current is deflected poleward, assuming conservation of absolute vorticity. Upper wind profiles in winter suggest that such splits occur in the westerly jet streams which encounter the Rocky Mountains and the Tibetan plateau-Himalaya (Chaudhury 1950), although in the latter case at least the cause is probably not simply a dynamical one. In winter, the Tibetan plateau is a high-level cold source which causes a strong baroclinic zone on its southern edge where the westerly subtropical jet tends to be anchored.

The question of how to treat mountain terrain and associated subgridscale processes in general circulation models (GCMs) merits discussion since knowledge of mountain topography and climates can provide some guide as to what aspects need to be adequately parameterized in order to obtain realistic model results and in their interpretation. Especially critical is the question of the scale of the topography in relation to the horizontal and vertical resolution of the model. In the vertical dimension most GCMs specify multiple levels. In early models these usually had an approximately regular height spacing (e.g. 3 km) whereas most models now use sigma coordinates (p/p_0) with greater resolution in the lower layers. GCM results are usually presented in terms of spatial fields at standard geopotential or sigma levels, or as vertical cross-sections. Actual surface conditions in mountain regions are not considered unless special fine-mesh submodels are nested with the GCM to examine specific problems such as orographic precipitation.

The horizontal resolution of current GCMs is typically of the order of 2.5°–5° latitude, with fine-mesh resolution of 1° latitude available for some purposes. Early models used latitude–longitude grids whereas many GCMs now treat the horizontal domain spectrally (Washington and Parkinson 1986). This causes the topography to be smoothed so that mean elevations are lower; adjacent to high mountains and plateaux there may be dips in the modelled surface extending below sea level. Recent studies suggest that the atmosphere responds more to the 'envelope' topography through the mountain peaks than to mean elevation because deep valleys and basins act as part of the terrain when they are filled with stagnant air. The truncation level adopted in spectral models, especially those used for most climate sensitivity experiments, causes mountains to be greatly smoothed. Therefore, adjustments have to be made to the mean surface heights in order to obtain adequate representation of the 'envelope' orography (Wallace *et al.* 1983). One procedure is to add twice the standard deviation of height values, determined over a fine resolution (10 minute spacing) terrain grid, to the mean for each grid box (Slingo and Pearson 1987).

The switch from latitude–longitude grids to spectral representations and also the increase in resolution of recent model formulations (e.g. from 5° to 2.5° resolution), has led to some significant problems. The latitudinal pressure gradient is increased in mid-latitudes, causing strong zonal flows – 'the westerly problem'. For example, the NCAR Community Climate Model with spectral truncation at n = 15, simulates the northern hemisphere winter circulation quite well, but is less successful with the more zonal circulation in summer and especially in the southern hemisphere. The n = 30 truncation in contrast, improves the southern hemisphere but gives excessively strong flow in the northern hemisphere. Similar errors have been found in the 11-layer UK Meteorological Office (UKMO) model (Palmer *et al.* 1986). They consider that the problem is caused by the increase in energetic transient eddies with the higher resolution. Deep cyclones are formed that fill only slowly at 50°–60°N, causing excessive westerlies over the eastern North Atlantic and Europe. The lower resolution models give a more reasonable simulation because of compensatory underestimation of both surface drag and poleward momentum flux.

Two approaches to correct this model deficiency have been tested. The first is to raise the mountains by using an envelope orography to increase surface drag (Wallace *et al.* 1983) and the second is to incorporate the effects of breaking orographic gravity-waves, mainly in the lower stratosphere (Palmer *et al.* 1986). Slingo and Pearson (1987) provide a comparative evaluation of both alternatives with the UKMO model. The envelope orography approach is not satisfactory because it does not remove the 'westerly problem' and introduces other biases in the circulation fields. Similar findings were made by Wallace *et al.* (1983) with the European Centre for Medium Range Weather Forecasting (ECMWF) model. Envelope orography is also likely to generate spurious elevated heat sources in summer over high mountains and plateaux. The UKMO simulations incorporating gravity-wave drag show substantial improvement in the circulation patterns in both seasons and in the storm track locations. However, the best treatment of orography in models is still an open question (Slingo and Pearson 1987). The solution may involve the use of mean elevations at high spatial resolution because this should result in better precipitation patterns. Another approach to estimating surface drag considers the silhouette of a mountain barrier as 'seen' by an approaching airstream (see Note 1, p. 193).

Synoptic-scale effects

While changes of the planetary flow are of major importance to global climate, modifications to synoptic systems are of more immediate consequence for conditions in the mountains themselves. Orography has two major effects on such systems; frontal cyclones crossing a mountain range undergo structural modification, and in the lee of the mountains cyclogenesis is enhanced. The first of these effects is the more important for mountain weather, although the second may be related to wind conditions downwind of the range.

The routine analysis of synoptic phenomena in mountain areas is greatly hindered by the limited availability of data and the pronounced mesoscale effects induced by the orography. Special analyses are necessary in such circumstances; Steinacker (1981) suggests the use of streamline analyses that omit areas lying above 750 m altitude, as illustrated in Figure 3.1, and also the analysis of stream functions on isentropic surfaces. Available mountain station data can be

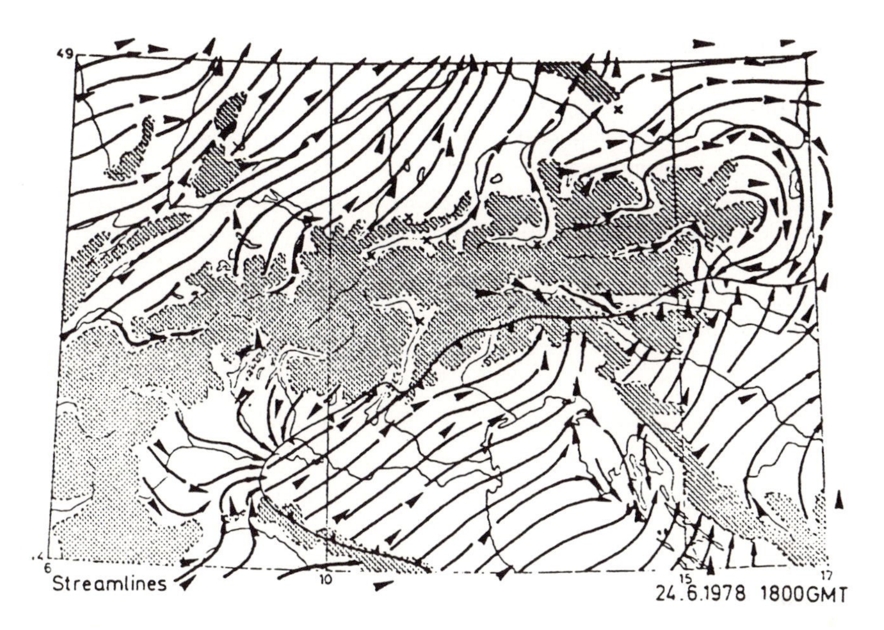

Figure 3.1 An example of a streamline analysis for the Alps, omitting areas exceeding 750m altitude, 24 June 1978
Source: From Steinacker 1981

incorporated into 850 and 700 mb analyses, if any systematic errors are first removed by comparison of their long-term mean values with those at surrounding stations. Rawinsonde data also need adjustment, when pressure tendencies are large, if there exists a time difference between the sounding and the standard synoptic observation.

There are several dynamic and thermodynamic mechanisms involved in the orographic modification of frontal characteristics. One category of effect is referred to as 'masking' (Godske *et al.* 1957: 610). Fronts crossing mountain systems with extensive intermontane basins may over-ride shallow cold air. This can diminish the low-level temperature contrast across a cold front and accentuate that across a warm front. In situations where pre-frontal föhn occurs, a warm front can also be masked due to the reduced temperature difference.

Dynamic and thermodynamic effects of orography result from the forced ascent of air over the barrier, which leads to distortion of the temperature structure through adiabatic processes. Since warm fronts have a typical slope of 1:100, subparallel to the usually steeper (1:20) windward slope of the barrier, air ahead of the front can become trapped, thus tending to retard the motion of the lower section of the front (Figure 3.2). This gives rise to prolonged cloud and rain on the windward slopes. The motion of the upper front is unimpeded and lee-slope föhn may cause it to separate from the lower front, with subsequent regeneration taking place downwind of the range. As a result of such retardation of warm fronts approaching the Scandinavian Mountains, or Greenland, from the west, the system may occlude with a new centre developing to the south where the warm front swings around the barrier (Godske *et al.* 1957: 613). Cold fronts, with a typical 1:20 slope, also tend to be slowed down by mountain barriers, since the wind component normal to the front is slowed first, and to a greater degree, at lower levels. This tends to lift the frontal surface through the accumulation of cold air near ground level (Radinovic 1965). In these situations, cold air may penetrate through major gaps in the mountain barrier, thus distorting the frontal profile. This is often apparent when cold fronts move south-east across Spain and France, encountering the Pyrenees, Massif Central and Alps.

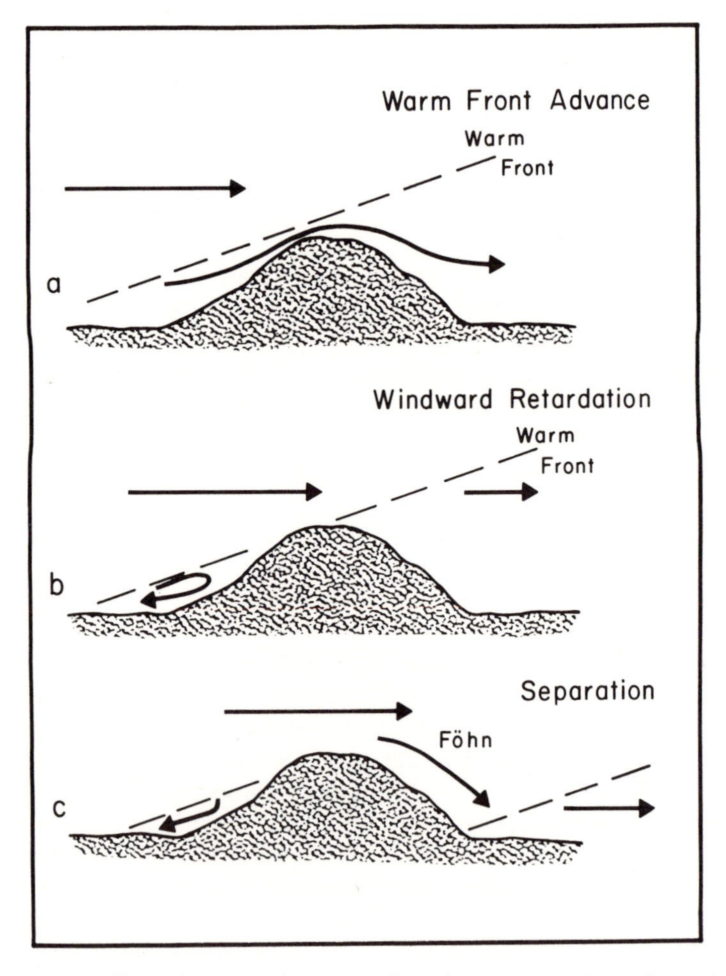

Figure 3.2 The effects of mountain barriers on frontal passages: (*a*) Warm front advance; (*b*) Windward retardation; (*c*) Separation.

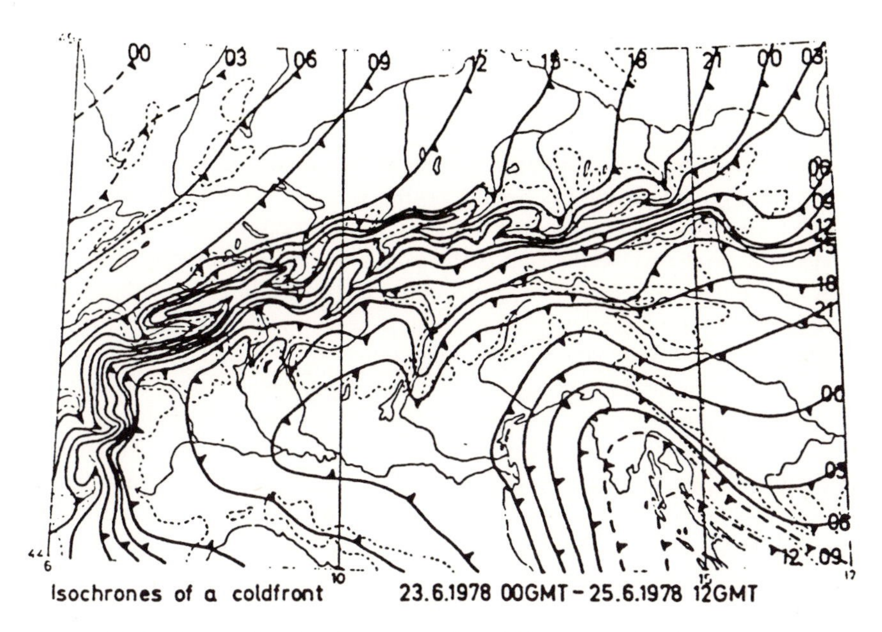

Figure 3.3 Isochrones showing the passage of a cold front over the Alps, 00.00 GMT 23 June to 12.00 GMT 25 June 1978
Source: From Steinacker 1981

Figure 3.3 illustrates a detailed analysis of isochrones of a cold front moving over the Alps (Steinacker 1981); note the lee side acceleration of the front. Other examples for central Asia are described by Lydolph (1977: 160–9). On the lee of a range, and in intermontane basins, the presence of residual pools of cold air can weaken the lower section of a cold front.

Retardation and distortion of fronts also cause modifications of the vertical motion field and therefore of cloud and precipitation patterns. For example, when uplift on the windward slope results in precipitation, fronts which cross the range resemble those in a dry atmosphere. This implies an absence of condensational heating and weakening of the frontogenetic tendencies (Palmén and Newton 1969: 265 and 344).

Mountain effects are most pronounced when fronts move in a direction perpendicular to the barrier, although, even then, the system may undergo some deflection towards one end of the range. The effects also depend on the intensity and speed of the system. The modification of strong, rapidly moving systems crossing the St Elias Mountains, Alaska-Yukon, for example, is limited, while fronts moving parallel to the range are scarcely affected (Taylor-Barge 1969). Studies during ALPEX illustrate considerable complexity (Smith 1986). As a cold front approaches the Alps from the north-west, the cross-front winds that advect the front at low levels like a density current are blocked or deflected by the barrier. A shallow prefrontal south föhn is often experienced in the northern

Alpine valleys associated with higher pressure to the south. This conjunction promotes the formation of an orographic low-level jet (Egger 1989). The cold air is accelerated eastward along the mountain front as a trapped density current since the mountains block the Coriolis acceleration to the right. This flow favours propagation of the front along the north side of the Alps producing a lateral 'nose'; cold air may penetrate around the mountains into Yugoslavia; similar features are observed for the Appalachians, and east of the Great Dividing Range in Australia where the cold outbreak is termed the 'Southerly Buster'. In each case, the cold air commonly forms a wedge banked up against the eastern slopes of the mountains.

The so-called 'barrier effect' of mountain ranges to air motion is most evident when they are high and continuous and when the movement of stable cold air masses is involved. An analysis of an outbreak of continental polar air from central Canada towards the Pacific coast of the state of Washington in January, 1940 illustrates this (Church and Stephens 1941). The Rocky Mountains, especially, and to a lesser extent the Cascades, only permitted air with higher potential temperatures above the summit levels to move westward. Subsidence occurred over a depth of at least 2000 m as the air moved westward with warming occurring at close to the dry adiabatic lapse rate. The barrier effect of the Himalaya ranges, in limiting northward movement of summer monsoon air, and of the Tibetan plateau on the southward movement of cold air from the shallow Siberian high in winter is well known in terms of average climatic conditions, although hardly investigated in detail.

Over high mountains and plateaux, weather conditions during the passage of synoptic systems reflect the fact that upper level features become 'surface' ones over high ground. Taylor-Barge (1969) notes that *tro*ughs of *w*arm *al*oft ('trowals' in the terminology of the Canadian Atmospheric Environment Service), for example, may affect the weather on the St Elias Mountains above 2500 m. Likewise, studies of weather on the Greenland ice sheet show that 700-mb synoptic analyses best describe frontal systems crossing the area (Hamilton, 1958a,b).

Apart from their effect on frontal systems, mountain barriers also modify the wind field through differential pressure effects. Observations across many mountain ranges show two characteristic types of flow disturbance: first, a pressure differential (of the order of 10 mb at the surface) between windward (high) and leeward (low) slopes of the range and second, an upstream deflection of airflow to the left in the northern hemisphere, in association with the orographically-disturbed pressure field (Smith 1982). Some of the pressure differential may be caused by horizontal temperature advection and latent heat effects; the typical 'föhn nose' evident on synoptic pressure maps during classical föhn situations in the Alps and Rocky Mountains illustrates such effects (Malmberg 1967; Brinkmann 1970). Pressure data from valley stations, however, are commonly biased by daytime heating and nocturnal ponding of cold air (cf. Walker 1967). The windward-side high pressure 'nose' in Smith's

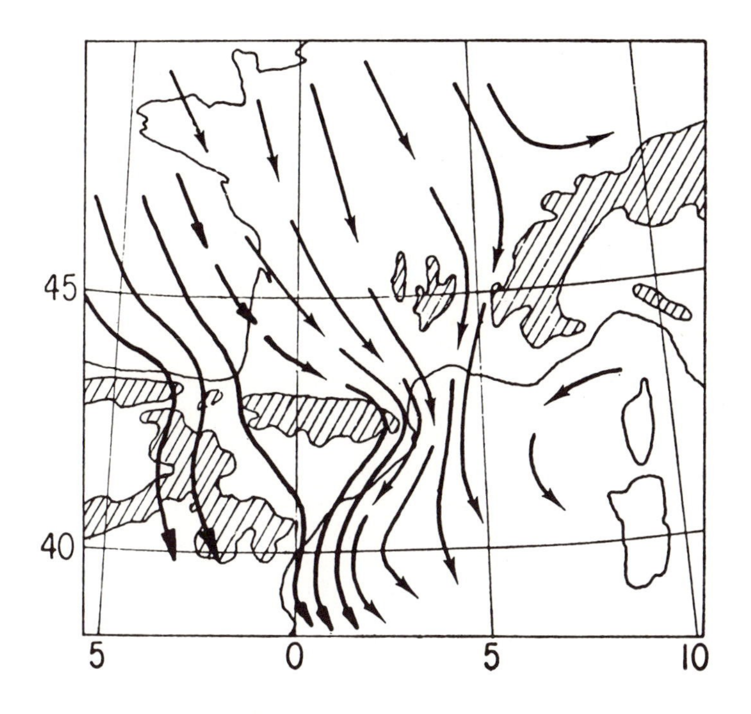

Figure 3.4 'Corner effect' on airflow east of the Massif Central (the Mistral), east of the Pyrenees (the Tramontane), and east of the Cantabrian Mountains of Spain.
Source: From Cruette 1976

analyses is primarily attributable to hydrostatic flow over the mountains; lifting causes mid-level adiabatic cooling creating a deeper layer of cool, dense air (Smith 1979a: 103), which hydrostatically increases the lower level pressure. In the lee of the mountain range, warm air descends to replace that deflected by the barrier. Smith (1982) shows that for steady hydrostatic flow, air approaching a mountain decelerates as it encounters the high pressure region; the Coriolis force is thereby decreased and the air turns to the left (in the northern hemisphere) because of the background pressure gradient. Air parcels that have been deflected to the left are at lower pressure and so move faster than air parcels on the right side of the flow. The essence of this theory is the alteration of horizontal trajectories of air parcels by the Coriolis force, whereas the pressure field is unaltered. In contrast, idealized geostrophically-balanced flow, with isentropic surfaces parallel to the terrain, requires an unrealistic 'mountain anticyclone' (Smith, 1979a, 1982).

Queney (1963) examined the results of the anticyclonic deformation of streamlines over a ridge. He noted that this sets up an intensified pressure gradient at the left end of the ridge viewed downwind (in the northern hemisphere), referred to as the *corner* effect (T. Bergeron, in Godske *et al.* 1957:

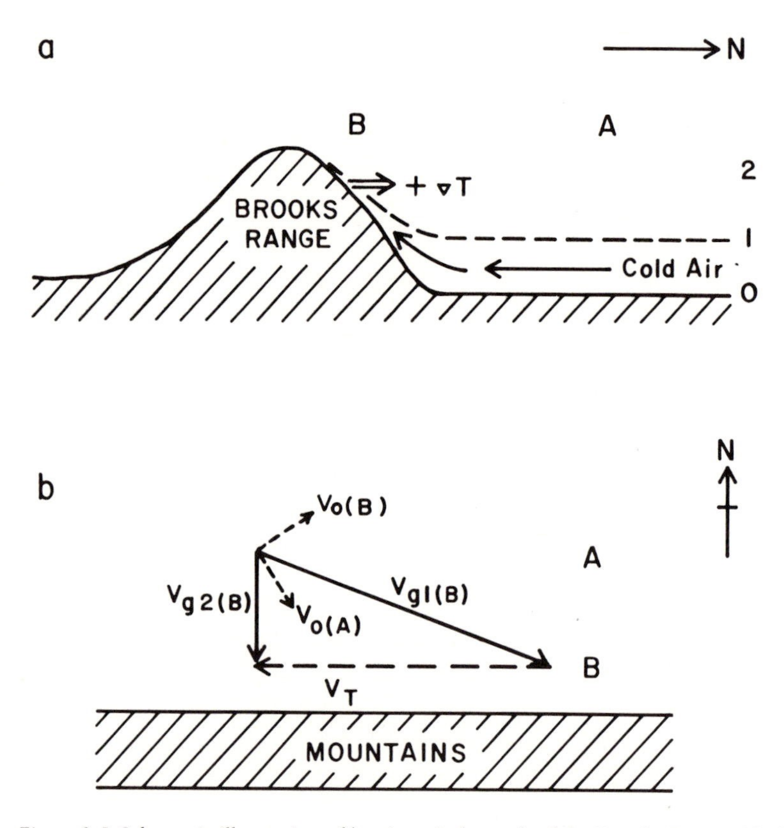

Figure 3.5 Schematic illustration of barrier winds north of the Brooks Range, Alaska, where northerly upslope flow is deflected to give westerly components at levels near the mountains:

Notes: (*a*) Vertical cross-section of stable air flowing towards the Brooks Range. Level 1 denotes the inversion; ∇T = the temperature gradient (cold to warm); (*b*) Plan view of the wind vectors: V_T = thermal wind; V_g = geostrophic wind; subscript numerals 0, 1 and 2 denote the level; $V_{O(A)}$ and $V_{O(B)}$ represent the surface winds at points A and B; they are deflected about 30° to the left of the corresponding geostrophic winds.

Source: After Schwerdtfeger 1975

606). The gradient causes a local wind maximum such as that observed during general northwesterly airflow east of the Pyrenees, known as the Tramontane, or the well-known Mistral, east of the Massif Central in France (Cruette 1976; Jansá 1987). These patterns are illustrated by the streamlines in Figure 3.4.

The degree of influence of the Alps on airflows depends on the flow direction (Reuter and Pichler 1964). For 500 mb northerly flow, the latitudinally-decreasing Coriolis term is balanced by orographically-induced anticyclonic curvature (see p. 109), resulting in no significant deflection. For southerly flow, the orographic and Coriolis terms have the same sign leading to reduced wave amplitude.

Mountain barriers exert substantial influence on shallow stable airstreams

(Dickey 1961; Schwerdtfeger 1975). In winter, north of the Brooks Range, Alaska, for example, stable cold air approaching the range from the north is deflected to the left, giving westerly *barrier winds* parallel to the mountains. The isobaric temperature gradient perpendicular to the north slope (with cold air against the slope) causes a thermal wind component parallel to the mountain from the east (Figure 3.5a). This is equivalent to a strong south to north pressure gradient, thus giving a *westerly* component of geostrophic wind at the base of the cold air on the slope. The thermal wind vector between the layers has to be subtracted from the upper layer geostrophic wind in order to obtain the lower layer geostrophic wind (Figure 3.5b). The magnitude of the thermal wind effect is determined by the slope of the terrain (and the cold air boundary), the inversion intensity, and the damming up of the cold air by the barrier. Similar low-level barrier jets occur along the western slope of the Sierra Nevada, California, in winter as stable air approaching from the west causes pressure excesses of ~ 5 mb (Parish 1982).

Lee cyclogenesis is of major importance downwind of barriers to the mid-latitude westerlies, such as the Western Cordillera of North and South America and the Tibetan Plateau, and in the lee of lesser topographic features such as the Alps. The topic merits discussion here in terms of the processes involved and also because the lee cyclone may affect weather conditions at least on the lee slopes.

The effect of the Rocky Mountains in causing cyclogenesis, especially in eastern Alberta and Colorado, is well known (Hess and Wagner 1948; McClain 1960; Hage 1961; Chung *et al.* 1976; Palmén and Newton 1969, Chapter 11). In westerly airflow there is a lee trough associated with low-level adiabatic warming in the downslope flow component and cyclonic vorticity generation east of the mountains. This standing pressure trough deepens when a Pacific low approaches the west coast and intensifies the flow across the mountains. In the subsequent stage of development, cold air advection offsets any warming due to abiabatic descent. The cyclogenetic tendencies continue, however, as a result of upper level advection of positive (cyclonic) vorticity with upper level divergence encouraging low-level convergence and rising air. Finally, the surface cold front of the Pacific system may move into the lee trough and orographic effects cease as the downslope component of flow disappears with a shift of the surface wind direction to the north-west.

Cyclones forming in the lee of the Alps may occur in any season, but are most frequent in autumn and the cold half-year generally. Two patterns leading to lee cyclogenesis have been identified from the ALPEX studies (Pichler and Steinacker 1987). They are:

1 blocking of cold northwesterly airflow by the Alps. A mistral wind in the Rhône Valley accompanies this pattern as cold air flows around the Alps. In this type, a cyclone is present upstream of the Alps; this weakens on the windward side and intensifies downwind of the mountains. The system develops and moves rapidly south-east causing limited effects within the Alps;

2 southwesterly upper-level flow ahead of an eastward-moving trough. Here, blocking and flow-splitting may generate a warm front over northern Italy and a cold front in the Gulf of Genoa.

Laboratory model studies simulating the orography of the Pyrenees and Alps by Boyer *et al.* (1987) indicate that, with westerly to northwesterly flows, flow splitting is generated west of the Alps with lee vortex development. This is illustrated in Plate 3.

The processes involved in the case of the Alps are complex (Speranza 1975; Buzzi and Tibaldi 1978; McGinley 1982; Radinović 1965, 1986). An analysis of eight cases of Alpine lee cyclogenesis during ALPEX (March–April 1982), shows that over a 27-hour period they develop from the surface upward to 300 mb (Radinović 1986). Half of these were most intense at the surface, the other half in the upper troposphere. The eight cases show that lee cyclogenesis is preceded by diffluent upper level flow over the Alps and western Mediterranean. Cyclogenesis is triggered by the orographic blocking of a meridional flow of cold air and an increase of baroclinicity in the lower troposphere (McGinley 1982). Development begins when an upper trough approaches the Alps and undergoes deformation.

A case study by Buzzi and Tibaldi (1978) illustrates the interaction between topography and a cold front advancing from the north-west, operating in conjunction with an intensifying upper-level baroclinic field over northern Italy (Figure 3.6a–d; p. 122). They show that it is important to distinguish between the low-level effects of the Alps on the pressure field and on frontal structures, on the one hand, and upper tropospheric adjustments on the other. Below 2 km, where the Alps form a 450 km barrier to westerly airflow (Egger 1972), interaction between the barrier and the airflow initiates a low-level pressure perturbation with anticyclonic vorticity, produced by vortex tube compression, over the mountains and cyclonic vorticity set up in the lee. The arcuate, convex shape of the western Alps causes distortion of the thermal field at low levels due to blockage of the cold air, and this, accentuated by cold air advection west of the Alps along the Rhône valley, modifies the upper tropospheric flow structure, as implied by the thermal wind relationship (Radinović, 1965). Thus, strong northwesterly jet streams across France tend to split with one branch curving cyclonically eastward, north of the Alps, and the other penetrating southward toward the Gulf of Lions (Buzzi and Rizzi 1975). According to Radinović, the effects of non-adiabatic heating (such as condensation release) on the thickness field changes are much less than those due to the blockage or retardation of advection in the lower layers. The 1000–500 mb thickness pattern typically shows a trough west of the Alps, that is deformed as it approaches the mountains, and a ridge in the concave lee area of the Alps related to cyclonic development at the surface. Cyclogenesis takes place on the eastern limb of a confluent thermal trough, associated with the southward penetration of cold air to the west of the Alps and the ridge south of the mountains. Illustrations of

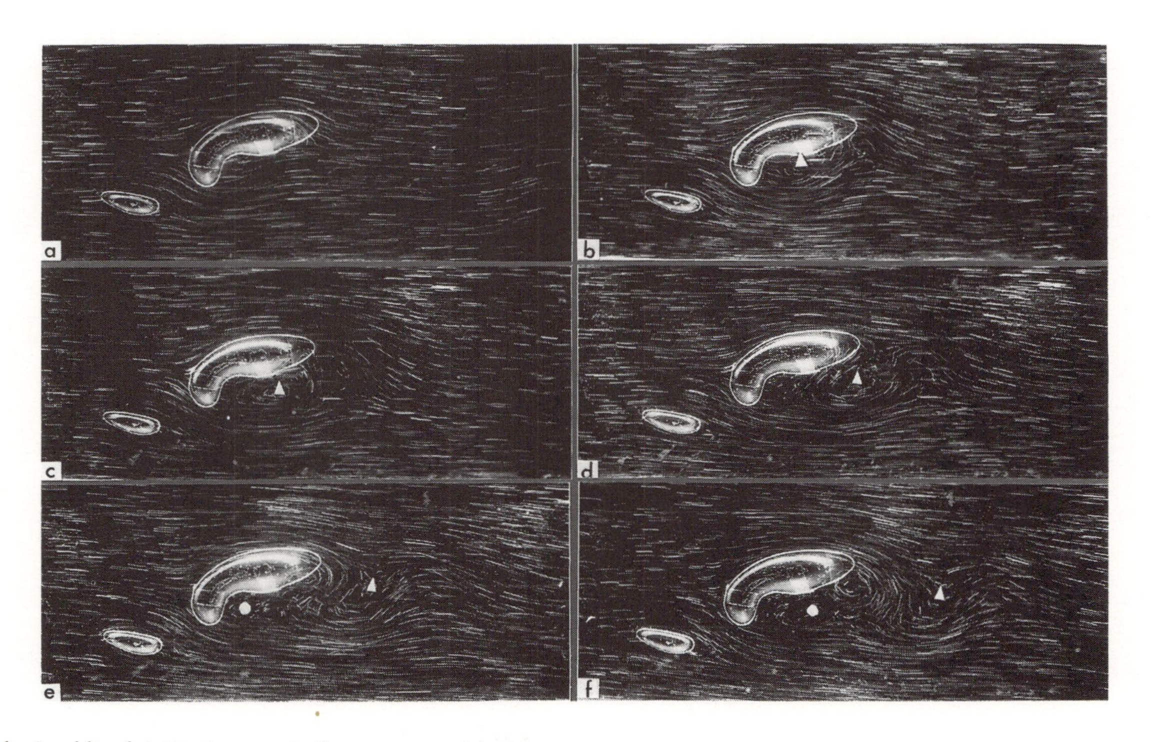

Plate 3 Cyclonic eddies forming in westerly flow past a model Alpine/Prenean topography (courtesy of Professor D.L. Boyer and the editor and publisher of *Meteorology and Atmospheric Physics*. From Boyer *et al.*, 1987). The laboratory experiment uses a towed obstacle in a linearly stratified, rotating water channel 2.4 m long, 0.4 m wide and 0.3 m high. The flow evolves from (*a*) 13 hours equivalent atmospheric time, through (*b*) 26h (*c*) 39h (*d*) 52h (*e*) 65h and (*f*) 78h, respectively. The triangle marks the centre of the first cyclone and the circle the second; flow splitting occurs at the western end of the Alps near the location of Geneva.

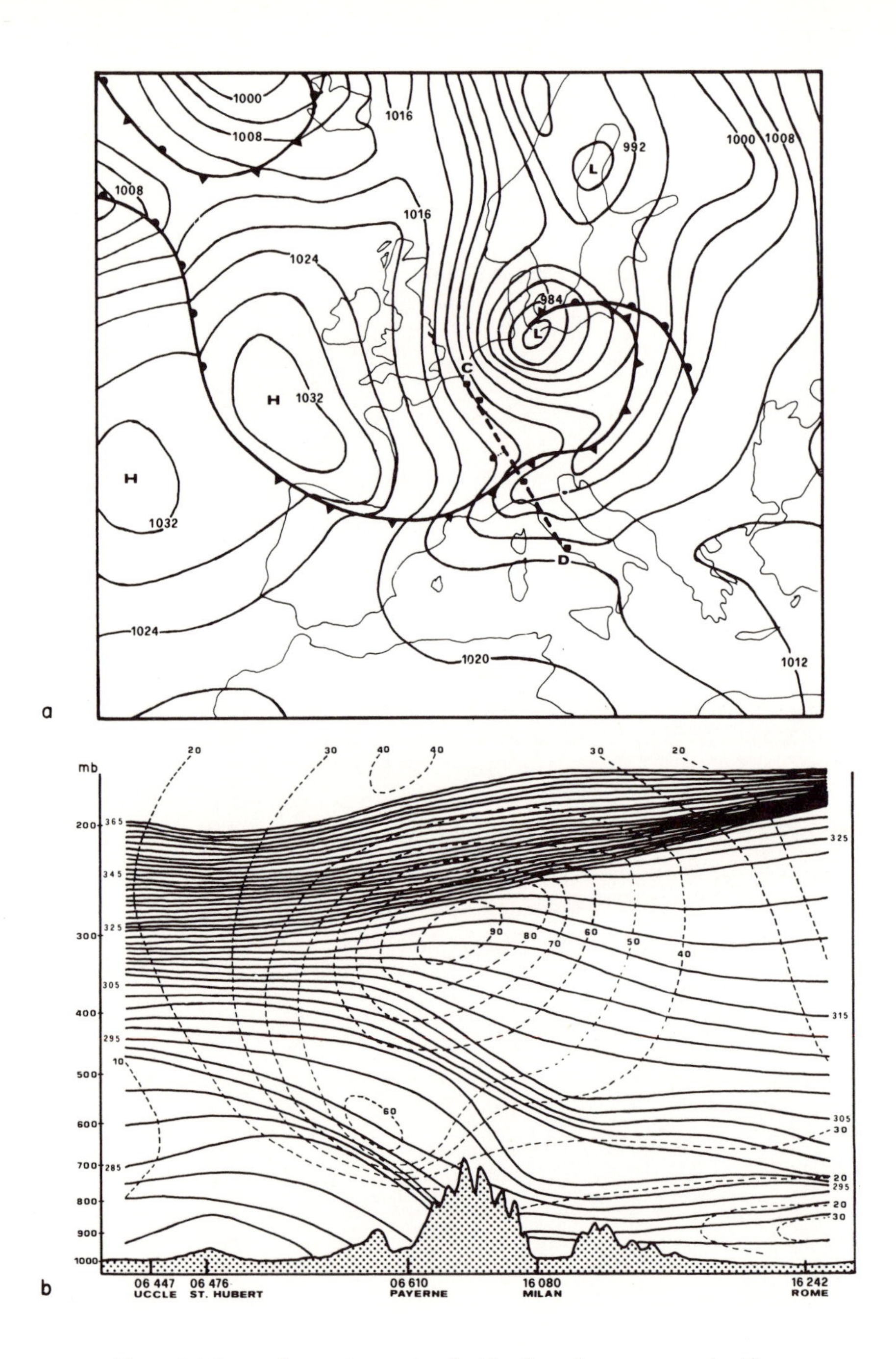

Figure 3.6 Lee cyclogenesis associated with a frontal passage over the Alps
Notes: (*a*) Surface pressure map, 3 April 1973, 00.00 GMT; (*b*) Cross-section along the line C–D of (*a*). Isentropes (K) and isotachs (knots); (*c*) As (a) for 12.00 GMT; (*d*) As (*b*) for 12.00 GMT along the line E–F.
Source: From Buzzi and Tibaldi 1978

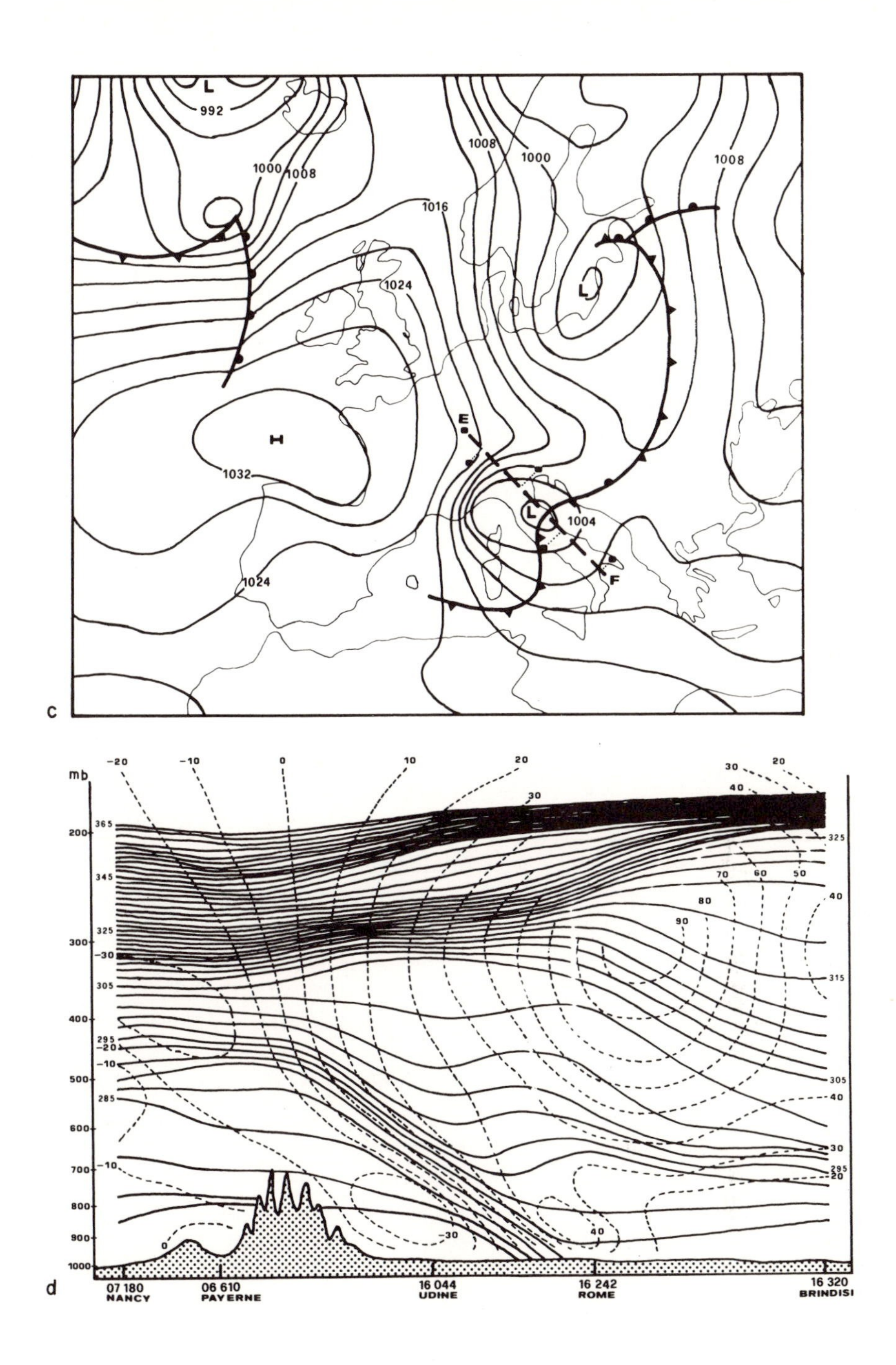
c
L
992
1000
1008
1016
1024
1032
H
1008
1000
1008
L
E
L
1004
F
1024
d
mb
200
300
400
500
600
700
800
900
1000
07 180
NANCY
06 610
PAYERNE
16 044
UDINE
16 242
ROME
16 320
BRINDISI

thickness patterns and cyclogenesis/anticyclogenesis are given by Sutcliffe and Forsdyke (1950); also Stringer (1972: 407).

Several studies indicate that in many cases an eddy forms in the lee of the Alps *prior* to the arrival of postfrontal cold air. This is related to low-level blocking of the cold air mass accompanied by flow splitting and frontal deformation. The maximum height of the barrier seems to control the blocking, not an 'average' mountain surface. The criteria for determining the likelihood of blocking and flow splitting by a mountain barrier are rather arbitrary and are not always reliable. Chen and Smith (1987) examined low-level trajectories of northerly and northwesterly flows of post-cold frontal air approaching the Alps. They defined a splitting parameter S as the length of the Alps (800 km) divided by the distance between the starting positions of two flow trajectories that just pass round the west and east ends of the barrier (near Lyon and Vienna, respectively). Blocking occurs for $S \sim 2$, whereas for $S \sim 1$ there is no blocking. They note, however, that in all eight cases analysed, the surface winds north of the Alps are steered parallel to the mountains. A second key element is the advection of upper-level vorticity across the mountains in propagating jet streaks (Mattocks and Bleck 1986). Such high velocity wind zones are associated with intense cyclones moving directly over the Alps, rather than passing to the north.

In the example shown in Figure 3.6 there is strong cold air advection on the windward side of the Alps in the northern and western sectors of the cyclone and above mountaintop level over the developing surface cyclone. The orographically-induced pressure perturbation (Figure 3.6a,b) is initially sub-synoptic scale (250 km). When cold air begins to flow over the mountains, and around them (via the Rhône valley) into the western Mediterranean, the upper trough moves over the low level cyclone. This causes positive vorticity advection aloft that triggers baroclinic instability. Rapid pressure falls associated with this readjustment are then followed by more moderate deepening and enlargement of the system as a synoptic-scale upper level trough deepens with the intensification of the thermal gradient parallel to the convex front of the Alps and the deflection of the upper jet system (see Figure 3.6d). Vertical coupling between the low-level pressure perturbation and the upper trough then allows for normal development by baroclinic instability forming a cut-off low.

A cold front crossing a mountain barrier is also stretched near the southern end of the range (Smith 1982). This helps to accentuate cross-front gradients, which may serve to promote baroclinic instability. Most of the cyclones forming in the lee of the Alps move out over the eastern Mediterranean and have a life time of about four days.

In an attempt to provide a comprehensive theory of lee cyclogenesis, Buzzi *et al.* (1987) stress the direct interaction between orography and pre-existing baroclinic waves. The spatial structure of the waves is modified by the large-scale slopes of the mountains so that the wave amplitude intensifies downwind on the warmer side of the mountain and weakens on the colder upstream side. The modifications occur on a scale comparable with the Rossby radius of deformation (*i.e.*, the horizontal distance beyond which rotational effects exceed buoyancy effects). A case study for the Alaskan coast ranges (cyclogenesis in the Gulf of Alaska during northwesterly air flow parallel to the mountains) and model simulations for cyclogenesis in the lee of the Alaska ranges, the Rocky Mountains, and the Tibetan Plateau, support the theoretical concepts. For cyclogenesis west of the mountains bordering the Gulf of Alaska, cyclogenetic tendencies are enhanced because the thermal wind is parallel to the mountains. However, the approach of an upper trough from the north-west sets off the development. Similar pre-existing lows are found in the other two areas. Lee cyclogenesis occurs as lows approach the Rocky Mountains from the Pacific, or the Tibetan Plateau from the west.

Local airflow modification

Wave phenomena

Airflow over mountains involves motions with a horizontal scale of 1–100 km, apart from the major long-wave features discussed earlier (p. 105) and these perturbations to the flow are of great importance to weather in the immediate area.

The behaviour of airflow over an obstacle depends principally on (1) the vertical wind profile, (2) the stability structure, and (3) the shape of the obstacle. First, we can examine the effects of a simple long ridge perpendicular to the airflow for the case of a stable atmosphere, where the potential temperature increases with height. For these conditions, three basic types of flow have been distinguished by Förchgott (1949), according to the vertical profile of wind speed (Figure 3.7). With light winds which remain essentially constant with height, the air flows smoothly over the ridge in a shallow wave (a) and there are only weak vertical currents. This is known as *laminar streaming*. When the wind speeds are somewhat stronger and show a moderate increase with height, the air overturns on the lee side of the barrier forming a *standing eddy* (b). With a more intense vertical gradient of wind speed, the oscillation caused by the mountains sets up a train of *lee waves* (c) and wave clouds for 25 km or more downwind (Plate 4). These are stationary gravity waves, provided that the flow conditions do not change. Lee waves usually form only when there is a deep airflow directed within about 30° of perpendicular to the ridge line, with little change in wind direction with height. The wind speed must increase upward, with a minimum horizontal velocity of about 7 m s^{-1} at the crest for low ridges (1 km) and 15 m s^{-1} for ranges 4 km high (Nicholls 1973).

In b, c and d of Figure 3.7, the streamlines show separation of the airflow from the surface. When this occurs, rotor motion develops forming individual vortices below the wave crests (Scorer 1955). Rotors are highly turbulent zones which present a severe hazard to aircraft. Sometimes, the vortex may attain mountain-sized dimensions as illustrated for Mt Fuji, Japan (Soma 1969) and

Plate 4 Wave clouds showing both lenticular and 'pile of plates' forms over the Front Range of the Rocky Mountains, Boulder, Colorado, 7 March 1966. Light southerly airflow in polar continental air at the surface was overlain by strong northwesterlies (courtesy of Robert Bumpas, National Center for Atmospheric Research, Boulder, Colorado).

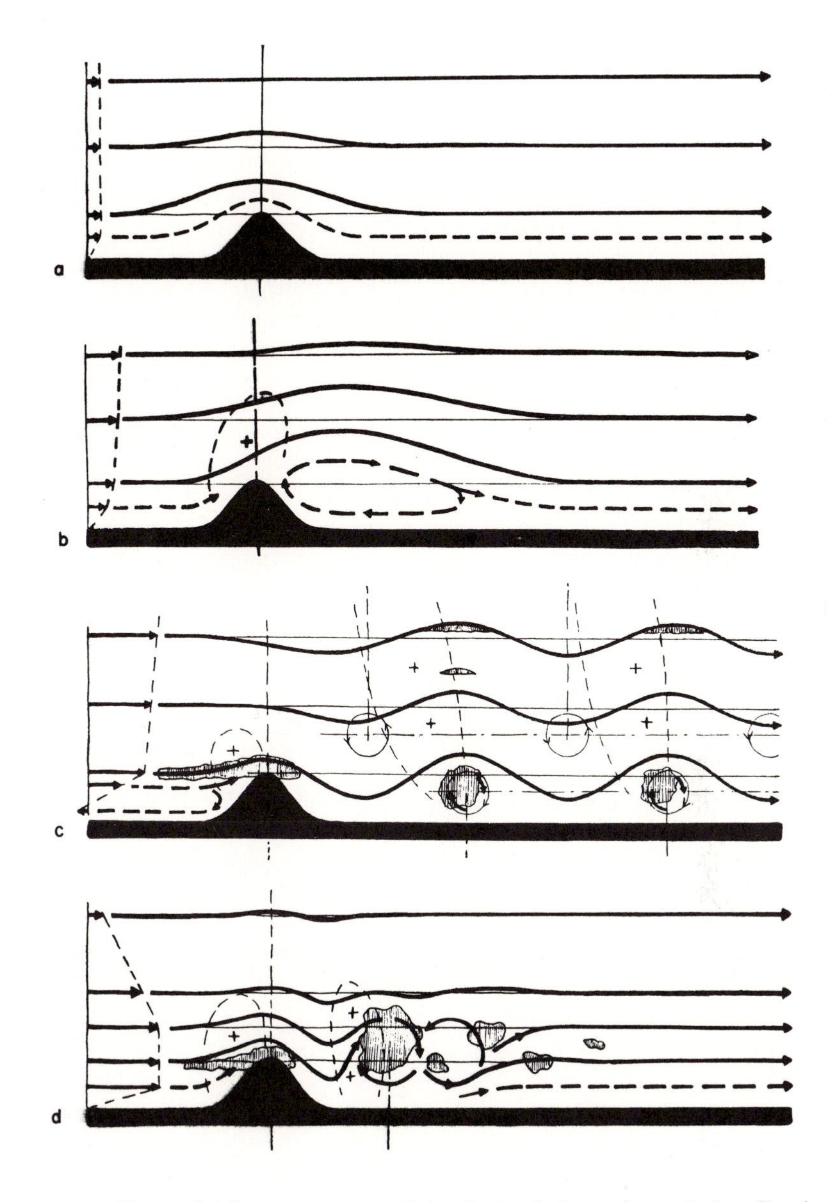

Figure 3.7 Types of airflow over a mountain barrier in relation to the vertical profile of wind speed

Notes: (*a*) Laminar streaming; (*b*) Standing eddy streaming; (*c*) Wave streaming, with a crest cloud and downwind rotor clouds; (*d*) Rotor streaming.

Source: From Corby 1954

the Little Carpathian Ridge, Czechoslovakia (Förchgott, 1969). In katabatic drainage (see p. 158), separation is suppressed since the flow follows the slope, whereas under convective conditions especially on the lee slope, the likelihood of separation is enhanced. Generally, separation occurs when the air in the overlying layers cannot maintain the necessary steady motion (see p. 72). However, its occurrence reduces the effect of the mounains on the higher level airflow, by lessening the lee wave amplitude (Scorer 1955).

Förchgott's scheme is supported by the conclusions of an observational programme carried out in the French Alps by Gerbier and Bérenger (1960). They distinguish three general cases according to the strength of the wind component normal to the ridge:

1 For winds < 8 m s^{-1}, any waves are weak and shallow. Turbulent flow predominates with one lee rotor. The mountain effect is limited to the layer of speed discontinuity 100–200 m above the ridge crest.
2 For winds of 8–15 m s^{-1}, the flow is more turbulent with a succession of low-level rotors in phase with the ridges. Wave forms increase in amplitude with height if the speed is constant, and are stronger if the atmosphere is stable.
3 For winds > 15 m s^{-1} the features are similar to case (2). The lee wavelength and their vertical amplitude and the occurrence of rotors depend on the stability and vertical gradient of wind speed. Rotors tend to be weak if the stratification is stable and the wind speed increases regularly with height.

Theories of airflow over mountains are mathematically complex and only the essential points can be given here. A topographic barrier initiates a vertical displacement in air crossing it, and on the lee side this is counteracted by the restoring force of gravity. The air commonly 'overshoots' the equilibrium position and thereby develops vertical oscillations as it flows downwind. If the atmosphere is stable and winds are light, the oscillation period is short (i.e. high frequency) whereas, in situations with low stability and strong winds, slow oscillations of long wavelength are formed. A stable atmosphere favours the formation of short-wavelength, large-amplitude waves by facilitating the restoring action of the force of gravity on the air motion; the occurrence of a shallow inversion near the ridge crest is especially effective (Corby and Wallington 1956). The natural airstream oscillations are disturbed over mountains, with *resonance* occurring if a topographically forced wave amplifies the 'free' waves (Beer 1976).

The natural frequency of vertical oscillation for a compressible medium, in the absence of frictional and pressure effects, is referred to as the Brunt–Väisälä frequency; it has a magnitude of the order of 10^{-2} s^{-1}, and is given by

$$N = \frac{1}{2\pi}\left[\frac{g(\Gamma - \gamma)}{T}\right]^{1/2} = \sqrt{S}/2\pi,$$

where $S =$ the static stability parameter,

Γ = the adiabatic lapse rate (defined as positive for a decrease of temperature with height),
γ = the environmental lapse rate ($= \partial T/\partial z$).

Wilson (1968) shows that the maximum vertical displacement of an air parcel from a given reference level is $w/\sqrt{S}$, where w = vertical velocity. For an isothermal atmosphere ($\partial T/\partial z = 0$), where $T = 270°K$, $1/\sqrt{S}$ is 53 seconds; in this case, the displacement ranges from 0.5 m for $w = 1$ cm s^{-1} to 160 m for $w = 3$ m s^{-1}. Greater (smaller) displacements occur in more unstable (stable) conditions.

In the presence of pressure gradients a vertically-displaced air parcel can oscillate along a path tilted from the vertical by angle θ; in this case the oscillations have a reduced frequency N cos θ (Durran 1990).

In a stratified atmosphere, flowing over sinusoidal ridges, gravity waves occur when the intrinsic frequency ($n_i = \bar{U}k$) of the motion is less than the Brunt-Väisälä frequency; here $k = 2\pi$/wavelength of the terrain ridges. This type of gravity wave is *vertically propagating*, implying that the disturbance does not decay with height. The phase lines for such waves tilt upstream with height, associated with the propagation of energy vertically upward (the group velocity). (Figure 3.8). It is assumed that no components of the flow can radiate energy downward (the 'radiation condition'). The energy flux is parallel to the parcel trajectories in the gravity wave. The pressure and velocity perturbations are zero where the buoyancy perturbations have extreme (maximum and minimum) values, as shown in Figure 3.8. Wave fronts, or lines of constant phase, propagate perpendicular to the energy flux and air parcel trajectories. It should be noted that a steady airflow crossing sinusoidal ridges can set up waves that are aligned vertically with the ridges, and decay exponentially with height (evanescent waves), if $\bar{U}k > N$. This is because buoyancy forces cannot be supported at frequencies in excess of the Brunt-Väisälä frequency (Durran 1990).

A second basic type of wave is the *trapped lee wave* which occurs downstream of the barrier. The development of lee waves is favoured when there is a vertical decrease of stability, and/or an increase in wind speed. This implies a decrease with height of l^2; l is a stability factor known as Scorer's parameter (Scorer 1949). Neglecting effects due to large shear the vertical wind profile (measured by the quantity $\partial^2 U/\partial z^2 \sim 0$), l^2 can be expressed in terms of the Brunt-Väisälä frequency:

$$l^2 \simeq \frac{1}{U^2}\left\{\frac{g}{T}(\Gamma - \gamma)\right\} = \frac{S}{U^2} = \frac{4\pi^2 N^2}{U^2},$$

where U = the horizontal wind speed perpendicular to the barrier, and the other terms are defined above.

A threshold condition exists for the formation of waves; if l_1^2 is specified for a lower layer of thickness z_1 and in a higher layer l_2^2 is determined, then this condition is $l_1^2 - l_2^2 > (\pi/2z_1)^2$.

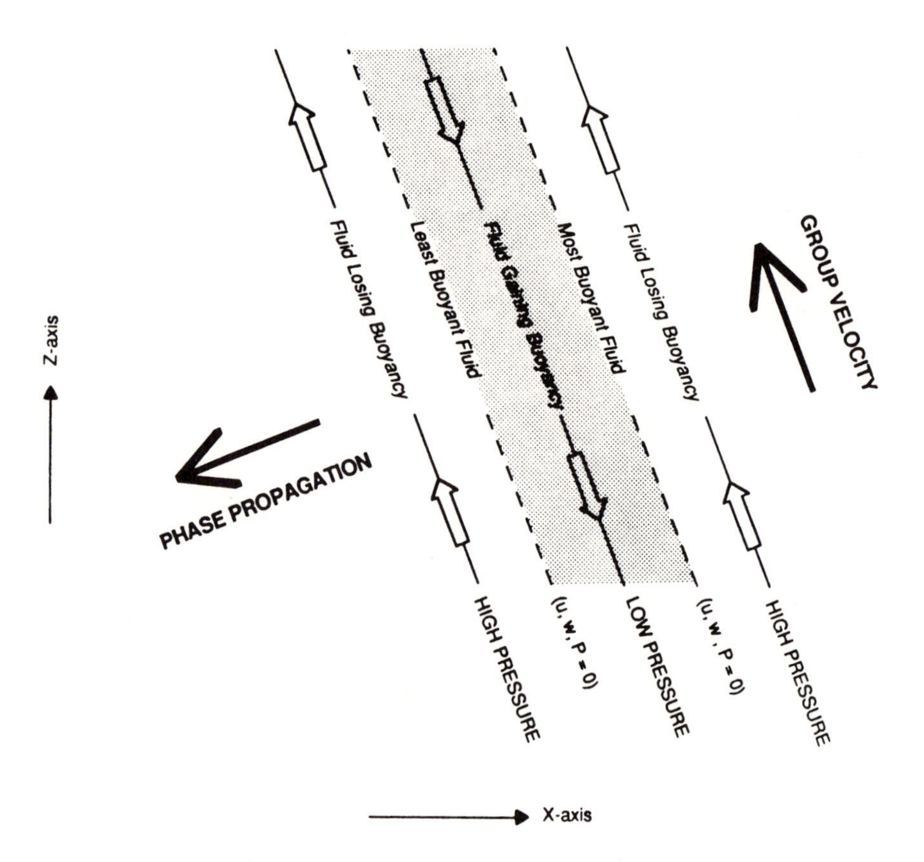

Figure 3.8 The distribution of velocity, pressure and buoyancy perturbations in an internal gravity wave in the $x-z$ plane. The phase of the wave is constant along the slanting lines; along the solid lines buoyancy perturbations are zero, while velocity (u,w) and pressure (P) perturbations have extrema; along the dashed lines, buoyancy perturbations have extrema, while velocity and pressure perturbations are zero. Small arrows indicate perturbation velocities

Source: From Durran 1990

Synoptic situations favourable for wave formation include those with a jet stream aloft (increasing wind speed with height) and near fronts with middle level stability and a large change of density (Corby 1957). Seasonally, waves are more common in winter due to increased low-level stability and stronger circulation.

Typical values of l ($\times 10^6$) may range in the vertical from ≈ 1 to < 0.05 km^{-1}. A scale for determining values of l from a tephigram plot of an appropriate upper-air surrounding has been developed by Scorer (1953) and Wallington (1970). Corby (1957) provides numerous examples of atmospheric soundings and profiles of l^2, while Casswell (1966) gives graphs for computing wave dimensions and velocities from l.

The wavelength (λ) of the dominant lee waves is related to the mean horizontal component perpendicular to the barrier ($\bar{U}$) and is inversely proportional to stability. If vertical accelerations are neglected, (Lyra, 1943; Scorer, 1949):

$$\lambda \sim 2\pi\bar{U}\left\{\frac{T}{g(\Gamma - \gamma)}\right\}^{1/2} = 2\pi/l$$

For $\bar{U} = 10$ m s^{-1}, $\partial T/\partial z = 6.5$ km^{-1} and $T = 260$ K, $l = 0.0035$ m^{-1} and $\lambda = 1.8$ km, for example.
The wavelengths tend to increase with a daytime reduction in stability in the lower layers, because l decreases. In the evening hours, conversely, the wavelength may gradually decrease (Scorer 1953).

Examples of vertical temperature and wind profiles and the calculated wave development are illustrated in Figure 3.9. In (a) the streamlines are tilted upstream near the ridge crest and there is strong downward flow over the lee slope. In the vertical, the pattern repeats with a wavelength of $2\pi/l$; the descending flow is displaced some 2–3 km eastward at 10 km. In figure 3.9b there are low-level lee waves as well as vertically propagating waves aloft with tilted phase lines. When l changes rapidly with height, wave energy may be partially or totally reflected. Computations by Sawyer (1960) for soundings from various actual airstreams show general agreement with observations made at the time. The lee flow shown in Figure 3.9a is quite characteristic, but the relationships between wave amplitude over the ridge, or in its lee, and profile of wind, temperature and l^2 are complex and variable (see Cox 1986).

It is worth noting that the wind profile in Figure 3.9 implies l^2 increases with height; hence smooth waves cannot form, as shown by Förchgott's observations. Severe turbulence occurs in such cases up to two to three times the height of the barrier (Corby 1957).

A useful approximate relationship for the wavelength (λ, in km) of lee waves in the lower troposphere is:

$$= 0.5\bar{U}$$

where $\bar{U}$ = mean tropospheric wind speed (m s^{-1}), assuming an average temperature lapse rate of 5°C km^{-1} (see Corby 1954). The relationship between λ and $\bar{U}$ is closely linear. However, the *first* lee wave crest downstream of a mountain range is approximately 0.75λ downwind. Observations of lee waves indicate a wavelength range of 5–30 km with a modal value of about 10 km. The horizontal extent of the lee waves downwind of topographic barriers is inversely proportional to the thickness of the stable layer according to Cruette (1976). Her results are based on 226 cases over western Europe and North Africa in 1966–8 examined via satellite photography and aircraft measurements. The stable layer traps the energy of the mountain perturbation acting as a 'wave guide'.

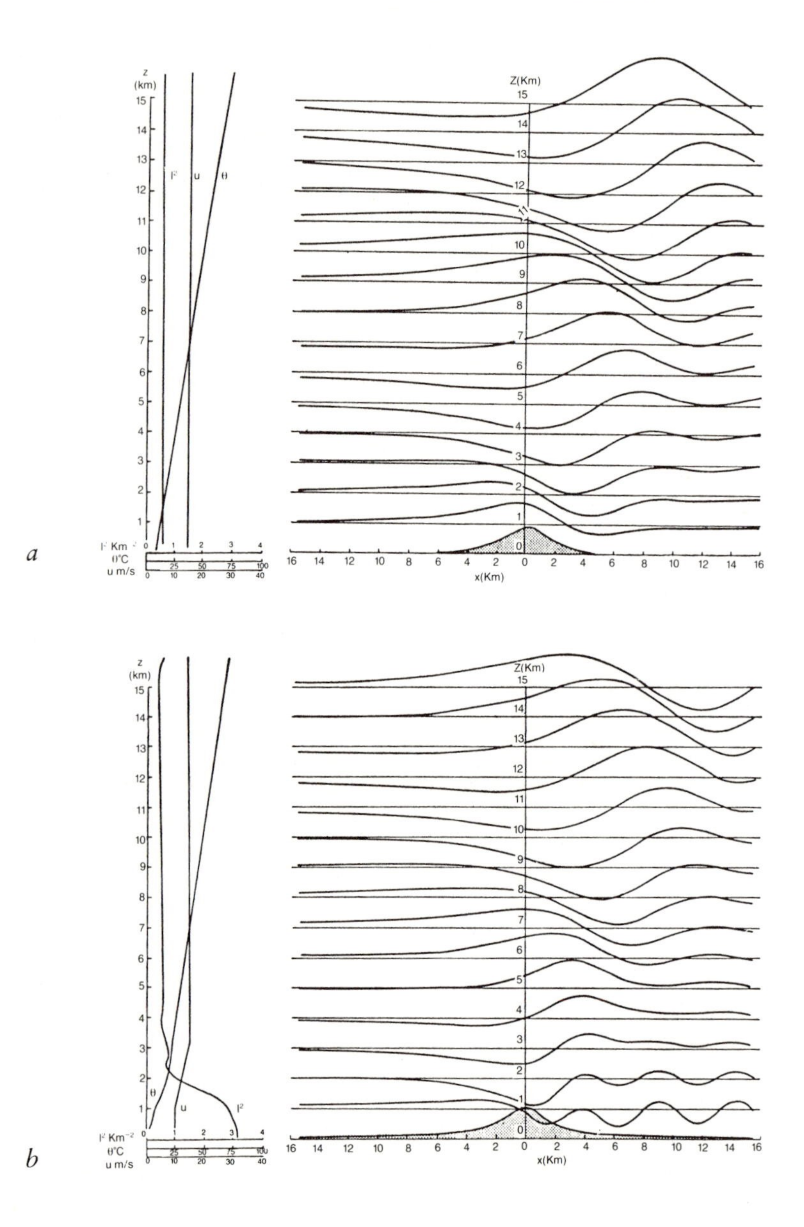

Figure 3.9 Calculated streamlines showing wave development over a ridge for two idealized profiles of wind speed (u) and potential temperature (θ). Values of l^2 (see text) are also shown.

Notes: (*a*) u and l^2 constant with height; (*b*) l^2 large in the lowest layer and then decreasing rapidly to low values.

Source: From Sawyer 1960

It is interesting to note why waves form only *downstream* of a mountain barrier. Smith (1979: 113) provides an explanation based on the fact that in fluid flow the speed of motion of wave crests (phase velocity) exceeds the rate of energy propagation (group velocity). The phase velocity in a standing wave has to be equal and opposite in direction to the mean wind speed, $\bar{U}$. Thus, $\bar{U}$ exceeds the group velocity, and advection by the mean wind dominates the transport of wave energy downstream away from its source at the obstacle. The wave energy is reflected up and down between the ground and an upper region where l^2 is small (see Figure 3.9b).

The lee wave amplitude depends on the decrease with height of l^2, the mountain height, and width. Analysis shows that the largest amplitude waves occur when the airstream characteristics meet the condition for waves by only a small margin (Corby and Wallington 1956). Moreover, maximum amplitudes tend to occur when there is a shallow inversion layer in the lower troposphere. For a given height of barrier, the amplitude is largest when the width of the range is adjusted to the natural wavelengths in a particular airstream; this occurs when λ/π is equal to the half-width of the barrier. Thus, the largest waves occurring in an airstream are *not* necessarily associated with the largest mountain ranges. The height of the maximum wave amplitude generally corresponds to the top of the wave clouds (if present) and this is also usually an inversion layer (Scorer 1967).

When the topography consists of a succession of ridges perpendicular to the airflow there may be superposition of several 'wave trains' that are set up independently by the different obstacles (Scorer 1967). This can lead to amplification or damping of the waves, depending on the spacing (the phase relationship) of the individual wave trains. An obstacle located one-half wave-length downstream can eliminate the lee waves set up by an upstream ridge. For this reason, air may sometimes descend while crossing some sections of a broad mountain range; Wallington (1961) has observed this during aircraft profiles across the Welsh mountains. Even over simple symmetrical ridges, however, there are commonly two or more different wave modes, with maximum amplitudes at different heights in the lower troposphere, as a result of the complexities of airstream characteristics.

The synoptic conditions which favour lee wave development are, of course, quite variable according to location. In the Sierras, lee waves may occur with westerly flow associated with an upper trough, or with the passage of a cold front or occluded front from the north-west (Alaka 1960: 38). The jet stream associated with the frontal zone is usually north of the area, and its occurrence is not a requirement for lee wave development. As implied by the earlier mathematical discussion, the characteristics of waves occurring in a specific situation depend critically on the particular airstream characteristics. Since these may change rapidly, and cannot always be specified by appropriate upper-air soundings, it has been suggested that radar observations may be the most feasible means of providing short-term aviation forecasts (Starr and Browning 1972).

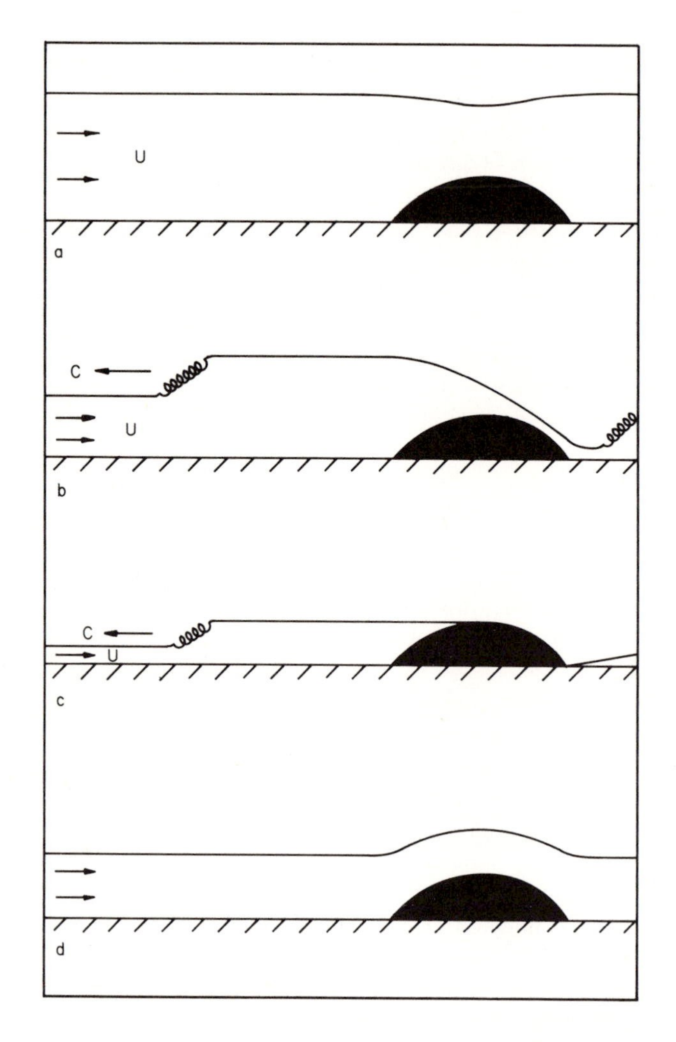

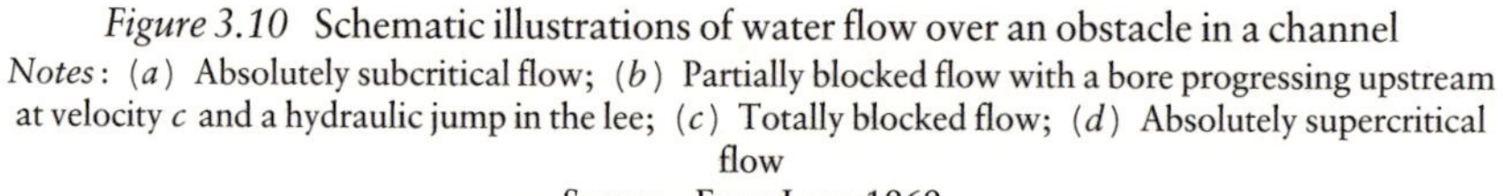

Figure 3.10 Schematic illustrations of water flow over an obstacle in a channel
Notes: (*a*) Absolutely subcritical flow; (*b*) Partially blocked flow with a bore progressing upstream at velocity *c* and a hydraulic jump in the lee; (*c*) Totally blocked flow; (*d*) Absolutely supercritical flow
Source: From Long 1969

Distortions of the air motion are detected by variable backscatter from layers of different density and humidity and, therefore, different refractive index.

The basic theories of wave motion are strictly applicable to disturbances of small amplitude, and they do not provide the necessary mechanisms for upstream blocking of airflow, rotor development, or 'hydraulic jump' type phenomena. Theories of large-amplitude disturbances were developed more slowly since in

Plate 5 Hydraulic jump illustrated by cloud on the eastern (lee) side of Elk Mountain, Wyoming, 1345 hours, 1 February 1973. The view is south-westward from the northeastern edge of Elk Mountain (courtesy of Dr August H. Auer, Jr, New Zealand Meteorological Service).

these circumstances the equations of motions become non-linear, except in special cases.

As discussed in the section on pp. 66–8 the flow of air (or water) over an obstacle may be supercritical ($F > 1$) or subcritical ($F > 1$). In the former case, the flow thickens and slows down as it crosses the barrier. Potential energy goes into creating kinetic energy. In the subcritical case, the opposite occurs; the flow thins and accelerates over the obstacle (Figure 3.10a). If the flow is partially blocked, with an increase in velocity as air ascends the obstacle and a decrease in thickness, a transition from sub- to supercritical flow may occur over the crest. The flow accelerates down the slope (shooting flow) and then jumps to a higher level (Figure 3.10b and Plate 5).

The *hydraulic jump* is a well-known feature of water flow in channels in the wake of large boulders, where there is a region of depressed flow which breaks down into turbulence downstream. In the atmosphere, hydraulic jumps are common occurrences in the strong katabatic flows in coastal Antarctica (p. 156). They may also be initiated when the airflow across a ridge is partially blocked

(Long 1970). A wave disturbance (a 'bore') formed by the barrier propagates *upstream*; the deeper flow is drawn down over the obstacle, becoming shallow on the lee side before jumping back to a higher level (Figure 3.10). Houghton and Isaacson (1970) show that hydraulic jumps may occur with high mountain ranges and a low upstream Froude number. A description of possible jump features during downslope winds in the lee of 600 m Mt. Lofty, near Adelaide, Australia, is reported by Grace and Holton (1990). For multi-layer fluids, an internal Froude number (F_1) is defined (Long 1954):

$$F_1 = \frac{U}{\left(g \dfrac{\Delta\rho}{\rho} H\right)^{1/2}}$$

where $\Delta\rho$ = the density difference between the top and bottom layers, and H = the total fluid depth.
Assuming an incompressible fluid

$$F_1 = \frac{U}{\left(g \dfrac{\Delta\theta}{\theta} H\right)^{1/2}}$$

where θ = potential temperature.

For a laboratory model resembling the atmospheric situation with a tropopause and a lower tropospheric inversion, lee jumps occur for $F_1 \geqslant 0.2$, approximately. This corresponds to a mean wind of 20 m s^{-1} and $(g\ \Delta\theta/\theta\ H)^{1/2} \simeq 10^4$ (Long 1954). In models with stratified flow over three-dimensional hills, jumps occur for $F_1 \geqslant 0.4$, but the types of flow are essentially those inferred for two-dimensional flow (see Plate 6) (Hunt and Snyder 1980).

Klemp and Lilly (1975) argue that the hydraulic jump mechanism is too restrictive in its assumption to account for many observed aspects of strong wave amplification and downslope windstorms. They propose instead that partial reflection of upward-propagating wave energy by a midtropospheric stable layer leads to wave amplification of the kind that is observed.

Recently, numerical solutions of the equations of motion have been used to examine non-linear flow behaviour over mountains. Peltier and Clark (1979) show that, for homogeneous stable flows over a two-dimensional barrier, non-linearities are related to the aspect ratio of the barrier. When the mountain height is comparable to the height of any inversion present or, in a continuously stratified atmosphere, to the vertical wavelength of hydrostatic waves ($= 2\pi U/N$), linear theory ceases to be applicable. This is typically the case for mountains exceeding 0.5–1 km. For inhomogeneous flows, Peltier and Clark find resonant lee waves and trapping and amplification of internal wave disturbances by reflection from a region of wave-breaking and turbulence in the lower stratosphere. Klemp and Lilly (1978) incorporate an upper dissipative boundary region to remove upward propagating wave energy before reflection. They also analyse a

Plate 6 Lee waves developed in simulated northwesterly flow over a model Alpine/Pyrenean topography (courtesy of Professor D.L. Boyer, and the editor and publishers of *Meteorology and Atmospheric Physics*. From Boyer *et al.*, 1987). This laboratory experiment is similar to that used in Plate 3. The sequence shows (*a*) no lee waves, (*b*) upper level lee waves, and (*c*) lee waves at all levels, in response to increasing Froude number (F_o). The transitions between cases (*a*)–(*b*) and (*b*)–(*c*) occur for $F_o \simeq 0.25$ and 0.37, respectively; corrresponding equivalent atmospheric free stream winds are 8.6 m s^{-1} and 12.8 m s^{-1}.

non-linear large amplitude case, simulating a downslope windstorm at Boulder, Colorado, on 11 January 1972. In this study, they find upstream influence with a stable layer upwind of the mountains being raised. However, in a subsequent study (Lilly and Klemp 1979) they demonstrate an *absence* of upstream effects and show that terrain shape has a significant effect on wave amplitude and mountain form drag. This is discussed further below. Long's results suggesting upstream influence (Figure 3.10) appear to be determined by the assumptions of two-dimensional flow and of a rigid-lid upper boundary condition (Smith, 1979).

The occurrence of rotors, as illustrated in Figure 3.7d (p. 127) is one of the most important aspects of mountain waves. The idea that they are related to hydraulic jumps has been proposed by Kuettner (1958), and Yoshino (1975: 403–6) outlines laboratory wind-tunnel experiments to examine such effects. However, Scorer (1967) shows that the important terms in the equation for the vertical displacement of air, w in standing waves in the xz plane are:

$$\frac{\delta^2 w}{\delta z^2} = (K^2 - l^2)w$$

where l = the Scorer parameter,
and K = streamline curvature (or wavenumber in the U direction); $2\pi/K$ = wavelength.

The criterion for rotors to occur is

$$\frac{\delta w}{\delta z} > 1 \text{ or } < -1.$$

which implies backward sloping streamlines such that the air has been overturned and has become statically unstable. Rotors therefore tend to develop when the wave amplitude increases where the slope of the w profile is largest. They are most common near the ground, but may occur elsewhere in lee wave troughs and crests depending on the profile of l^2 (Figure 3.9, p. 132).

When a barrier has steep slopes or bluffs, especially on the lee side, the flow may become highly turbulent. Smith (1977) shows analytically that steep lee slopes accentuate the forward steepening of mountain waves, causing earlier breakdown of the waves and increased downslope wind velocity. Gust speeds during the downslope windstorms on the lee slopes of the Rocky Mountains appear to be intensified by this factor at locations such as Boulder, Colorado (Lilly and Zipser 1972; Brinkmann 1974). In their numerical study of the effect of mountain cross-profile on wave motion, Lilly and Klemp (1979) find that the ratio of maximum/mean surface wind speed over a symmetrical mountain is 1.72, whereas over one with a steep lee slope and a gentle windward one it is 2.36. Separation of the flow from the ground is also accentuated by steep slopes, especially on the lee sides of a barrier. For stratified flow over low to moderate slopes ($< 45°$), the boundary layer flow regime is determined primarily by the ratio of the wavelength of the lee waves ($2\pi U/N$) to the total width of the barrier (W), not by its height (Hunt and Snyder 1980). When this ratio is close to unity,

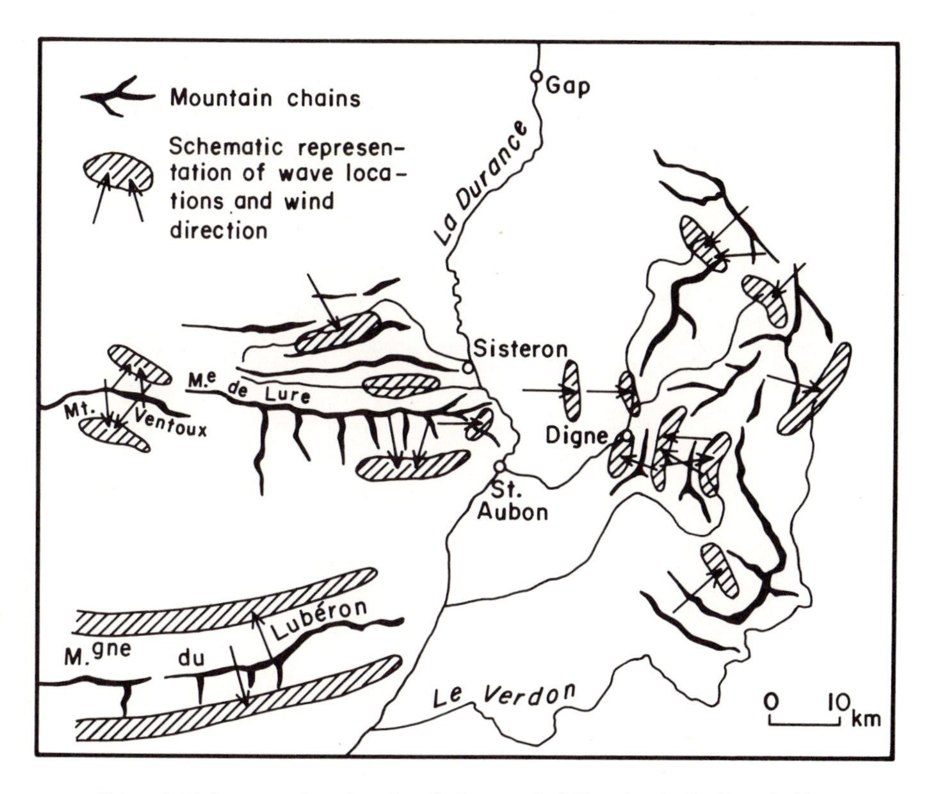

Figure 3.11 Lee wave locations in relation to wind direction in the French Alps
Source: After Gerbier and Bérenger 1961

separation is unlikely, but when $W \ll 2\pi U/N$ (i.e. $F \gg 1$), lee separation is induced by the boundary layer flow. In this case, separation tends to occur near the hill crest (or other location of maximum slope curvature) (see Figure 2.28, p. 73, for example).

Most theoretical analyses of mountain effects on airflow treat the problem as a two-dimensional one, but it is evident that in many cases there will be passage of air through mountain valleys and gaps, or around the ends of the barrier. As might be anticipated, isolated peaks cause the least vertical perturbation, although air tends to flow around any mountain range of limited length rather than rise over it as demonstrated in Figure 3.11. This effect is augmented when the crestline is convex upwind, whereas a crestline which is concave upwind accentuates any tendency to wave development.

The effects of isolated obstacles in an airflow are seen in satellite photography depicting lee wave clouds behind mountain ranges or peaks. Gjevik and Marthinsen (1978), for example, report trapped waves during inversion conditions in the lee of Jan Mayen, Bear Island and Hopen. The waves occur in a wedge-shaped wake behind the island, frequently in a diverging pattern with the

Plate 7 Parallel vortex sheets, identified by cyclonic and anticyclonic stratocumulus cloud eddies, extending 600–800 km downwind of the Canary Islands (28°N, 14°–18°W) as seen by a Defence Meteorological Satellite Programme visible band sensor 1136 GMT, 1 August 1978. (From transparencies archived at the National Snow and Ice Data Center, Boulder, Colorado for NOAA/NESDIS.) Light northeasterly surface winds were overlain by a subsidence inversion at 1100 m altitude and a further stable layer up to 2 km. The vortices are forming downwind of Gran Canaria and Tenerife, visible in the centre-right, and also La Palma, mainly cloud-covered, to the west. The lower island of Fuerteventura (upper right margin) appears to have little effect.

crests orientated outward from the wave centre. Less commonly, the wave clouds are transverse, with their crests perpendicular to the wind. Another meso-scale circulation which may occur in such locations is the *vortex street.* This is common in the lee of islands in the trade wind zone and other areas with low-level inversions (Plate 7). By analogy with von Karman's vortex street theory,

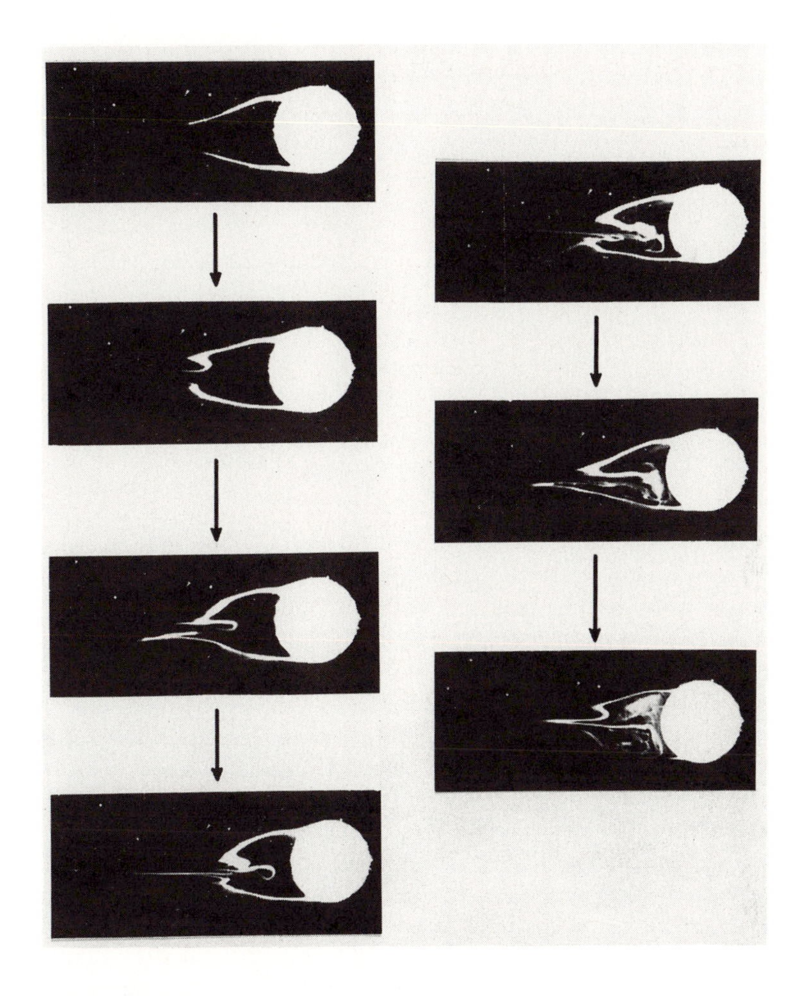

Plate 8 A laboratory model of double-eddy formation in the lee of a cylindrical obstacle for easterly flow in a rotating water-tank. Motion in the larger eddy at the right-hand edge of the obstacle, viewed downstream, circulates cyclonically, while motion in the smaller detached eddy at the left side is anticyclonic (courtesy of Professor D.L. Boyer and the Royal Society, London. From Boyer and Davies, 1982).

airflow drag over such high, steep-sided islands leads to eddies with a vertical axis being shed alternately on each side with a period of about 5–10 hours (Zimmerman 1969, Chopra 1973). In the case of the Canary Islands, which project above the trade wind inversion to 2–3 km elevation, the eddies are typically 10–30 km in size, and the downstream wakes are of the order of 50 km wide and 500 km long. Clouds below the inversion serve as tracers of the vortices, which in the easterly airflow are cyclonic from the northern sides and anticyclonic from the southern sides of the islands (Figure 3.12; Plate 7).

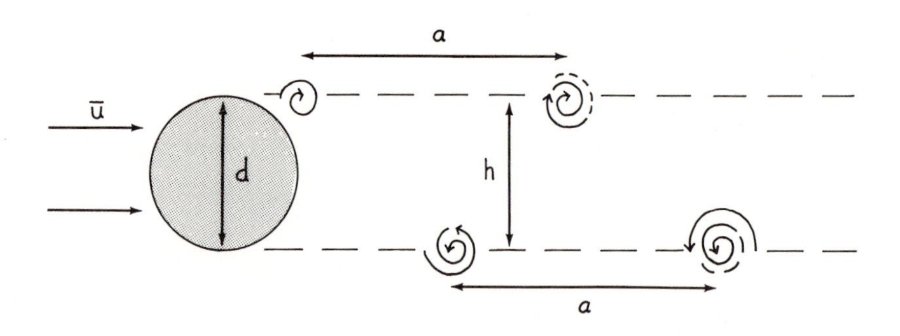

Figure 3.12 Schematic illustration of a von Karman vortex street in the lee of a cylindrical obstacle (diameter d). h = lateral street width; a = longitudinal spacing between successive vortices in each row. Empirically, $0.3 < h/a < 0.5$.
Source: After Chopra 1973

Laboratory simulations of flows on a β-plane (*i.e.* the Coriolis parameter increases with latitude) show that downstream of obstacles flow separation is suppressed (enhanced) for eastward (westward) flows (Boyer and Davies 1982). Eastward flows have downstream acceleration and a stationary Rossby wave, whereas westward flows exhibit eddy shedding resembling Figure 3.12, with stronger cyclonic than anticyclonic eddies (see Plate 8). Laboratory experiments are typically performed with hemispheric or conical obstacles whereas slopes of islands, where downstream eddies are observed, are seldom steeper than 1:10 to 1:15. The additional factor contributing to vortex street formation in many cases is the presence of a stably-stratified flow. Etling (1989) notes that this is present in the trade wind zones and during wintertime outbreaks of cold air off east Asia, for example. In such situations, there is an elevated inversion below the island summit. The air below the level of the 'dividing streamline' (see p. 67) goes around the obstacle and a train of vertical eddies forms downwind.

Experimental analysis of strongly stratified flow past a bluff obstacle by Brighton (1978) supports the idea of the key role of a strong low-level inversion in vortex shedding. His results demonstrate that flow around the obstacle remains nearly horizontal at low levels, but with intermittent vortices forming in the lee. Near the summit level, however, lee waves are present downstream and occasionally a 'cowhorn' eddy may develop, initially below the first lee wave crest, with the horns of this eddy pointing downstream.

Small vertical vortices ('mountainadoes') have occasionally been reported at Boulder during windstorms, as well as during more moderate westerly flow conditions (Bergen 1976). These may well represent vortex shedding in the lee of the locally sharp and irregular mountain front. Similar vortices, made visible as 'cloud spouts' extending 300 m below the cloud base, have been observed in the lee of Mount Washington, New Hampshire (Brooks 1949).

Plate 9 Cirriform clouds at 8.5 km over the Himalaya. Lhotse peak (8500 m; 28.6°N, 86.6°E) is visible in the centre. The cloud forms indicate winds probably in excess of 30 m s^{-1} over the summits (courtesy of Dr K. Steffen, Institute of Geography, Zurich).

Associated cloud forms

A detailed classification of orographic cloud was proposed from cinephotography of Mt Fuji, Japan, and studies of laboratory models by Abe (1941), but it is unnecessarily elaborate. Nevertheless, he notes that cumuliform cloud, stratocumulus, and turbulent fracto-forms, may all occur over or near mountains.

The mountain wave system is characterized by several distinctive cloud forms. All of them are *stationary* clouds which continually dissipate on the lee edge and reform on the upwind edge. The three principal categories are the cap cloud, lenticular cloud and rotor cloud, which may or may not occur together.

The cap (or crest) cloud. This forms over a ridge or isolated peak when forced ascent of air raises it to saturation level (see Plate 9). The cloud base is usually near or below the summit level for the term, 'cap cloud', to be applied. It has a smooth upper outline, but the lee side often appears as a wall (föhn wall, or chinook arch in the Rocky Mountains) with fibrous elements dissipating from it downward.

A further type of cap cloud form is the *banner cloud*. This forms in the lee of sharp isolated peaks, such as the Matterhorn (Douglas 1928). The pressure reduction caused by the flow of air around the mountain causes air to rise on the lee side. This form is sometimes hard to distinguish from streamers of snow blowing off the summit area.

Lenticular clouds. These are lens-shaped clouds forming in regularly spaced bands parallel to the mountain barrier on the lee side (Plate 4). The first descriptions of lee waves (Moazagotl) to the north of the Sudeten mountains on the Czech–Polish border were derived from study of such cloud forms (Küttner 1939a). The Moazagotl cloud occurs with southerly flow and typically extends 50–60 km from its windward edge just north of the 1200–1500 m mountain range. Within this zone, Küttner (1939b) has identified up to six lee waves extending 250 km cross-wind with cloud 1–4 km thick. The term 'great hill-wave cloud' has been proposed by Ludlam (1980) for such features related to major escarpments. In complex terrain, the wave pattern may not be clearly related to the underlying surface. Indeed, the 'hohe Föhnwelle' identified over the eastern Alps is an example of upper cloud related to the overall mountain effect, rather than a particular range (Krug-Pielsticker 1942). Due to a variable humidity stratification, these clouds sometimes occur in layers, one above another, forming a 'pile of plates', so that stratocumulus, altocumulus and cirriform cloud may be involved. Affronti (1963) describes such forms in the lee of Mt Etna where they are termed Contessa del Vento. On rare occasions, particularly in winter, the wave motion extends into the stratosphere and forms nacreous (mother-of-pearl) clouds at a height of 25–30 km. The cloud top in the wave tends to be sharply defined when a stable layer, which causes a marked upward decrease in humidity, is present.

The lee cloud forms at Mt Fuji are termed 'turusi' (suspended). However, they include vertical rotary forms which cannot be explained solely on the basis of wave motion (Abe 1941: 108). In the case of an isolated conical mountain, turusi commonly assume a V-shaped form, with the wings pointing downstream. Abe's photographs and laboratory model results show a clear resemblance between this form and the 'cowhorn eddy' described by Brighton (1978).

Rotor clouds. The first wave crest downwind of the mountains is commonly occupied by a rotor cloud band. At Crossfell, in the Pennines of northern England, this is the well-known 'Helm bar' which develops during strong easterly winds (Manley 1945). The rotation in these clouds is readily visible in time-lapse camera photography. Turbulence connected with the rotor usually causes the cloud outline to be ragged, such that the 'fracto-' designation is appropriate.

Superimposed on the forms of these three clouds, especially lenticular and rotor clouds, may be bands or striations known as *billows* (Ludlam 1967). These are caused by small-scale instabilities induced by the vertical shear of the wind which move through the larger-scale wave cloud. They may form in existing shallow layer clouds in the lower troposphere when radiative fluxes set up an unstable stratification. The overturning occurs transverse to the shear vector with a typical wavelength of about 1 km. Less commonly, irregular cirriform billows may occur near fronts or jet streams. A relationship with orographic features is apparent in the case of the lower tropospheric forms and even high-level cirrus billows can develop 20–30 km downward of particular hills and ridges.

'Fall' winds

When the synoptic situation is favourable, the mechanical and thermodynamic effects of topography on airflow can give rise to distinctive winds blowing down the lee slopes of a mountain range. These so-called 'fall' winds include the föhn (or chinook), the bora, and (meso-scale) katabatic winds. In the simplest terms, the föhn wind is defined with reference to a downslope wind that causes temperatures to rise and relative humidities to fall on the lee side of a mountain range, whereas the corresponding *bora* causes temperatures to fall. Both may be gusty. The föhn also has important desiccating effects on vegetation and soil moisture extending up to 50 km from the foothills of the Rocky Mountains in Colorado (Ives 1950; Riehl 1974). A katabatic wind is a gravity wind down any incline, but, in the present context, the reference is to wind systems on a scale affecting more than individual slopes (see p. 158). The generic term 'drainage wind' is also used, particularly to denote a downslope flow on a scale larger than on a single simple slope. It may apply to broad flows, not confined to valleys, over large uniform slopes (Sturman 1987).

Föhn

The recognition and study of föhn winds has a history of over 100 years in the European Alps, where the first broadly correct account of their origin was given by Hann (1866). The generic name derives therefore from the Alps, although the term *chinook* is used along the high plains east of the Rocky Mountains and there are many other local names throughout the world (Brinkmann 1971). The classical mechanism used to account for the föhn phenomenon begins with the forced ascent of moist air against a mountain range, causing cloud build-up and precipitation on the windward slope. The rising air cools at the saturated adiabatic lapse rate (*ca* 5–6°C km^{-1}) due to latent heat release by condensation above the cloud base, whereas on the lee slope the descending, cloud-free air warms at the dry adiabatic lapse rate of 9.8°C km^{-1}. Thus, potential temperatures are higher on the lee side (Figure 3.13b). In many instances, however, föhn may occur without moisture removal on the windward slope. This was first noted by Hann (1885) and has subsequently been widely demonstrated (Cook and Topil 1952; Lockwood 1962; Brinkmann 1973). Hence, it is sufficient for air to descend from the summit level to the surrounding lowland and undergo adiabatic compression, due to blocking of air at low levels by an inversion (Figure 3.13a). Typical effects during north- and south-föhn are illustrated for the Alps in Tables 5.4 and 5.5 (p. 324, 325).

Two further mechanisms producing föhn-type temperature fluctuations have been identified on the east slope of the Rocky Mountains by Beran (1967). One is

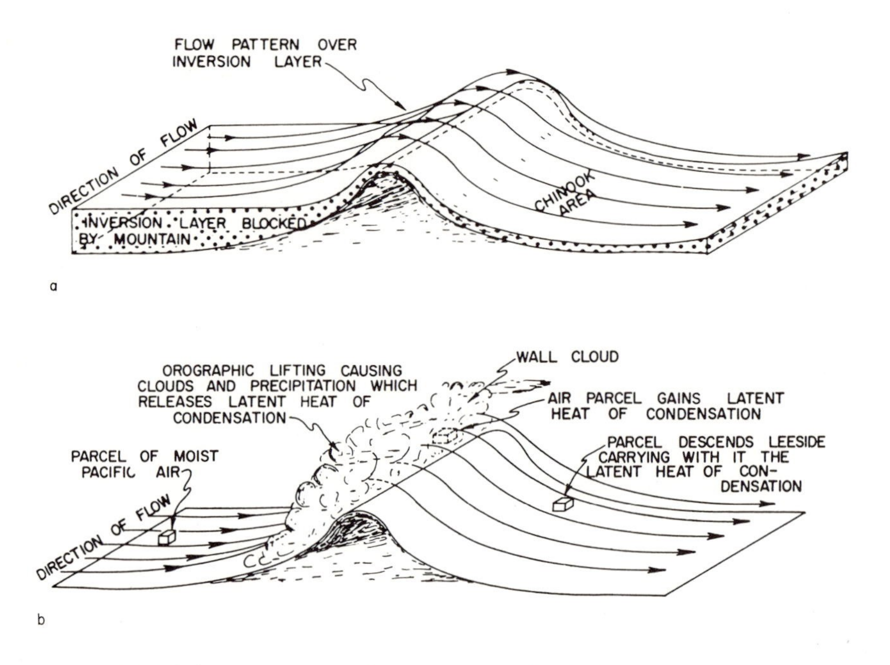

Figure 3.13 Adiabatic temperature changes associated with different mechanisms of föhn descent

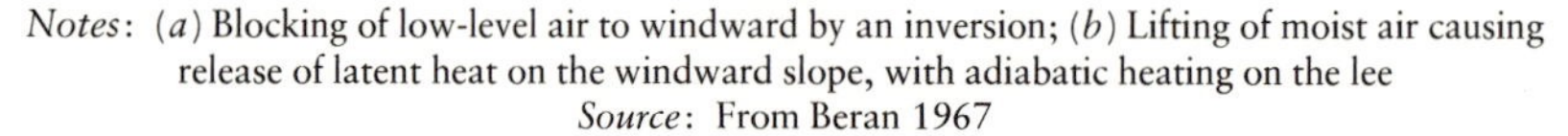

Notes: (*a*) Blocking of low-level air to windward by an inversion; (*b*) Lifting of moist air causing release of latent heat on the windward slope, with adiabatic heating on the lee
Source: From Beran 1967

a nocturnal feature which occurs when advection of (warm) air from the west sets up turbulence and prevents, or greatly reduces the normal radiational cooling trend. The second involves the displacement of a pool of shallow, cold polar air covering the lower east slopes by air of Pacific origin which crosses the mountains. The interface between these two air masses may develop minor waves roughly parallel to the mountain front causing pronounced temperature fluctuations to be recorded at locations near the interface. Both situations are effectively special cases of the basic downslope föhn wind.

A classification of föhn winds has been proposed by Čadež (1967), based on the temperature and pressure differences across the mountain barrier. His three types are illustrated schematically in Figure 3.14. Types (a) and (b) both occur in a cyclonic pressure field; the atmosphere is less stable in (b) and the leeward temperature rise is correspondingly greater. The pressure gradient across the Alps may be up to 10 mb/100 km during föhn situations, compared with a mean value of only 2 mb/100 km (Hoinka 1985b). In type (c) there is damming of cold air on the windward side due to an anticyclonic inversion. In this connection it may be mentioned that Bilwiller (1899) first introduced the term 'anticyclonic föhn' to refer to dynamic warming by large-scale anticyclonic subsidence over the

Alpine area. Other meteorologists have subsequently used the term 'free föhn' in the same sense (e.g. Flohn 1942). Such cases are not in the category of fall winds.

The problem of definition is one of some importance since the criteria adopted determine the frequency of föhn conditions calculated for particular locations (Brinkmann 1970, 1971). For a study in Alberta, for example, Longley (1967) used an arbitrary maximum temperature of $\geqslant$ 4.4°C in winter months. He showed that over a wide area of southern Alberta chinooks occur on 15 per cent or more of winter days. Commonly, three criteria are used at lee stations. They are: surface winds blowing from the direction of the mountains, an abrupt temperature rise, and simultaneous drop in relative humidity (Osmond 1941; Frey 1957). Ives (1950) used a definition in accord with the original thermodynamic theory of föhn requiring precipitation occurrence on the windward slope and higher potential temperatures on the lee side. Ives noted that about a third of these cases were not recognized as chinooks by the plains dwellers of Colorado because there was no temperature rise. Conversely, about half of the warm wind events recognized as chinooks by the plainsmen did not satisfy the meteorological criteria! A further possible criterion is the existence of isentropic conditions (dry adiabatic lapse rate) between mountain summit and leeward valley stations. In this connection, Schütz and Steinhauser (1955) assumed a lower limit of 7°C km^{-1} since a pair of stations may not be exactly on the same streamline. Brinkmann (1970) also attempted to use synoptic criteria. For chinooks, the upper flow should be perpendicular to the mountains and the surface pressure field should show a 'föhn nose', or ridging, over the mountains. These cases were compared with 'non-chinook' periods of westerly surface winds, defined by upper airflow parallel to the mountains and no föhn nose in the surface isobars over the Canadian Rockies. Analysis based on the three most common climatic criteria (surface wind speed, temperature and relative humidity) used to identify chinooks indicated that nearly 50 per cent of the cases were misclassified by discriminant analysis. The föhn nose pattern is not usually well marked over the western United States, probably due to the occurrence of stable cold air pools in the Great Basin, according to Brinkmann (1971).

Synoptic conditions for the Santa Ana winds in southern California, south and west of the San Bernardino Mountains, have also been examined (Sommers 1978). Here the range is orientated between north–south and west–north-west/east–south-east directions. The typical Santa Ana situation involves northerly flow, associated with an upper trough to the east and a ridge in the eastern Pacific. The presence of large-scale subsidence on the windward side of the mountains giving strong stability, together with an inversion near summit level, are important controls. Sommers agrees with Brinkmann's results for Colorado that temperatures on the lee slopes with these föhn situations may increase or decrease. Relative humidities may fall below 10 per cent, creating major forest fire risks, and wind gusts of 10–50 m s^{-1} are observed.

A different approach has been developed from the theory of lee waves. According to Scorer and Klieforth (1959), upstream blocking of the airflow by a

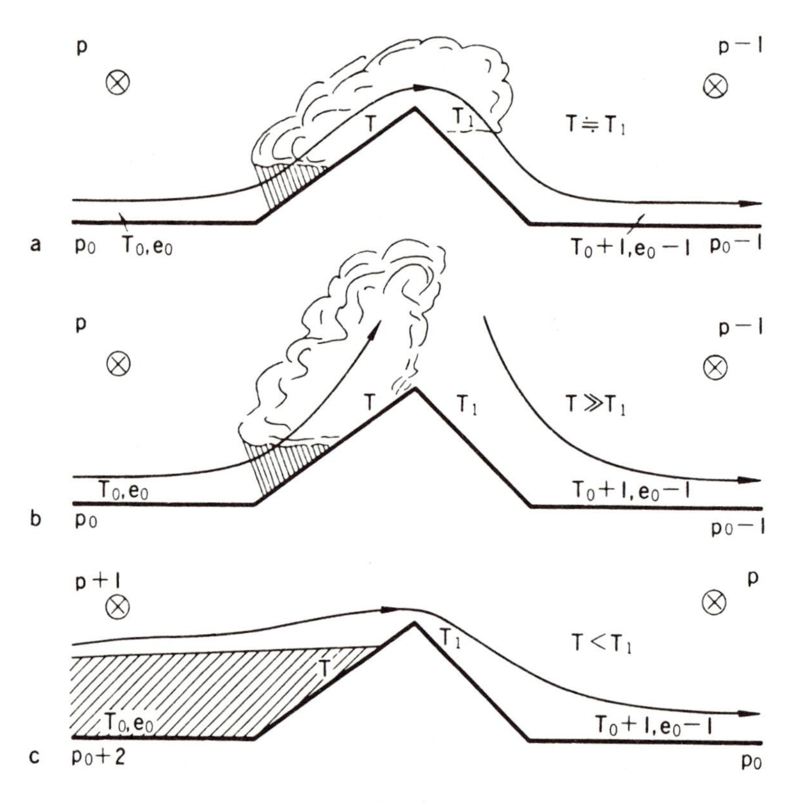

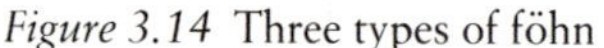

Figure 3.14 Three types of föhn

Notes: (*a*) Cyclonic föhn in a stable atmosphere with strong winds; (*b*) Cyclonic föhn in a less stable atmosphere; and (*c*) Anticylonic föhn with a damming up of cold air: *T*, air temperature; *p*, air pressure; and *e*, vapour pressure; Suffix 0, values at the ground level; and 1, values on the leeside slope

Source: After Cadež 1967; from Yoshino 1975

mountain barrier occurs if the ridge height $h > \pi/l$, where l = Scorer's parameter (see p. 129). This criterion is equivalent to $F[= U/\sqrt{(hS)}] < \pi^{-1}$. Lockwood (1962) found that in four out of five cases of föhn winds in the British Isles, π/l slightly exceeded the mountain height. In Colorado, Beran (1967) found it to be a necessary but not a sufficient criterion for predicting chinook occurrence in the lee of the Rocky Mountains. A small-scale three-dimensional numerical model has been used by Vergeiner (1978) to analyse föhn winds in the vicinity of Innsbruck, Austria. The equations are linearized, but they are used in time-dependent form with several arbitrary airstreams. The topography is idealized with a 2.5 km horizontal and 250 m vertical grid. The model results show: (1) forced orographic waves at low levels, with wavelengths of 10–15 km, tilting backward with height and (2) 'free' resonant lee waves of shorter wavelength generally at higher levels. These latter are trapped and propagate wave energy downstream (*cf.* the discussion on p. 129). The wind fields are similar to those observed during northerly and southerly föhn with strong föhns recurring in preferred locations, according to the assumed airflow characteristics, and 'blocking' effects due to the slopes facing the wind direction, causing flow reversal. Vergeiner also finds evidence of strong ascent over the crest of the northern (Karwendel) Alps (termed rohe Föhnwelle – high föhn wave) which is related to the orographic effect of the Alps as a whole, rather than that of individual ranges. While such models are useful diagnostically, input data are in most cases lacking for predictive use.

In summary, there are various approaches to defining and predicting föhn occurrence, none of them wholly satisfactory. In part, this reflects the different situations that may give rise to 'föhn conditions'. Föhn and chinook winds can be expected to exhibit some differences because the Alps form a lower barrier than the Rocky Mountains and, especially during southerly föhns, the air is considerably more humid; this is likely to modify the wave dynamics (Hoinka 1985b). Additionally, a south föhn in the Alps is usually pre-frontal.

Bora

The cold, dry and gusty winds that blow in winter over the Dalmatian Mountains of Yugoslavia towards the Adriatic Sea give their name to this type of fall wind. Subsequently, it has been widely used for similar downslope winds on the Black Sea coast of the Crimea (Arndt 1913) and elsewhere in the USSR (see Yoshino 1972), in the Apennines of Italy (Georgii 1967), the fjords of northern Norway (Köppen 1923; Mook 1962) and along the east slope of the Colorado Front Range (Brinkmann 1974b). Analogous winds with local names also occur at Crossfell in the northern Pennines of England – the northeasterly helm wind (Manley 1945), and in the Kanto Plain, inland of Tokyo, Japan – the *oroshi* (Yoshino 1975: 368–72).

In its type area, the bora occurs along the eastern shore of the Adriatic in the vicinity of Trieste, Italy, and southward for nearly 500 km between Rijeka and Dubrovnik. Strong northeasterly gusts may be encountered up to 50–60 km offshore (Yoshino *et al.* 1976). The bora is generally more intense in winter, with gusts in excess of 40 m s^{-1} frequently recorded at Trieste and inland in the Ajdovscina basin. Velocities show a nocturnal maximum, peaking between about 5 and 8 a.m. Each event lasts 12–20 hours, on average, but spells of 6–7 days or more with bora usually occur at least once each winter. Temperatures are around freezing on the coast (Table 3.1) and relative humidities may fall below 40 per cent during anticyclonic bora situations.

The vertical structure of bora conditions 130 km to windward (Zagreb) and leeward (Split) of the Dinaric Alps is illustrated in Table 3.2 showing differences between these stations for 142 winter cases. The wind is only stronger at Split near the surface, but temperature and humidity effects extend to about 3 km.

Aerial observations over the Dinaric Alps (Smith 1987) do not support the idea of a 'pure' fall wind model for the bora. Internal hydraulic mechanisms are shown to be important, since the initial cross-mountain flow is often weak. The mountains modify the upstream flow conditions and the development of a turbulent layer in the middle troposphere helps to decouple the descending air from the upper flow. Numerical modelling of mountain waves shows several important factors in the dynamics of bora events (Klemp and Durran 1987). 'Shooting flow' resembling that described by hydraulic concepts can be produced during bora conditions. Strong lee slope flow can arise in several ways. Vertical energy propagation may be restricted by a 'critical layer' (located near the top of the inversion) where the cross-mountain wind reverses direction. Alternatively, wave overturning may occur in the continuously-stratified cold air beneath the inversion. In several respects there appears to be considerable similarity between bora events and other severe downslope winds (discussed on p. 155).

Aircraft and sounding data collected during ALPEX confirm that the bora is a cold air flow trapped beneath an inversion. Pettre (1984) applies hydraulic theory through the equations of motion for shallow water (used earlier for study of katabatic winds in Antarctica by F.K. Ball; see p. 157). Pettre shows that the flow becomes supercritical in the lee of the Dinaric Alps; over the sea where the flow is strong the cold air is only about 500 m deep.

Four synoptic pressure patterns giving rise to bora conditions have been distinguished by Yoshino (1971). In winter these mainly involve a cyclone over the Mediterranean or an anticyclone over Europe. In summer, cyclonic patterns are less common and the anticyclone may be located further west. With each pattern the gradient wind is easterly to northeasterly. Yoshimura (1976) identifies a wider range of synoptic conditions associated with bora and shows that cyclonic cases have shallow bora winds (approximately 1 km) whereas in the anticyclonic patterns they may extend to 3 km. However, these distinctions may be of limited value. Development and maintenance of the bora requires the combination of suitable pressure gradient and the damming up of cold air east of the mountains with its overflow transforming geopotential into kinetic energy (Petkovsěk and Paradiž 1976). The bora is best developed where the Dinaric Alps are narrow and close to the coast as at Split. This accentuates the coastal–inland temperature gradient and the fall wind effect. The Alps rise to 1000 m or more and lower passes, such as that inland of Senj, also favour locally strong bora. On days with bora, an inversion layer is typically present between 1500–2000 m windward of the mountains and at lower levels to leeward (Yoshimura 1976; Hoinka 1985b).

On a hemispheric scale, Januaries with frequent bora days on the Adriatic Coast also have frequent oroshi winds in the Kanto Plain of Japan (Tamiya 1975). The mean 500 mb pattern features a dominant two-wave mode which favours outbreaks of cold polar air in these two sectors.

Yoshino (1976) notes that bora and oroshi winds may show either bora or föhn characteristics in terms of temperature change. The possible difficulties of distinguishing bora and föhn can be illustrated with reference to Figure 3.14c. If

Table 3.1 Mean conditions during bora in January at Senj, Yugoslavia

	Air temperature (°C)	*Relative humidity (%)*	*Wind speed (m s^{-1})*
Cyclonic bora	−0.1	61	15.6
Anticyclonic bora A[a]	1.5	49	13.4
Anticyclonic bora B[b]	−0.2	55	18.8
Non-bora	7.4	64	5.3

Notes: [a]High pressure over western Europe; [b]High pressure over the eastern Atlantic.
Source: From a study by the Hydrometeorological Institute, Zagreb, after Yoshino 1976

the cold air dammed up on the windward side of the mountain range deepens sufficiently to flow across the mountains, then bora conditions will replace föhn. Adiabatic warming through descent and disruption of surface inversions may cause other complexities. As noted by Suzuki and Yabuki (1956), the temperature characteristics on the lower slopes may be masked by local heating/cooling effects. In a detailed study of 20 downslope windstorms at Boulder, Colorado, Brinkmann (1974b) found that almost half were genetically cold, i.e. cold advection lowered the potential temperature on the mountains slopes (at approximately 650–750 mb). Only 4 out of 20 cases showed warm air advection and the remaining 8 cases were indifferent. However, at the foot of the mountains only 4 cases showed bora characteristics with temperature decreases of up to 15°C, while 15 appeared as föhn winds; 5 of these would be considered 'cold air föhn', 6 as indifferent, and only 4 as true föhn.

Windstorms. Considerable insight into causal mechanisms has been derived through study of violent downslope windstorms on lee slopes (Lilly and Zipser 1972; Brinkmann 19754a). Theoretical and laboratory model analyses demonstrate the critical role of an inversion (or stable) layer just above the mountain top level in triggering such conditions (Suzuki and Yabuki 1956; Arakawa 1968; Long 1970; Klemp and Lilly 1975; Durran 1990). Figure 3.15

Table 3.2 Climatological characteristics for winter bora days expressed as differences (Split–Zagreb)

Level (mb)	*Air temperature (°C)*	*Relative humidity (%)*	*Wind speed (m s^{-1})*
500	−0.1	+ 2.4	−1.5
700	+1.8	− 0.8	−2.7
850	+5.4	−10.2	−2.0
Surface	+7.4	−26.7	+3.9

Notes: Positive values indicate Split (leeward) > Zagreb (windward).
Source: From Yoshino 1976

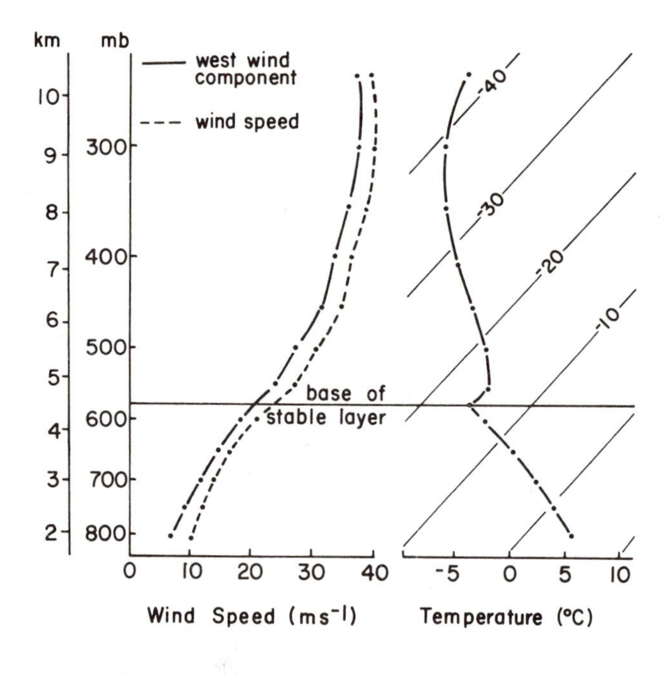

Figure 3.15 Composite soundings for times of windstorms in Boulder, Colorado:
Notes (*a*) Upwind sounding (west of the Continental Divide); (*b*) Downwind soundings (Denver) for storms in Boulder or on the slopes just to the west
Source: From Brinkmann 1974a

illustrates a composite up-wind sounding, using data at Grand Junction, Salt Lake and Lander according to the upper airflow trajectory and a composite Denver sounding, about the time of onset of wind storms in Boulder and on the slopes. A modal stable layer is indicated with a base at 575 mb; this served as a reference point for averaging the other sounding data (Brinkmann 1974a). Surface gusts during the analysed windstorms at Boulder averaged 36 m s^{-1}, well above the mean wind speeds up to 450 mb. The high speeds at low levels are generated by a small-scale lee trough which is located, hydrostatically, beneath a region of high potential temperature set up by a large-scale lee wave (Figure 3.16). This local trough, which may be superimposed upon a synoptic-scale pressure minimum, accelerates the low-level airflow towards it. Consequently, a narrow zone of very high winds is usually observed parallel to the mountain front (Aanensen 1965; Whiteman and Whiteman, 1974); on rare occasions this may extend tens of kilometres from the mountain foot (Lester 1978).

A numerical simulation of the 11 January 1972 windstorm using a hydrostatic model with a parameterization of turbulent kinetic energy suggests that surface friction plays a significant role in delaying the onset of strong surface wind and in

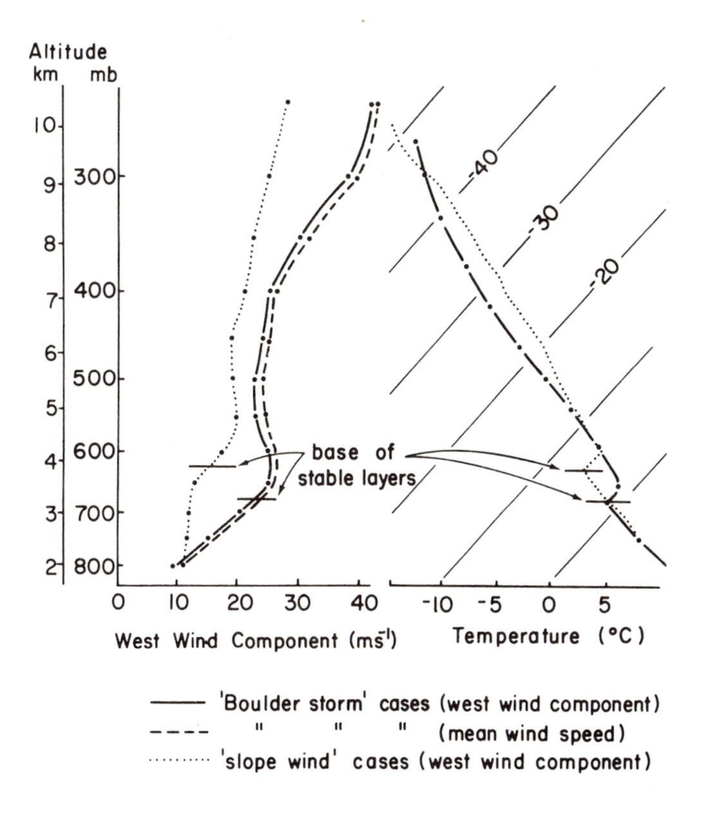

Figure 3.15b

preventing the zone of maximum winds from propagating downstream (Richard *et al.* 1989). Their results are consistent with the hydraulic theory that a transition from subcritical flow upstream to supercritical flow takes place over the mountain.

The fine structure of a moderate windstorm event at Boulder, Colorado, has recently between determined by Doppler lidar observations (Neiman *et al.* 1988). A low-level wind maximum exceeding 42 m s^{-1} was observed near 700 mb. As a result of surface drag, the surface winds averaged 25 m s^{-1}, with gusts over 30 m s^{-1}. The observed pressure fall of 4.7 mb during the windstorm corresponds to a windspeed of 32 m s^{-1} according to the Bernoulli equation. The lidar also indicated a hydraulic jump-like reversal of the flow along the eastern edge of the mountain wave some 5 km east of Boulder. Propagating wind gusts, advected by the mean wind, were identified with periods of 4–5 and 14 minutes.

The theoretical basis available for predicting severe downslope winds is still incomplete (Smith 1985). A key feature of such events appears to be the existence of a zone of weak winds with strong turbulence in the middle troposphere, overlying the strong low-level wind maximum, with weaker waves at higher

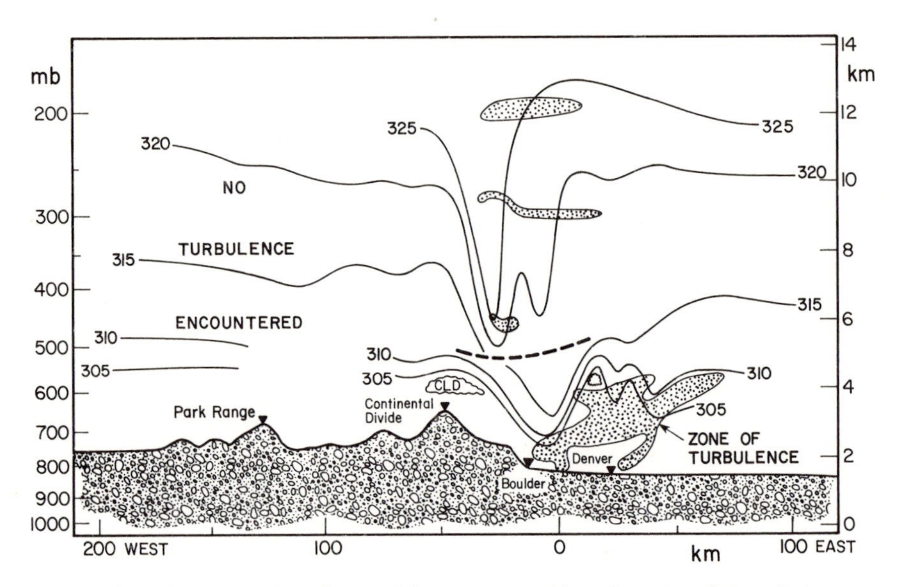

Figure 3.16 A cross-section of potential temperature K based on aircraft data during a windstorm in Boulder on 11 January 1972. The dashed line separates data collected at different times. The three bands of turbulence above Boulder were recorded along horizontal flight paths and are probably continuous vertically; no turbulence was observed at these levels further east.
Source: After Lilly 1978

levels *cf* (Figure 3.16). Strong mountain wave development may be a response to partial resonance produced by upstream conditions and tropopause reflection, or to wave reflection from the turbulent region. Alternatively, the severe wind mode may be triggered by the mountain top level inversion causing hydraulic acceleration and a jump feature, as observed in the lidar case study. Similarities between hydraulic-like shooting flows during windstorms along the eastern slope of the Rocky Mountains and during bora events on the western slopes of the Dinaric Alps are apparent (Klemp and Durran 1987). However, the chinook windstorms involve high speed and deep cross-mountain airflow.

The general conditions necessary for violent, gusty flow according to Yabuki and Suzuki (1967) are:

1 Ascending air on the windward slope subject to the Bernoulli effect enhanced by an inversion just above the ridge line.
2 Strong winds forced to descend from the summit level by the wave structure. The evidence from Brinkmann's (1974a) study for Boulder, and analysis of damaging winds at Sheffield, England, by Aanensen (1965), strongly suggests the importance of long wave-length (20–30 km) lee waves in such situations.
3 Topography with a steep lee slope.

The last point is supported by model calculations of Smith (1977), Lilly and

Klemp (1979) and Hoinka 1985a) (see p. 138). Arakawa (1968) notes that strong winds are also more likely at the surface if the inversion level downstream of the range is lower than over the summit. In fact, Brinkmann's data support this argument (Figure 3.15).

The criteria for windstorm are quite subtle. An analysis of wind profiler and stability data from Grand Junction, Colorado, and Lander, Wyoming (both upstream) during windstorms in Boulder shows that windstorm/no windstorm conditions are not separable based on wind direction, upper tropospheric wind shear or the vertical phase shift (Bower and Durran 1986). They suggest that information on the location and strength of inversions is particularly necessary for prediction purposes.

East of the Rocky Mountains, where there have been numerous studies of downslope windstorms, the zone of occurrence extends from about Colorado Springs to Cheyenne. Based on newspaper reports since the 1860's in Boulder, Colorado, as well as more recent scientific analysis, about 85 per cent of cases occur between November and March, with a strong January maximum, a preference for the night hours, and lasting an average of eight hours (Julian and Julian 1969; Whiteman and Whiteman 1974). The 100-year record suggests a frequency of about 1.5 storms per year, but more complete reporting since 1945 indicates that three or four damaging storms per year cause property losses in Boulder associated with peak wind speeds of about 30 m s^{-1}.

Severe windstorms may differ in important ways. During an event on 24 January 1982, isolated damaging gusts from an easterly direction occurred during otherwise light easterly flow in Boulder, while 45–50 m s^{-1} westerly winds affected the foothills to the west (Zipser and Bedard 1982). Evidently, a large rotor was present over Boulder. A week earlier, a closely similar large-scale synoptic situation produced a nine-hour downslope windstorm, with west winds that gusted to 61 m s^{-1} on the roof of the National Centre for Atmospheric Research, and caused $10 million damage in Boulder. Prediction of such severe storms is still unreliable despite improved theoretical understanding.

In reviewing the available evidence and model simulations, Durran (1990) finds strong support for the hydraulic jump mechanism in supercritical flow. Several simulations are inconsistent with the linear resonance mechanism of wave amplification proposed by Peltier and Clark (1979). However, there is an important linkage between the 'breaking' of vertically-propagating waves set up by high mountain barriers and hydraulic wave amplification. Non-linear effects of the partial reflection of vertically propagating waves by a layer interface become significant with large values of pressure drag. Durran notes that three rather different conditions may favour a transition from subcritical to supercritical flow, namely:

1 wave breaking forced by a high mountain barrier;
2 a two-layer atmosphere in terms of the Scorer parameter, for mountains too small to force wave breaking,

3 an atmosphere capped by a mean-state critical layer (see Note 2, p. 193) above mountain top level, as in the Yugoslavian bora, forcing wave breaking.

Windstorms in Boulder, Colorado, and Owens Valley, California, lack condition 3. They occur in deep cross-mountain flow when the wind is nearly perpendicular to the mountains with summit speeds exceeding some terrain-dependent threshold, and a stable layer or inversion located near the summit level (Durran 1990). The gustiness may be a result of competition between wave build-up by gravity wave forcing and wave breakdown through convective instability (Clark and Farley 1984; Seinocco and Pettier, 1989).

Recent research in the lower Rhône Valley, between the Massif Central and the Alps (Figure 3.4, p. 117), shows that some *mistral* events have several similarities to other 'fall' winds. The valley has a maximum depth of about 500 m; its slope is only 2.5 per cent, but the valley width decreases rapidly near Valence (Pettre 1982). During cold northerly flows, a sharp transition is observed between tranquil flow in the northern part of the valley and the presence of wave motions downstream. If the inversion height is much lower at the exit from the narrows than at their entrance, a hydraulic jump may be present downwind. Violent winds can occur in the lower valley, even as far south as the Mediterranean coast depending on the specific conditions.

Katabatic winds

This class of fall wind is distinguished from the nocturnal downslope drainage of cold air slopes (see p. 158) by its scale and consequent involvement of additional forces, including Coriolis accelerations. On the polar ice sheets of Antarctica and Greenland, air drainage may originate on remote ice domes and the extensive high cold plateaux. This motion is modified by the occurrence of a sloping, low-level inversion which sets up thermal wind components, such that the winds in the interior of the ice sheets are not properly katabatic flows (Schwerdtfeger 1970, 1972). They involve a balance between gravitational acceleration, Coriolis acceleration, friction and inversion strength. The flow adjusts more rapidly to gravitational forcing than to Coriolis effects (Gosink 1982); the continuity of flow cannot be maintained under equilibrium conditions when the trajectory is $>$ 10–100 km in length.

However, the main interest here is with the coastal zones, where the 'katabatic winds' are fall winds that may display föhn or bora characteristics and irregular fluctuations in velocity (Streten 1963).

Reports of extreme wind conditions at Cape Dennison (67°S, 143°E) during Douglas Mawson's 1912–13 expedition were not at first generally accepted. During a 12-month period, the 24-hour average wind was $\geqslant$ 18 m s^{-1} on 64 per cent of the days, and a high constancy of speed and direction was reported throughout the area. Subsequently, similar extremes have been reported at other coastal locations in eastern Antarctica. The katabatic zone is at least 150 km wide and extends inland 300 km from Cape Dennison. The coastal topography does not cause these localized winds, but diagnostic analysis of the flow regime demonstrates that large-scale topography inland determines their strength and persistence through a channelling of radiatively cooled air from the interior (Mather and Miller 1967; Parish 1980).

At Mawson (67°S, 63°E), where cyclone activity may be responsible for the frequent strong winds (Streten 1968), the katabatic is only one component and three types of wind profile can be identified. Streten (1963) distinguishes a 'normal' katabatic regime when synoptic control is weak. Here the mean wind is about 10 m s^{-1} in the lowest 300 m. Sometimes katabatic flow over-rides the lower layers, with mean speeds of 10 m s^{-1} at 1200–1500 m. Finally, blizzard winds occur with mean speeds of about 30 m s^{-1} between 300 and 1000 m when synoptic controls augment any katabatic effect.

The onset and cessation of strong coastal winds tends to be very abrupt at Cape Dennison, and Ball (1957) interprets this as a standing jump phenomenon. Between Cape Dennison on the coast and Charcot Station (69°S, 2400 m a.s.l.) a strong temperature gradient enhances the basic gravity flow of cold air from the Antarctic Plateau. At the 2400 m altitude, there is a 17°C difference in annual mean temperature between these two locations which creates (assuming it is an isobaric temperature gradient) a density difference of about 7 per cent (Loewe 1972). A thermal wind component related to the surface-based temperature inversion is also probably involved since the winds are usually a few hundred metres deep. The jump feature is usually just out to sea, but if it shifts inland the regime of strong winds (shooting flow) upstream of the jump gives way to near calm conditions in a deepened layer of cold air (*cf* Figure 3.10b). Ball shows that the typical conditions in this area are conducive to a jump since the Froude number is far in excess of unity. Near Davis (68°S, 78°E), standing jumps are commonly marked by a 30–100 m high wall of drifting snow (Lied 1964). Between 30 May–14 November 1961, 31 standing jumps were seen or heard (from the roar of the wind) at the Davis station. Lied reports their typical occurrence several hours after the development of a katabatic regime.

Katabatic winds in Greenland are less extreme than those in Adelie Land, although gusts in excess of 50 m s^{-1} have been recorded on the coast. Fjord gales in East Greenland, which are primarily katabatic in origin, occur on about 20 per cent of days (Manley 1938; Putnins 1970).

In a wholly different environment, jump features are reported by Lopéz and Howell (1967) from the Cauca Valley in equatorial Colombia. Here cool, moist Pacific air crosses the Western Cordillera and, as a result of a potential temperature difference of 2–4°C with the air to the east, descends as a katabatic wind. Speeds of 16 m s^{-1} are observed at Cali in late afternoon, compared with about 5 m s^{-1} at the pass at 700 m a.s.l. Hydraulic jump type phenomena are observed in the longitudinal valleys as the katabatic flow descends and is heated. Lopéz and Howell show that the speed of the katabatic wind is much more sensitive to the difference in potential temperature than to the depth of the overflow.

THERMALLY INDUCED WINDS

The topographically induced modifications to airflow discussed in the last section are basically due to mechanical effects of mountain barriers. In addition to these influences, the thermal patterns of the topography also give rise to characteristic systems of air motion, especially when the regional pressure gradients are weak. The primary forcing agents are elevational differences in potential temperature, causing vertical motion, and differential heating/cooling along slopes which may set up air circulations with horizontal and vertical components. In some locations, such systems are sufficiently frequent and pronounced in their effects as to create distinctive and semi-permanent topoclimatic patterns. Such is the case, for example, in the deep valleys of the Himalayan ranges.

Thermally driven wind systems include land-sea breezes, which are not discussed here, as well as the more complex mountain-valley winds. The basic dynamical processes that are involved are (i) an *antitriptic wind* component directed towards low pressure, when the Coriolis effect is small and (ii) *a gravity wind* component directed downslope, in the absence of any general pressure gradient (Flohn 1969). It is appropriate to begin by considering the nature and mechanisms of slope winds.

Slope winds

In general, downslope movement of cold air at night is referred to as *katabatic* flow and upslope movement during the day is termed *anabatic* flow.

Katabatic winds in the strict sense are *local* downslope gravity flows caused by nocturnal radiative cooling near the surface under calm clear-sky conditions. The extra weight of the stable layer, relative to the ambient air at the same altitude, provides the mechanism for the flow. Conversely, upslope flow is associated with daytime slope heating and buoyancy induced by this. The basic patterns of slope flow associated with potential temperature gradients are illustrated schematically in Figure 2.22 (p. 59). The cold air thickness over mountain slopes and alpine glaciers is typically 20–50 m with maximum speeds of 2–3 m s^{-1} at 20–40 m (Defant 1949). The large-scale katabatic-type flows over Antarctica, which may be an order of magnitude deeper and stronger, are discussed separately (p. 156). Figure 3.17 shows pilot ballon observations of wind profiles on the Nordkette, near Innsbruck, on a 42° slope. The maximum speed for both upslope and downslope winds occurred at a height of 27 m. As shown, upslope systems are generally stronger than their nocturnal counterparts since, in fine weather, the large daytime radiative exchanges encourage strong buoyancy effects. The entrainment of air from over the valley into the slope circulation also makes the upslope system somewhat deeper. In general, upslope winds are best developed on south-facing slopes, although even there they tend to be concentrated in gullies.

There are some important differences between katabatic slope winds and the weak ($\sim$ 1 m s^{-1}) small-scale drainage of air which causes cold air pockets and 'frost hollows'. While both arise from radiative cooling and density differences, a minimum slope of about 1:150 to 1:100 seems to be necessary for katabatic airflow (Lawrence 1954), whereas small-scale air drainage does not set up any significant compensating currents, primarily as a result of the limited horizontal scale of surface irregularities. From observations on the Niagara Peninsula, Ontario, Thompson (1986) suggests that cold hollows are a result of the early cessation, or decrease of turbulent heat transfer in sheltered locations rather than of cold downslope airflow. Nevertheless, small closed basins can give rise to some dramatic local temperature inversions. The famous Gstettneralm sink hole (a limestone depression) near Lunz, Austria, which was extensively investigated by W. Schmidt, has recorded many minimum temperature readings below −40°C, for example (see Geiger 1965: 398–401). It is located at 1270 m, with a relief of 100–150 m, and may show inversions of 20–30°C or more.

Several expressions have been developed for estimating the wind speed associated with slope drainage. For small-scale movements, the equation of Reiher (1936) seems appropriate. This relates the speed (v) to the temperature difference between the cold slope air (T) and the surrounding air (T'):

$$v = \left(\frac{2gh(T' - T)}{T'}\right)^{1/2}$$

where g = acceleration due to gravity (9.81 m s^{-2}),
h = height above the surface of v.

For $h = 500$ cm, $T = 273.2$K and $T' = 275.8$K, Reiher calculated a speed of 1 m s^{-1} compared with an observed value of 1.2 m s^{-1}. In an experimental study on artificial slopes, Voights (1951) obtained results which also agreed with this expression.

The basic processes at work on slopes involve density gradients caused by heating/cooling that produce buoyancy differences, downslope advection of momentum, and frictional drag. For example, heating of air over a slope sets up a horizontal density gradient with the ambient air and this generates a horizontal pressure gradient directed towards the slope. The presence of the valley side prevents an opposite gradient force acting on the parcel, thus the air moves upward along the slope. For slope cooling, the air moves downslope through the component of gravitational acceleration (g) parallel to the slope, while the component of g perpendicular to the ground is balanced by the pressure gradient force (Mahrt 1982). From the thermodynamics viewpoint, slope flows are determined by very localized, almost instantaneous equilibria (Vergeiner and Dreiseitl 1987). Thus, upslope flows may cease abruptly when clouds obscure the sun, cutting off the direct beam radiation on a slope. Vergeiner and Dreiseitl propose that the mass flux is proportional to the thermal forcing by the available sensible heat and inversely proportional to the static stability:

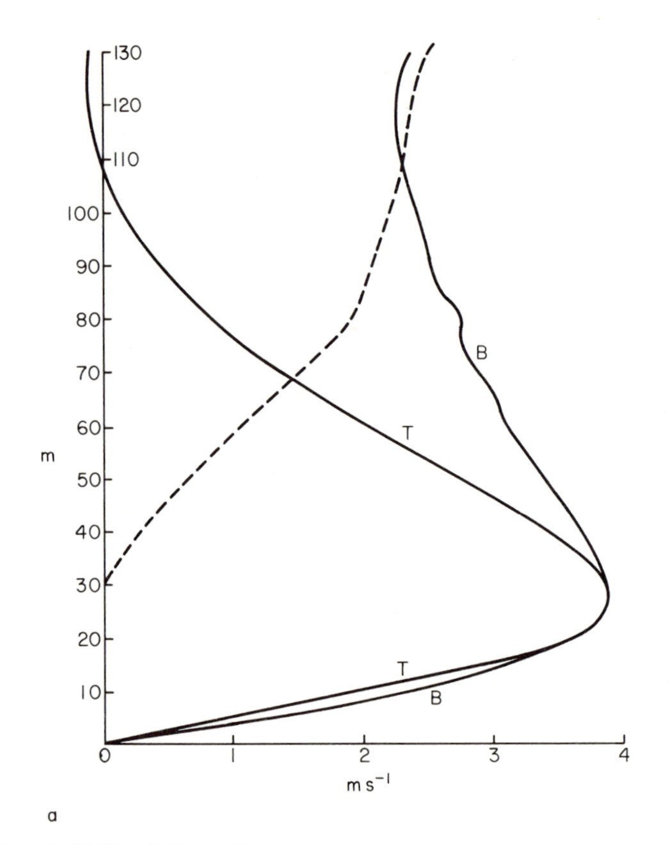

Figure 3.17 Pilot-balloon observations (B) and theoretical (T) slope winds on the Nordkette, Innsbruck

Notes: (*a*) Upslope (11 cases); (*b*) Downslope (5 cases). The dashed lines show the difference between theory and observation.

Source: From Defant 1949

$$V\delta = \frac{(H/\tan s)\,(1 - Q)}{c_p \rho d\theta/dz}$$

where $V\delta$ = mass flux per unit width of slope
H = sensible heat to the slope layer
δ = slope layer thickness
Q = sensible heat transferred into the valley atmosphere.

They also suggest that anabatic slope flows comprise two separate cells on the upper and lower segments of the slope, separated by an inversion layer.

Three theoretical approaches to the analysis of katabatic flows can be distinguished (Manins and Sawford 1979a). The earliest is based on hydraulic considerations of the downslope flow of a cooled air layer (Defant 1933). This

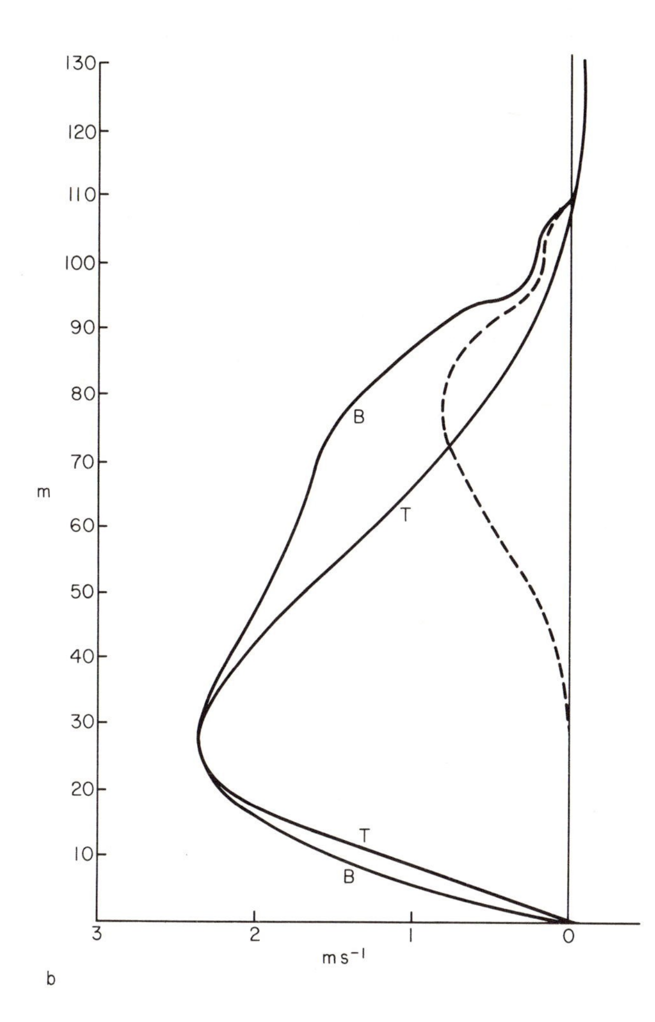

first approach has been followed subsequently by Fleagle (1950) and by Petkovsěk and Hočevar (1971). Internal (vertical) structure of the flow is ignored, but time variations are considered. In the second approach, Prandtl (1952), Defant (1949, 1951) and Holmgren (1971) analyse the vertical structure of the temperature and wind velocity above the slope by regarding the flow as steady and invariant down the slope. This approach ignores effects of advection and assumes small slopes and constant exchange coefficients. The third approach allows inclusion of advection through numerical solution of the primitive equations of motion (Thyer 1966). The mixing processes are parameterized.

Defant (1933) attempted to formulate a general expression for katabatic flow on slopes incorporating slope angle (s), friction effects, and the depth of the cold air involved (ΔZ):

$$v = \left(g \frac{\Delta Z}{C_d} \frac{(\theta_2 - \theta_1)}{\theta_2} \sin s \right)^{1/2},$$

where θ = potential temperature of an upper layer (θ_2) and a lower layer (θ_1) (θ_1)

and $C_d = 0.0025$, a dimensionless friction coefficient.

However, the direct relationship between wind speed and slope angle is not in accord with many observations of katabatic airflow. Lawrence (1954) suggests that slope *length* is probably important, since short slopes can supply little cold air. He proposes that the expression:

$$v = \left(\frac{(2gl \sin s)\,(T' - T)}{T'} \right)^{1/2}$$

may be representative of the wind speed during the developmental stage of katabatic flow.

Subsequently, Petkovsěk and Hočevar (1971) developed an expression for the downslope wind by assuming that, in the steady-state case, downslope acceleration of the cooled slope air due to gravity is balanced by friction. Adiabatic warming due to this motion is also taken into account. The speed is given by:

$$v = \left(\frac{C}{(\Gamma - \gamma') \sin s} \right) \left[1 - \exp \left(- \frac{g}{KT'} (\Gamma - \gamma') \sin^2 st \right) \right],$$

where $C = \frac{1}{c_p} \frac{dL_n}{dt}$, the mean radiational cooling of the layer,

c_p = specific heat of dry air at constant pressure,
L_n = effective long-wave radiation loss,
γ' = the original lapse rate of the surroundings,
K = a friction coefficient,
T' = temperature of the ambient warmer air,
s = slope angle,
t = time.

The maximum speed theoretically occurs as

$$t \rightarrow \infty$$

$$V_{max} = C/[(\Gamma - \gamma') \sin s]$$

and it is proportional to the net radiational cooling and inversely proportional to slope angle and the lapse rate. Calculations illustrating the wind speeds for different lapse rates and friction coefficients with a slope angle of 11.5° are shown in Figure 3.18. Winds on the McCall Glacier, Alaska, are in general accord with calculations based on this model according to Streten *et al* (1974). Before the steady state is reached, v is proportional to the lapse rate in the surrounding air and inversely proportional to the friction coefficient as well as to the slope angle. In fact, high speeds are observed in katabatic flows on the margins of the Greenland and Antarctic ice sheets, but it must be noted that the model takes no account of possible effects due to a limited extent of slope.

The theoretical treatment of Fleagle (1950) expresses *mean* downslope flow in terms of slope angle, frictional effects and the non-adiabatic cooling rate of the slope air. The solution for equilibrium velocity is again proportional to the net outgoing radiation and inversely proportional to slope angle and to the thickness of the layer which is cooling. In addition, the model indicates that the flow is initially periodic and that during this time the speed is proportional to the cotangent of the slope angle. This fluctuating flow is attributed to the accelerating and divergent downslope flow of air weakening the initial pressure gradient. The adiabatic heating then exceeds the radiational cooling and causes an upslope pressure gradient which slows the airflow. The cycle is repeated when the radiational cooling rebuilds the downslope pressure gradient. Mcnider (1982) proposes that the fluctuations in downslope flow depend on the tempera-

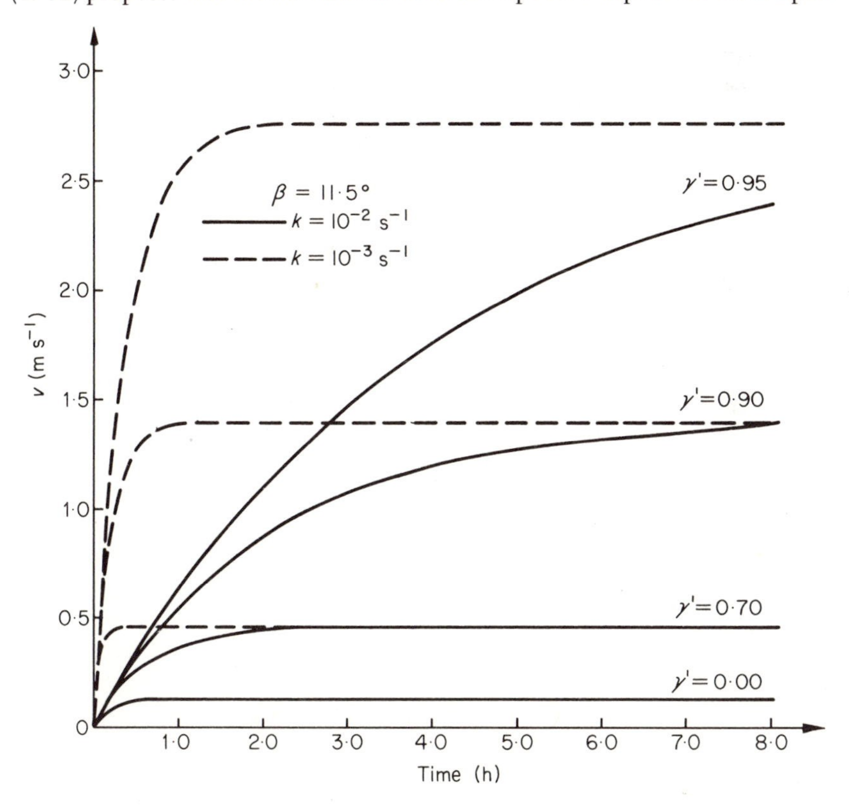

Figure 3.18 Drainage wind velocity (v) on an 11.5° slope as a function of time, for different values of lapse rate ($\gamma = 10^{-2}$°C m^{-1}) and friction coefficient (solid lines $K = 10^{-2}\ s^{-1}$, dashed lines $K = 10^{-3}\ s^{-1}$)
Source: From Petkovsěk and Hočevar 1971

ture difference between the katabatic layer and the valley atmosphere. The oscillation period becomes determined by slope angle, ambient vertical temperature structure and a friction coefficient; the period is shorter as stability increases. Doran and Horst (1981) found temperature and windspeed fluctuations with a period of about 1.5 hr in a valley in the Geysers areas of northern California.

Surges of air are commonly observed on slopes greater than 10°. Küttner (1949) referred to these as 'air avalanches'. He noted a regular 5-minute interval between gusts throughout two successive nights at a point 100 m below the 2800 m Höllenthal summit in the Wetterstein Mountains.

The final steady velocity, indicated by the model of Fleagle, is that which compensates for the cooling rate, maintaining equilibrium between the pressure gradient and frictional forces. He notes that, since the time required to reach the final velocity on gentle slopes is so long, changes are likely to occur in the regional pressure field or in the diurnal regime, modifying the katabatic system and thereby removing the apparent paradox of v increasing without limit as $s \rightarrow 0$.

The Prandtl-Defant approach relating to the vertical structure of the flow, will not be detailed here since later work appears to invalidate their assumptions, at least for the katabatic case.

Manins and Sawford (1979a) have extended the hydraulic approach to include time dependence, downslope modification due to turbulent entrainment of the ambient air into the katabatic flow, surface stress, and radiational cooling of the katabatic flow. A key conclusion, which contradicts most previous theoretical studies, is that katabatic flows with strong stratification near the ground are dynamically isolated from the surface so that *surface drag is negligible*. The dominant retardation of the flow is by turbulent mixing across the interface with the ambient air. In Manins and Sawford's study, entrainment of air from the environment into the katabatic flow is parameterized by making the velocity of inflow proportional to the velocity scale (u) of the layer,

$$W_H = -EU$$

where W_H = velocity normal to the slope at height H,
the volume flux Uh = $\int$HO udn,
h = thickness
n = direction normal to the slope,
$E = A/S_1 Ri$,
where S_1 = 0.5 (a constant)
Ri = a layer Richardson number which is a function of the slope
($Ri = \Delta h \cos s/U^2$)

$S_1 Ri$ is an inverse internal Froude number of the flow and for katabatic flows is of order unity, $A \sim 2 \times 10^{-3}$ (a constant). E represents the change in potential energy of the katabatic flow by incorporation of ambient air. It is proportional to the turbulent kinetic energy produced by shear at the interface between the layer and the environment.

The model of Manins and Sawford shows that a katabatic flow grows in thickness linearly with distance downslope and increases in speed at a decreasing rate. The buoyancy of the flow decreases gradually through entrainment of ambient air. For a stably stratified environment with a buoyancy frequency $N = 10^{-2}\ s^{-1}$, the flow properties at 4, 8 and 12 km downslope from the crest are as follows:

	4 km	*8 km*	*12 km*
Speed ($m\,s^{-1}$)	3.0	3.3	3.0
Thickness (m)	45.0	95.0	186.0
Deficit of potential virtual temperature (°C)	1.6	0.71	0.19

As the cooled air moves downslope it meets denser ambient air and a greater proportion of the cooling is needed to maintain a buoyancy deficit in the layer. A balance between layer cooling and the flux of entrained cool ambient air is established. The initial cooling involves surface heat fluxes, but as thermal stratification develops radiative divergence in the layer probably dominates. In Veracruz, Mexico, reduced nocturnal cooling and opposing trade winds can delay katabatic onset by 4–8 hours (Fitzjarrald, 1984).

A field study by Manins and Sawford (1979b) in southern Australia confirms the inapplicability of previous one-dimensional models. Their observations show that surface friction effects are restricted to a 'skin' only a few metres thick. The katabatic flow reaches a maximum in a layer 40 m thick and there is a deep 'interface' to the ambient air above 160 m. The onset of the katabatic flow on a 4.5° slope about 4 a.m. was related to increasing thermal stability with a gradient of 130K km^{-1} in the lowest 40 m. The inversion extended to 120 m above the surface. Other studies on a simple slope (Rattlesnake Mountain, Washington) confirm that inversion depth increases with distance downslope and the elevation of the wind maximum rises as the flow strengthens and deepens (Horst and Doran 1986); however, the inversion strength did not diminish downslope.

By way of summary, the various theories of downslope flow can be categorized according to the terms of the momentum equation that are included (Mahrt 1982). The principal ones, following Mahrt are:

1 *advective-gravity flows* where the downslope advection of weaker momentum balances the buoyancy acceleration. The equation for a gravity flow on a snow surface developed by Businger and Rao (1965) states that

$$u = \left[g \frac{(\theta_0 - \theta)}{\theta_0} (\sin s)\, x \right]^{1/2}$$

i.e. the velocity increases as the square root of the distance down the slope (x) and as the square root of the temperature deficit ($\theta_0 - \theta$).

2 *equilibrium flows*, where buoyancy acceleration is balanced by turbulent-stress divergence. Defant (1949) adopts an eddy diffusivity and heat diffusion balanced by temperature advection. Here the velocity is linearly proportional to the temperature deficit; for constant temperature deficit and flow depth, the speed is constant. In Petkovsěk and Hočevar's work (1971) the flow is linearly proportional to buoyancy deficit and the slope angle.
3 *shooting flows* where downslope momentum advection and turbulent transport balance the buoyancy deficit. These flows have a Froude number > 1 and are commonly characterized by the occurrence of a hydraulic jump. Such flows were first noted in coastal Antarctica (see p. 157). The flow adjusts to the equilibrium case at a rate proportional to the stress divergence; *i.e.* more rapidly for greater drag. Drag can arise through entrainment of momentum at the upper boundary of the downslope flow (Manins and Sawford 1979a).
4 *Ekman-gravity flows* involve the additional effect of Corolis acceleration because of large slope length and long time scales. On large ice sheets the terrain shape also introduces a thermal wind term when the flow is deep.

With regard to upslope circulations, it is worth pointing out that slope winds as such may not develop if the lapse rate is unstable or even neutral. Model calculations by Orville (1964) for a 1000 m-high mountain with 45° slopes indicate that, in a neutral environment, convection bubbles move up and away from the slope. With a stable environment, however, upslope motion occurs with columnar convection over the mountain crest. Over the broad South Park Basin, Colorado, upslope flows develop in a shallow convective boundary layer on the lee (eastern) side of a heated mountain slope. The presence of a nocturnal inversion layer seems to be a requirement for their development (Banta 1986) and their duration is inversely related to the westerly winds above the ridgetops. By late morning/early afternoon the slope winds come to resemble the ridgetop winds through strong turbulent mixing in the convective boundary layer (Banta and Cotton 1981). During this transition, a local lee convergence zone may form at the upwind edge of a pool of cold air in the basin. This convergence generates cumulus cloud that later may grow into cumulonimbus (Banta 1984).

For the nocturnal downslope regime, several studies have shown the nature of the return flow from the valley or adjacent lowland. Observations between 515 and 830 m on the south-eastern slope of Mt Bandai, Japan (37°36′N, 140°04′E), together with special soundings, allowed Mano (1956) to develop a model of such flows. Figure 3.19 illustrates the circulations and the area with higher nocturnal temperatures on the slope, where air ascends over the inversion layer towards the mountain. This is termed the 'thermal belt' (see p. 217) and Mano notes that it moves with a regular see-saw motion with a period of about four hours. Similar downslope winds and return circulations over a shallow slope near Aryk-Balyk (53°N, 68°E), Kazakhstan, USSR are reported by Vorontsov (1958).

There, the depth of the 'cold air lake' at the foot of the slope is approximately 0.20–0.25 of the relative relief.

Mountain and valley winds

One of the first descriptions of mountain-valley wind systems was provided by Fournet (1840). He noted that, in the valleys of Savoie, there are daytime up-valley winds, especially in summer, whereas at night the wind is down-valley (a *mountain wind*), especially in winter. Theoretical explanations began with Julius von Hann (1879) and were greatly elaborated by Wagner (1938), Ekhart (1934, 1944) and Defant (1951), based on aerological studies in the Alps, and modified more recently by Buettner and Thyer (1966), McNider and Pielke (1984) and Vergeiner and Dreiseitl (1987).

Defant argued that upslope winds develop prior to the daytime up-valley wind and, at night, katabatic drainage currents feed the mountain wind. His familiar diagram of the successive stages is widely reproduced, although several other investigations have demonstrated that the various component winds develop almost concurrently. In an alpine valley near Davos, Urfer-Henneberger (1967) found that in over 90 per cent of cases the downslope breeze ends at sunrise ± 20 minutes and the mountain wind about 25 minutes after sunrise. The upslope breeze begins between sunrise and 40 minutes later and the valley wind

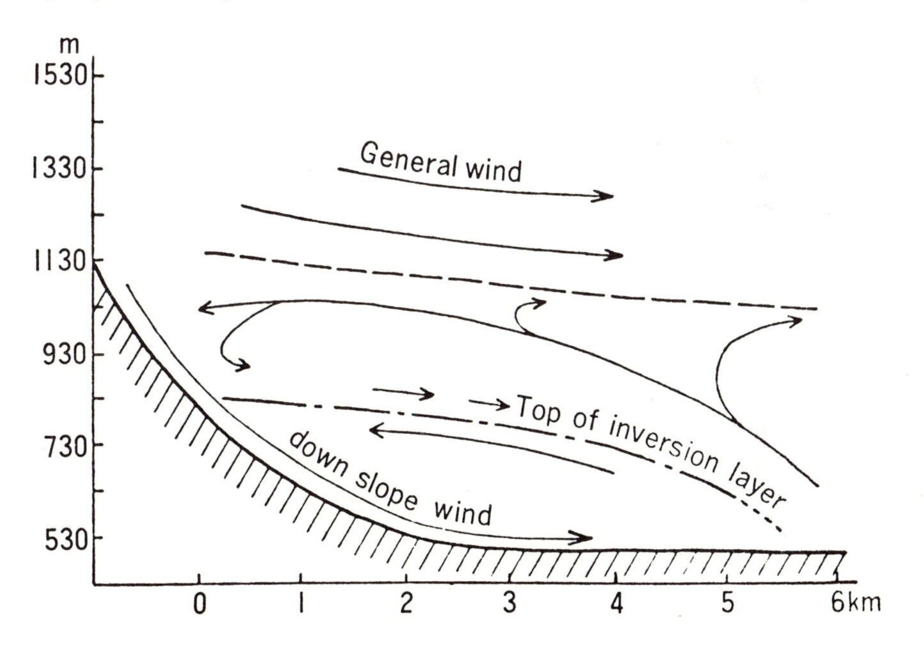

Figure 3.19 Model of nocturnal air circulation on the slope of Mt. Bandai, Japan. Below the top of the inversion layer is a cold air lake.
Source: After Mano 1956, from Yoshino 1975

about one hour after sunrise. Thus, the transitions occupy less than an hour in each case. In a later study (Urfer-Henneberger 1970), she stresses that the valley wind sets in synchronously along the length of the valley, although in the foothills of the Drakensberg, South Africa, the onset of the mountain wind occurs as a local front of cold air moving down valley (Tyson 1968b).

These diurnal regimes appear to be most regular in narrow mountain valleys. Work by F. Steinhauser in the Rauris Valley north of the Sonnblick (cited by Flohn 1969), shows the nocturnal mountain wind to have a frequency exceeding 70 per cent in all months except May, whereas the valley wind has similar high frequencies only from about 1000–1400 hours from May through September. However, even in broad valleys such as the Salzach, at Salzburg, Austria, up- and down-valley circulations are apparent (Ekhart 1953). The flows along the valley axis are often stronger than the slope winds, despite the gentler topographic gradient, in line with the theories discussed above. For example, in the Inn Valley, maximum speeds average 3–4 m s^{-1} in the mountain wind and occur about 400 m above the valley floor (Defant 1951).

In addition to these low-level diurnal systems, continuity requirements necessitate the occurrence of 'anti-winds' (compensation currents) above the ridge crests. These are also more difficult to detect due to the interaction with the general upper airflow.

The daytime up-valley wind is primarily a feature of fine summer weather. The rising air over the valley slopes is replaced mainly by longitudinal flow from the plains, supplemented in *large* valleys by cross-valley flow with subsidence along the valley axis, as reported by Wagner (1938) (see also Urfer-Henneberger 1970: 38). The primary controlling factor is the pressure gradient between the plains and valley, which extends up to about the mean ridge height (the 'effective ridge altitude', Wagner, 1938).

Steinacker (1984) and Vergeiner and Dreiseitl (1987) show that the valley geometry, specifically the area-height distribution in the valley as a percentage of the total area, determines the difference in diurnal temperature amplitude between the air in the valley and that over the adjacent plains. The valley area is calculated from the valley length × width at the ridge height. The Inn Valley near Innsbruck has a valley: plains volume ratio of 1:2.1, corresponding closely to the 2.2:1 ratio of diurnal amplitude of virtual mean temperature between Innsbruck and Munich. For individual valley cross-sections, the ratio of valley width at ridge height to area of the vertical cross-section is compared to corresponding measures for the plain (McKee and O'Neal 1989). For example, a valley with convex sides has a smaller ratio of air volume unit area than one of the same depth with concave sides. For the ideal forms in Figure 3.20 where width is twice the depth the so-called 'topographic amplification factor' is 1.27 for a concave U-shaped valley, 2.0 for the V-shaped valley, and 4.66 for the convex shape compared to unity over the plain.

The driving force behind the geometric effect of the valley arises from the volume of air being heated and the associated diurnal temperature range and

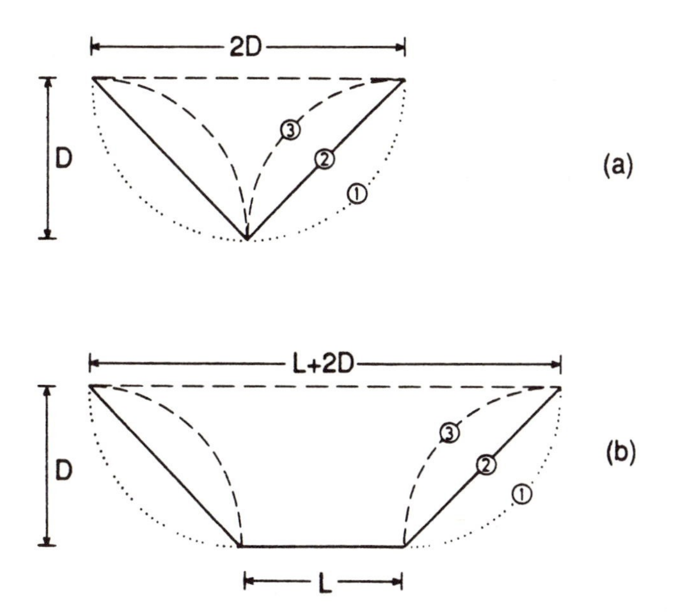

Figure 3.20 Depth – width and shape relationships in idealized valley cross-sections for a valley without and with a horizontal floor. See text.
Source: Müller and Whiteman 1988

pressure fluctuations in the air column between the plain (or valley floor) and ridge height. Vergeiner and Dreiseitl note that valley temperatures (as an average of 0700 and 1600, local time) are slightly higher than on the plains, but the diurnal range is 2–3 times greater in the valleys. Correspondingly, there is a plains–valley pressure difference in fair-weather afternoons of ≈ 5 mb, which disappears above the mean crest height. Interestingly, the valley wind is found to adjust rapidly to changes in pressure gradient attaining a local balance with frictional effects. Vergeiner and Dreiseitl also find that even on days with weak pressure gradients, less than half of them show a well-developed valley wind regime.

A second factor that interacts with valley geometry is the stability structure of the valley air (Steinacker 1984). If energy is input mainly to the narrow, lower parts of a valley, where air volumes are small, the temperature change will be large and conversely if the energy is confined to the larger volume in the upper parts the change is small. Less attention has so far been paid to this effect.

This geometric, topographic amplification effect, augmented when there is stratification associated with an inversion, gives a reaction time of about 30 minutes for the valley wind system to reach an equilibrium between pressure gradient and frictional forces (Nickus and Vergeiner 1984). The pressure differences, typically 1–2 mb/100 km (Whiteman 1990) have been studied along the Gudbrandsdalen, which transects the mountains of southern Norway, by Sterten and Knudsen (1961). Using pressure anomalies at 6-hour intervals from a

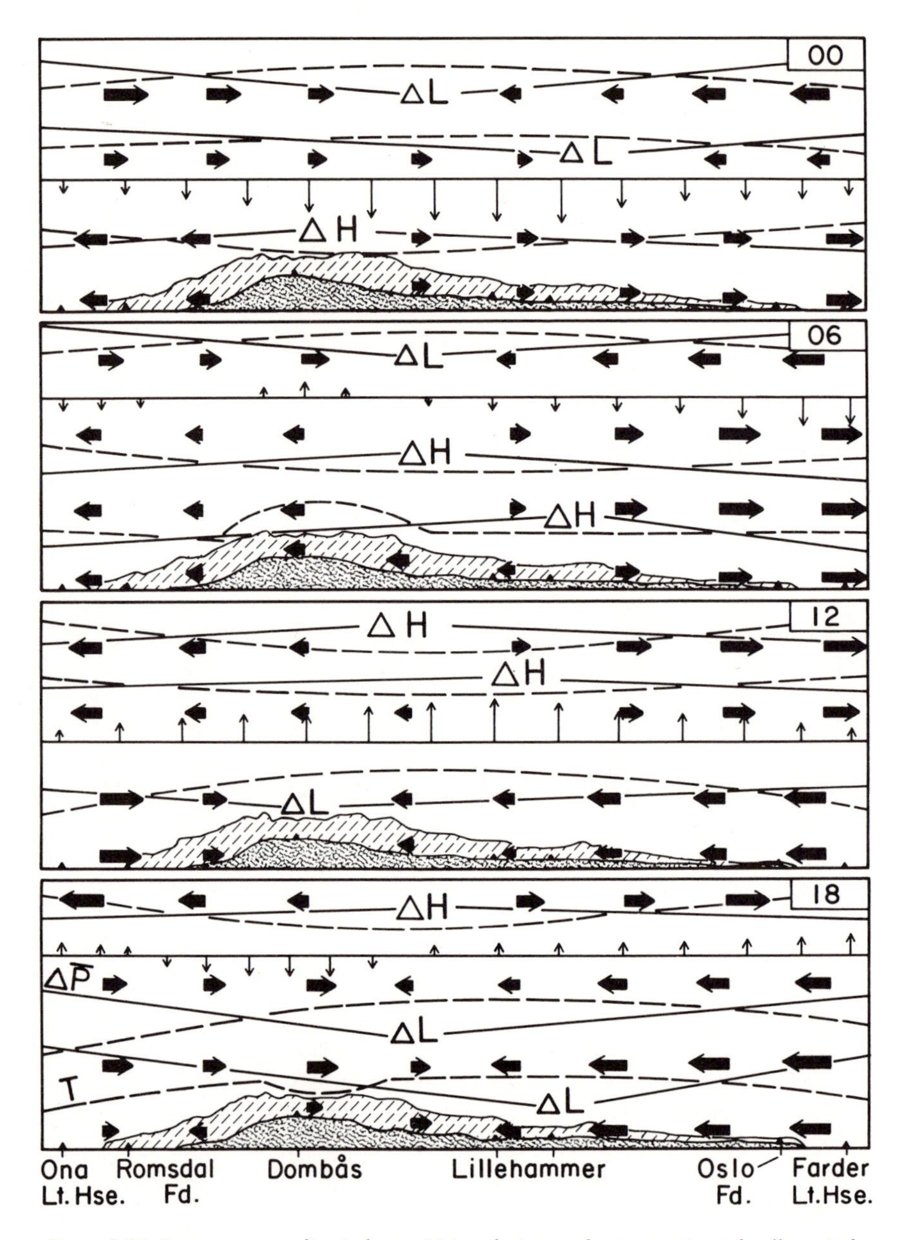

Figure 3.21 Pressure anomalies (schematic) in relation to the mountain and valley wind system at 6-hourly intervals along the Gudbrandsdalen, Norway. Farder (59°S) is at the mouth of Oslo Fjord, Ona (~ 63°N, 7°E) is on the west coast. The land rises to 650 m at Dombäs.

Source: After Sterten and Knudsen 1961

centred 24-hour mean, they show typical fluctuations of ± 1.5 mb on days with at least three stations recording mountain valley winds in summer, and ±0.7 mb in winter. The pressure anomalies and their tendencies are shown schematically in Figure 3.21.

Intensive field measurements, including balloon and aircraft observations, were carried out in the Carbon River Valley on the north-west side of Mt Rainier, Washington, by Buettner and Thyer (1966) during the summers of 1957–60. They interpreted the data by plotting time–height and longitudinal sections of wind structure at several valley stations. Figure 3.22a shows a longitudinal section in the upper Carbon River Valley for 8–10 August 1960 and Figure 3.22b a cross-section of the same valley on 9 July 1959. The maximum speeds are located at about a quarter to a third of the valley–ridge height difference above the valley floor, due to friction near the surface and the smaller amplitude of the temperature perturbation with increasing elevation. The anti-winds above have a similar vertical extent to the main system below.

Observations on four nights in September 1984 in the 25 km-long Brush Creek Valley, Colorado, with tethered balloons and Doppler lidar (Clements *et al.* 1989) indicate a nocturnal mountain wind maximum of 5–6 m s^{-1} at a height of 0.2 × valley depth (80–100 m above the floor). There is a surface-based inversion averaging 3.2°C/100 m in the lowest 200 m, above which the air is isothermal to above ridge-top height (Figure 3.23). The depth of the drainage flow and volume flux are inversely related to the strength of external geostrophic winds. For Brush Creek, an upvalley wind at ridge level > 8 m s^{-1} will totally override drainage flow (Barr and Orgill 1989). Other studies in the same valley show complex interactions between the main valley drainage and that of small tributaries. These may cause 8–16 minute wind oscillations and also changes during the course of the night due to changes in cooling rates of the tributary sidewalls and upper ridge slopes (Porch *et al.* 1989). The large *sky view factor* (solid angle of visible sky/2π) of upper slopes permits greater radiation and cooling than lower slopes, which also receive downward infra-red radiation from valley walls (Whiteman 1990).

The theoretical basis of the valley–mountain wind system is the circulation theorem of V. Bjerknes which relates to the thermally driven circulation set up by a horizontal temperature gradient. At its simplest, one can envisage two vertical air columns, one in the upper valley, the other on the plain. The former is heated more by radiation on the adjacent slopes and the air expands vertically, aided by the upslope winds. Consequently, airflow develops near the surface from the plain towards the valley head and an upper outward return flow forms. This type of mechanism was proposed by Wenger (1923) and evaluated for a simple case. A complex numerical model incorporating not only this effect, but also the slope winds, has been developed by Thyer (1966). The mathematical formulation involves the three-dimensional equation of motion and of heat transfer, the equation of continuity of mass, the equation of state and Poisson's thermodynamic equation. These were applied to a symmetrical V-shaped valley 400 m

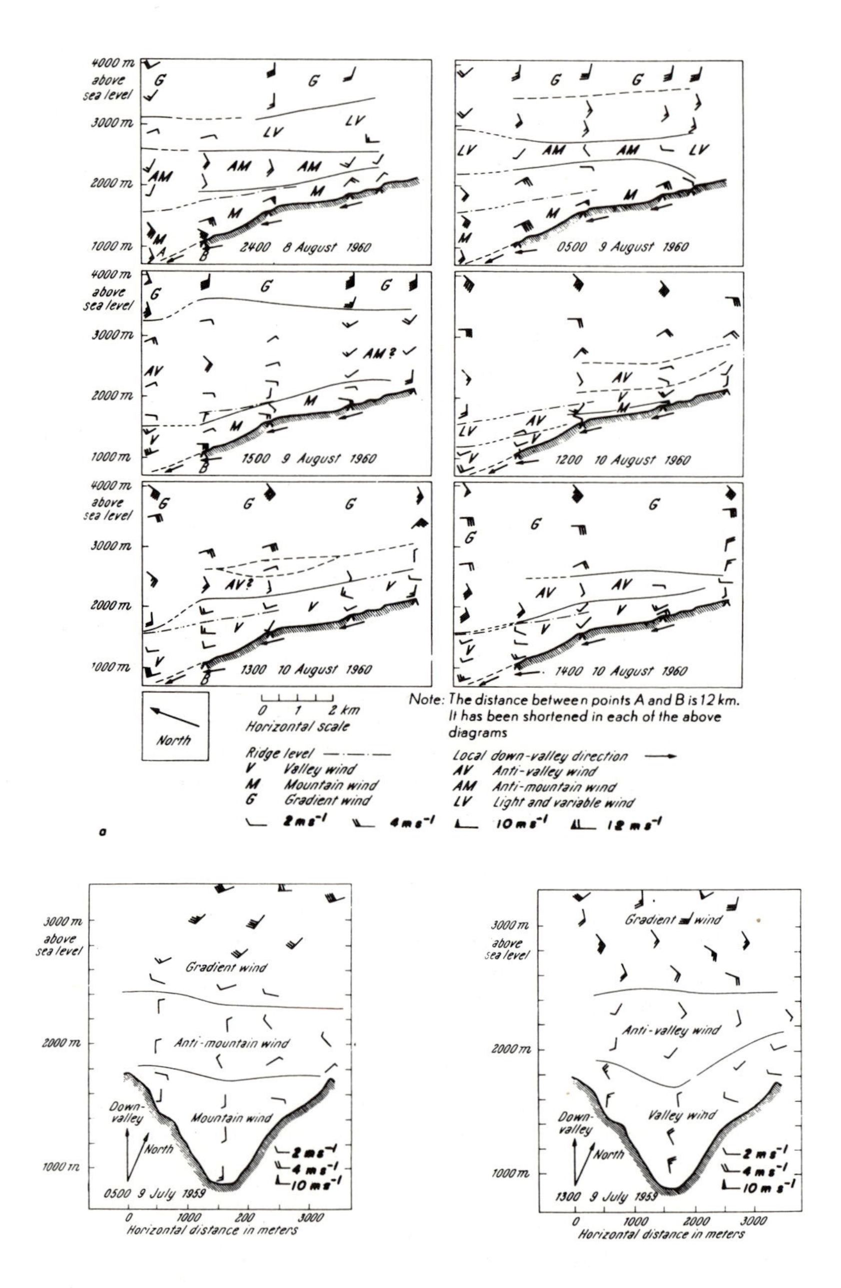

Figure 3.22 Mountain and valley winds in the vicinity of Mount Rainier, Washington
Notes: (*a*) Longitudinal section in the upper Carbon River Valley, 8–10 August 1960; (*b*) Cross-section in the same valley, 9 July 1959.
Source: From Buettner and Thyer 1966

long and 1000 m high with 45° slopes. Separate experiments successfully simulated both the shallow slope winds and the longitudinal circulations with their compensation currents. A reduction of slope angle to 26.5° and ridge height to 200 m reduced the wind speeds considerably by comparison with the standard run. The experiments represent only the earlier phases of these circulations, since the computer runs were short, although Thyer points out that a steady state does not occur in the valley wind system in nature. An earlier theoretical analysis by Gleeson (1953) suggests that the Coriolis effect is of some importance, at least in determining the vertical phase of the wind changes, although Thyer's model did not take this into account.

A three-dimensional numerical mesoscale model, forced by a surface energy budget, has been used to simulate mountain and slope winds for an idealized valley and an actual Colorado valley by McNider and Pielke (1984). Realistic downslope flows and cold air pooling were obtained. The model shows that the deep cooling of valley air is not directly due to downslope flow, but is determined by upward motion over the valley axis. The convergence and turbulent mixing of the downslope flows also redistribute radiatively cooled air in the valley bottom.

Valley orientation and therefore slope exposure may considerably modify these theoretical systems as demonstrated by Urfer-Henneberger's (1970) study of the Dischma Valley near Davos; this 15 km long, 4 km wide, 800–100 m deep valley runs SSE–NNW.

Temperature data collected on both slopes display the anticipated time asymmetry of slope heating during fine weather. As a result, the circulations are complicated by lags between the onset of the slope winds and by cross-valley

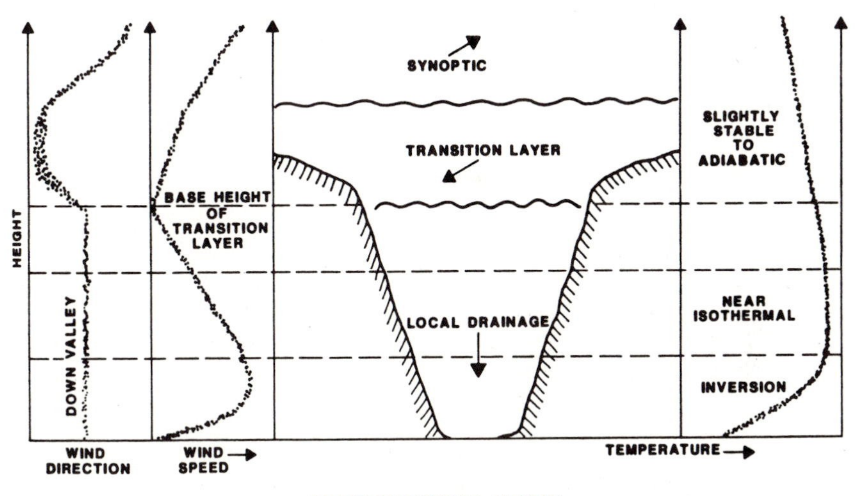

Figure 3.23 Schematic model of the structure of a nocturnal mountain (drainage) wind observed in western Colorado
Source: From Barr and Orgill 1989

components. The eight diurnal phases that she identified are shown schematically in Figure 3.24 and differ markedly from the original scheme of Defant (1951). Urfer-Henneberger also questions some of the mechanisms inferred by Defant (1949). In particular, air temperatures at the slope stations in the Dischma Valley do not show a nocturnal temperature inversion, although she acknowledges the possibility of stronger ground level cooling. Three-dimensional

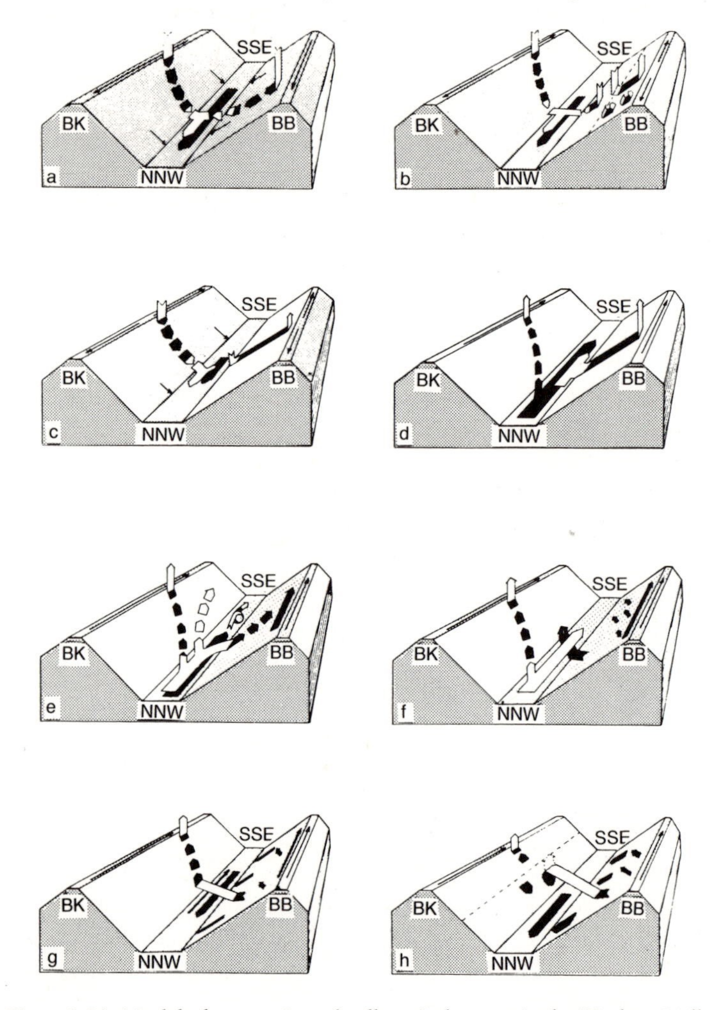

Figure 3.24 Model of mountain and valley wind system in the Dischma Valley, Switzerland

Notes: (*a*) Midnight to sunrise on the E-facing slope; (*b*) Sunrise on the upper E-facing slope; (*c*) Whole E-facing slope in sunlight; (*d*) Whole valley in sunlight; onset of valley wind; (*e*) W-facing slope receiving more solar radiation than E-facing slope; (*f*) Solar radiation only tangential to E-facing slope; (*g*) Sunset on E-facing slope and valley floor; (*h*) After sunset on the lower W-facing slope

Source: From Urfer-Henneberger 1970

observations in the Dischma Valley using ground stations, balloons and motor gliders during project DISKUS amplify the earlier investigations. Hennemuth and Schmidt (1985) show that deviations of the slope winds from the downslope gradients are produced by the superimposed valley/mountain winds and the cross-valley components induced by differential slope heating.

The concept of accounting for the geometry of a drainage basin to explain mountain/valley winds, developed by Steinacker (1984, 1987) and Vergeiner and Dreiseitl (1987) for the Inn Valley, has been evaluated by a programme of field measurements. During spring 1982, the Mesoscale Experiment in the Region Kufstein-Rosenheim (MERKUR) took place in a 37 km long, 8 km wide section of the Inn Valley that contains major side valleys and terraces. Freytag (1987) estimates the heating rate for the total valley air volume to that of the adjacent foreland is about 2.3:1. Mean mass fluxes up to 1000 m above the surface of the valley have been calculated for the 2.2-day observational period (Table 3.3). The mean mass budget equation can be expressed:

$$\Delta M_x + M_s + M_y + M_r = 0$$

where ΔM_x = horizontal mass flux difference
M_s = flux into/out of the slope layer
M_y = flux in/out of the side valley
M_r = residual ascent/subsidence.

Table 3.3 can be used to distinguish four major phases. Following Freytag (1987), these are as follows:

1 upslope winds and continuation of the mountain wind (0500–1100 GMT). Upslope transport (M_s) accounts for most of the residual flux associated with air subsiding at about 2 cm s^{-1} at ridge height. There is a positive horizontal mass flux down valley (ΔM_x) as a result of mountain winds converging from the side valleys ($-M_y$), implying that there is a 'local' compensation of the upslope flow within any particular valley cross-section;
2 upslope winds and valley wind (1100–1500 GMT). The data show that the valley wind increases up valley ($\Delta M_x > 0$). This results from greater heating of air in the side valleys, leading to lateral components of valley wind. These must be compensated by subsidence into the foreland end of the main valley ($-M_r$). The main valley narrows upvalley and the side valleys are smaller and less numerous. Hence, the merging upslope and valley winds here produce a mass accumulation and rising air. This phase comprises three superimposed scales of circulation: slope winds with local compensation by subsidence; valley wind components entering the side valleys with subsidence over the main valley; and the longitudinal valley wind with compensating descent over the adjacent foreland;
3 continuing valley wind, with downslope winds (1500–2100 GMT). Downslope winds ($-M_s$) now contribute to the valley wind. Less subsidence is required to offset the flow into the side valleys;

4 downslope winds and mountain wind (0100–0500 GMT). There is an increasing mountain wind along the Inn Valley ($\Delta M_x > 0$) due to the combined side valley and downslope flows. There may be weak ascent in the centre of the valley when the mountain winds are weak. This is the only phase when there is a single overall direction of circulation.

A related field and modelling study in the Dischma Valley provides an energy balance for the valley atmosphere. Hennemuth and Köhler (1984) show that observed mid-day heating rates of 0.9 K hr^{-1} agree well with the rate calculated from the convergence of sensible heat flux from the surface. This energy is advected upward during the afternoon and appears to be transported via subsidence into the main Davos valley which the secondary Dischma Valley joins (Hennemuth 1985). However, associated studies of the specific humidity suggest that only a small fraction of the daytime surface evaporation leaves the valley by vertical transport. Rather, upslope winds export moisture from the Dischma Valley in the morning while the valley winds import moisture in the afternoon (Hennemuth and Neureither 1986). These findings emphasize the importance of scale transfers from side valleys into main valleys.

The valley geometry concept of Steinacker (1984, 1987) assumes that

Table 3.3 Mean mass budgets[a] of mountain and valley winds during MERKUR

Time (GMT) 25–26 March 1982	05.00–11.00	11.00–15.00	15.00–21.00	21.00–01.00	01.00–05.00
Phase		Heating			Cooling
Wind	Mountain		Valley		Mountain
Horizontal mass flux difference, ΔM_x	7.9	9.5	11.3	12.4	10.1
Flux into the slope layer, M_s	6.3	5.8	−8.5	−7.0	−4.4
Flux into side valleys, M_y	−5.0	7.6	8.6	0.3	−7.3
Residual ascent/ subsidence, M_r	−9.2	−22.9	−11.4	−5.7	+1.6
Mean vertical wind, 0–1 km	−1.8	−4.0	−0.8	−1.0	0.1

Notes: [a]Mass budget of the valley segment is $\Delta M_x + M_s + M_y + M_r = 0$.
Main budget components for 05.00–11.00 GMT: $M_s \simeq -M_r$; $\Delta M_x \simeq -M_y$; for 11.00–15.00 GMT: $M_s + M_y + \Delta M_x \simeq -M_r$; for 15.00–21.00 GMT: $-M_s \simeq \Delta M_x$; $M_y \simeq -M_r$; for 01.00–05.00 GMT: $\Delta M_x \simeq -(M_y + M_s)$. Fluxes: 10^6 kg s^{-1}; vertical wind: cm s^{-1}.
Source: From Freytag 1987

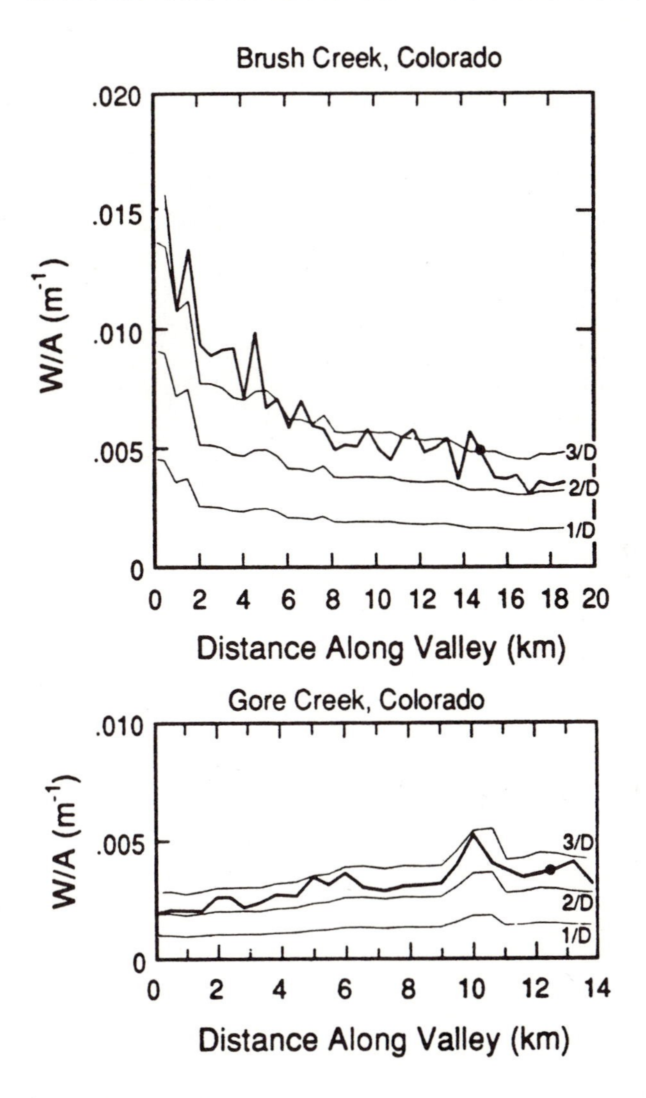

Figure 3.25 Plots of valley width/area (W/A) ratios (m^{-1}) along Brush Creek, Colorado (draining) and Gore Creek, Colorado (pooling)
Source: From Whiteman 1990, after McKee and O'Neal 1989

mountain valleys have immediately adjacent lowlands. In much of the interior of the western United States, however, many valleys are cut into plateaux remote from plains. McKee and O'Neal (1989) extend the topographic analysis by considering variations in valley cross-section geometry along its length. They compare plots of valley width/area against distance downvalley for Brush Creek and Gore Greek, Colorado (Figure 3.25) and show that in the former, a 'draining' valley for nocturnal airflow, the width/area ratio decreases down-

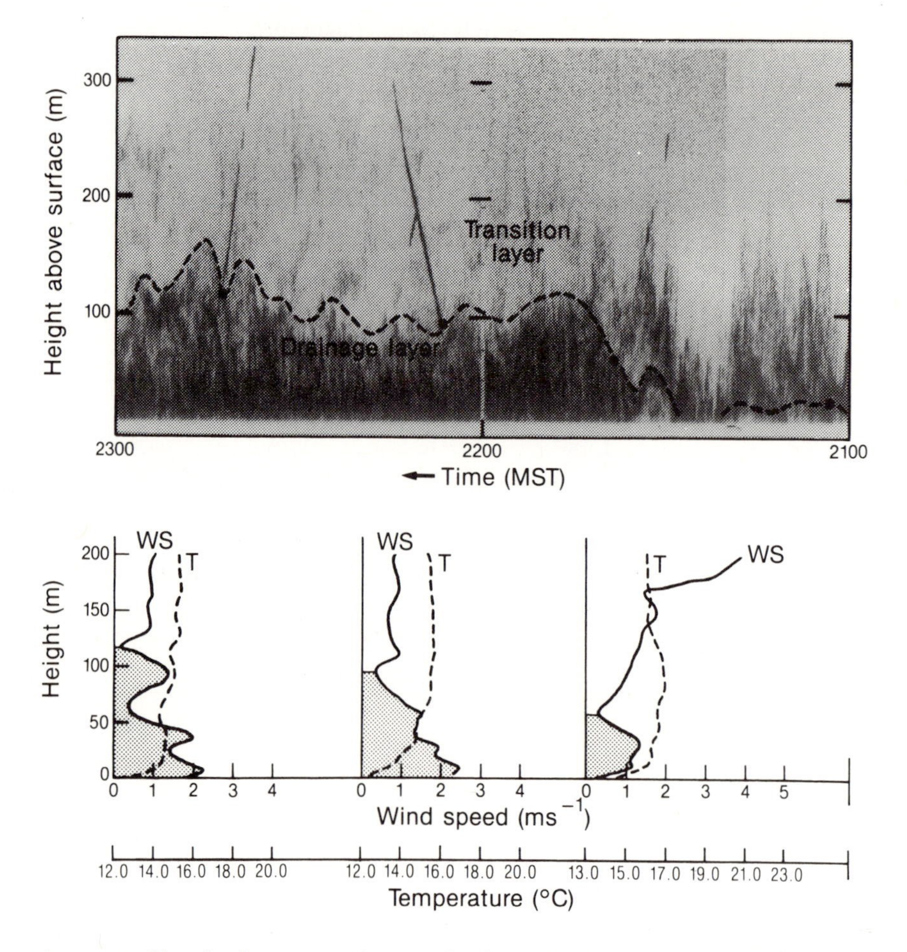

Plate 10 Profiles of sodar returns (above) and tethersonde observations (below) during drainage initiation in Willy's Gulch, western Colorado, 16 September 1986 (courtesy of Dr W.D. Neff and C.W. King, NOAA Environmental Research Laboratories, and the American Meteorological Society. From Neff and King, 1989). The arrows denote the transit direction of the tethersonde balloon; the point at approximately 2140MST marks the initiation of northerly drainage flow, with a decrease of southerly ridge top winds from about 7 m s^{-1} to 3–4 m s^{-1}. The drainage layer thickens to 100 m or more (dashed line). The wind speed (WS) and temperature (T) profiles in the lower part correspond to the three soundings shown on the sodar record; the drainage layer is stippled. Note that time runs from right to left.

valley. Theoretical nocturnal cooling rates decrease from 1.0–1.5°C hr^{-1} at the head of Brush Creek to 0.4°C hr^{-1} 18 km downvalley. In contrast, Gore Creek is a 'pooling' valley because the cooling rates increase in response to an increasing width/area ratio. Tethersonde profiles of temperature and wind speed and sodar echoes of boundary-layer structure in Brush Creek (see Plate 10) and its eventual outlet into the Colorado River east of Grand Junction, Colorado, illustrate the pooling effect of basins (Neff and King 1989; see 2, p. 193). Nocturnal air drainage from Brush Creek, and others joining the southeastward flowing Roan Creek, is blocked 50 km to the south by a 300 m escarpment. The De Beque basin fills with drainage flows over several hours, despite air exiting through De Beque Canyon on the Colorado River. The depth of this cold pool (in excess of 300 m) can modify the tributary drainage in Brush Creek and causes drainage flows from tributary valleys to enter the basin as elevated 'jets'. Geometric characteristics of basins described by Petkovsěk (1978, 1980) can be related to the accumulation and outflow of such cold air pools, but more work on this is needed.

The effect of cross-valley wind components due to differential slope heating has been analysed theoretically by Gleeson (1951). He demonstrates that inertia effects, which arise through the cumulative role of solar radiation differences between slopes, have considerable importance for winds dynamics. Such effects are ignored, for example, in Fleagle's equilibrium approach to slope drainage. Changes in friction and in the Coriolis parameter are also shown to be important. As slope angles increase, the maximum speed of the cross-valley winds increases due to the enhanced contrast in slope heating. Likewise, cross-valley winds tend to be stronger in summer than winter. Changes in slope angle and valley orientation also affect the phase of maximum cross-valley winds speeds. For example, if the angle of west-facing slopes is increased, the sun rise occurs later on the slope, giving a maximum speed later in the morning. If the valley orientation is shifted from north/south to north-east/south-west, the wind speed maxima in both morning and afternoon occur later, as illustrated by Urfer-Henneberger's results.

There have been some studies of flows in mountain basins. In central Colorado, Banta and Cotton (1981) identify three wind regimes in a broad mountain basin. There is nocturnal drainage downslope, a shallow upslope flow in the daytime convective boundary layer, and transitional flows related to strong turbulent mixing, in the late morning or afternoon; the transitional flows resemble the general westerly flows above the ridgetops.

Several of the cited studies report periodic fluctuations in the mountain wind. Buettner and Thyer (1966) found a 25-minute periodicity in the Carbon River Valley with speed fluctuations of 1.5 to 6.5 m s^{-1}. In mountain valleys of 300–350 m depth in the Drakensberg foothills near Pietermaritzburg, South Africa, Tyson (1968a) reports surges in the mountain wind at about 90–150 m above ground level varying between 45 and 75 minutes with the maximum speed ranging between 1.6 and 3.8 m s^{-1}. He also shows that the profile can be approximated by the Prandtl model where maximum speeds occur at 0.75 of the depth of the downvalley air. In another study in Bushmans valley in the Drakensberg, however, periodicities ranged between 2 hours and 4 hours (Tyson 1968b).

Two special cases of mountain wind systems merit note. One is the persistent downvalley, *glacier wind*, which occurs in alpine valleys where there is a glacier or snowfield. It is generally strongest and deepest (50–300 m) in early afternoon when the temperature difference between the cold surface and the air is at a maximum (Tollner 1931; Hoinkes 1954; see p. 165). The glacier wind is not restricted to fine weather conditions although by day it does not reach far down-valley and may be overlain by a valley wind. It is commonly gusty and turbulent, like nocturnal slope winds and the mountain wind. Hoinkes reported double maxima (before sunrise and before sunset) and minima (before noon and before midnight) in the Alps. Streten and Wendler (1968) found the same pattern in central and southern Alaska, but in northern Alaska there is a single nocturnal maximum and late afternoon minimum (Streten *et al.* 1974). They suggest that this is due to the strong 'nocturnal' inversion and the rather weak temperature gradient between the ice and its surroundings during the day.

The second special regime is known as the *Maloja wind*, from its type location between the Engadine and Bergell valleys, Switzerland (Defant 1951). This is a 'mountain wind' blowing from the south-west down the upper Engadine valley during the day. The valley wind in the Bergell valley to the south-west climbs 300 m over the Maloja Pass between the two valleys to flow northeastward along the Engadine towards St Moritz. Frequently associated with the Maloja wind, especially on summer afternoons, is an unusual 'snake'-like tube of stratocumulus cloud 100–500 m wide and 30–300 m above the surface. Sometimes it fills most of the pass (Holtmeier 1966; Gross 1984). It occurs during dry anticyclonic weather with a northeasterly gradient wind, but the phenomenon requires a moist low-level counter-current from the south-west – the valley wind in the Bergell valley. Cloud forms as this flow rises 300 m to the pass and extends northeastward into the Engadine (Gross 1984) (c.f. Plate 11). Valley channelling with anti-winds is also reported by Wipperman (1984) in the middle Rhine in Germany.

Despite the numerous studies of mountain-valley circulations, several aspects of these systems remain to be clarified. These include the characteristics of the atmospheric structure at the times of wind reversal and the role of cross-valley gradient winds. These questions are considered next.

New instrument systems can now be used to investigate the atmospheric structure in valleys. Whiteman and McKee (1977), for example, used a tethered balloon profiler in the 600 m deep Gore River Valley, western Colorado, to examine the development and decay of the nocturnal inversion in December. In their case study, a 100 m deep ground-based inversion developed by 1600 hours decoupling the surface air flow from the continuing up-valley winds. By 21.45 the inversion (of 3°C) had deepened to 225 m and winds became down valley throughout the 400 m sounding. The inversion had intensified to 11°C over 444 m by 08.30 next morning with continuing light mountain winds, but strong up-valley winds developed shortly afterwards just above the inversion. Subsequently, the inversion level descended at 120 m hr^{-1} with both adiabatic and sensible heating occurring above the inversion. Shallow up valley winds also

Plate 11 A tubular stratocumulus cloud observed in early September (late morning) in a pass near Kremmling, Colorado. The motion is from the north-west (left) where the valley drops steeply about 200 m. The topographical setting and cloud forms are analogous to the well-known Maloja-Schlange (Maloja Snake) in Maloja Pass, Switzerland described in the text (courtesy of Dr G. Kiladis, CIRES, University of Colorado).

developed at the surface in a superadiabatic layer caused by radiative heating of the ground. The authors suggest that solar heating of the slopes sets up convective plumes which penetrate the stable layer of cold air. This entrains cold air from the base of the stable layer, removing it upslope, thereby causing the inversion top to descend through mass continuity, as illustrated in Figure 3.26. They point out that the wind data show that the cold stable air did not drain down the valley and that the stable structure above the inversion rules out erosion by turbulent mixing from above.

The idealized model developed from observations in western Colorado by Whiteman (1982) has been successfully reproduced in a numerical model by Bader and McKee (1983). However, studies in the Inn Valley, Austria, indicate that erosion of the stable core can be more complex. Brehm and Freytag (1982) attribute the erosion to upvalley flows developing above and below the decaying mountain wind. Warming of the valley atmosphere (calculated for two days in October 1978) occurred through the combined effects of subsidence (45 per cent) basal convection (40 per cent), and horizontal advection and warming of the slope wind layer (15 per cent).

The role of cross-valley winds on the circulation within a valley has been examined both theoretically (Tang 1976; Wippermann 1984) and observationally (MacHattie 1968; Yoshino 1957). MacHattie, for example, found that winds in the main Kananaskis Valley, Alberta, are more subject to modification by the synoptic-scale gradient winds than is the case in adjoining sub-valleys. Also, daytime valley winds are more affected in this way than the nocturnal system due to the more effective decoupling of the circulations at night. Even a broad and shallow valley, such as the upper Rhine in Germany, causes winds crossing it to be channelled so as to flow almost along the valley direction. Such channelling occurs for winds crossing at any angle (Wippermann 1984). Moreover, a counter-current often develops; in the south-north Rhine Valley, geostrophic winds from 115° to 170° give rise to northerly valley winds while geostrophic winds from 295° to 350° produce southerly valley winds. These counter-currents exist, according to Wippermann, when the topographically-influenced part of the v component of wind (V_t) is contrary to and greater than the undisturbed part (V_r). The pressure gradient along the valley has the direction of V_r, but since it cannot be balanced by the Coriolis acceleration in the valley, V_t has the opposite direction. Thus, for instance, with a westerly geostrophic wind and lower pressure to the north, a southerly flow develops in the Rhine Valley. However, with a south-southeasterly geostrophic wind and lower pressure to the south-west, a northerly valley flow is observed.

A recent analytical treatment by Tang (1976) specifically considers the interaction between a prevailing cross-valley wind and the slope winds; Coriolis terms are neglected. Calculations for a daytime case show that a separated circulation cell develops over the lee slope due to friction, reinforced by upslope winds associated with the typical differential heating on the slope. Downward motion occurs over the valley centre (cf. Figure 28c, p. 73) and high above the lee slope due to a standing wave tilted upstream from the windward slope. Conversely, at night a separated cell is formed above the windward slope. Tang provides observational evidence from Vermont and near Innsbruck, Austria, showing similar features. Observations by Yoshino (1957) in a small V-shaped valley in Japan also show lee slope eddies in cross-valley wind conditions. He found periodic development of the eddies, associated with upvalley airflow, occurring over an interval of about eight minutes.

Regional-scale interactions

The final element in thermally induced topographic wind systems is the nature of large-scale interactions. These involve both regional circulations set up by extensive mountainous areas in relation to surrounding lowlands, as well as synoptic-scale wind systems. Model studies for a circular plateau with radial valleys show that regional circulations are driven by the heating distribution and its diurnal variations (Egger 1987). Valley winds make a significant contribution to the plateau circulation. Over the Alps, Burger and Ekhart (1937) identified a regional compensation flow moving radially away from the mountains by day

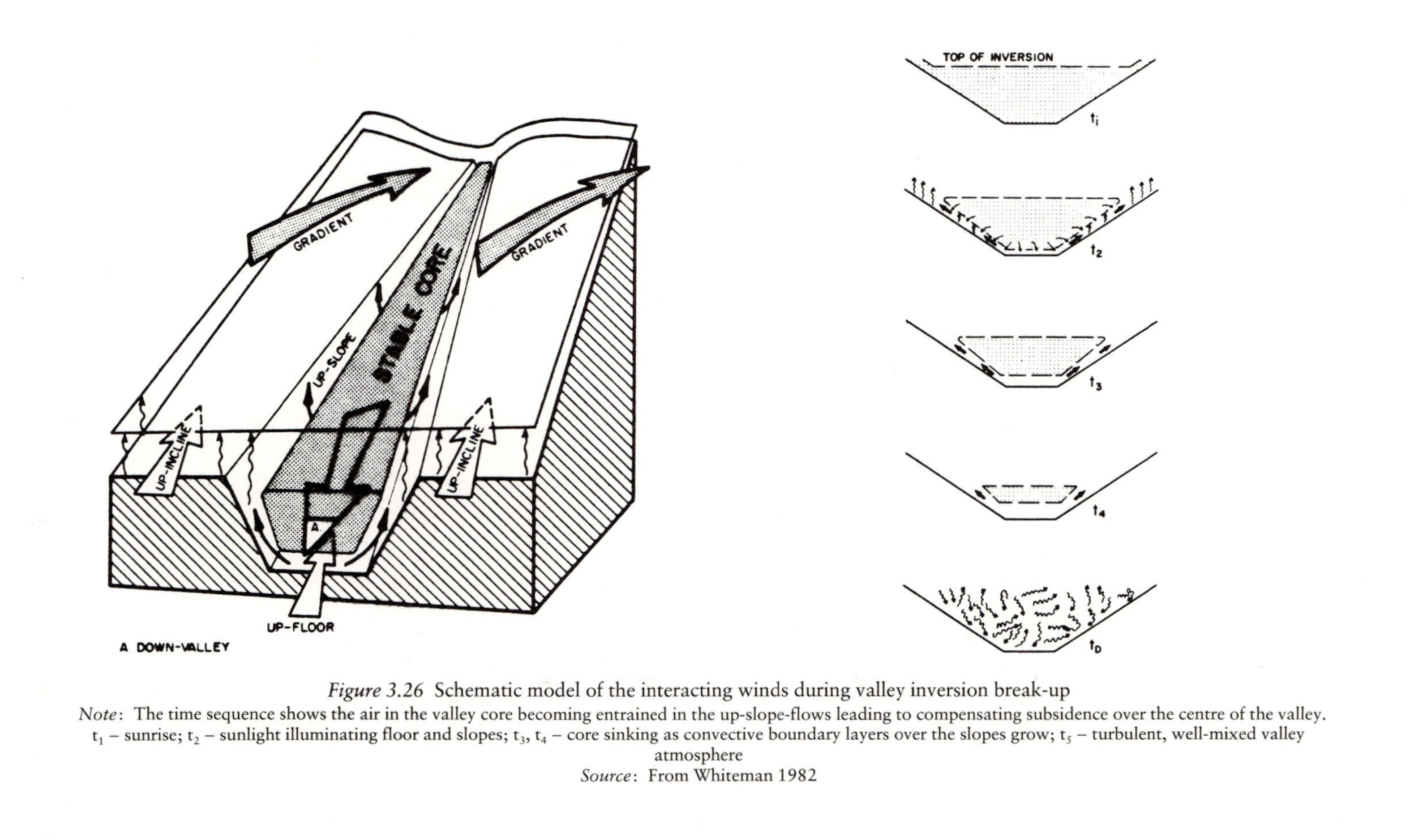

Figure 3.26 Schematic model of the interacting winds during valley inversion break-up

Note: The time sequence shows the air in the valley core becoming entrained in the up-slope-flows leading to compensating subsidence over the centre of the valley. t_1 – sunrise; t_2 – sunlight illuminating floor and slopes; t_3, t_4 – core sinking as convective boundary layers over the slopes grow; t_5 – turbulent, well-mixed valley atmosphere

Source: From Whiteman 1982

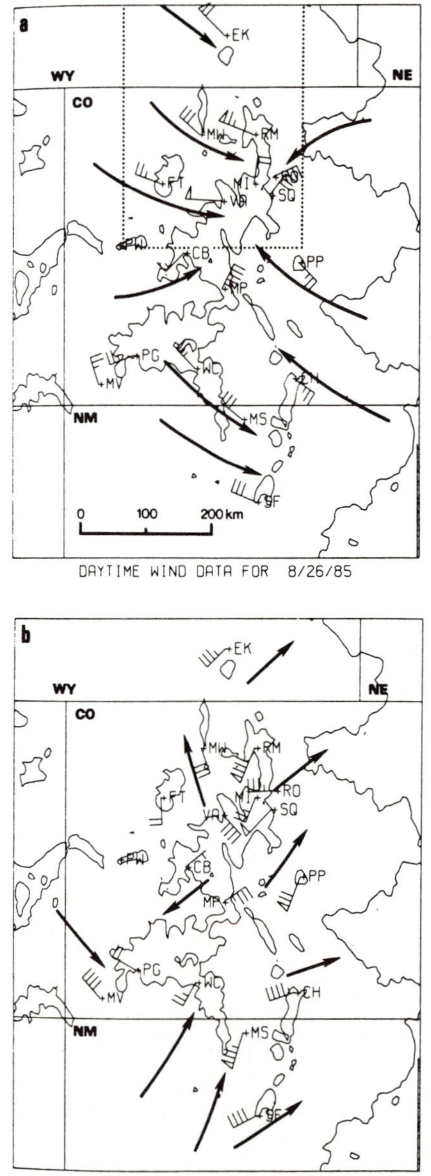

Figure 3.27 Regional-scale diurnal wind reversals over the Rocky Mountains, Colorado, based on mountain-top observations of average resultant wind for (*a*) 12.00 to 15.00 MST for 26 August 1985 and (*b*) 00.01 to 03.00 MST for 27 August 1985

Notes: Each barb represents 1m s^{-1} and the triangle 5m s^{-1}. The 1500 and 3000m height contours are shown

Source: From Bossert *et al*. 1989

and subsiding over the plains. Regional circulations were measured over the Inn Valley during the MERKUR project (see p. 175). Freytag (1987) reports nocturnal mountain winds reaching 30 km into the foreland and estimates that they force a general ascent of about 6 cm s^{-1}. For a 1500 m-deep layer over a drainage area 40 km in width, the anti-mountain wind would have a mean velocity of 40 cm s^{-1}. Yet as Urfer-Henneberger (1970) points out, the level of this flow (c. 4000 m or more) is such that it is invariably a component of the gradient winds.

Similar regional-scale diurnal flows have been identified from summit wind observations in the Colorado Rocky Mountains, using a network of mountain-top sites during four summers (Bossert *et al.* 1989). Two regional-scale diurnal wind regimes are identified (Figure 3.27). During weak synoptic situations on clear summer days with strong solar radiation, the radiatively-forced wind regime exhibits a simple daytime inflow and nocturnal outflow pattern. There are slow transitions lasting 6–7 hours in the evening and 4–5 hours in the morning. The

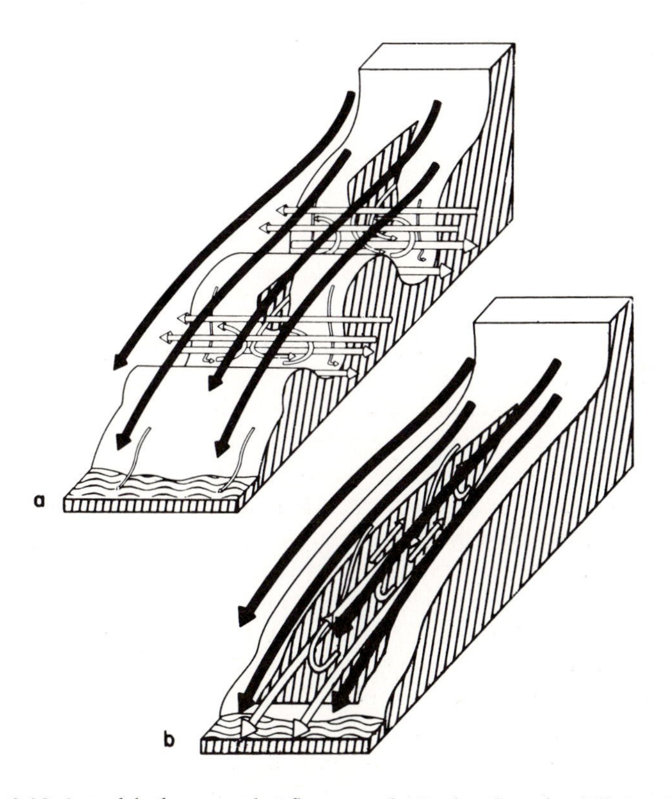

Figure 3.28 A model of nocturnal airflow over the Drakensberg foothills in winter. Daytime patterns are reversed.

Notes: (*a*) Valleys at right angles to mountains and coastline; (*b*) Valleys parallel to the slope

Source: From Tyson and Preston-Whyte, 1972

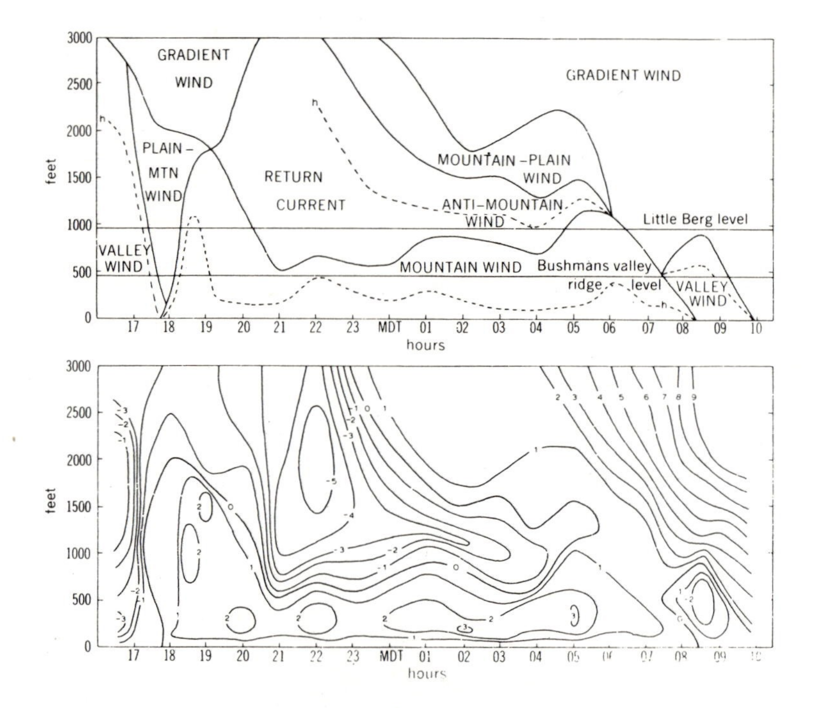

Figure 3.29 Time-section of local winds (m s^{-1}) in and above Bushmans Valley, Drakensberg Mountains, 12–13 March 1965
Source: From Tyson 1968b

nocturnal outflow occurs in a shallow layer at mountaintop level above a deep stable valley air layer caused by radiative cooling. In contrast, during periods with convective thunderstorm activity and latent-heat forcing, the gradual morning transition from outflow to inflow is disrupted in mid-afternoon by a shift from inflow to outflow, and by a further reversal in late afternoon – early evening. Data over four years indicates that these shifts occur on 27 per cent of days. Over lower ranges or single mountains, the return flows occur much lower and can be considered as strictly anti-valley or anti-mountain winds.

This problem has been highlighted by the investigations of Tyson (1968b; Tyson and Preston-Whyte 1972) on the south-east side of the Drakensberg escarpment, South Africa, spanning a 180 km transect to the Indian Ocean at 30°S. The escarpment edge is at 3000 m, falling sharply to a sloping plateau at 950 m which is dissected by valleys 250–550 m deep. While the major valleys trend NW–SE, others are at right angles to the general southeastward slope of the plateau. Tyson and Preston-Whyte propose schematic models of the nocturnal wind systems in relation to this topography (Figure 3.28). The mountain–valley winds are overlain by a regional air movement away from the Lesotho Massif and the Drakensberg at night (the 'mountain–plain' wind) and its

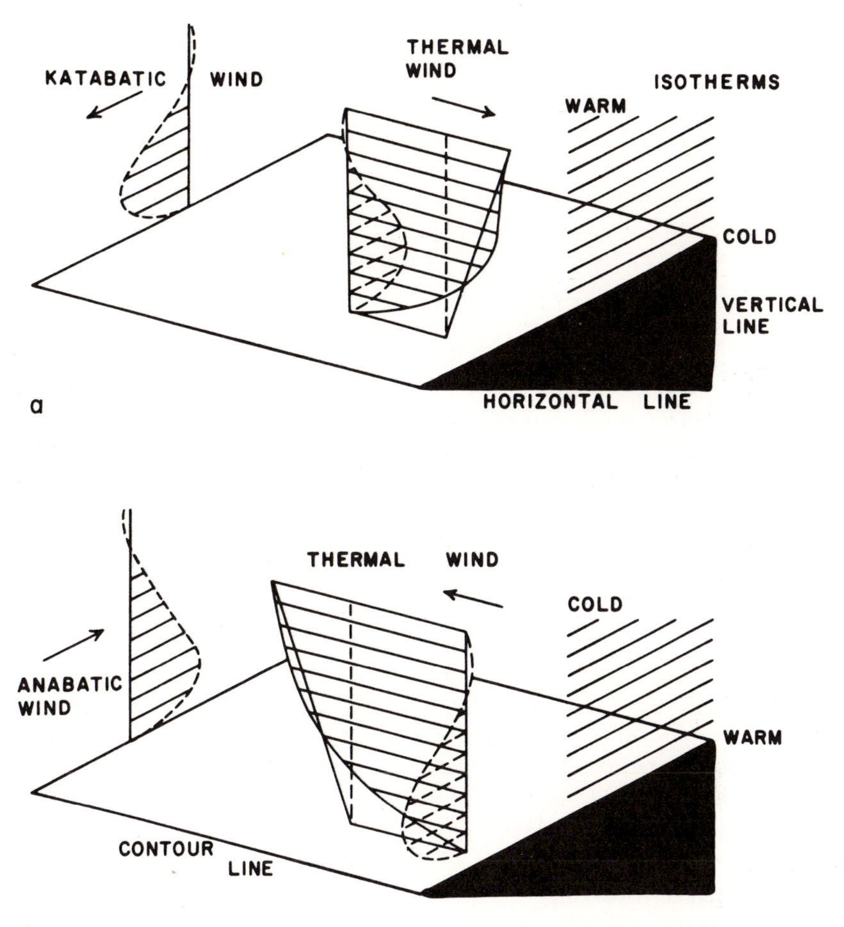

Figure 3.30 Thermal winds developed in the steady-state boundary layer over sloping terrain in the northern hemisphere. The nocturnal and daytime phases of temperature stratification and associated geostrophic wind shear are shown. The assumed large-scale geostrophic flow is parallel to the terrain contours, so that the frictionally induced cross-isobaric components of motion counteract the katabatic flow in case (a) and the anabatic flow in case (b). The reverse cases of augmentation are readily inferred.

Notes: (*a*) Nocturnal phase; (*b*) Daytime phase

Source: From Lettau 1967

daytime converse. A specific case is illustrated for the Bushmans Valley (Figure 3.29). The valley– and plain–mountain winds are best developed in summer when the combined mountainward flow may be 1000–1250 m deep. They are seldom related to the general pressure gradients whereas the northwesterly mountain–plain winds may be an integral part of the large-scale circulation.

In the context of extensive plateau slopes it is appropriate to mention the thermo-tidal wind theory of Lettau (1978). He shows that large-scale mountain slopes can produce a thermal wind parallel to the terrain contours due to heating effects. For example, if high ground is relatively warm compared to its surroundings (by day/summer), an anticyclonic circulation develops provided the normal lapse situation prevails. Conversely, over cold highlands, given inversion conditions, a cyclonic circulation will occur. The former is well illustrated over the Tibetan and Bolivian highlands (see p. 57). In addition, the diurnal heating cycle generates a harmonic forcing function in the boundary layer, giving rise to a 'thermo-tidal' wind. Schematic patterns of steady-state solutions are indicated in Figure 3.30. Lettau discusses the apparent contribution of such systems to the nocturnal low-level jet east of the Rocky Mountains and to the shore-parallel winds over the western foothills of the Andes in Peru. The nocturnal boundary layer jet east of the Rocky Mountains and in similar locations elsewhere has been identified in both observational (Paegle 1984) and modelling studies (McNider and Pielke 1981; Garratt 1985). The nocturnal stratification enhances the momentum coupling according to Mason and Sykes (1978), helping to establish a supergeostrophic wind maximum in the lowest kilometre or so. Paegle suggests that the time-averaged jet can be accounted for by western boundary dynamics related to the meridional variation of the Coriolis parameter, while its diurnal velocity fluctuations involve buoyancy oscillations above a sloping surface as proposed by Lettau (1967) and Holton (1967). Lettau's theory shows that the low-level wind speed maximum occurs at different times of day according to latitude. This results from the appearance in the equation for boundary-layer tidal flow of the term $f/(f^2 - \omega^2)$ where $f =$ Coriolis parameter ($= 2\omega \sin \phi$), $\phi =$ latitude angle, $\omega =$ frequency of the tidal force (which here corresponds to earth's angular velocity), and $(f^2 - \omega^2) = \omega^2 (4 \sin^2 \phi - 1)$. The critical latitude is where $f = \pm \omega$, which is latitude 30°. Lettau (1967) shows that the signs of the two parameters involved are:

	30°–90°N	*0°–30°N*	*0°–30°S*	*30°–90°S*
f	+	+	–	–
$f^2 - \omega^2$	+	–	–	+

As a result of the implied phase differences, the low-level jet occurs about midnight in middle and high latitudes, whereas equatorward of latitude 30° it occurs about midday. This accords with the observed nocturnal maximum east of the Rocky Mountains and the daytime one over the Peruvian Andean foothills. Further general support for this theory is presented by Lettau (1978) for South America, although Tyson (1968b) notes an absence of such winds parallel to the general terrain contours in Natal.

In Namib, south-west Africa (23°S), a large-scale northwesterly plains–mountain wind exceeding 1500 m depth dominates the summer circulation both day and night, in contrast to the diurnally-oscillating regime in Natal (Lindsay and Tyson 1990). In winter, however, there is a nocturnal southeasterly mountain–plains wind.

MODELS OF THE OROGRAPHIC WIND FIELD

The dynamic modifications to airflow due to topography, superimposed thermal effects setting up local circulations, and basic altitudinal effects on wind velocity, must obviously combine to make the observed wind field highly complex and variable in mountainous terrain. The individual roles of these factors have already been detailed, but it will be useful to close this chapter by some consideration of interactions in the total wind field. In complex terrain, with several topographic scales represented, the airflow may show different effects at different levels in the atmosphere. Nappo (1977) illustrates this in the 50–60 km wide Tennessee River Valley between the Cumberland Plateau (1000 m) to the north-west and the Great Smoky Mountains (2000 m) to the south-east. Three distinct layers can be identified. Below about 200 m in the valley the flow shows no large-scale terrain effects although there is channelling by minor ridges (100–150 m) with variations in speed and direction occurring virtually independently of stability. Above this layer, up to about 800 m above the valley floor (which is approximately half the height of the major terrain features), the wind speed profile is still like that for a rough plain surface, whereas wind direction is affected by local topography and stability. At higher levels, speeds increase but the directions tend to remain constant into the free air.

For modelling purposes, the wind field can be regarded as a result of the combined effects of three major factors: synoptic-scale forcing; topographic blocking or channelling; and thermal effects. The synoptic-scale pressure field itself can also be greatly modified by the dynamic and thermodynamic effects of large-scale topography as shown earlier (section on pp. 108–25).

If the gradient winds are strong, mountain winds essentially show only kinematic adjustments to the terrain. For southern California, where this situation is represented during Santa Ana conditions, an empirical model has been developed by Ryan (1977). Wind direction is expressed by a diversion factor (F_D) which is a function of the downwind slope (s_d, in per cent), and the angle between the slope azimuth (A) and the wind direction (V_0):

$$F_D = -\,0.255\, s_d \sin\,[2(A - V_o)]$$

F_D is a maximum of 22.5° for s_d = 100 per cent (45° slope) and $(A - V_0) = 45°$. $F_D < 0$ denotes that the wind is backed (turned counterclockwise) if A is less than 90° clockwise from V_0. Ryan also estimates 'sheltering effects' due to topography based on the upwind slope between the site and the horizon. The sheltering factor is:

$$F_s = k \arctan\,(0.17\, s_u)/100,$$

where s_u = the upwind slope (per cent) (limited arbitrarily to 100 per cent),

k = unity, except under Santa Ana conditions where it is subjectively adjusted to unity at 600 m and zero at 1250 m and above.

At the other extreme, when the synoptic control is weak, the surface winds in deep mountain valleys are decoupled from the gradient flow and are determined by the thermal forcing (Schwabl 1934). In southern California this involves the additional complication of sea breeze effects (Sommers 1976). The airflow at ridge levels under these conditions remains closely related to the gradient pattern. Thermal influences obviously change diurnally and seasonally, with changes in the effects of diurnal season heating on slopes and on lapse rate structure in the lower atmosphere.

Many efforts are now under way to develop meso-scale dynamical models of boundary layer flow over complex terrain, especially in view of its importance to such problems as air pollution dispersal in mountain valleys and the spread of forest fires in mountainous terrain. The low-level airflow over complex terrain can be modelled in a variety of ways. Relatively simple models suitable for diagnostic analyses or forecasting applications can be grouped into three types according to the approach adopted: mass conservation; one-layer vertically-integrated primitive equations of motion; or one-level primitive equations (Mass and Dempsey 1985).

Mass conservation models (e.g. Dickerson 1978) assume a well-mixed (constant density) layer beneath a low-level inversion where mass is conserved by applying the continuity equation:

$$\frac{\partial h}{\partial t} + \frac{\partial (hu)}{\partial x} + \frac{\partial (hv)}{\partial y} + w = 0$$

where h = height of the inversion base above the ground surface
w = vertical outflow through the inversion base (upper boundary)
u,v = horizontal components of mean velocity in the mixed layer.

The same approach can be extended for three-dimensional wind fields (Sherman 1978; Ross *et al.* 1988). The variational analysis procedure is used whereby a solution is obtained to an integral function that minimizes the variance of the difference between observed and analysed variable values, subject to imposed constraints e.g., observed horizontal fluxes must satisfy the continuity equation, and a non-divergent three-dimensional analysed wind field (Sherman 1978). A mass continuity model has been used by Tesche and Yocke (1978) to examine wind flow patterns in the Lake Tahoe basin and the Donner Pass area of the Sierra Nevada, California.

Examination of the effect of diurnal heating cycles on air flow over a mountain has been performed numerically using hydrostatic meso-scale models incorporating a detailed boundary layer formulation (Mahrer and Pielke 1975, 1977). Surface heat and momentum fluxes are parameterized. Analysis with the two-dimensional version shows that, over the windward slope, diurnal heating leads to a decrease in daytime horizontal wind components and an increase at night, while the opposite is found over the lee slope. With respect to vertical motion, the mountain acts as an elevated heat source by day, with stronger ascent over the windward slope, but weaker descent over the lee slope. At night, the descent on the lee side is increased, aided by the accelerating effect of a nocturnal pressure maximum due to cooling at the crest. The planetary boundary layer is predicted to be 600–900 m lower on the downwind side of the mountain than on the windward side. Further simulations with the three-dimensional version (Mahrer and Pielke 1977) for the area of the Sacramento Mountains and White Sands, New Mexico, illustrate effects of deflection around the mountains, stronger downslope winds on the lee side, augmented winds upwind of ridges, and valley and slope wind effects.

Fosberg *et al.* (1976) have developed a numerical model based on a combination of their earlier work on thermally induced winds and work by Anderson (1971) on terrain-induced airflow. The procedure is based on a one-layer model of boundary layer flow applied to areas of about 2500 km^2 with grid spacings of the order of 1–6 km (finer grids with increasing relative relief). The gradient wind pattern is first transferred into the terrain-modified throughflow, determined from a steady-state mass continuity equation (Anderson 1971). A rigid lid, about 1.5–2 km above the spatially smoothed terrain, is applied. Modification of the wind field by thermal and frictional effects are then taken into account based on the divergence and vorticity equations, neglecting advection. The approach is to transform the computed divergence and vorticity into stream functions and velocity potentials (see Barry and Perry 1973: 32) and then determine horizontal wind components from these. After the calculation of background values, representing the influence of large-scale terrain features on the flow, divergence is calculated for the actual terrain and frictional effects are taken into account. The data requirements of the model are: temperature, pressure, elevation and surface roughness, length at each grid point and values for the area specifying static stability, geostrophic wind velocity, eddy viscosity, latitude, and large-scale vorticity. Tests of the model for summer situations in Oregon and California showed good agreement with observed meso-scale flow patterns. Fosberg *et al.* comment that flow perturbations due to divergence effects are about 2.5–4 times as large as those due to rotational effects. They also note that thermal control on the flow vanishes and terrain influences become weak when the background wind speed reaches 10 m s^{-1}.

The second simplified model approach uses the primitive equations of motion, vertically integrated for a well-mixed boundary layer (Lavoie 1974; Overland *et al.* 1979). These models consider a surface, constant stress, layer, a well-mixed boundary layer, an inversion layer, and the overlying free atmosphere. Mass is conserved in the mixed layer, but not the layers above. The simulated flow may be blocked and deflected when the mixed layer is forced upward by convergence on a hill slope and cooler air in the mixed layer replaces warmer air above the subsidence inversion (Mass and Dempsey 1985).

The third type of model uses the primitive equations for one level without a continuity equation. For example, Danard's (1977) model requires the geostrophic wind at the surface and 850 mb, the lower tropospheric lapse rate and surface air temperature. It integrates to a steady state the tendency equations at the surface only for wind, pressure and potential temperature. Mass and Dempsey (1985) extend that approach to calculate surface wind and temperature using equations for horizontal momentum and temperature tendency in sigma coordinates. The mass field is determined by the vertical temperature structure. Thermally-induced circulations due to diabatic forcing can also be included. The results of model runs for the coastal mountain area of southern British Columbia, western Washington state and north-western Oregon with different synoptic-scale forcings indicate that stably stratified air is deflected around high terrain and converges on the lee sides over a zone several times the diameter of the obstacle. Such channelling and deflection is shown to be due to the effects of adiabatic cooling (as air rises) and warming (as it descends), which results in redistributions in the mass field. In view of the modest requirements of such models for initial data and for computing resources, they are useful for analysis and forecasting of wind, temperature, and pollutant transport in areas of mountainous terrain. Limited area, fine resolution prognostic models that resolve mesoscale features are also available (see, for example, Pielke, 1984).

It is not possible to detail such models here. However, it is worth noting that a current trend is to embed mesoscale models capable of resolving regional detail in general circulation models. An illustration for the western United States is provided by Giorgi and Bates (1989). The Pennsylvania State University/National Center for Atmospheric Research (NCAR) mesoscale model is embedded in the NCAR Community Climate Model (CCM) and is driven by initial conditions and lateral boundary conditions for January 1979. The mesoscale model has a 60 km grid for a 3000×3000 km^2 area; there are five levels in the lowest 1.5 km and a detailed surface biosphere parameterization. A CCM multiyear simulation provides a set of representative weather patterns for analysis; statistical relations are developed between the climatic patterns given by the regional model and predictions from the global model. These relationships are then applied to the long-term global simulations to obtain regional climate descriptions. The simulated regional climate patterns for the western United States in January, including precipitation and snow depth, are reasonably realistic in relation to major topographic features although certain model biases in the different variables are noted (Giorgi and Bates 1989).

In a further study using the mesoscale model nested in the NCAR CCM, Giorgi (1990) shows that with forcing from a spectral triangular expansion truncated at wave number 42 (corresponding to $2.89° \times 2.89°$ resolution), the 60 km resolution of the mesoscale model provides regional detail of temperature, precipitation and snowcover distribution that agree well with observations. This type of approach will clearly be useful for climate assessments and scenarios of changed forcings.

Notes

1 Roughness length (z_0) is estimated from 0.5 h A/S, where A = silhouette area of obstacle, S = surface area, h = relief; A/S is typically ≤ 0.1 (Mason, 1985).
2 A critical layer occurs where the phase speed of a wave = the mean flow velocity.
3 Descriptions of acoustic sounder observations are given by Neff (1988) and the full suite of boundary-layer remote sensing techniques (lidar, sodar, radar) is detailed in Lenschow (1986).

REFERENCES

Planetary- and synoptic-scale effects

Alaka, M.A. (ed.) (1960) *The Airflow over Mountains*, Tech. Note no.34, Geneva, World Meteorological Organization.
Beer, T. (1976) 'Mountain waves', *Sci. Prog.*, 63, 1–25.
Bolin, B. (1950) 'On the influence of the earth's orography on the character of the westerlies', *Tellus*, 2, 184–95.
Boyer, D.L. and Davies, P.A. (1982) 'Flow past a circular cylinder on a β-plane', *Phil. Trans. R. Soc. Lond.* A306, 533–66.
Boyer, D.L. and Chen, R.R. (1987) 'Laboratory simulation of mountain effects on large-scale atmospheric motion systems: the Rocky Mountains', *J. Atmos. Sci.*, 44, 100–23.
Boyer, D.L., Chen, R. and Davies, P.A. (1987) 'Some laboratory models of flow past the Alpine/Pyrenees mountain complex', *Met. Atmos. Phys.*, 36, 187–200.
Brinkmann, W.A.R. (1970) 'The chinook at Calgary (Canada)', *Arch. Met. Geophys. Biokl.*, B, 18, 269–78.
Buzzi, A. and Rizzi, R. (1975) 'Isentropic analysis of cyclogenesis in the lee of the Alps', *Rivista Ital. Geofis.* (XIII Cong. Int. Met. Alpina), 1, 7–14.
Buzzi, A., Speranza, A., Tibaldi, S. and Tosi, E. (1987) 'A unified theory of orographic influences upon cyclogenesis', *Met. Atmos. Phys.*, 36, 91–107.
Buzzi, A. and Tibaldi, S. (1978) 'Cyclogenesis in the lee of the Alps: A case study', *Q.J.R. Met. Soc.* 104, 271–87.
Chaudhury, A.M. (1950) 'On the vertical distribution of wind and temperature over Indo-Pakistan along the meridian 76°E in winter', *Tellus*, 2, 56–62.
Chen, W.D. and Smith, R.B. (1987) 'Blocking and deflection of airflow by the Alps', *Mon. Wea. Rev.*, 115, 2578–97.
Chung, Y.S., Hage, K.D. and Reinelt, E.R. (1976) 'On lee cyclogenesis and airflow in the Canadian Rockies and the east Asian mountains', *Mon. Wea. Rev.*, 104, 879–91.
Church, P.E. and Stephens, T.E. (1941) 'Influence of the Cascade and Rocky Mountains on the temperature during the westward spread of polar air', *Bull. Am. Met. Soc.*, 22, 25–30.
Colson, D. (1949) 'Airflow over a mountain barrier', *Trans. Am. Geophys. Union.* 30, 818–30.
Cruette, D. (1976) 'Experimental study of mountain lee-waves by means of satellite photographs and aircraft measurements', *Tellus*, 28, 499–523.
Dickey, W.W. (1961) 'A study of topographic effect on wind in the Arctic', *J. Met.*, 18. 790–803.
Dickinson R.E. (1980) 'Planetary waves: theory and observation', in Hide, R. and White, P.W. (eds), *Orographic Effects in Planetary Flows*, GARP Publ. Series no. 23, pp. 1–49, Geneva, WMO-ICSU Joint Scientific Committee, World Meteorological Organization.

Egger, J. (1972) 'Incorporation of steep mountains into numerical forecasting models', *Tellus*, 24, 324–35.

Egger, J. (1989) 'Föhn and quasi-stationary fronts', *Beitr. Phys. Atmos.*, 62, 20–9. GARP-ALPEX (1987) 'The alpine experiment', *Met. Atmos. Phys.* 36, 1–296.

Godske, C.L., Bergeron, T., Bjerknes, J. and Bundgaard, R.C. (1957) *Dynamic Meteorology and Weather Forecasting*, Boston, Mass., American Meteorological Society.

Hage, K.D. (1961) 'On summer cyclogenesis in the lee of the Rocky Mountains', *Bull. Am. Met. Soc.,* 42, 20–33.

Hamilton, R.A. (1958a) 'The meteorology of northern Greenland during the midsummer period,' *Q.J.R. Met. Soc.*, 84, 142–58.

Hamilton, R.A. (1958b) 'The meteorology of north Greenland during the midwinter period,' *Q.J.R. Met. Soc.*, 355–74.

Hess, S.L. and Wagner, H. (1948) 'Atmospheric waves in the northwestern United States,' *J. Met.*, 5, 1–19.

Jansá, A. (1987) 'Distribution of the mistral: a satellite observation', *Met. Atmos. Phys.*, 36, 201–14.

Kasahara, A., Sasamori, T. and Washington, W.M. (1973) 'Simulation experiments with a 12-layer stratospheric global circulation model. I. Dynamical effects of the earth's orography and thermal influence of continentality', *J. Atmos. Sci.*, 30, 229–50.

Kutzbach, J.E., Guetter, P.J., Ruddiman, W.F. and Prell, W.L. (1989) 'Sensitivity of climate to Late Cenozoic uplift in southern Asia and the American West: numerical experiments', *J. Geophys. Res.*, 94(D15), 18, 393–407.

Lydolph, P.E. (1977) *Climates of the Soviet Union*, pp. 160–9, 193–5, Amsterdam, Elsevier.

McClain, E.P. (1960) Some effects of the Western Cordillera of North America on cyclonic activity. *J. Met.* **17**, 104–15.

McGinley, J. (1982), 'A diagnosis of Alpine lee cyclogenesis', *Mon. Wea. Rev.*, 110, 1271–87.

Malmberg, H. (1967) 'Der Einfluss der Gebirge auf die Luftdruckverteilung am Erdboden', *Met. Abhand.*, 71(2).

Manabe, S. and Broccoli, A.J. (1990) 'Mountains and arid climates of middle latitudes', *Science* 247, 192–5.

Manabe, S. and Terpstra, T.B. (1974) 'The effects of mountains on the general circulation of the atmosphere as identified by numerical experiments' *J. Atmos. Sci.*, 31, 3–42.

Mason, P.J. (1985) 'On the estimation of orographic drag', in *Physical Parameterization for Numerical Models of the Atmosphere*, 2, pp. 139–65, ECMWF, Reading, UK.

Mattocks, C. and Bleck, R. (1986), 'Jet streak dynamics and geostrophic adjustment processes during the initial stages of lee cyclogenesis', *Mon. Wea. Rev.*, 114, 2033–56.

Nicholls, J.M. (1973) *The Airflow over Mountains. Research, 1958–1972*, WMO Tech. Note no. 127, Geneva, World Meteorological Organization.

Palmén, E. and Newton, C.W. (1969) *Atmospheric Circulation Systems*, pp. 344–50, New York, Academic Press.

Palmer, T.N., Shutts, G.J. and Swinbank, R. (1986) 'Alleviation of a systematic westerly bias in general circulation and numerical weather prediction models through an orographic gravity wave drag parameterization', *Q.J.R. Met. Soc.*, 112, 1001–39.

Parish, T.R. (1982) 'Barrier winds along the Sierra Nevada mountains', *J. appl. Met.*, 21, 925–30.

Pichler, H. and Steinacker, R. (1987) 'On the synoptics and dynamics of orographically induced cyclones in the Mediterranean', *Met. Atmos. Phys.*, 36, 108–17.

Queney, P. (1948) 'The problem of airflow over mountains: a summary of theoretical studies,' *Bull. Am. Met. Soc.*, 29, 16–26.

Queney, P. (1963) 'Etat actuel de la dynamique des courants aériens près des montagnes', *Geofis. Met.* (7 Cong. Int. Met. Alpina), 11, 1–11.
Radinović, D. (1965) 'Forecasting of cyclogenesis in the West Mediterranean and other areas bounded by mountain ranges by a baroclinic mode', *Arch. Met. Geophys. Biokl.* A, 14, 279–99.
Radinović, D. (1986), 'On the development of orographic cyclones', *Q.J.R. Met. Soc.*, 112, 927–51.
Reiter, E.R. (1963) *Jet Stream Meteorology*, Chicago, University of Chicago Press.
Reuter, H. and Pichler, H. (1964) 'On the orographic influences of the Alps', *Tellus*, 16, 40–2.
Ruddiman, W.F. and Kutzbach, J.E. (1989) 'Forcing of late Cenozoic northern hemisphere climate by plateau uplift in southern Asia and the American West', *J. Geophys. Res.*, 94(D15), 18, 409–27.
Ruddiman, W.F., Prell, W.L. and Raymo, W.E. (1989) 'Late Cenozoic uplift in southern Asia and the American West: Rationale for general circulation modelling experiments', *J. Geophys. Res.*, 94(D15), 18, 379–91.
Schwerdtfeger, W. (1975) 'Mountain barrier effects on the flow of stable air north of the Brooks Range,' in Weller, G. and Bowling, S.A. (eds) *Climate of the Arctic*, pp. 204–8, Fairbanks, Geophysics Institute University of Alaska.
Slingo, A. and Pearson, D.W. (1987) 'A comparison of the impact of an envelope orography and of a parameterization of orographic gravity-wave drag on model simulations', *Q.J.R. Met. Soc.*, 113, 847–70.
Smith, R.B. (1979a) 'The influence of mountains on the atmosphere', *Adv. Geophys.*, 21, 87–230.
Smith, R.B. (1979b) 'Some aspects of the quasi-geostrophic flow over mountains', *J. Atmos. Sci.*, 36, 2385–93.
Smith, R.B. (1982) 'Synoptic observations and theory of orographically disturbed wind and pressure', *J. Atmos. Sci.* 39, 60–70.
Smith, R.B. (1986) 'Mesoscale mountain meteorology in the Alps', in *Scientific Results of the Alpine Experiment (ALPEX), Vol. II*, Geneva. WMO/TD No. 108, GARP Publ. Series no. 27, pp. 407–23. World Meteorological Organization.
Speranza, A. (1975) 'The formation of basic depressions near the Alps', *Ann. Geofis.*, 28, 177–217.
Steinacker, R. (1981) 'Analysis of the temperature and wind field in the Alpine region', *Geophys. Astrophys. Fluid Dynamics*, 17, 51–62.
Stringer, E.T. (1972) *Foundations of Climatology*, p. 407, San Francisco, W.H. Freeman.
Sutcliffe, R.C. and Forsdyke, A.G. (1950) 'The theory and use of upper air thickness patterns in forecasting', *Q.J.R. Met. Soc.*, 76, 189–217.
Taylor-Barge, B. (1969) *The summer climate of the St. Elias mountain region*, Arctic Institute of North America Res. Pap. No. 53, Montreal.
Walker, J.M. (1967) 'Subterranean isobars,' *Weather*, 22, 296–7.
Wallace, J.M., Tibaldi, S. and Simmons, A.J. (1983) 'Reduction of systematic forecast errors in the ECMWF model through the introduction of an envelope orography', *Q.J.R. Met. Soc.*, 109, 683–717.
Washington, W.M. and Parkinson, C.L. (1986) *An Introduction to Three-Dimensional Climate Modeling*, University Science Books, Mill Valley, CA, 422 pp.

Local airflow modification: wave phenomena

Alaka, M.A. (ed.) (1960) *The Airflow over Mountains*, Tech. Note no. 34, Geneva, World Meteorological Organization.
Beer, T. (1976) 'Mountain waves', *Sci. Prog.*, 63, 1–25.

Bergen, W.R. (1976) 'Mountainadoes: a significant contribution to mountain windstorm damage?' *Weatherwise*, 29, 64–9.

Boyer, D.L. and Davies, P.A. (1982) 'Flow past a circular cylinder on a β-plane', *Phil. Trans. R. Soc. Lond.* A306, 533–66.

Brighton, P.W.M. (1978) 'Strong stratified flow past three-dimensional obstacles', *Q.J.R. Met. Soc.*, 104, 289–307.

Brinkmann, W.A.R. (1974) 'Strong downslope winds at Boulder, Colorado', *Mon. Wea. Rev.*, 102, 596–602.

Brooks, F.A. (1949) 'Mountain-top vortices as causes of large errors in altimeter heights', *Bull. Am. Met. Soc.*, 30, 39–44.

Casswell, S.A. (1966) 'A simplified calculation of maximum vertical velocities in mountain lee waves', *Met. Mag.* 95, 68–80.

Chopra, K.P. (1973) 'Atmospheric and oceanic flow patterns introduced by islands', *Adv. Geophys.*, 16, 297–421.

Corby, G.A. (1954) 'The airflow over mountains: a review of the state of current knowledge', *Q.J.R. Met. Soc.*, 80, 491–521.

Corby, G.A. (1957) *Air Flow Over Mountains*, Met. Rep. no. 18 (vol. 3, no. 2), Meteorological Office, London, HMSO.

Corby, G.A. and Wallington, C.E. (1956) 'Airflow over mountains: the lee-wave amplitude', *Q.J.R. Met. Soc.*, 82, 266–74.

Cox, K.W. (1986) Analysis of the Pyrenees lee-wave event of 23 March 1982. *Mon. Wea. Rev.* 114, 1146–66.

Cruette, D. (1976) 'Experimental study of mountain lee-waves by means of satellite photographs and aircraft measurements', *Tellus*, 28, 499–523.

Durran, D.R. (1990) 'Mountain waves and downslope winds', in W. Blumen (ed.) *Atmospheric Processes Over Complex Terrain, Met monogr*, 23(45), 59–81, Boston, Mass., Amer. Met. Soc.

Etling, D. (1989) 'On atmospheric vortex streets in the wake of large islands', *Met. Atmos. Phys.*, 41, 157–64.

Förchgott, J. (1949) 'Wave streaming in the lee of mountain ridges', (in Czech), *Met. Zpravy*, 3, 49–51.

Förchgott, J. (1969) 'Evidence for mountain-sized, lee eddies', *Weather*, 24, 255–60.

Gerbier, N. and Bérenger, M. (1961) 'Experimental studies of lee waves in the French Alps', *Q.J.R. Met. Soc.*, 87, 13–23.

Gjevik, B. and Marthinsen, T. (1978) 'Three-dimensional lee-wave pattern', *Q.J.R. Met. Soc.*, 104, 947–58.

Grace, W. and Holton, I. (1990) 'Hydraulic jump signatures associated with Adelaide downslope winds', *Austral. Met. Mag.*, 38, 43–52.

Houghton, D.D. and Isaacson E. (1970) 'Mountain winds', *Studies in Num. Anal.*, 2, 21–52.

Hunt, J.C.R. and Snyder, W.H. (1980) 'Experiments on stably and neutrally stratified flow over a model three-dimensional hill', *J. Fluid Mech.*, 96, 671–704.

Klemp, J.B. and Lilly, D.K. (1975) 'The dynamics of wave-induced downslope winds', *J. Atmos. Sci.*, 32, 320–39.

Klemp, J.B. and Lilly, D.K. (1978) 'Numerical simulation of hydrostatic mountain waves', *J. Amos. Sci.*, 35, 78–107.

Kuettner, J.P. (1958) 'The rotor flow in the lee of mountains', *Schweiz. Aero Revue*, 33, 208–15. (Also: (1959) *Geophys. Res. Dir. Res. Notes, no.* 6, Cambridge, Mass., US Air Force, Cambridge Research Center.)

Lilly, D.K. (1978) 'A severe downslope windstorm and aircraft turbulence induced by a mountain wave', *J. Atmos. Sci.*, 35, 59–77.

Lilly, D.K. and Klemp, J.B. (1979) 'The effect of terrain shape on non-linear hydrostatic

mountain waves', *J. Fluid Mech.*, 95, 241–61.
Lilly, D.K. and Zipser, E.J. (1972) 'The front range windstorm of 11 January 1972 – a meteorological narrative', *Weatherwise*, 25, 56–63.
Long, R.R. (1954) 'Some aspects of the flow of stratified fluids. II. Experiments with a two-fluid system', *Tellus*, 6, 97–115.
Long, R.R. (1969) 'Blocking effects in flow over obstacles', Tech. Rep. 32, WB-ESSA, Baltimore, Md., Johns Hopkins University.
Long, R.R. (1970) 'Blocking effects in flow over obstacles', *Tellus*, 22, 471–80.
Lyra, G. (1943) 'Theorie der stationären Leewellenströmung in freier Atmosphäre, *Zeit. angew. Math. Mechan.*, 23, 1–28.
Nicholls, J.M. (1973) *The Airflow over Mountains, Research 1958–1972*, WMO Tech. Note No. 127, Geneva, World Meteorological Organization.
Peltier, W.R. and Clark, T.L. (1979) 'The evolution of finite-amplitude mountain waves. Part II. Surface wave drag and severe downslope windstorms', *J. Atmos. Sci.* 36, 1498–529.
Sawyer, J.S. (1960) 'Numerical calculation of the displacements of a stratified airstream crossing a ridge of small height', *Q.J.R. Met. Soc.*, 86, 326–45.
Scorer, R.S. (1949) 'Theory of waves in the lee of mountains', *Q.J.R. Met. Soc.* 74, 41–56.
Scorer, R.S. (1953) 'Forecasting mountain and lee waves', *Met. Mag.*, 82, 232–4.
Scorer, R.S. (1955) 'Theory of airflow over mountains. IV. Separation of airflow from the surface', *Q.J.R. Met. Soc.* 81, 340–50.
Scorer, R.S. (1967) 'Causes and consequences of standing waves', in Reiter, E. and Rasmussen, J.L. (eds.) *Symposium on Mountain Meteorology*, Atmos. Sci. Pap. no. 122, pp. 75–101, Fort Collins, Colorado State University.
Scorer, R.S. (1978) *Environmental Aerodynamics*, Chichester, Ellis Horwood.
Smith, R.B. (1977) 'The steepening of hydrostatic mountain waves', *J. Atmos. Sci.*, 34, 1634–54.
Smith, R.B. (1979) 'The influence of mountains on the atmosphere', *Adv. Geophys.* 21, 87–230.
Soma, S. (1969) 'Dissolution of separation in the turbulent boundary layer and its application to natural winds', *Pap. Met. Geophys*, 20, 111–74.
Starr, J.R. and Browning, K.A. (1972) 'Observations of lee waves by high-power radar', *Q.J.R. Met. Soc.*, 98, 73–85.
Wallington, C.E. (1961) 'Airflow over broad mountain ranges: a study of five flights across the Welsh mountains', *Met. Mag.* 90, 213–22.
Wallington, C.E. (1970) 'A computing aid to studies of airflows over mountains,' *Met Mag.* 99, 157–65.
Wilson, H.P. (1968) 'Stability waves', Meteorological Branch. Tech. Mem. 703, Toronto, Department of Transport.
Wilson, H.P. (1974) 'A note on meso-scale barrier to surface airflow', *Atmosphere*, 12, 118–20.
Yoshino, M.M. (1975) *Climate in a Small Area*, pp. 355–407, Tokyo, University of Tokyo Press.
Zimmerman, L.I. (1969) 'Atmospheric wake phenomena near the Canary Islands', *J. appl. Met.*, 8, 896–907.

Associated cloud forms

Abe, (1941) 'Mountain clouds, their forms and connected air currents, Part II', *Bull. Cent. Met. Observ.*, Japan, 7, 93–145.
Affronti, di F. (1963) 'Le nubi d'onda sull'Etna con flusso occidentale', *Geofis. Met.* (7 Congr. Int. Met., Alpina), 11, 75–80.

Brighton, P.W.M. (1978) 'Strongly stratified flow past three-dimensional obstacles', *Q.J.R. Met. Soc.*, 104, 289–307.

Douglas, C.K.M. (1928) 'Some alpine cloud forms', *Q.J.R. Met. Soc.*, 54, 175–7.

Krug-Pielsticker, U. (1942) 'Beobachtungen der hohen Föhnwelle an den Ostalpen', *Beitr. Phys. frei. Atmos.*, 27, 140–64.

Küttner, J. (1939a) 'Moazagotl und Föhnwelle', *Beitr. Phys. frei Atmos.*, 25, 79–114.

Küttner, J. (1939b) 'Zur Entstehung der Föhnwelle', *Beitr. Phys. frei Atmos.*, 25, 251–99.

Ludlam, F.H. (1967) 'Characteristics of billow clouds and their relation to clear-air turbulence,' *Q.J.R. Met. Soc.*, 93, 419–35.

Ludlam, F.H. (1980) *Clouds and Storms*, pp. 369–80, University Park, Pa., Pennsylvania State University Press.

Manley, G. (1945) 'The Helm wind of Crossfield, 1937–9', *Q.J.R. Met. Soc.*, 71, 197–219.

'Fall' winds

Aanensen, C.J.M. (1965) 'Gales in Yorkshire in February 1962', *Geophys. Mem.*, 14(3), No. 108.

Arakawa, S. (1968) 'A proposed mechanism of fall winds and Dashikaze', *Pap. Met. Geophys.*, 19, 69–99.

Arndt, A. (1913) 'Uber die Bora in Noworossisk', *Met. Zeit.*, 30, 295–302.

Ball, F.K. (1957) 'The katabatic winds of Adelie Land and King George V Land', *Tellus*, 9, 201–8.

Beran, D.W. (1967) 'Large amplitude lee waves and chinook winds', *J. appl. Met.*, 6, 865–77.

Bilwiller, R. (1899) 'Uber verschiedene Entstehungsarten und Erscheinungsformen des Föhns', *Met. Zeit.* 16, 204–15.

Bower, J.B. and Durran, D.R. (1986) 'A study of wind profiler data collected upstream during windstorms in Boulder, Colorado' *Mon. Wea. Rev.*, 114, 1491–500.

Brinkmann, W.A.R. (1970) 'The chinook at Calgary (Canada)', *Arch. Met. Geophys. Biokl.*, B. 18, 279–86.

Brinkmann, W.A.R. (1971) 'What is a foehn?' *Weather*, 26, 230–9.

Brinkmann, W.A.R. (1973) *A Climatological Study of Strong Downslope Winds in the Boulder Area*, Inst. Arct. Alp. Res., Occas. Pap. no. 7, Boulder, University of Colorado.

Brinkmann, W.A.R. (1974a) 'Strong downslope winds at Boulder, Colorado', *Mon. Wea. Rev.*, 102, 596–602.

Brinkmann, W.A.R. (1974b) 'Temperature characteristics of severe downslope winds in Boulder, Colorado', *Zbornik Met. Hidrol. Radova*, 5, 143–7.

Čadež, M. (1967) 'Uber synoptische Probleme in Südostalpinen Raum', Veröff. *Schweiz. Met. Zentralanstalt*, 4, 155–75.

Clark, T.L. and Farley, R.D. (1984) 'Severe downslope windstorm calculations in two and three spatial dimensions using an elastic interactive grid nesting: A possible mechanism for gustiness', *J. Atmos. Sci.* 41, 329–50.

Cook, A.W. and Topil, A.G. (1952) 'Some examples of chinooks east of the mountains in Colorado', *Bull. Am. Met. Soc.*, 33, 42–7.

Durran, D.R. (1990) 'Mountain waves and downslope winds', in W. Blumen (ed.) *Atmospheric Processes over Complex Terrain, Met. monogr.*, 23 (45), 59–81, Boston, Mass., Amer. Met. Soc.

Flohn, H. (1942) 'Häufigkeit, Andauer and Eigenschaften des "freien Föhns" auf deutschen Bergstationen', *Beitr. Phys. frei, Atmos.*, 27, 110–24.

Frey, K.K. (1957) 'Zur Diagnose des Föhns', *Met. Rdsch.*, 2, 276–80.

Georgii, W. (1967) 'Thermodynamik und Kinematik des Kaltluftföhns', *Arch. Met.*

Geophys. Biokl. A, 16, 137–52.

Gosink, J. (1982) 'Measurements of katabatic winds between Dome C and Dumont d'Urville', *Pure Appl. Geophys.* 120, 503–26.

Hann, J. (1866) 'Zur Frage über den Ursprung des Föhns', *Zeit. Osterreich Ges. Met.*, 1 (17), 257–63.

Hann, J. (1885) 'Einige Bermerkungen zur Entwicklungs-Geschichte der Ansichten über den Ursprung des Föhns', *Met. Zeit.*, 2, 393–9.

Hoinka, K.P. (1985a) 'A comparison of numerical simulations of hydrostatic flow over mountains with observations', *Mon. Wea. Rev.*, 113, 719–35.

Hoinka, K.P. (1985b) 'Observation of the airflow over the Alps during a foehn event', *Q.J.R. Met. Soc.*, 111, 199–224.

Ives, R.L. (1950) 'Frequency and physical effects of chinook winds in the Colorado high plains region', *Ann. Ass. Am. Geog.*, 40, 293–327.

Julian, L.T. and Julian, P.R. (1969) 'Boulder's winds', *Weatherwise*, 22, 108–12, 126.

Klemp, J.B. and Durran, D.R. (1987) 'Numerical modelling of bora winds', *Met. Atmos. Phys.*, 36, 215.

Klemp, J.B. and Lilly, D.K. (1975) 'The dynamics of wave-induced downslope winds', *J. Atmos. Sci.* 32, 320–39.

Köppen, W. (1923) 'Die Bora in nordlichen Skandinavien', *Ann. Hydrogr. Marit. Met.*, 51, 97–9.

Lester, P.F. (1978) 'A severe chinook windstorm', in *Conference on Sierra Nevada Meteorology. Preprints*, pp. 104–8, Boston, American Meteorological Society.

Lied, N.J. (1964) 'Stationary hydraulic jumps in a katabatic flow near Davis, Antarctica, 1961', *Aust. Met. Mag.* 47, 40–51.

Lilly, D.K. (1978) 'A severe downslope windstorm and aircraft turbulence induced by a mountain wave', *J. Atmos. Sci.*, 35, 59–77.

Lilly, D.K. and Klemp, J.B. (1979) 'The effect of terrain shape on non-linear hydrostatic mountain waves', *J. Fluid Mech.*, 95, 241–61.

Lilly, D.K. and Zipser, E.J. (1972) 'The Front Range windstorm of 11 January 1972 – a meteorological narrative', *Weatherwise*, 25, 56–63.

Lockwood, J.G. (1962) 'Occurrence of föhn winds in the British Isles,' *Met. Mag.*, 91, 57–65.

Loewe, F. (1972) 'The land of storms', *Weather* 27, 110–21.

Long, R.R. (1970) 'Blocking effect in flow over obstacles', *Tellus*, 22, 471–80.

Longley, R.W. (1967) 'The frequency of winter chinooks in Alberta', *Atmosphere*, 5, 4–16.

Lopéz, M.E. and Howell, W.E. (1967) 'Katabatic winds in the equatorial Andes', *J. Atmos. Sci.*, 24, 29–35.

Manley, G. (1938) 'Meteorological observations of the British East Greenland Expedition, 1935–1936, at Kangerdlugssuak, 68°10′N, 31°44′W', *Q.J.R. Met. Soc.*, 64, 253–76.

Manley, G. (1945) 'The helmwind of Crossfield, 1937–9,' *Q.J.R. Met. Soc.*, 71, 197–219.

Mather, K.B. and Miller, G.S. (1967) 'The problem of katabatic winds on the coast of Terre Adelie,' *Polar Record*, 13, 415–32.

Mook, R.H.G. (1962) 'Zur Bora an einem nordnorwegischen Fjord', *Met. Rdsch.*, 15, 130–3.

Neiman, P.J., Hardesty, R.M., Shapiro, M.A., Cupp, R.E. (1988), 'Doppler lidar observations of a downslope windstorm', *Mon. Wea. Rev.*, 116, 2265–75.

Osmond, H.W. (1941) 'The chinook wind east of the Canadian Rockies', *Can. J. Res.*, A, 19, 57–66.

Parish, T.R. (1980) *Surface Winds in East Antarctica*, Madison, Dept. of Meteorology, University of Wisconsin.

Peltier, W.R. and Clark, T.L. (1979) 'The evolution of finite-amplitude mountain waves. Part II. Surface wave drag and severe downslope windstorms', in *J. Atmos. Sci.*, 36, 1498–529.

Petkovsěk, Z. and Paradiž, B. (1976) 'Bora in the Slovenian coastal region', in Yoshino, M.M. (ed.) *Local Wind Bora*, pp. 135–44, Tokyo, University of Tokyo Press.

Pettre, P. (1982) 'On the problem of violent valley winds', *J. Atmos. Sci.*, 39, 542–44.

Pettre, P. (1984) 'Contribution to the hydraulic theory of bora wind using ALPEX data', *Beitr. Phys. Atmos.*, 57, 536–45.

Putnins, P. (1970) 'The climate of Greenland', in Orvig. S. (ed.) *Climates of the Polar Regions*, pp. 3–128, Amsterdam, Elsevier.

Richard, E., Mascart, P. and Nickerson, E.C. (1989) 'On the role of surface friction in downslope windstorms', *J. Appl. Met.*, 28, 241–51.

Riehl, H. (1974) 'On the climatology and mechanisms of Colorado chinook winds', *Bonn. Met. Abhandl.*, 17, 493–504.

Schütz, J. and Steinhauser, F. (1955) 'Neue Föhnuntersuchungen aus dem Sonnblick. *Arch. Met. Geophys. Bioklim.*, B, 6, 207–24.

Schwerdtfeger, W. (1970) 'The climate of the Antarctic,' in Orvig. S. (ed.) *Climates of the Polar Regions*, pp. 253–355, Amsterdam, Elsevier.

Schwerdtfeger, W. (1972) 'The vertical variation of the wind through the friction-layer over the Greenland ice cap', *Tellus*, 24, 13–16.

Scinocco, J.F. and Peltier, W.R. (1989) 'Pulsating downslope windstorms', *J. Atmos. Sci.*, 46, 2885–914.

Scorer, R.S. and Klieforth, H. (1959) 'Theory of mountain waves of large amplitude', *Q.J.R. Met. Soc.*, 85, 131–43.

Smith, R.B. (1977) 'The steepening of hydrostatic mountain waves,' *J. Atmos. Sci.*, 34, 1634–54.

Smith, R.B. (1985) 'On severe downslope winds', *J. Atmos. Sci.*, 42, 2597–603.

Smith, R.B. (1987) 'Aerial observations of the Yugoslavian bora', *J. Atmos. Sci.*, 44, 269–97.

Sommers, W.T. (1978) 'LFM forecast variables related to Santa Ana wind occurrences', *Mon. Wea. Rev.*, 106, 1307–16.

Streten, N.A. (1963) 'Some observations of Antarctic katabatic winds', *Aust. Met. Mag.*, 42, 1–23.

Streten, N.A. (1968) 'Some characteristics of strong wind periods in coastal East Antarctica', *J. appl. Met.*, 7, 46–52.

Sturman, A.P. (1987) 'Thermal influences on airflow in mountainous terrain', *Prog. Phys. Geog.*, 11, 183–206.

Suzuki, S. and Yakubi, K. (1956) 'The air-flow crossing over the mountain range' *Geophys. Mag.*, 27, 273–91.

Tamiya, H. (1972) 'Chronology of pressure patterns with bora on the Adriatic coast', *Climat. Notes* (Tokyo), 10, 52–63.

Tamiya, H. (1975) 'Bora and oroshi: their synoptic climatological situation in the global scale', *Jap. Progr. Climatol.*, 29–34.

Vergeiner, I. (1978) 'Foehn flow in the Alps-three-dimensional numerical simulations on the small- and meso-scale', *Arbeiten, Zentralanstalt Met. Geodynam.*, 32, (63), 1–37.

Whiteman, C.D. and Whiteman, J.G. (1974) 'A historical climatology of damaging downslope windstorms at Boulder, Colorado', Tech. Rep. ERL-336-APCL 35, Boulder, Colorado, NOAA.

Yabuki, K. and Suzuki, S. (1967) 'A study on the airflow over mountain', *Bull. Univ. Osaka Prefecture*, B, 19.

Yoshimura, M. (1976) 'Synoptic and aerological climatology of the bora day', in Yoshino, MM. (ed.) *Local Wind Bora*, pp. 99–111, Tokyo, University of Tokyo Press.

Yoshino, M.M. (1971) 'Die Bora in Yugoslawien. Eine synoptisch-klimatologische Betrachtung', *Ann. Met.*, N.F., 5, 117–21.

Yoshino, M.M. (1972) 'An annotated bibliography on bora', *Climat. Notes*, 10, 1–22.

Yoshino, M.M. (1975) *Climate in a Small Area*, Tokyo, University of Tokyo Press.

Yoshino, M.M. (ed.) (1976) *Local Wind Bora*, Tokyo, University of Tokyo Press.

Yoshino, M.M., Yoshino, M.T., Yoshimura, M., Mitsui, K., Urushibara, K., Ueda, S., Owada, M. and Nakamura, K. (1976) 'Bora regions as revealed by wind-shaped trees on the Adriatic Coast', in Yoshino M.M. (ed.) *Local Wind Bora*, pp. 59–71, Tokyo, University of Tokyo Press.

Zipser, E.J. and Bedard, A.J. (1982) 'Front Range windstorms revisited', *Weatherwise*, 35, 32–5.

Thermally induced winds

Bader, D.C. and McKee, T.B. (1983) 'Dynamical model simulation of morning boundary layer development in deep mountain valleys', *J. Clim. appl. Met.*, 22, 341–51.

Banta, R.M. (1984) 'Daytime boundary-layer evolution over mountainous terrain. Part 1. Observations of the dry circulations', *Mon. Wea. Rev.*, 112, 340–56.

Banta, R.M. (1986) 'Daytime boundary-layer evolution over mountainous terrain. Part 2. Numerical studies of upslope flow duration', *Mon. Wea. Rev.* 114, 112–30.

Banta, R. and Cotton, W.R. (1981) 'An analysis of the structure of local wind systems in a broad mountain basin', *J. appl. Met.*, 20, 1255–66.

Barr, S. and Orgill, M.M. (1989) 'Influence of external meteorology on nocturnal valley drainage winds', *J. appl. Met.*, 28, 497–517.

Bossert, J.E., Sheaffer, J.D. and Reiter, E.R. (1989) 'Aspects of regional-scale flows in mountainous terrain', *J. appl. Met.*, 28, 590–601.

Brehm, M. and Freytag, C. (1982) 'Erosion of the night-time thermal circulation in an alpine valley', *Arch. Met. Geophys. Biokl.*, B31, 331–52.

Buettner, K.J.K. and Thyer, N. (1966) 'Valley winds in the Mount Rainier area', *Arch. Met. Geophys Biokl.*, B, 14, 125–47.

Burger, A. and Ekhart, E. (1937) 'Uber die tägiche Zirkulation der Atmosphäre im Bereiche der Alpen', *Gerlands Beitr. Geophys.*, 49, 341–67.

Businger, J.A. and Rao, K.R. (1965) 'The formation of drainage wind on a snow-dome', *J. Glaciol.* 4, 833–41.

Clements, W.E., Archuleta, J.A. and Hoard, D.E. (1989) 'Mean structure of the nocturnal drainage flow in a deep valley', *J. Appl. Met.* 28, 457–62.

Defant, A., (1933) 'Der Abfluss schwerer Luftmassen auf geneigten Boden, nebst einigen Bemerkungen zur Theorie stationärer Luftstrome', *Sitz. Berichte Preuss. Akad. Wiss.* (Phys. Math. Klasse), 18, 624–35.

Defant, F. (1949) 'Zur Theorie der Hangwinde, nebst Bemerkungen zur Theorie der Bergund Talwinde', *Arch. Met. Geophys. Biokl.*, A, 1, 421–50. (translated as 'A theory of slope winds, along with remarks on the theory of mountain winds and valley winds', in Whiteman, C.D. and Dreiseitl, E. (eds) (1984) *Alpine Meteorology*, PNL-5141, ASCOT-84-3, pp. 95–120, Richland, WA: Baltelle, Pacific Northwest Laboratory.

Defant, F. (1951), 'Local winds', in Malone, T.F. (ed.) *Compendium of Meteorology*, pp. 655–72, Boston, American Meteorological Society.

Doran, J.C. and Horst, T.W. (1981) 'Velocity and temperature oscillations in drainage winds', *J. appl. Met.*, 20, 361–4.

Egger, J. (1987) 'Valley winds and the diurnal circulation over plateaus', *Mon. Wea. Rev.*, 115, 2177–86.

Ekhart, E. (1934) 'Neuere Untersuchungen zur Aerologie der Talwinde: Die periodischen Tageswinde in einem Quertale der Alpen', *Beitr. Phys. fr. Atmos.*, 21, 245–68.

Ekhart, E. (1944) 'Beiträge zur alpinen Meteorologie', *Met. Zeit.*, 61, 217–31 (translated as 'Contributions to alpine meteorology', in Whiteman, C.D. and Dreiseitl, E. (eds) (1984) *Alpine Meteorology*, PNL-5141, ASCOT-84-3, pp. 45–72, Richland, WA, Baltelle, Pacific Northwest Laboratory.

Ekhart, E. (1953) 'Uber den täglichen Gang des Windes im Gebirge', *Arch Met. Geophys. Biokl.*, B, 4, 431–50.

Fitzjarrald, D.R. (1984) 'Katabatic winds in opposing flow', *J. Atmos. Sci.*, 41, 1143–58.

Fleagle, R.G. (1950) 'A theory of air drainage', *J. Met.*, 7, 227–32.

Flohn, H. (1969) 'Local wind systems', in Flohn, H. (ed.) *General Climatology, World Survey of Climatology*, vol. 2, pp. 139–71, Amsterdam. Elsevier.

Fournet, M.J. (1840) 'Des brises de jour et de nuit autour de montagnes', *Ann. Chim. Phys.*, 74, 337–401.

Freytag, C. (1987) 'Results from the MERKUR Experiment: Mass budget and vertical motions in a large valley during mountain valley wind', *Meteorol. Atmos. Phys.*, 37, 129–40.

Garratt, J.R. (1985) 'The inland boundary layer at low latitudes: Part I. The nocturnal jet', *Boundary-Layer Met.*, 31, 307–27.

Geiger, R. (1965) *The Climate near the Ground*, pp. 393–417, Cambridge, Mass., Harvard University Press.

Gleeson, T.A. (1951) 'On the theory of cross-valley winds arising from differential heating of the slopes', *J. Met.*, 8, 398–405.

Gleeson, T.A. (1953) 'Effects of various factors on valley winds', *J. Met.*, 10, 262–9.

Gross, G. (1984) 'Eine Erklärung des Phänomens Maloja-Schlange mittels numerische Simulation', Dissertation, Fachbereich Mechanik, Darmstadt, Technische Hochschule.

Hann, J. von (1879) 'Zur Meteorologie der Alpengifel', *Wien Akad. Wiss. Sitzungsberichte* (Math.-Naturwiss. Klass.). 78 (2), 829–66.

Hennemuth, B. (1985) 'Temperature field and energy budget of a small alpine valley', *Contrib. Atmos. Phys.*, 58, 545–59.

Hennemuth, B. and Köhler, U. (1984) 'Estimation of the energy balance of the Dischma Valley', *Arch. Met. Geophys. Biokl.*, B34, 97–119.

Hennemuth, B. and Neureither, I. (1986) 'Das Feuchtefeld in einem alpinen Endtal.', *Met. Rdsch.*, 39, 233–9.

Hennemuth, B. and Schmidt, H. (1985) 'Wind phenomena in the Dischma Valley during DISKUS' *Arch. Met. Geophys. Biokl.*, B35, 361–87.

Hoinkes, H. (1954) 'Beiträge zur Kenntnis des Gletscherwindes', *Arch. Met. Geophys. Biokl.* B, 6, 36–53.

Holmgren, H. (1971) 'Climate and energy exchange on a sub-polar ice cap in summer. Part C. On the katabatic winds on the northwest slope of the ice cap. Variations of the surface roughness', *Uppsala Univ. Met. Inst.* Meddel. 109.

Holtmeier, F.K. (1966) 'Die "Malojaschlange" und die Vorbreitung der Fichte', *Wetter u. Leben*, 18, 105–8.

Holton, J.R. (1967) 'The diurnal boundary layer oscillation over sloping terrain', *Tellus* 19, 199–205.

Horst, T.W. and Doran, J.C. (1986) 'Nocturnal drainage flow on simple slopes', *Boundary-Layer Met.* 34, 263–86.

Küttner, J. (1949) 'Periodische Luftlawinen', *Met. Rdsch*, 2, 183–4.

Lawrence, E.N. (1954) 'Nocturnal winds', *Prof. Notes. Met. Office* (London), 7(111), 1–13.

Lenschow, D. (ed.) (1986) *Probing the Atmospheric Boundary Layer*, Boston, Mass, American Meteorological Society.

Lettau, H.H. (1967) 'Small to large-scale features of boundary layer structure over mountain slopes', in Reiter, E.R. and Rasmussen, J.L. (eds) *Proceedings of the Symposium on Mountain Meteorology*, Atmos. Sci. Pap. no. 122, pp. 1–74, Fort

Collins, Colorado State University.
Lettau, H.H. (1978) 'Explaining the world's driest climate', in Lettau, H.H. and Lettau, K. (eds) *Exploring the World's Driest Climate*, Rep. 101, Inst. Environ. Studies, Univ. of Wisconsin, Madison, 182–248.
Lindsay, J.A. and Tyson, P.D. (1990) 'Thermo-topographically induced boundary layer oscillations over the central Namib, southern Africa', *Int. J. Climatol.*, 10, 63–77.
MacHattie, L.B. (1968) 'Kananaskis Valley winds in summer', *J. appl. Met.*, 7, 348–52.
McKee, T.B. and O'Neal, R.D. (1989) 'The role of valley geometry and energy budget in the formation of nocturnal valley winds', *J. Appl. Met.*, 28, 445–56.
McNider, R.T. (1982) 'A note on velocity fluctuations in drainage flows', *J. Atmos. Sci.*, 39, 1658–60.
McNider, R.T. and Pielke, R.A. (1981) 'Diurnal boundary-layer development over sloping terrain', *J. Atmos. Sci.*, 38, 2198–212.
McNider, R.T. and Pielke, R.A. (1984) 'Numerical simulation of slope and mountain flows', *J. Clim. appl. Met.*, 23, 1441–53.
Mahrt, L. (1982) 'Momentum balance of gravity flows', *J. Atmos. Sci.*, 39, 2701–11.
Manins, P.C. and Sawford, B.L. (1979a) 'A model of katabatic winds', *J. Atmos. Sci.*, 36, 619–30.
Manins, P.C. and Sawford, B.L. (1979b) 'Katabatic winds: a field case study', *Q.J.R. Met. Soc.*, 105, 1011–25.
Mano, H. (1956) 'A study on the sudden nocturnal temperature rise in the valley and the basin', *Geophys. Mag.*, 27, 169–204.
Mason, P.J. and Sykes, R.I. (1978) 'On the interaction of topography and Ekman boundary layer pumping in a stratified atmosphere', *Q.J.R. Met. Soc.* 104, 475–90.
Müller, H. and Whiteman, C.D. (1984) 'Breakup of a nocturnal temperature inversion in Dischma Valley during DISKUS', *J. appl. Met.*, 27, 188–94.
Neff, W.D. (1988) 'Observations of complex terrain flows using acoustic sounders: Echo interpretation', *Boundary-Layer Met.*, 42, 207–28.
Neff, W.D. and King, C.W. (1989) 'The accumulation and pooling of drainage flows in a large basin', *J. appl. Met.*, 28, 518–29.
Nickus, U. and Vergeiner, I. (1984) 'The thermal structure of the Inn valley atmosphere', *Arch. Met. Geophys. Biokl.*, A, 33, 199–215.
Orville, H.D. (1964) 'On mountain upslope winds', *J. Atmos. Sci.*, 6, 622–33.
Paegle, J. (1984) 'Topographically bound low-level circulations', *Riv. Met. Aeronaut.*, 44, 113–25.
Petkovsěk, Z. and Hočevar, A. (1971) 'Night drainage winds', *Arch. Met. Geophys. Biokl.*, A, 20, 353–60.
Petkovsěk, Z. (1978) 'Relief meteorologically relevant characteristics of basins', *Met. Zeit.*, 28, 333–40.
Petkovsěk, Z. (1980) 'Additional relief meteorologically relevant characteristics of basins', *Met. Zeit.*, 28, 333–40.
Porch, W.H., Fritz, R.B., Coulter, R.L. and Gudiksen, P.H. (1989) 'Tributary, valley, and sidewall air flow interactions in deep valley', *J. appl. Met.*, 28, 578–89.
Prandtl, L. (1952) *Essentials of Fluid Dynamics* (transl. of 1949 edn of Führer durch die Strömungslehre), pp. 422–5, New York, Hafner Publishing Co.
Reiher, M. (1936) 'Nächtlicher Kaltluftfluss an Hindernissen', *Bioklim. Beiblätter* (Braunschweig); 3, 152–163.
Steinacker, R. (1984) 'Area-height distribution of a valley in its relation to the valley wind', *Contr. Atmos. Phys.*, 57, 64–71.
Steinacker, R. (1987) 'Zur Ursache der Talwindzirkulation', *Wetter u. Leben*, 39, 61–4.
Sterten, A.K. (1965) 'Alte und Berg- und Talwindstudien', *Carinthia II, Sonderheft*, 24, pp. 186–94, Vienna.

Sterten, A.K. and Knudsen, J. (1961) *Local and Synoptic Meteorological Investigations of the Mountain and Valley Wind System*, Kjeller-Liuestrom, Forsvarets Forskingsinstitutt, Norweg. Def. Res. Establ. Internal Rep. K-242.

Streten, N.A., Ishikawa, N. and Wendler, G. (1974) 'The local wind regime on an Alaskan glacier', *Arch. Met. Geophys. Biokl.*, B, 22, 337–50.

Streten, N.A. and Wendler, G. (1968) 'Some observations of Alaskan glacier winds', *Arctic*, 21, 98–102.

Tang, W. (1976) 'Theoretical study of cross-valley wind circulation', *Arch. Met. Geophys. Biokl.*, A, 25, 1–18.

Thompson, B.W. (1986) 'Small-scale katabatics and cold hollows', *Weather*, 41, 146–53.

Thyer, N.H. (1966) 'A theoretical explanation of mountain and valley winds by a numerical method', *Arch. Met. Geophys. Biokl.* A, 15, 318–47.

Tollner, H. (1931) 'Gletscherwinde in den Ostalpen', *Met. Zeit.*, 48, 414–21.

Tyson, P.D. (1968a) 'Velocity fluctuations in the mountain wind,' *J. Atmos. Sci.*, 25, 381–4.

Tyson, P.D. (1968b) 'Nocturnal local winds in a Drakensberg valley', *S. Afr. Geog. J.*, 50, 15–32.

Tyson, P.D. and Preston-Whyte, R.A. (1972) 'Observations of regional topographically-induced wind systems in Natal', *J. appl. Met.*, 11, 643–50.

Urfer-Henneberger, C. (1967) 'Zeitliche Gesetzmässigkeiten des Berg und Talwindes', *Veröff. Schweiz. Met. Zentralanstalt*, 4, 245–52.

Urfer-Henneberger, C. (1970) 'Neurere Beobachtungen über die Entwicklung des Schönwetterwindsystems in einem V-förmigen Alpental (Dischma bei Davos'), *Arch. Met. Geophys. Biokl.* B, 18, 21–42.

Vergeiner, I. and Dreiseitl, E. (1987) 'Valley winds and slope winds – observations and elementary thoughts', *Met. Atmos. Phys.*, 36, 264–86.

Voights, H. (1951) 'Experimentalle Untersuchungen über den Kaltluftfluss in Bodennähe bei verschiedenen Neigungen und verschiedenen Hindernissen', *Met. Rdsch.*, 4, 185–8.

Vorontsov, P.A. (1958) 'Nekotorie voprosi aerologicheskikh issledovanii po granichnogo sloia atmosferi (Some questions on aerological observations of the atmospheric boundary layer)', *Sovremenie Problemy Meteorologii Prizemnogo Sloia Vozdukha: Sbornik Statei*, pp. 157–79, Leningrad, Glav. Geofiz Observatory.

Wagner, A. (1938) 'Theorie und Beobachtungen der periodischen Gebirgswinde', *Gerlands Beitr. Geophys.*, 52, 408–49 (translated as 'Theory and observation of periodic mountain winds', in Whiteman, C.D. and Dreiseitl, E. (eds) (1984) *Alpine Meteorology*, PNL-5141, ASCOT-84-3, pp. 11–43, Richland, WA: Baltelle, Pacific Northwest Laboratory).

Wenger, R. (1923) 'Zur Theorie der Berg-und Thalwinde', *Zeit Met.*, 40, 193–204.

Whiteman, C.D. (1982) 'Breakup of temperature inversions in deep mountain valleys. Part I. Observations', *J. appl. Met.*, 21, 270–89.

Whiteman, C.D. (1990) 'Observations of thermally developed wind systems in mountainous terrain', in W. Blumen (ed.) *Atmospheric Processes over Complex Terrain, Met. Monogr.* 23(45), pp. 5–42, Boston, Mass, Amer. Met. Soc.

Whiteman, C.D. and McKee, T.B. (1977) 'Observations of vertical atmospheric structure in a deep mountain valley', *Arch. Met. Geophys. Biokl.*, A, 26, 39–50.

Wippermann, F. (1984) 'Airflow over and in broad valleys: channelling and counter-current', *Contrib. Atmos. Phys.*, 57, 92–105.

Yoshino, M.M. (1957) 'The structure of surface winds over a small valley', *J. Met. Soc. Japan*, 35, 184–95.

Yoshino, M.M. (1975) 'Climate in a small area', *An Introduction to Local Meteorology*, Tokyo, University of Tokyo Press.

Models of the orographic wind field

Anderson, G.E. (1971) 'Meso-scale influences on wind fields', *J. appl. Met.*, 10, 377–86.

Barry, R.G. and Perry, A.H. (1973) *Synoptic Climatology: Methods and Applications*, Methuen, London, 555 pp.

Danard, M. (1977) 'A simple model for mesoscale effect of topography on surface winds', *Mon. Wea. Rev*, 105, 572–81.

Dickerson, M.H. (1978) 'MASCON – A mass consistent atmospheric flow model for regions with complex terrain', *J. appl. Met.*, 17, 241–53.

Fosberg, M.A., Marlatt, W.E. and Krupnak, L. (1976) *Estimating Airflow Patterns over Complex Terrain*, Fort Collins, US Dept. of Agriculture, Forest Service, Res. Pap. RM-162.

Giorgi, F. (1990) 'Stimulation of regional climate using a limited area model nested in a general circulation', *J. Climate* 3, 941–63.

Giorgi, F. and Bates (G.T. (1989) 'On the climatological skill of a regional model over complex terrain', *Mon. Wea. Rev.*, 117, 2325–47.

Lavoie, R.L. (1974) 'A numerical model of the trade-wind weather on Oahu', *Mon. Wea. Rev.*, 102, 630–7.

Mahrer, Y. and Pielke, R.A. (1975) 'A numerical model of the airflow over mountains using the two-dimensional version of the University of Virginia meso-scale model', *J. Atmos. Sci.*, 32, 2144–55.

Mahrer, Y. and Pielke, R.A. (1977) 'A numerical study of the airflow over irregular terrain', *Beitr. Phys. Atmos.*, 50, 98–113.

Mass, C.F. and Dempsey, D.P. (1985) 'A one-level, mesoscale model for diagnosing surface winds in mountainous and coastal regions', *Mon. Wea. Rev.*, 113, 1211–27.

Nappo, C.J. Jr. (1977) 'Meso-scale flow over complex terrain during the eastern Tennessee trajectory experiment-(ETTEX)', *J. appl. Met.*, 16, 1186–96.

Overland, J.E., Hitchman, M.H., Han, Y-J. (1979) 'A regional surface wind model for mountainous coastal areas', NOAA-TR-ERL 407, Pacific Marine Environmental Lab. 32, Seattle: NOAA.

Pielke, R. (1984) *'Mesoscale Meteorological Modeling'*, Orlando, Florida, Academic Press.

Ross, D.G. Smith, I.N., Marius, P.C. and Fox, D.G. (1988) 'Diagnostic wind field modeling for complex terrain: Model development and testing', *J. Clim. appl. Met.*, 27, 785–96.

Ryan, B.C. (1977) 'A mathematical model for diagnosis and prediction of surface winds in mountainous terrain', *J. appl. Met.*, 16, 571–84.

Schwabl, W. (1934) 'Zur Kenntnis der Beeinflussung der Allgemeinströmung durch ein Gebirgstal,' *Met. Zeit*, 51, 342–5.

Sherman, C.A. (1978) 'A mass-consistent model for wind fields over complex terrain', *J. appl. Met.*, 17, 312–19.

Sommers, W.T. (1976) 'Mascon – A mass-consistent atmospheric flux model for regions with complex terrain', *J. appl. Met.*, 17, 241–53.

Tesche, T.W. and Yocke, M.A. (1978) 'Numerical modeling of wind fields over mountain regions in California', in *Conference on Sierra Nevada Meteorology* (Preprint Volume), pp. 83–90. Boston, American Meteorological Society.

4

CLIMATIC CHARACTERISTICS OF MOUNTAINS

The basic factors and processes affecting mountain climate have been discussed in Chapters 2 and 3. When climatic elements such as temperature or precipitation are considered, their temporal and spatial characteristics in mountain areas are inevitably determined by the total complex of these factors – latitude, altitude and topography – operating together. In this chapter, therefore, some general climatic characteristics of mountain areas are examined for individual climatic elements. We begin by considering energy budgets and slope temperature profiles. This is followed by a discussion of cloudiness, precipitation, other hydrometeors, and evaporation. The ways in which altitudinal and topographic effects, in particular, interact to create orographic patterns in the spatial and temporal distribution of each climatic element are illustrated.

ENERGY BUDGETS

It was noted in Chapter 2 that mountain sites were of special importance to early research on solar radiation, but there has been a general lack of modern radiation and energy budget studies in the mountains. An adequate level of information on the spatial and temporal distribution of radiation exists only for the European Alps. This material provided the basis for the generalizations on altitudinal effects presented in Chapter 2 (pp. 29–44) and all that can be usefully added here is to illustrate the types of work carried out in a few other mountain areas and some of the findings.

The intensive radiation studies in the Alps show that, up to about 3 km, global solar radiation increases by about 7–10 per cent km^{-1} under clear skies and 9–11 per cent km^{-1} under overcast, with the most rapid increase occurring in the lower levels, due to the vertical distribution of water vapour (p. 28). Data in the Valais of the south-western Alps for global solar radiation, show a small increase of only 2.6 per cent km^{-1} between 2 and 3.5 km in summer, whereas downward IR radiation decreases by 10 per cent km^{-1} (Müller, 1985).

Cloud conditions significantly affect the theoretical rates of altitudinal increase, particularly since cloud cover may be more frequent on mountain slopes. This is the case on the east slope of the Front Range, Colorado, where there is no altitudinal increase on an annual basis (Barry 1973). However, in individual months there may be an increase or decrease with height, depending on the frequency of cloud cover. This is apparent in Table 4.1. Concurrent cloud observations are not available to evaluate the specific cloud-radiation relationships, but Greenland (1978) illustrates the magnitude of probable differences due to cloud cover by comparison of two synoptic regimes. During 16 days of upslope flow in 1977, a daily average of 142 W m^{2} was recorded at 3480 m, compared with 105 W m^{-2} in Boulder, whereas for the same number of days of downslope flow the averages were 165 W m^{-2} and 188 W m^{-2}, respectively. Upslope weather, with more cloud at low elevations on the east slope of the Front Range, is common in spring, while downslope weather, which brings cloud mainly to the high elevations, is characteristic of winter months. In summer, cumuliform clouds predominate, with greater development occurring earlier in the day over the mountains. The radiation data in Table 4.1 illustrate some of these seasonal differences.

Table 4.1 Daily averages of global solar radiation on the east slope of the Front Range, Colorado ($W m^{-2}$)

	Altitude (m)				
	1590	*2591*	*3048*	*3480*	*3750*
	Station				
	Boulder	*Sugarloaf*	*Como*	*Niwot Ridge (tree line)*	*Niwot Ridge*
	Distance E of Divide (km)[a]				
Period	*37*	*22.5*	*9.7*	*4.5*	*2.6*
January–February 1965[b]	–	98	117	–	112
March–May	–	184	210	–	221
June–September	–	229	185	–	199
October–December	–	119	111	–	103
January–February 1977[c]	96	–	–	98	–
March–May	220	–	–	222	–
June	262	–	–	248	–

Notes: [a]The Continental Divide in this area is approximately 4000 m; [b] Bimetallic actinographs, calibrated against a 50 junction Eppley pyranometer; [c] Eppley pyranometers.
Source: After Greenland 1978

At high elevation stations, 5000 m and above, cloud layers are frequently thin and have high transmissivities. Thus, for Mt. Logan (5365 m), Yukon, the difference between clear and cloudy conditions (which averaged about 7/10 cover) was only 18 per cent in July 1970 (Brazel and Marcus 1979). Data for July 1968–70 show that, whereas solar radiation at Kluane Lake (787 m) averaged only 43 per cent of possible, the figure was 83 per cent for Mt. Logan (Marcus and LaBelle 1970; Marcus and Brazel 1974). The difference implies an average altitudinal increase of 8.7 per cent km^{-1}; the rate is greatest at low levels decreasing to about 6 per cent km^{-1} above 2650 m.

Complete energy budget studies in mountain areas are rare and there are very few measurements in the absence of snow cover. In the equatorial Andes, Korff (1971) showed that net radiation for 12 days in July on Cotapaxi at 3570 m, with a surface albedo of 0.22, averaged 60 per cent (53 per cent on clear days) of the incoming short-wave. These figures are in line with data given by Voloshina (1966) for the Caucasus at 3000–3500 m (see Figure 2.13, p. 43). At 4750 m on Mt. Everest (28°N) the net radiation on 9 days in April 1963 represented 55 per cent of the incoming short-wave with a surface albedo of 0.16 (Kraus 1971).

The albedo of alpine tundra surfaces shows considerable temporal and spatial variability – of the order of 25 per cent of the mean values for densely vegetated meadow and shrub tundra (0.18), sparsely vegetated fellfield (0.27) and krummholz (0.15) (Goodin and Isard 1989). Temporal variations are caused particularly by varying moisture conditions; spatial variations by the heterogeneous surfaces. The latter can be modelled in terms of canopy structure (foliage inclination, canopy strata and plant area) as illustrated for light extinction in alpine plant communities by Tappenier and Cernusca (1989).

Most work has been performed on glacier surfaces during the summer ablation season. Table 4.2 summarizes some of the available literature results, but these are insufficient to form a reliable basis for general conclusions. Data for glacier surfaces in the Caucasus show a distinct decrease in the ratio of net radiation to incoming solar radiation with altitude (Figure 2.13, p. 43). In fact, as the Quelccaya and Mt. Logan data in Table 4.2 show, over snow surfaces at high altitudes net radiation is frequently negative. The difference in net radiation over snow/ice and snow-free surfaces is discussed by Voloshina (1966). For 20 days at 3250 m on Karachaul Glacier (Mt. Elbrus), the net radiation is 119 W m^{-2} on bare ground (albedo of 0.10) compared with 111 W m^{-2} on ice (albedo of 0.37) only 200 m away. While the absorbed solar radiation is much greater on the bare ground, the effective back = radiation is at least two times greater than over the ice, leading to the near equality.

A major component of the energy budget in snow-covered areas is the summer snow melt. Snow melt rates are of particular significance for forecasting runoff from alpine basins. In the early melt phase, meltwater is retained in the snow until the free (liquid) water content of the pack (about 2 per cent by volume) is attained. Nocturnal refreezing may also take place at this stage (Martinec 1989). Measurements for a complete melt season (9 May–15 July 1985) made at the

Table 4.2 Selected energy budget data

Location	*Altitude*	*Month*	*S* $(W\,m^{-2})$	α	R_n $(W\,m^{-2})$	*H*	*LE* (% of R_n)	G	*References*
Everest (28°N)	4750	April	291	0.16	161	60	36	4	Kraus 1971
Niwot Ridge (39°N)	3650	July	252	0.17	144	50	38	12	LeDrew and Weller 1978
Turkestan Mts (41°N)	3150	September	–	0.14	169	71	14	14	Aizenshtat 1962
Turkestan Mts (31° south slope)	3150	September	–	0.20	70	59	28	13	Aizenshtat 1962
Turkestan Mts (31° south slope)	3150	September	–	0.15	206	71	18	11	Aizenshtat 1962
Austrian Alps (47°N)									
Hohe Mut	2560	July	367	~0.23	183	28	64	8	Rott 1979
Obergurgl	1960	July	331	~0.20	157	9	89	2	Rott 1979
Quelccaya Ice Cap (14°S)	5645	July	244	0.80	2	–	–	–	Hastenrath 1978
Mt. Logan (61°N) (snow surface)	5365	July	373	0.84	57[a]	10	90[b]	~0	Brazel and Marcus 1979
McCall Glacier (69°N)									
ice surface	1730	July	226	0.33	72	(78)[c]	7	13	Wendler and Ishikawa 1973
moraine	1740	July	226	0.19	71	49	43	8	Wendler and Ishikawa 1973
snow surface	2140	July	226	0.59	43	(59)[c]	63	1	Wendler and Ishikawa 1973

Notes: [a] *H* and *LE* here are towards the surface; [b] Includes sublimation; [c] *H* here is towards the surface.

Weissflujoch, Switzerland (2540 m), indicate snowmelt of about 30 cm water in both May and June and 40 cm in July. Martinec shows that net radiation accounts for 60 per cent of melt, sensible and latent heat for 40 per cent. Evaporation losses from the snow cover were estimated at 4 per cent.

One of the most extensive series of alpine measurements is that of LeDrew (1975; LeDrew and Weller 1978). Averages for the growing season at 3650 m on Niwot Ridge, Colorado, show large sensible heat fluxes due to the generally strong advection of cool westerly airstreams. The Bowen ratio ($\beta = H/LE$) ranges between 4 and 6 during daylight hours, indicating much higher moisture stresses than implied by the mean value in Table 4.2. Surface canopy temperatures on 41 days around solar noon averaged 28°C compared with 14°C at screen level (1.5 m). Measurements at Mount Werner (3200 m), Colorado, during August 1984 show a similar range to those of LeDrew for Niwot Ridge (Sheaffer and Reiter 1987) although the ground heat flux is much smaller and the Bowen ratio at that site averaged only 0.6.

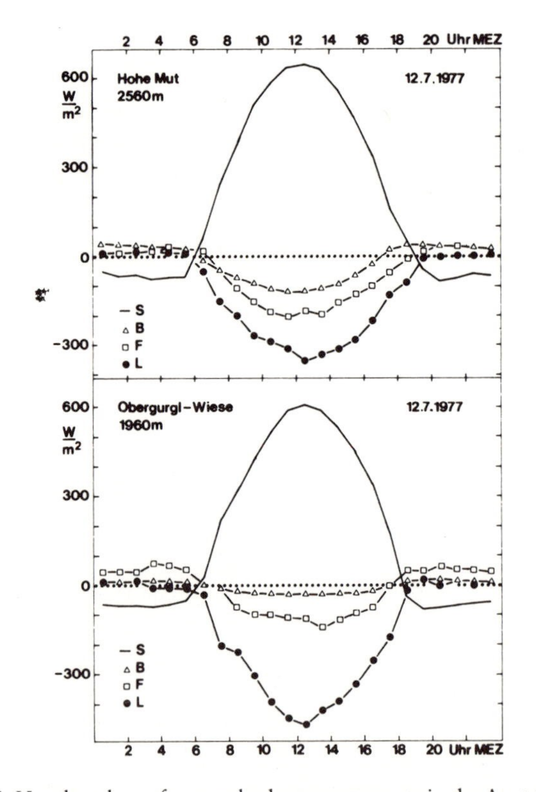

Figure 4.1 Hourly values of energy budget components in the Austrian Tirol, 12 July 1977, at Obergurgl–Wiese (1960 m) and Hohe Mut (2560 m)

Notes: S = net radiation; F = sensible heat flux; B = soil heat flux; L = latent heat flux; MEZ = Central European Time

Source: From Rott 1979

In the Austrian Alps, Rott (1979) provides comparative data at two elevations. Figure 4.1 illustrates his results on a clear day in July. The solar incoming and net radiation values are, respectively, about 10 per cent and 15 per cent greater at Hohe Mut (2560 m) than at Obergurgl (1960 m) for cloudless conditions (Table 4.2). This is attributed to greater horizon screening at Obergurgl, as well as to the larger atmospheric extinction effect. Latent heat accounts for a larger fraction of the available energy at the lower station, over a meadow, than over alpine grass heath at the higher station. During night-time hours, under cloudless skies, the negative radiation balance averaging about 70–75 W m^{-2} is largely offset by sensible heat flux from the air at Obergurgl, but by both ground heat flux and sensible heat flux at Hohe Mut (Rott 1979). For the Dischma Valley, near Davos, the energy balance has been determined using observations at three stations, measurements of surface temperature from an airborne infra-red thermometer, a map of the vegetation and a digital elevation model (Hennemuth and Köhler 1984). Net radiation is shown to be determined mainly by topographic exposure and latent heat flux by vegetation cover.

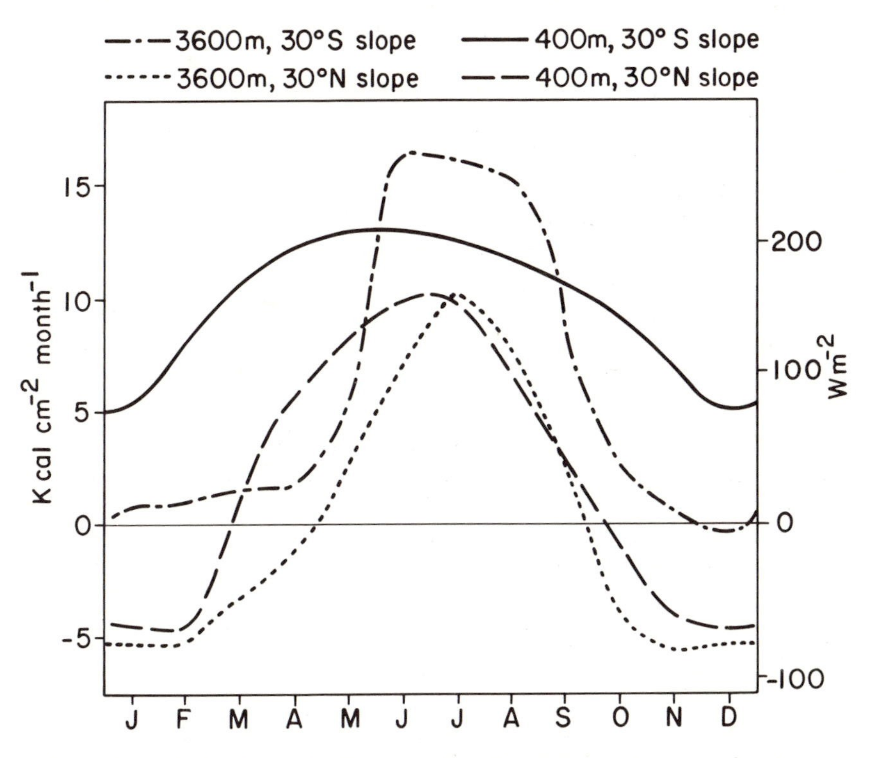

Figure 4.2 Calculated net radiation on north- and south-facing slopes of 30° in the Caucasus at 400 m and 3600 m

Source: Based on Borzenkova 1967

In view of the paucity of observational data for mountain regions, several sources have provided calculations of net radiation and energy fluxes. Figure 4.2 illustrates calculated clear-sky net radiation on north and south-facing slopes at 400 m and 3600 m in the Caucasus, based on temperature and albedo data at Tbilisi and Kasbegi and equations of Kondratyev and Manolova (Borzenkova 1967). Mean albedos are 17–21 per cent at Tbilisi and range from 75 per cent in March–April to 20 per cent in August at Kasbegi, due to snow cover (Borzenkova 1965). On an annual basis, net radiation decreases 15–16 per cent km^{-1} for both cloudless and overcast skies. In June, however, there is a decrease of 7 per cent km^{-1} for cloudless skies, but an increase of almost 4 per cent km^{-1} with overcast conditions. This may not be wholly real since the calculations assume low cloud layers whereas the cloud at 3600 m is certainly less opaque than at low levels. Nevertheless, there is observational evidence from the Alps for a summer increase of net radiation with height (Table 2.11, p. 49).

For the same general area, Borzenkova (1965) also provides calculated annual values of turbulent heat fluxes based on the methods of M. Budyko (Table 4.3). In both the Greater Caucasus and the Armyanski Mountains, the sensible heat flux decreases steadily with elevation, whereas the latent heat flux reaches a maximum at 2500–3000 m. The sensible heat loss is slightly greater in the drier, more southerly range.

Similar estimates on a monthly basis have been prepared for Croatia (Pleško and Šinik 1978) also using equations of Budyko. The stations range from 120–1594 m, at approximately 45°–46°N. Figure 4.3 shows the seasonal variation with height of R_n, H and LE. There is virtually no height variation for any of the components in winter. In summer, on the other hand, R_n is more or less constant to about 700 m and then decreases slightly above that height. LE increases likewise, perhaps due to higher rainfall, and then decreases above about 800 m possibly as a result of increased infra-red radiative cooling and lower surface temperatures, according to Pleško and Šinik. The sensible heat flux is slightly

Table 4.3 Annual turbulent fluxes in the Caucasus ($W m^{-2}$)

	Greater Caucasus			*Armyanski Mts*		
	Sensible heat	*Latent heat*	*Bowen ratio*	*Sensible heat*	*Latent heat*	*Bowen ratio*
Surface	44	27	1.65	–	–	–
1 km	29	44	0.67	32	24	1.33
2	19	55	0.34	23	28	0.81
3	9	44	0.21	13	35	0.38
3.5	4	28	0.14	–	–	–

Source: After Borzenkova 1965

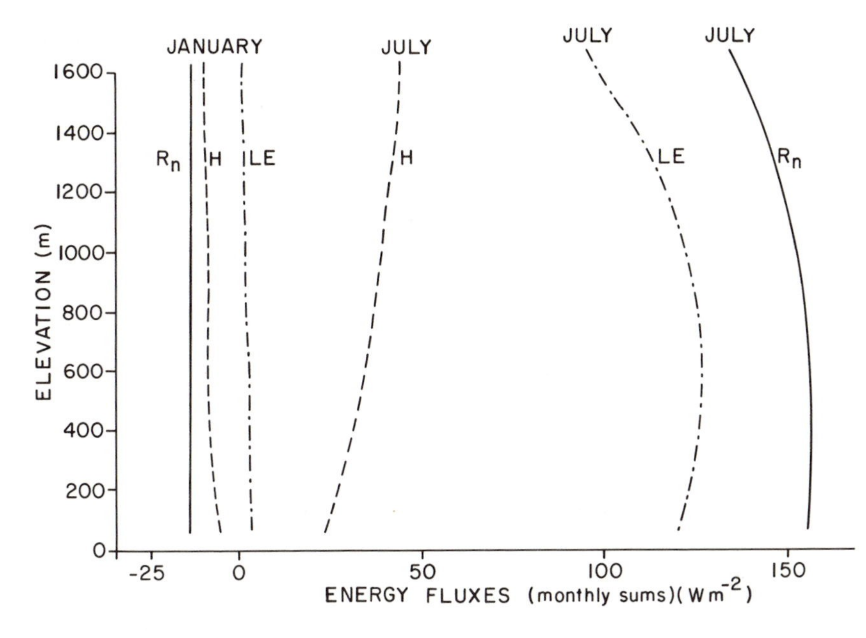

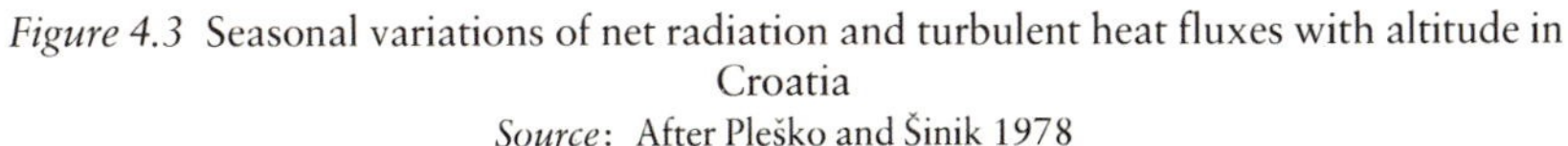

Figure 4.3 Seasonal variations of net radiation and turbulent heat fluxes with altitude in Croatia
Source: After Pleško and Šinik 1978

negative in winter, in association with winter temperature inversions. In summer, *H* increases with height, apparently in response to increased advection effects and a greater air–surface temperature gradient. This pattern is the opposite of that calculated by Borzenkova for the Caucasus area. The results for Croatia seem consistent, although over snow cover, of course, *H* will be zero or negative.

TEMPERATURE

Slope profile

Temperature is undoubtedly the single most important aspect of mountain climates. It has been widely observed in most mountain areas of the world and there are many statistical studies dealing with its altitudinal variation. This variation is a particular problem for climatic atlases, due to the sharp temperature gradients over short distances and their seasonal variability (Steinhauser 1967). Several investigations of mountain temperatures use regression analysis to relate temperatures to altitude and to separate between effects of slope-aspect and inversions, for example, Douguédroit and de Saintignon (1970) and de Saintignon (1976). Pielke and Mehring (1977) use linear regression analysis of mean monthly temperatures as a function of elevation in an attempt to improve the spatial representation of temperature for an area in north-western Virginia.

They show that the correlations are greatest ($r = -0.95$) in summer, as is commonly the case in middle latitudes. Inversions at low levels in winter introduce greater variability and better estimates may be obtained by fitting polynomial functions, or alternatively by the use of potential temperatures (Hennessy 1979). Numerous regression equations have been worked out in a similar fashion for the West Carpathian Mountains (Hess *et al.* 1975) in order to produce topoclimatic maps. For this purpose, separate regressions are used according to the slope profile, as described in Chapter 2, pp. 40–2. Remarkably, there have been few attempts to describe mountain temperature variations with any more general statistical model.

Although altitudinal and aspect effects determine the *mean* seasonal values of temperature (and other variables) in a mountainous area, there is some indication that short-term deviations, associated with synoptic weather events, are not so influenced. Furman (1978) shows that daily maximum temperatures (T_1) in summer at stations on a forested ridge in Idaho can be described by a second-order autoregressive model $T_i = a_1 T_{i-1} + a_2 T_{i-2}$. For seven sites over a 1000 m height range, with north and south aspects, a_1 was between 0.795 and 0.980, indicating a strong dependence on the previous day's value (T_{i-1}), whereas a_2 was between −0.004 and −0.176. The differences in coefficients between sites were shown to be a result of sampling variations, and the probability distribution of the residuals also showed no effects of location. It remains to be determined how widely these results are applicable in situations where the mountain scale has a greater range. Also, whereas daily maxima bear a close relationship to global radiation totals, minimum temperatures are much more site-dependent, especially with respect to cold air drainage.

Most studies of slope lapse rates rely on data from a summit and a base station, although in a few cases attempts have been made to provide detailed transects. C.L. Wragge and assistants, for example, made daily ascents of Ben Nevis in the summers of 1882 and 1883, taking observations at eight slope stations (Buchan 1890). However, such transects suffer from the problem of time differences between the observations, requiring averages to be taken over the ascent and descent.

Wagner (1930) provided a specimen continuous transect on the Nordkette, north of Innsbruck, by use of the cable car. Temperatures were determined every five seconds with an aspirated thermometer suspended from the car. Figure 4.4 shows four profiles: the first and third upward, between 08.00 and 08.50 hours and between 15.00 and 15.50 hours, the second and fourth downward, between 10.15 and 11.00 hours and between 18.05 and 18.35 hours. The morning was mainly overcast, clearing completely before noon, with cumulus development in the afternoon and a clear sky again by evening. The morning profile shows only a 4°C change over the 1700 m interval, but by mid-morning there is a marked warming on the steeper south-facing slopes – below the Hafelkar, and at lower levels below the Hungerburg (although here the thermometer was only about 1 m above the track). There is also a considerable increase in micro-scale irregu-

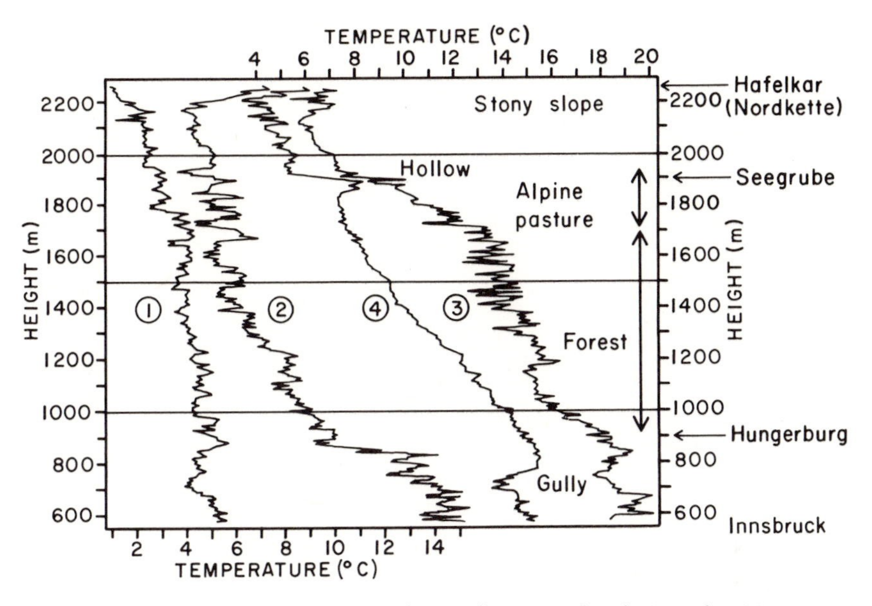

Figure 4.4 Slope temperature on the Nordkette, Innsbruck, 2 April 1930
Source: After Wagner 1930 and see text

larity between 08.30 and 10.30 and this feature is even more pronounced in the afternoon, when it is concentrated in the forested section (1300–1700 m). Wagner suggests that this irregular pattern is due to the canopy acting as a primary heating surface. The sharp cooling just above Seegrube, evident in plots 3 and 4, identifies a hollow where the cable car is up to 70 m above ground. The effect of a shady gully is also evident just below 750 m (except at mid-morning). The final transect, when the whole slope was in shadow, shows a markedly smooth profile. The relative warmth of the upper forest section has now gone and the mean temperature gradient between 900 and 2200 m is 0.85°C/100 m, almost adiabatic.

Another transect by Wagner on 23 November 1929 illustrated the descent of a föhn layer over a surface-based inversion. The latter persisted through a layer only some 60 m deep for eight hours during the day before the föhn reached the valley floor.

In recent work on the Zugspitze, a similar approach has been adopted by Reiter and Sladkovic (1970). Lapse rates determined from the cable car are compared with the exchange coefficient for eddy diffusivity based on radon (RaB) measurements. For days without fog or precipitation and $\leqslant 5/10$ high cloud, they find correlations of about -0.7 under all conditions in the layer between 1800 and 3000 m. Similar values were also obtained under non-stable conditions between 700 and 1800 m, but there is no correlation between lapse rate and exchange coefficient in this layer during stable conditions. Their analysis provides information that is of value in assessing the potential for aerosol

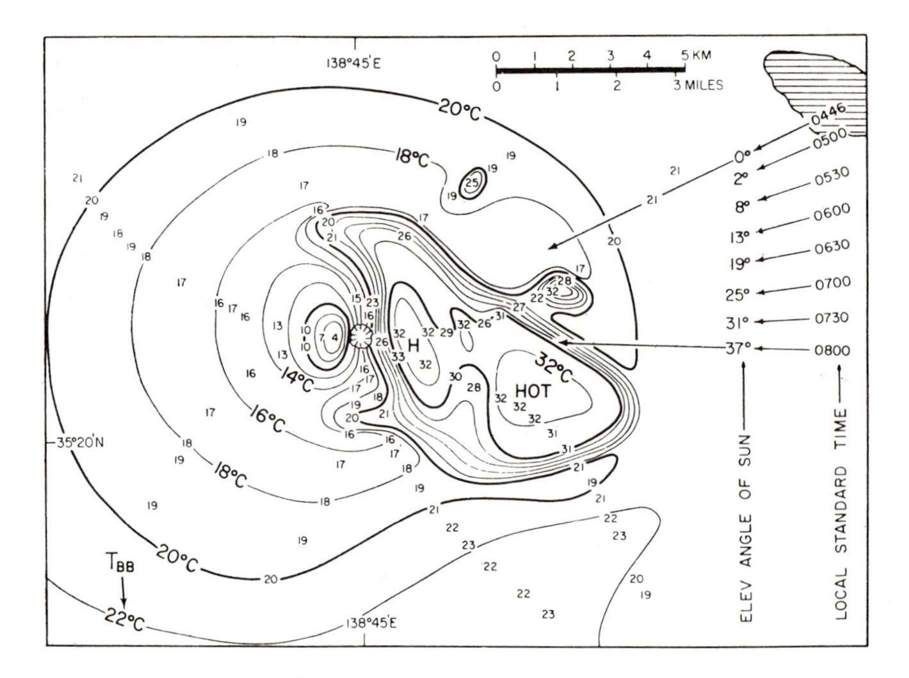

Figure 4.5 Equivalent black-body temperatures (T_{BB}) measured by radiation thermometer from an aircraft over Mt. Fuji, 07.43–07.56 JGT, 28 July 1967. Solar altitude and azimuth is shown since sunrise.
Source From Fujita *et al.* 1968

dispersion in mountain environments (see Chapter 6, pp. 258–63).

Such case studies provide useful insight into small-scale spatial and temporal variability and illustrate the need for care in siting stations and in interpreting slope lapse-rates based on data at only a few stations at fixed times. Aerial surveys using a radiation thermometer (Fujita *et al.* 1968) offer another means of determining 'apparent' (radiative) temperatures over mountain slopes, but such data are expensive to acquire and are not readily converted to absolute values. Also, the spot size viewed by the instrument is likely to be of the order of several hundred metres across and vary irregularly as the aircraft crosses the terrain. Nevertheless, the work of Fujita *et al.* over Mt. Fuji provided the interesting observation that solar heating of rocky slopes can cause nearly identical temperatures at the ground surface with negligible altitudinal differences (Figure 4.5). Independent slope measurements under similar conditions showed that, in contrast to the rapid surface heating, the rise in screen temperatures was very gradual.

Observed screen temperatures on slopes can be adjusted to estimate 'free air' temperature in a valley atmosphere if supplementary information is available. Dreiseitl (1988) uses a four-year record of hourly readings at six stations on the north side of the Inn Valley, Austria, to develop correction functions for each station, between 580 m and 2260 m altitude, that take account of day length, weather type, and cloudiness. The lower slopes are several degrees warmer than the valley air in the afternoon and slightly cooler in the morning. The seasonal values of corrected daily temperature range agree well with those derived from diurnal pressure fluctuations at Innsbruck and the summit station.

A technique for predicting surface temperature in mountainous terrain using 12 and 24-hour regional forecast data from the US National Weather Service Limited-Area Fine Mesh (LFM) model has been developed by McCutchan (1976). Diurnal temperature variation (T) is represented by the first two harmonic terms:

$$T_t = A\left[\left(a_0 + a_1 \cos\frac{\pi t}{12} + b_1 \sin\frac{\pi t}{12} + a_2 \cos\frac{\pi t}{6} + b_2 \sin\frac{\pi t}{6}\right) - B_t\right]$$

where A = an aspect pararameter relating to solar radiation,
a_0 = mean temperature,
a_1, b_1 = coefficients of the first harmonic,
a_2, b_2 = coefficients of the second harmonic,
and B_t = a bias condition function which depends on time, synoptic weather category, and elevation.

Five synoptic weather categories relating to temperature and moisture conditions were determined separately by discriminant analysis. The model first calculates each of the Fourier coefficients independently using stepwise regression analysis based on observed temperature, dew point and wind speed at the surface, 850 mb chart data, weather class, the sine and cosine of the Julian day, and 12-hour and 24-hour predictions from the LFM. On this basis, McCutchan developed surface temperature predictions up to 36 hours in advance for the San Bernardino Mountains, southern California. In areas where there are less detailed forecasts, other techniques would have to be evolved, but, in any case, a key factor would be the siting of suitable reference stations.

Thermal belts

The effect of nocturnal radiation on downslope air drainage has been examined in an earlier section (p. 158). The result of such conditions, in clear calm weather, is the formation of a pond of cold air in valley bottoms with high temperatures on the slope: the latter zone is referred to as the *thermal belt* (Figure 4.6). The first description of such zones is attributed to Silas McDowell, a farmer in the southern Appalachian Mountains of North Carolina in 1861 (Chickering 1884; Dunbar 1966). In this area the thermal belt is centred, on average, about 350 m above the valley floors (Cox 1923). The reduced risk of spring frosts in this zone makes it of importance to agriculture and horticulture.

Extensive studies of thermal belts and inversion heights on valley slopes have

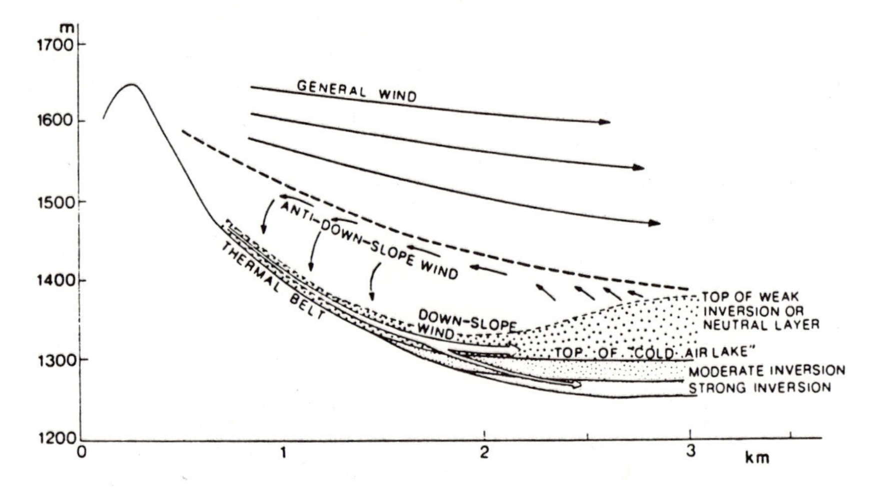

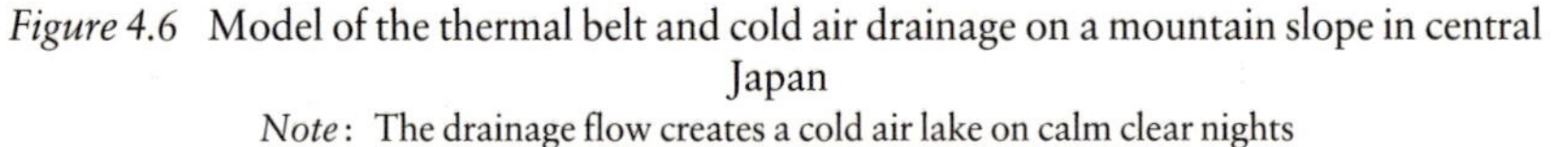

Figure 4.6 Model of the thermal belt and cold air drainage on a mountain slope in central Japan
Note: The drainage flow creates a cold air lake on calm clear nights
Source: From Yoshino 1984

subsequently been performed in many parts of the world. In Bavaria, R. Geiger and colleagues investigated the vertical temperature structure on Grosse Arber (1447 m) and demonstrated the more moderate diurnal temperature range on the slope, particularly during clear weather in spring. Slope and summit locations also have a corresponding lower range of relative humidity than the valley floor (Geiger 1965: 432–7). In the same area, detailed climatological and phenological studies have been made by Baumgartner (1960–2) on Grosse Falkenstein (1312 m). The 'growing season' (defined phenologically) is shown to be 1–2 weeks longer in the thermal belt than on the valley floor 200 m below, or 100 m or so higher up the slope. An effect is also apparent in earlier snow melt (Waldemann 1959). Hence, this mainly clear weather phenomenon is of sufficient magnitude to be manifested in a climatological sense. Observations in the Ötztal near Obergurgl, Austria, by Aulitsky (1967) from June 1954 to May 1955 show that the inversion intensity averages 3°C in winter and 1.5°C in summer for mean monthly minimum temperatures but only 1°C and 0.2°C, respectively, for monthly mean temperatures (Figure 4.7).

There has been considerable discussion as to the elevation of the thermal belt with respect to the valley floor. Obrebska-Starkel (1970) summarizes data from valleys in Europe on the mean *upper* limit of the inversion layer, capping the thermal belt, which approximates the centre of the belt (see also Yoshino 1975: 434). These figures indicate that in hilly terrain (the Mittelgebirge) with a relative relief of 500 m or less, the centre of the thermal belt is typically 100–400 m above the valley floor. In high mountains, Aulitsky's work in the Ötztal suggests that it is centred about 350 m above the valley floor in summer and 700 m above

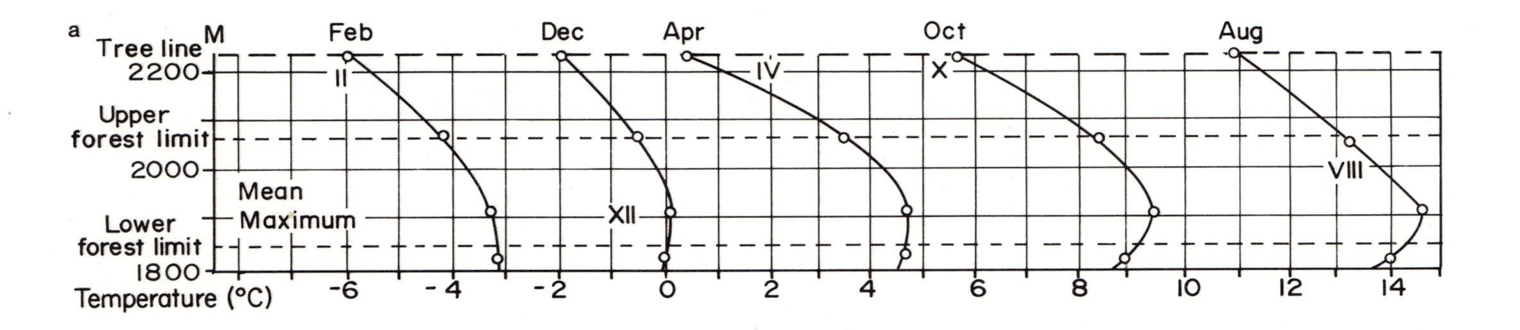

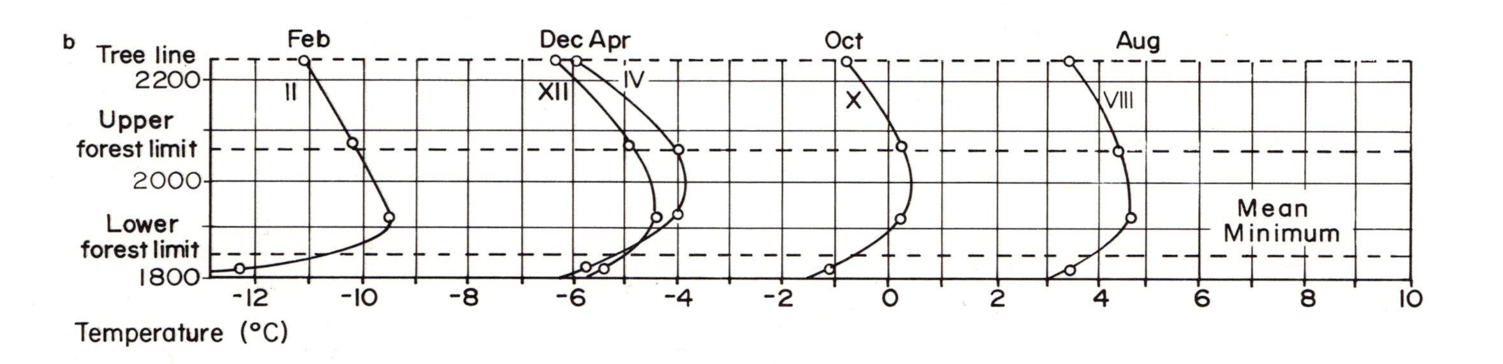

Figure 4.7 Mean monthly maxima (a) and minima (b) of air temperature (1954–55) near the forest limits on a WNW slope near Obergurgl
Source: After Aulitsky 1967

in winter. This difference is attributable to the deeper cold air in winter with the possibility of more stable inversion conditions. However, investigations by Koch (1961) show that terrain *profile* is more important than relative relief or absolute elevation in determining the height of the thermal belt. It is usually located at the steepest section of the valley side since the slope exercises a controlling effect on air drainage.

CLOUDINESS

Cloud cover is, in general, more frequent and thicker over mountains than over the surrounding lowlands. Mechanical uplift when an airstream encounters a topographic barrier is a primary cause, although this may be augmented by convective effects due to slope heating. When cloud is already present, slowing down of the air by the mountain barrier also leads to an increase in cloud water-content (Pedgley 1971). If an air parcel is forced to rise it expands, due to the lower ambient air pressure, and therefore cools. This assumes no exchange of heat between the parcel and its surroundings, i.e. an *adiabatic process*. The rate of cooling, for unsaturated air, can be determined as follows. The First Law of Thermodynamics states that the heat (dQ) supplied to a unit mass of gas must be balanced by an increase in the internal energy of the gas and the work done externally by the gas. For an adiabatic process $dQ = 0$, and the changes in internal energy and work done by a parcel of gas also sum to zero. This can be expressed:

$$c_p dT - V dp = 0,$$

where c_p = specific heat of air at constant pressure ($\sim 1.0 \times 10^3$ J kg^{-1} K^{-1})
V = specific volume of gas.

By substitution for dp from the hydrostatic equation,

$$\frac{dp}{dz} = -g\rho$$

where $g = 9.81$ m s^{-2} and ρ = air density,

and since $V\rho = 1$, the changes in the rising air parcel can also be expressed as

$$c_p dT + g dz = 0.$$

Hence, the lapse rate is

$$\frac{dT}{dz} = -\frac{g}{c_p}.$$

This *dry adiabatic lapse rate* ($\Gamma = dT/dz$) has a value of 9.8 K km^{-1}, and is negative for increasing height. If the air rises above its condensation level, cloud begins to form. Further ascent now causes cooling at some lesser rate, known as the saturated adiabatic lapse rate, due to the release of latent heat which partially offsets the cooling.

A hypothetical path curve for a rising air parcel plotted on a tephigram chart is illustrated in Figure 4.8a. This chart has coordinates of temperature and $p^{0.288}$. It may be noted that $T/p^{0.288}$ is a constant value and the potential temperature, θ, is determined from $T\,(1000/p)^{0.288}$ Figure 4.8b also shows the *lifting condensation level* – the intersection of a dry-adiabat (or θ line) through the air temperature (T_A) with a saturation mixing ratio line through the dew-point temperature (T_D); T_A and T_D are both at screen level. If we assume a daytime increase in T_A from T_0 to T_3, the *convective condensation level*, similarly determined, indicates a higher cloud base. This is characteristically observed in the course of diurnal heating. A useful empirical relationship states that the base of cumulus cloud (in metres) triggered by surface heating is approximately 120 (T_A – T_D), where the screen temperatures are in °C. Fliri (1967) has expressed the same relationship in terms of temperatures and relative humidity.

The effect of diurnal heating on the build-up of mountain cumulus in the Tirol has been described by Tucker (1954). In the morning, small fractocumulus develop over south-facing slopes in side valleys adjacent to the major valley troughs (Figure 4.9). Later, as diurnal heating raises the general condensation level, the main cumulus growth occurs over the ridge crests between the major valley systems. This lifting of the general condensation level is dramatically evidenced in the central highlands of Papua New Guinea where the nocturnal stratiform cloud in the valleys at about 1800–2000 m usually dissipates within an hour of sunrise and is replaced by fractocumulus on valley slopes. Within 2–3 hours, cumulus may extend up the higher slopes above 3500 m towards the valley heads.

Although the parcel method is a useful approximation, when an airstream crosses a mountain barrier it is, in practice, lifted as a slab. In the process of such uplift the lapse rate (γ) of the entire layer is modified: if $\gamma > \Gamma$ (a superadiabatic lapse rate), ascent decreases γ, stabilizing the layer; if $\gamma < \Gamma$ (a subadiabatic lapse rate), ascent increases γ, leading to destabilization of the layer.

Also, if the layer has a higher moisture content at lower levels, it reaches saturation more quickly for a given amount of uplift than the top of the layer. Consequently, with further lifting the lapse rate in the layer becomes greater. This is referred to as *potential* (or *convective*) *instability*.

Cloud type in mountain areas is primarily determined by air mass characteristics and is therefore related to the regional climate conditions. Special mountain cloud forms due to meso-scale air motion have been discussed already in a previous section, p. 143. On west coasts in middle latitudes, stratiform cloud is common, particularly in winter, and often occurs as hill fog, enveloping high ground. The mountains of north-western Europe and western North America experience such conditions frequently (see p. 328). If stratus cloud is already present in an air mass, due to turbulent mixing, orographic uplift tends to lower the cloud base. It is not always easy to distinguish orographically produced cloud

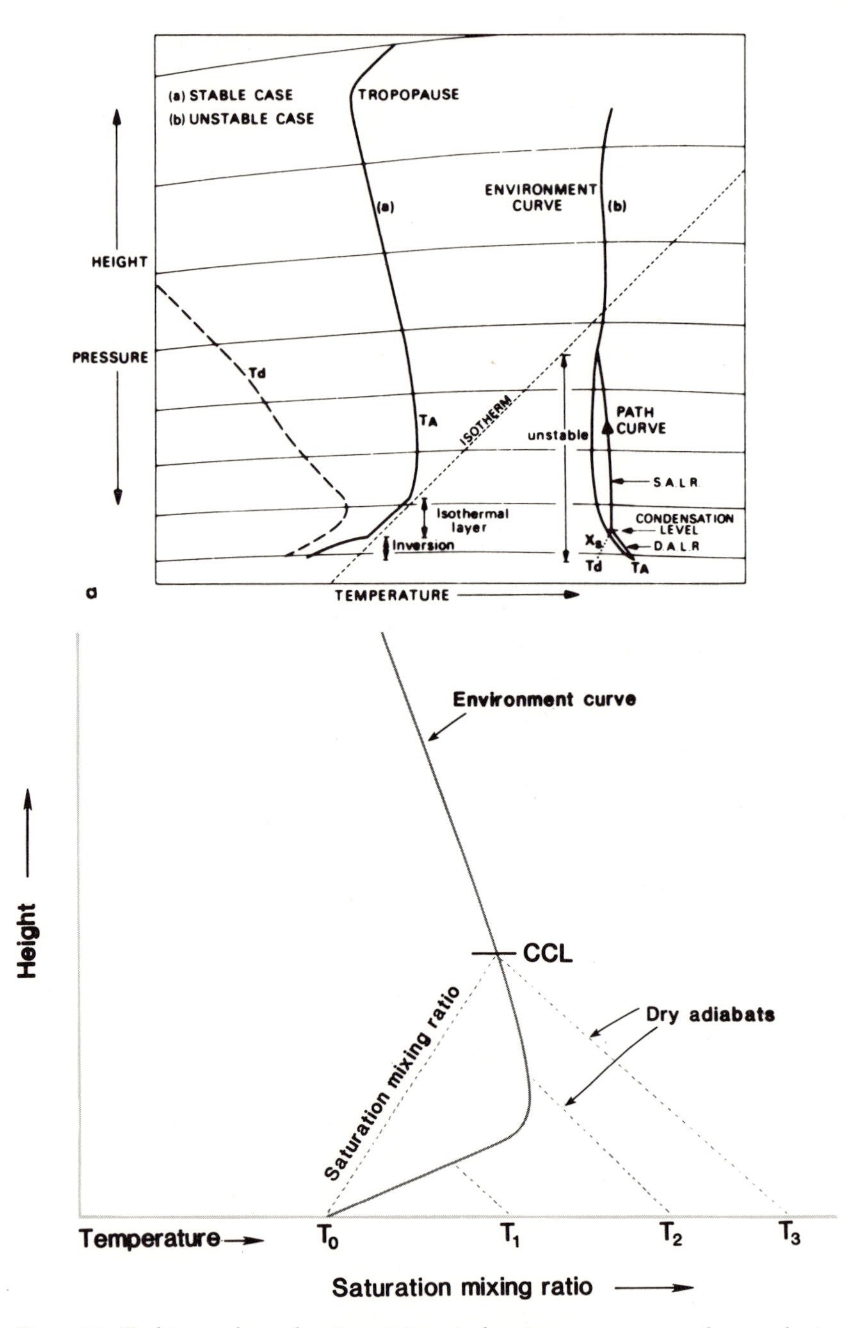

Figure 4.8 Tephigram charts showing; (*a*) typical environment curves and atmospheric stability. The path of a rising air parcel and condensation level are also shown; (*b*) cloud formation following surface heating
Source: From Barry and Chorley, 1987

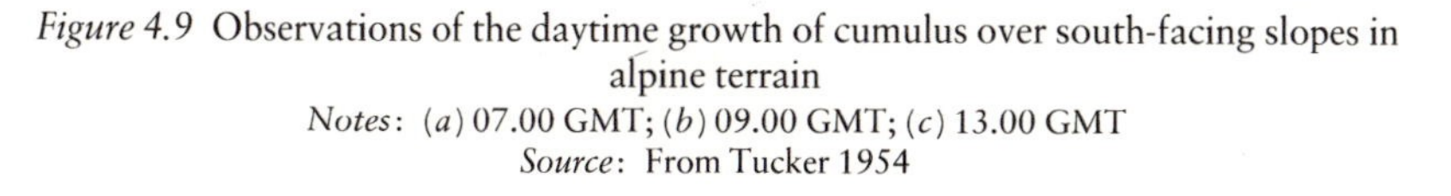

Figure 4.9 Observations of the daytime growth of cumulus over south-facing slopes in alpine terrain
Notes: (*a*) 07.00 GMT; (*b*) 09.00 GMT; (*c*) 13.00 GMT
Source: From Tucker 1954

from that occurring in a cyclonic system. Rawinsonde data from studies over the San Juan Mountains of south-western Colorado (Hjermstad 1975) indicate that, in winter orographic cloud systems, there is usually a moist layer extending some 1500–2000 m above the general summit level (3500–3800 m) and capped at around 500 mb by a stable dry layer. In contrast, deep cyclonic storm clouds may extend up to the 300 mb level.

In mid-latitudes in summer, and generally in continental and subtropical–tropical areas, the predominant cloud form is convective. The spatial distribution of convective upcurrents (*thermals*) in mountain regions shows some pronounced effects of topography. For example, there may be strong contrasts between shaded and sunny slopes. Fujita *et al.* (1968) reported rapid cumulus build-up on the slope of Mt. Fuji between 08.45–09.15 in July as the solar altitude increased from 47° to 53° and surface temperatures on the rocky slopes exceeded 30°C (see Figure 4.5). Surface temperatures on mid-latitude mountains during summer afternoons tend not to differ much from those in adjacent valleys since the change of net radiation with height is small (Scorer 1955; MacCready 1955). Consequently, potential temperatures are higher in the mountains. In Idaho, for example, MacCready found an average potential temperature gradient on summer afternoons of 2.9 K km^{-1} between 700 and 1700 m, with maximum rates of 5.5 K km^{-1}. Therefore, thermals start more readily over high ground, although because of the higher potential temperatures, cloud bases over these locations also tend to be higher. In such terrain, the height difference between the bases of cumuli over valleys and over hilltops is about half of the valley–summit relative relief (assuming cumulus actually forms at the lower level).

This pattern of stronger convective motion over high ground is confirmed by observations of Silverman (1960) in the Santa Catalina Mountains, Arizona, which rise 2000 m above the surrounding terrain. However, the wind field determines the cloud locations with respect to the ridge line and, in most cases, it is difficult to separate the effects on convection of barrier heating on the one hand and lifting, on the other (Hosler *et al.* 1963; Orville 1965a, b, 1968; Kuo and

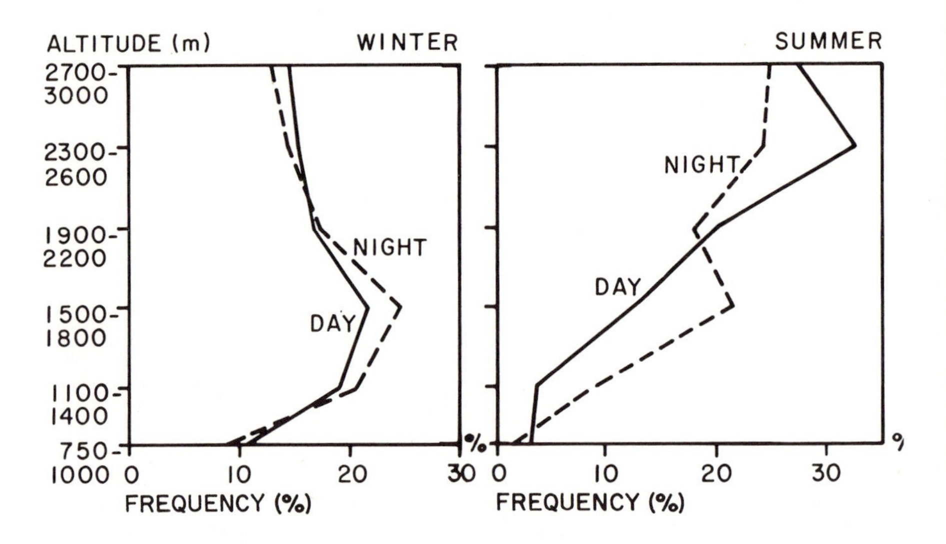

Figure 4.10 The diurnal frequency distributions with altitude of the tops of stratiform cloud observed from the Zugspitze in summer and winter, 1939–48
Source: After Hauer 1950

Orville 1973). The role of lee convergence in thunderstorm initiation in relation to ridgetop wind direction in the Colorado Rocky Mountains is discussed in Chapter 5, p. 314.

Fujita (1967) also demonstrated that a mountain mass, such as the San Francisco Peaks of Arizona, can cause a convective shower system to split due to downwind wake effects with cyclonic vorticity on the right and anticyclonic vorticity on the left, viewed downwind. Aircraft measurements during fair-weather conditions over the San Mateo Mountains, west-central New Mexico, suggest a scale interaction between 3–4 km wide convective eddies and a 20-km diameter toroidal 'heat island' type of circulation (Raymond and Wilkening 1980). Turbulent heat is transferred upward by the eddies to 200–1400 m above the summit levels creating a positive temperature anomaly. Mean sensible heat fluxes of 400 W m^{-2} or more are common at mountain-top level. This buoyancy generated by the mesoscale eddies supports lower-level inflow and higher level outflow. Given moisture availability, conditional instability in such situations would favour the initiation of cumulonimbus growth and thunderstorm activity.

A special problem relating to cloud data from mountain stations is the frequent obscuration of view by cloud at station level (i.e. hill fog). Moreover, cloud tops, especially in winter, may be below the station (Plate 12). Observations for 1939–48 from the Zugspitze show that the tops of stratiform cloud limited by an inversion layer have a maximum frequency about 1500–1800 m altitude, day and night in winter (Figure 4.10), for a total of 5650 cases. For April–September, the tops of stratiform cloud are most frequently about 2600 m

Plate 12 Low level stratus filling the Swiss lowlands and valleys. This is a typical autumn phenomenon when low-level inversions develop during anticyclonic weather conditions. The mountain in the foreground is the Rigi, near Lucerne (courtesy of Dr K. Steffen, Institute of Geography, ETH, Zurich).

with a secondary nocturnal maximum at 1500–1800 m for 1865 cases. Hauer (1950) notes that the explanation of the nocturnal distribution, with cloud tops frequent above 2700 m, probably lies in the persistence of a stratocumulus layer derived from the spreading out of daytime convection cloud in the proximity of the mountain.

Some practical problems with cloud reporting at high mountain stations results from the fact that the low cloud category, which includes stratocumulus and cumulus, has a height range of 0–2 km for the cloud base (World Meteorological Organization 1975), whereas over extensive mountain areas the base may lie at an altitude typical of that for medium cloud according to the synoptic reporting code. In addition, various cloud combinations and forms occur in mountain areas, such as a chaotic 'föhn sky', and fracto-types of cloud, which are not adequately covered by the code categories (Küttner and Model 1948; World Meteorological Organization 1956).

PRECIPITATION

Precipitation processes

The influence of mountain barriers on precipitation distribution and amount has been a subject of long-standing debate and controversy (e.g. Bonacina 1945). It is a problem that is compounded by the paucity of high-altitude stations and the additional difficulties of determining snowfall contributions to total precipitation, especially at windy sites. As recognized by Salter (1918) from analysis of British data, the effect of altitude on the vertical distribution of precipitation in mountain areas is highly variable in different geographical locations. To gain adequate understanding of these variations we must consider the basic condensation processes and the ways in which mountains can affect the cloud and precipitation regimes.

As pointed out by Fliri (1967), the primary distinction is between convective and air mass situations. The controls involved are the vertical profiles of moisture content and wind speed. The atmospheric water vapour content decreases quite rapidly with height in the lower troposphere (see p. 28), so that, at 3 km, the figure is typically about one third of that at sea level. From this standpoint, precipitation amounts might be expected likewise to decrease upward, but the vapour flux convergence, cloud-water content and vertical wind profile are the dominant influences.

In a simple convective cloud system (with only vertical motions), the precipitation maximum should be located close to the cloud base where the maximum size and number of falling drops occur before beginning to evaporate. In cumulonimbus clouds with strong vertical updraughts the drops tend to be transported upwards so that the zone of maximum precipitation may even be somewhat above the cloud base.

A convective pattern of vertical precipitation distribution is widely found in the tropics where the cloud base is typically about 500–700 m in coastal areas and 600–1000 m inland. As noted by Weischet (1969), these areas characteristically have a rainfall maximum between 1000 m and 1500 m. This pattern is especially pronounced in the trade-wind inversion belt where the air is very dry above the inversion. On Hawaii, for example, more than 550 cm falls at 700 m on the eastern slopes of Mauna Loa, whereas the summit (3298 m) receives only 44 cm. Similar trends are apparent on windward slopes of the coastal mountains of Central America (Hastenrath 1967). Flohn (1974) states that, in the area of the intertropical convergence, precipitation amounts on mountains above 3000 m are only 10–30 per cent of those in the maximum zone (e.g. Mt. Kenya and Mt. Cameroon).

In the middle latitudes, precipitation in the winter half-year at least is predominantly derived from advective situations. The large-scale forced ascent of air over a mountain barrier leads to a lifting condensation level (p. 221). This ascent may intensify the general vertical motion in a cyclonic system or it may release conditional instability and shower activity, especially in polar maritime airstreams (Smithson 1970).

Whereas in the tropical easterlies, wind speeds decrease with height, the westerlies in middle latitudes generally increase upwards. This increase more than compensates for the vertical decrease in moisture content up to at least 700 mb, according to evaluations for stations in the Alps by Havlik (1968). He shows that the totals increase up to the highest stations around 3500 m. Only a small fraction of this increase can be attributed to an increased frequency of days with precipitation. Instead, days with 30 mm day^{-1} at mountain stations account for almost half of the excess. These are commonly warm-sector situations with south-westerly airflow at 500 mb. Days with large vertical increases in precipitation amounts are shown to have three to four times the mean annual vapour flux between 850 and 500 mb. Nevertheless, an analysis of sounding data on the windward side of the Park Range, Colorado, suggests that only 2–5 per cent of the moisture flux is precipitated. Hindman (1986) estimates a vapour inflow of 3.20×10^6 (± 0.74)g hr^{-1} for several orographic storms, an outflow of 2.96×10^6 (± 0.68)g hr^{-1} of vapour, 0.14×10^6 of liquid water and 0.01×10^6g hr of ice particles. The precipitation averaged only 0.1×10^6 g hr^{-1}.

The amount of orographic precipitation depends on three factors operating on quite different scales (Sawyer 1956). They are (1) air mass characteristics and the synoptic-scale pressure pattern; (2) local vertical motion due to the terrain; and (3) microphysical processes in the cloud and the evaporation of falling drops. These are examined in turn.

Air mass characteristics of major importance are the stability and moisture content of the air; the pressure field determines the wind speed and direction. Heavy orographic precipitation is most likely in Britain when winds are strong and perpendicular to an extensive mountain range, the air is already moist and cloudy, and the lapse rate is near neutral, facilitating the release of conditional instability through uplift (Douglas and Glasspoole 1947). Such conditions are common in the warm sectors of frontal cyclones. Poulter (1936) noted that the orographic increase at warm fronts was about two-fold over windward slopes of similar inclination to the front (1:100), whereas cold fronts (slope 1:50) resulted in smaller increases of about 50 per cent. Radar studies (Browning *et al.* 1974, 1975) show that, over Britain, orographic effects are negligible at surface cold fronts, where precipitation is heavy in any case. Ahead of the front, in the warm sector, orographic effects vary according to the existence, first, of a low-level jet maintaining high liquid water contents in the low-level 'feeder clouds' and, second, of seeding particles falling from higher-level 'seeder' clouds as illustrated in Figure 4.11 (Bergeron 1949; Storebö 1968; Andersson 1980; Browning and Hill 1981). Such seeding augments rainfall rates through the washout of droplets, or by enhancing snow crystal growth, in orographic cloud over low hills. Convective meso-scale precipitation areas (MPAs) within cyclone warm sectors (Hobbs 1978; Smith 1979) may also be accentuated by orography (see Figure 4.12). They can persist 6 hours or more and contribute much of the heavy

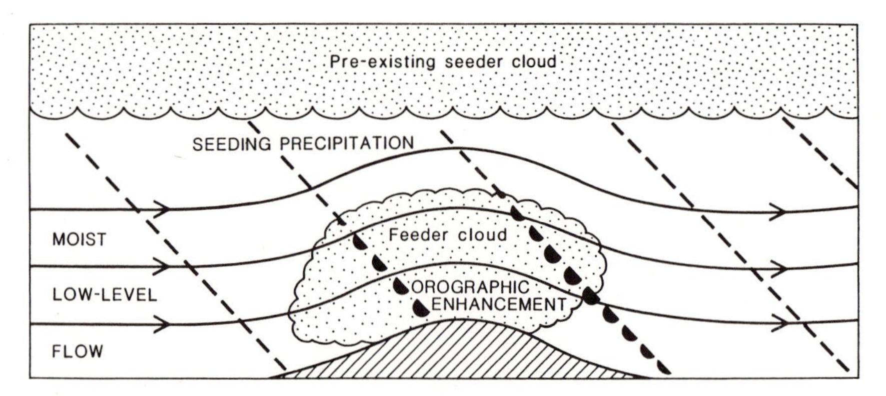

Figure 4.11 Schematic illustration of seeder-feeder clouds and enhancement of precipitation over a hill
Source: From Browning and Hill 1981

precipitation. Radar analyses show that most of the augmentation occurs in the lowest 1500 m associated with pre-existing rain areas, high humidities and winds ≥ 20 m s^{-1} (Hill *et al.* 1981). These newer studies appear to confirm earlier ideas of Bonacina (1945) on the role of convective instability and the need for 'pre-conditioning' of the upstream airflow, but the relative roles of (1) existing stratiform-derived precipitation giving an orographic seeder-feeder effect, (2) existing convective instability released by forced ascent, or (3) existing baroclinicity releasing instability through orographic blocking or frontal deformation, remains uncertain (Smith 1982). According to Carruthers and Choularton (1983), for wide hills with half-width ≥ 20 km the precipitation maximum occurs on the upwind side, but at the summit or in the lee side of narrow hills due to wind drift.

The vertical motion induced by the terrain is determined by the relief and shape as well as by airmass stability, wind speed and direction (see Chapter 3, pp. 112–39). For the apparently simple case of southwesterly monsoon airstreams approaching the Western Ghats, India, the evidence suggests that smooth orographic uplift is an insufficient explanation (Smith and Lin 1983). The rainfall occurs in wet spells of 5–10 days' duration. Deep cumulus convection is apparently triggered by localized mesoscale lifting that is driven by latent heating. Smith and Lin argue that the Ghats merely anchor the rainfall system. However, low-level flow interactions with the mountains and the extent of upstream influence is still debated (Smith 1985). For low hills, assuming no deflection or blockage of the flow, the vertical velocity (w_s) of the low-level air can be obtained from:

$$w_s = u \tan s$$

where u = component of the wind normal to the barrier
s = the corresponding slope angle.

Orographic precipitation models are discussed below.

The amount of condensation depends on the lifting depth, the amount of air lifted, and the moisture content of the air. In a stable air mass, the lifting depth may be only a few kilometres, whereas, in a cloudy airmass with strong winds the lifting may be up to 6 km. The condensation rate (c) for saturated air subject to orographic lifting can be expressed:

$$c = -\int_z \rho w \frac{\partial r_s}{\partial z} dz,$$

where r_s = saturation mixing ratio, and w = vertical velocity. This implies that c decreases with a decrease in temperature (which determines r_s) and with increasing altitude. For given amounts of uplift for specific mountain cross-sections, an equation of this form can be solved by assuming appropriate profiles of w and r_s (Fulks 1935). Pedgley (1970) indicates that, for each kilometre of uplift, the condensation from saturated air is about 1.5 g kg^{-1} (1.5 g m^{-3}) of liquid water.

The fallout of raindrops (or snow flakes) depends on the growth rate of cloud droplets, their coalescence efficiency, and their fall speed in relation to wind velocity. About one hour is required for droplet growth to raindrop size by coalescence, compared with about half as long for snow flakes by the Bergeron–Findeisen process in clouds with both ice crystal and supercooled droplets present. In this process, ice crystals grow at the expense of supercooled droplets, since the saturation vapour pressure is lower over an ice surface than over a water surface.

An analysis (Pedgley 1970) of six cases of heavy precipitation associated with strong southwesterly airflows over Snowdonia, north Wales, in summer 1966, suggests that there would be insufficient time for raindrops to grow in the airstream across these mountains. Pedgley calculates that if droplets of 2×10^{-3} cm radius were already present, and if the winds were <40 km h^{-1}, small drizzle drops (10^{-2} cm radius) could grow within half an hour. The centre of Snowdonia (averaging 600 m) is about 20 km from the margins of the elevated dome. Heavy rainfalls apparently require a scouring process by droplets of 0.1–0.5 mm radius falling from upper cloud layers. Large drops have a high fall speed and so gain little by accretion. However, a droplet of 0.5 mm may grow to 1.0 mm radius (a volume increase of 8) in about 10 minutes while falling through 3 km of cloud with a liquid water content of 0.5 g m^{-3}. Such raindrop growth is therefore feasible for winds <120 km h^{-1}. Observations indicate that wind speeds are typically 80 km h^{-1}. Also, the heavy rains are commonly of the prolonged 'fine rain' type with intensities of about 6 mm h^{-1}. Half of the liquid water in a cloud column is removed in about 6 minutes by which time the cloud has been displaced 5–10 km. Sawyer (1956) suggests that, in favourable conditions, most of the available water is precipitated, but as little as 30–50 per cent of the available water may fall out if only shallow moist layers are present. This example illustrates the importance of taking account of the microphysical processes in the clouds.

Climatic characteristics of mountains

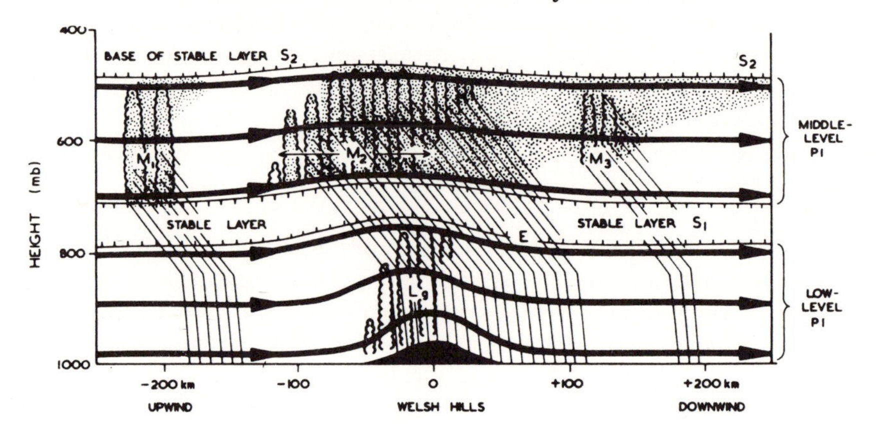

(Streamline arrow) Mean streamlines within the strong west-south-west flow crossing the Welsh hills, drawn to be consistent with the observed pattern of precipitation development; although the precise form of the streamlines is arbitrary, notice that the middle-level air begins to ascend far upwind of the hills.

S_1 Layer with rather high static stability separating the potentially unstable air at low levels from potentially unstable air at middle levels.

S_2 Base of the region of high static stability that extends throughout the upper troposphere.

(Convection symbol) Small scale convection occurring where the low-level or middle-level potential instability (PI) is realized by general ascent.

(Stippling) Ice crystal (anvil) 'cloud' resulting from the middle-level convection and perhaps also, above 500 mb, from stable ascent over the hills.

(Inclined lines) Precipitation trajectories relative to the ground, strongly inclined because of the high winds; the change in slope occurs at the melting level at about 840 mb.

M_1 Middle-level convection within isolated MPAs due to areas of mesoscale ascent that occur in the warm sector even over the sea.

←M_2→ Abundant middle-level convection triggered by orographic uplift over the hills, occurring as fresh outbreaks within and between existing MPAs.

M_3 Decaying middle-level convection mainly associated with MPAs previously in existence far upwind of the hills (i.e. M_1).

L_g Rapid low-level growth of precipitation falling from aloft, producing a large increment in rainfall rate tied closely to the hills.

E Evaporation in the lee of the hills, decreasing the amount of precipitation from middle-levels that reaches the ground over central England; however, because of the enhanced generation of precipitation over the hills (M_2) widespread rain continues to fall up to 100 km downwind of the hills.

Figure 4.12 Model of warm sector rainfall based on radar studies over the Welsh hills showing the role of potential instability in meso-scale precipitation areas (MPAs) and orography

Source: From Browning *et al.* 1974

Soundings and aircraft data for wintertime storms over the Colorado Rocky Mountains indicate some findings that contrast with conditions over the lower and less extensive British mountains. Marwitz (1980) identifies four stages that occur in most cyclonic storms over the San Juan Mountains. First, moist westerly flow does not extend above mountain top level (3.5–4 km) and so is blocked and diverted. This stable atmosphere situation changes to a neutral one as a deep storm extends throughout the troposphere. However, a baroclinic zone is usually only present in the upper half of the troposphere. In stage three, a convective cloud line develops above a low-level horizontal convergence zone at the base of the mountains and upwind. Finally, in stage four, dissipation begins with subsidence occurring at mountain top level. Microwave and radar measurements of storms in the northern Colorado Rocky Mountains show that liquid water content is related to cloud top temperature (i.e. higher, colder clouds), but there is an inverse relation between liquid water content and precipitation rate (Rauber *et al.* 1986). The influence of the extensive barrier of the Sierra Nevada on wintertime Pacific storm systems has also received much attention. Two deep orographic storms that caused heavy snowfalls have been analysed with aircraft and Doppler radar by Marwitz (1987). Precipitation rates averaged 4 mm hr^{-1} with forced ascent rates of 0.4 ms^{-1} in the very stable storm case. This blocking reduced the wind component normal to the barrier. The stable orographic clouds were found to have concentrations of cloud droplets that are typically marine above the freezing level, *i.e.*, mean diameters of 20–30 μm. The ice crystals concentrated near the −5°C level were predominantly needles indicating rime-splintering. The cloud droplets are efficiently accreted by ice crystals, depleting the condensate and producing ice multiplication. Marwitz notes that little super-cooled water will pass over the crest line of the Sierras.

Lee effects are highly variable for individual storms. For example, Pedgley (1971) demonstrates for Snowdonia that 'rain-shadow' patterns are weak when the airstream already has substantial cloud cover, with precipitation falling upwind of a range and strong winds. The same is true also when there is little orographic lifting due to light winds and the airstream has only shallow moist layers.

In general, precipitation maxima over mid-latitude mountain ranges are closely related to the 'smoothed' topography (Pedgley 1970). Thus, gauges in small valleys within a mountain complex may record totals more characteristic of the surrounding ranges. Large, deep valleys, however, set up their own wind systems that can cause very different precipitation distributions (see p. 207). Convective activity may also cause erratic patterns of precipitation distribution. Towering cumulus generated by thermal convection and valley wind effects over the Rocky Mountains, for example, is typically carried eastward by the general airflow and may cause afternoon–evening thunder showers if continued convection over the adjacent high plains permits cumulonimbus to form in the lee of the mountains (Henz 1972).

Altitudinal characteristics

Following these general considerations of precipitation processes, let us now examine the empirical evidence of altitudinal effects. During the late nineteenth century, several studies were made of the vertical distribution of precipitation in the Himalaya (Hill 1881) and the Alps. Subsequently, many computations have been made of the rate of precipitation change with height and its geographical variation. For example, Salter (1918) found increases of 8–15 cm 100 m^{-1} in southern England and about 12–30 cm 100 m^{-1} on windward slopes in western England. However, he noted that the rate of increase was much lower where there are high ranges to windward. On lee slopes the increase with height was found to be larger, especially on the lower slopes, due to the frequent occurrence of descending air motion and the removal of moisture upwind. Salter (1918: 54) also noted that stations in narrow valleys in mountain areas typically record much larger annual totals than would be expected from their relatively low elevation. For the eastern Pennines, mean elevation over an 8 km radius from a gauge site appears to be a better predictor of annual precipitation than station height (Chuan and Lockwood 1974).

Several studies demonstrate that the altitudinal increase is due to the combined effect of higher intensities and greater duration of precipitation (Atkinson and Smithson 1976). In north Wales, for example, average daily rainfall rates in Snowdonia are nearly twice those on the Irish Sea coast but, in addition, there are almost twice as many hours with rain falling in the mountains. This reflects the complex make-up of orographic precipitation discussed above.

The most complete global survey of vertical precipitation profiles has been carried out by Lauscher (1976a) using data for 1300 long-term stations grouped into three major categories: below 1 km (1029 stations), 1–2 km (222) and 2–3 km (43), for 10° latitude–20° longitude sectors between 35°S and 55°N, from 130°E westward to 110°W. He distinguishes five general types as shown in Figure 4.13. These are: 'tropical' (T) with a clear maximum at about 1.0–1.5 km; 'equatorial' (E) where there is a general decrease with height above a maximum close to sea level; a 'transition' type (Tr) in the subtropics where there is either little height dependence, or conditions vary considerably locally; a 'mid-latitude' type (M) which shows a strong increase with height; and a 'polar' type (P) where higher totals tend to occur near sea level, at least in the vicinity of open water. These altitudinal patterns reflect the processes discussed above.

Despite these useful generalities, many local or regional complications occur. For example, Lauer (1975) shows that the southern slopes of Mt. Cameroon, west Africa, have a maximum at their foot due to the monsoon regime but, on the north-east side, where trade wind influences are dominant, it occurs at 1500 m. On the Caribbean slopes of the Mexican meseta, the typical tropical maximum occurs between 600 and 1400 m but there is a weak secondary maximum around 3000 m. This results from convective heating over the high basins. Lauer reports a similar phenomenon in Ethiopia. The low-level, warm

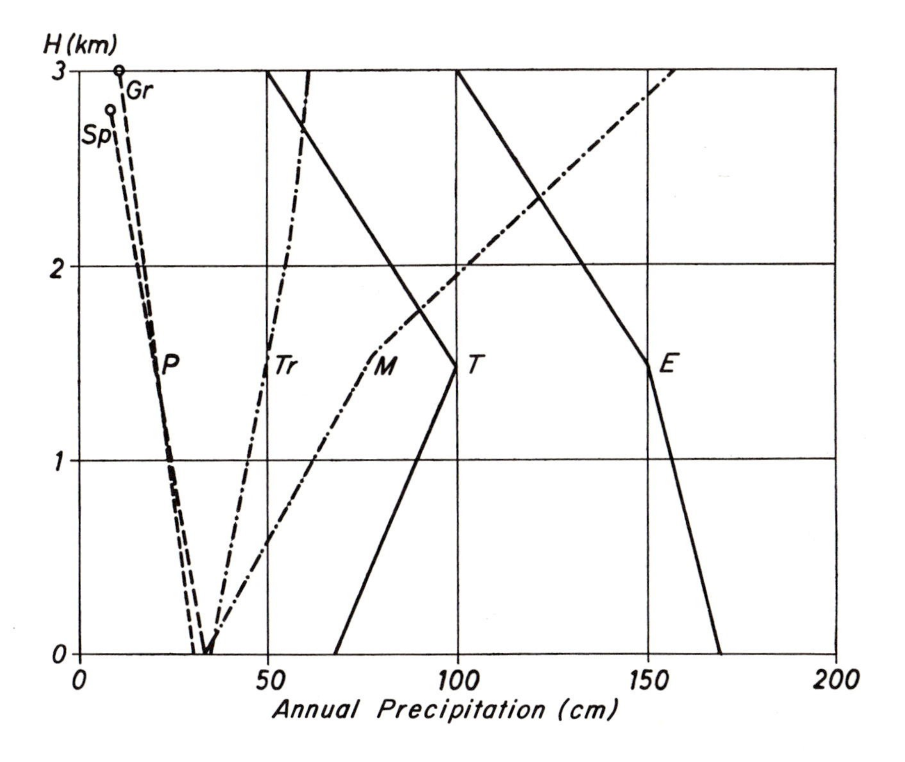

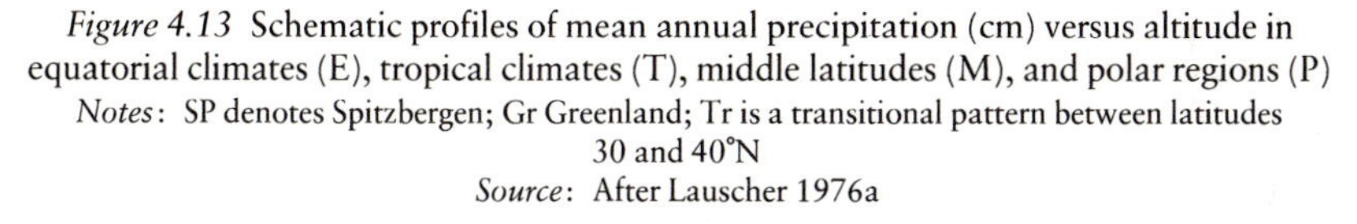
Figure 4.13 Schematic profiles of mean annual precipitation (cm) versus altitude in equatorial climates (E), tropical climates (T), middle latitudes (M), and polar regions (P)
Notes: SP denotes Spitzbergen; Gr Greenland; Tr is a transitional pattern between latitudes 30 and 40°N
Source: After Lauscher 1976a

moist monsoon air reaches the plateaux along the valleys where it enhances convective activity causing maximum precipitation to occur between 2000 and 2500 m. Similar intense heating occurs over the subtropical deserts so that in the Sahara, for example, convection is set off within disturbances in the tropical easterlies and gives rise to a precipitation maximum at 2500 m altitude in the Hoggar. Lauer summarizes this variety of equatorial and tropical patterns of precipitation profile (see Figure 4.14), showing that the maximum zone in these regions tends to rise with decreasing annual total. Locally, of course, altitudinal gradients may be the reverse of those anticipated due to the presence of regional gradients caused by atmospheric circulation effects or moisture sources. An illustration for New Guinea is provided in Figure 5.1 (p. 300).

In the equatorial Andes of Peru (1°S), annual totals on the western slopes show a tropical pattern with maximum values around 1500 m, corresponding to a greater number of rain days, whereas on the eastern slopes values decline from 600 m to about 2500 m, with a secondary maximum about 3250 m, again

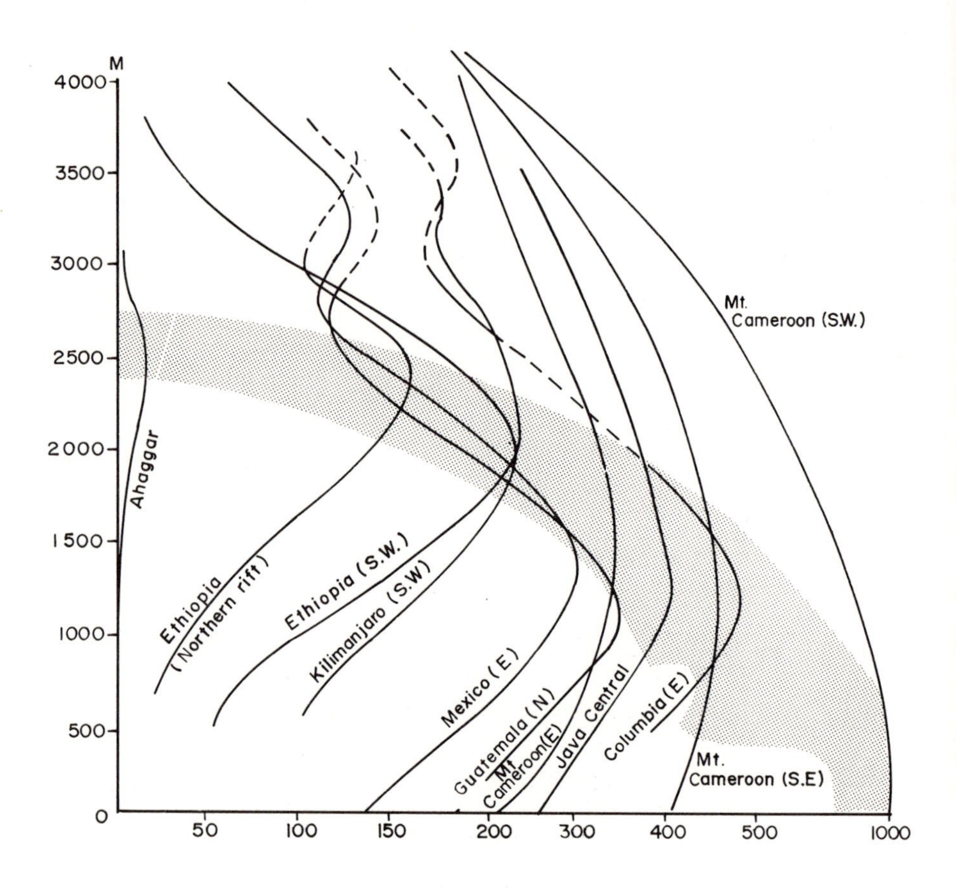

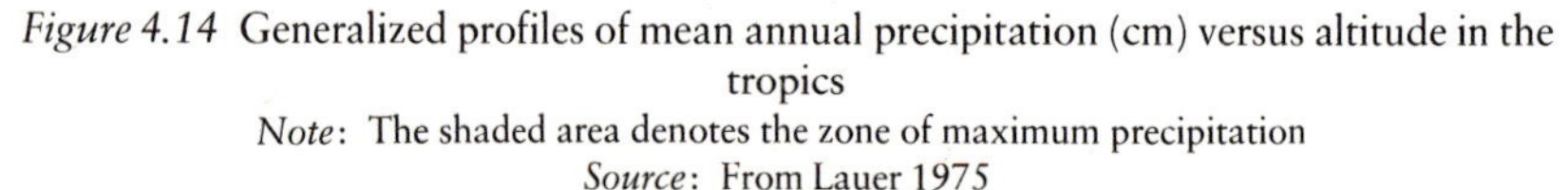

Figure 4.14 Generalized profiles of mean annual precipitation (cm) versus altitude in the tropics

Note: The shaded area denotes the zone of maximum precipitation

Source: From Lauer 1975

corresponding to the frequency of rain days (Frère *et al.* 1975). The annual regime is bimodal on the eastern slopes, but unimodal with precipitation during November–May on the western slope. At 16°–18°S in Bolivia, where there are no trade winds, amounts decline almost linearly with elevation from about 900 mm at 500 m to around 200–250 mm at 3500–4000 m altitude. There is little change in the numbers of raindays, but the amount per rainday decreases. Amounts per rainday in Bolivia and western Ecuador decrease from around 13 mm/rainday at 500–800 m to 4–5 mm/rainday at 3500 m; in eastern Ecuador, amounts fall off rapidly to about 5 mm/rainday at 2000 m.

In middle latitudes, the general tendency for increased precipitation with height, often to the highest levels of observations, is modified considerably by a leeward or windward slope location. This is illustrated for the eastern Alps in Austria by Lauscher (1976a). The profile for the lee situation of the Ötz valley

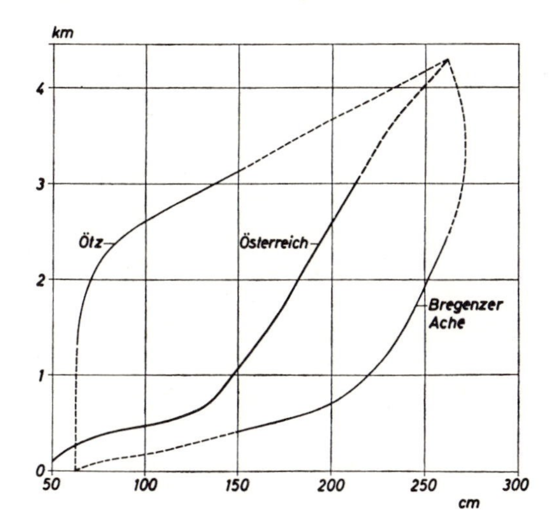

Figure 4.15 The altitudinal profile of mean annual precipitation (cm) for Austria as a whole, for the Ötztal in a lee situation and the Bregenz area in a windward situation
Note: The dashed lines are hypothetical
Source: From Lauscher 1976a

contrasts strongly with that for the windward exposure of the Bregenz district (Figure 4.15). Similar differences are observed in the Swiss Alps. In the sheltered Valais (Wallis), the rate of increase averages 27 mm/100 m from 470 m to 1700 m altitude and 99 mm/100 m from 1700 m to 3810 m, whereas in the north-eastern Alps the opposite is found: 85 mm/100 m from 380 m to 1700 m and 57 mm/100 m from 1700 m to 3810 m. In both areas, however, the overall average from 400 m to 3800 m altitude is close to 75 mm/100 m (Lang 1985).

In the Coast Ranges of western North America there is an increase usually up to the summit level, (as in the Olympic Range, Washington, *c.* 2200 m; Schermerhorn, 1967), whereas the maximum appears to occur on the western slopes around 1600 m in the Sierra Nevada further inland (Armstrong and Stidd 1967). In the Rocky Mountains, patterns are complicated by the Pacific origin of winter precipitation and the Gulf of Mexico origin of precipitation in the summer halfyear (see Chapter 5, pp. 312–14). On the western slopes of the central Colorado Rockies, winter precipitation at 3200 m is almost six times that at the base of the slopes (1750 m) according to an analysis by Hjermstad (1970) (see Table 5.2, p. 315). Hanson (1982) showed strong linear relations exist between annual amounts and elevation in south-west Idaho both on upwind (south-western) slopes and lee (northern) slopes for elevations between 1200 and 2100 m. However, lee sites received more precipitation than upwind ones at equivalent elevations. The annual pattern is determined by the winter season altitudinal gradient associated with cyclonic storms whereas summer thunderstorm rainfall shows only a small altitudinal effect.

For polar regions, data on altitudinal effects on precipitation relate primarily to snow accumulation records on the Greenland and Antarctic ice sheets. These have been summarized by Sugden (1977) who notes that accumulation in Antarctica and north Greenland increases inland to about 1500–1600 m altitude and thereafter decreases. In Antarctica, however, latitudinal effects introduce complications, in south-east Greenland and eastern Antarctica the maximum occurs about 750 m. Using precipitation and accumulation data for Greenland, however, Ohmura (1991) shows the existence in western Greenland of a zone of maximum occurring around 2400 m at 69°N and descending northward to about 1500 m at 76°N. In eastern Greenland, the highest values are along the coast.

The altitude of maximum precipitation (Pmax) for an ideal bell-shaped mountain is determined primarily by the lapse rate (Γ) and the height of the mountain according to an analytical expression developed by Alpert (1986). This altitude, and the amount of precipitation, decrease as Γ increases. For a saturated atmosphere with $\Gamma = 6.5°\text{C km}^{-1}$, the limiting altitude of P max is shown to be about 3.8 km. Large-scale synoptic uplift is a less important variable than lapse rate, according to Alpert's model. In contrast to these results for high mountains, however, microphysical processes and mesoscale flow characteristics in the boundary layer play a much more significant role for precipitation over low hills, as shown above.

In addition to geographical patterns there are also distinctive patterns on seasonal and shorter time scales. For example, Erk (1887) noted a winter maximum at 700 m in the Bavarian Alps, but a summertime increase up to 1600 m. This shift reflects the predominance of cyclonic or convective precipitation types. Using monthly data for 1956–75 to construct profiles on the northern and southern sides of the Sonnblick Observatory in Austria, Lauscher (1978) shows that the typical mid-latitude pattern, with a summit maximum, occurs in 73 per cent of months on the south side, and 66 per cent on the north side. This pattern also characterizes the average profiles for the 12 wettest months and the 12 driest months. Nevertheless, there is a maximum around 2300 m on the north side in 26 per cent of months and in 15 per cent on the south side. This 'tropical' pattern obviously reflects the predominance of particular storm types.

Storm 'types' and synoptic flow patterns can also introduce major differences in orographic effects. For example, Peck (1964, 1972b) shows that in the Wasatch Mountains, Utah cold lows have a lower ratio (2.7:1) of summit/base precipitation than non-cold low storms (4:1 ratio). In the San Juan Mountains, Colorado, winter storms from the south-east and north-west are generally associated with light precipitation, due to high upper air temperatures, whereas storms from the west or south-west generally give rise to more precipitation over the mountains. Such differences are of great significance for winter cloud-seeding operations (Rangno 1979).

Evaluating the orographic component

The precipitation falling on a mountainous area consists of an amount that would occur in the absence of the mountains as a result of convection and cyclonic convergence, and an orographic component due to the intensification of these processes over the mountains, as well as to the forced uplift of air by the terrain. Usually the total precipitation only is discussed, especially when statistical prediction is attempted by regression methods, but there have also been some attempts to distinguish the specific contribution of orographic effects.

For the Rocky Mountains in Alberta, Reinelt (1970) uses a statistical approach. The mean monthly precipitation is first calculated for each of five zones located parallel to and east of the mountains. Each zone is 100–150 km wide. These averages show that, between October and April, precipitation decreases eastward since most of it is caused by the upslope flow of shallow arctic air which generates widespread stratiform cloud. There is also a strong orographic effect in May–June when deep unstable air in cold lows sets off large-scale precipitation. Between July and October, thermal convection largely outweighs orographic influences so that the plains receive larger totals. A harmonic analysis is performed on the monthly zone-averages and the seasonal regime, as expressed by the phase angle and the amplitude of the first three harmonic components, is shown to differ markedly between the mountains and the other four zones. Precipitation in zone 5, 500–600 km east of the mountains, is assumed to be free of orographic influences and its monthly averages are therefore used as a reference datum. By subtracting the amounts in zone 5 from the corresponding averages for each of the other four zones, Reinelt shows that the orographic component for the Rocky Mountains in Alberta averages 37 per cent of the total annual precipitation and exceeds 50 per cent during September through to April. In the foothills zone around Calgary and Lethbridge, the annual orographic component is still 18 per cent.

The orographic enhancement over Britain has been estimated for different airflow directions by stratifying daily data according to the ratio of the contribution of falls with rates ≥ 0.05 mm hr^{-1} to the mean annual precipitation (Hill 1983). For hills exposed to maritime winds, enhancement factors are 1.5 to 3.5. Similar enhancements are estimated for the Hawaiian Islands by Nullet and McGranaghan (1988), who suggest that mountain shape is more important than its size.

A specific case-study of precipitation during the passage of an occluded front over the Cascade Mountains, Washington, shows a similar magnitude of orographic augmentation (Hobbs *et al.* 1975). There is also evidence of a 20-hour period of light precipitation on the windward slopes, after the frontal passage, attributable to orographic effects. In contrast, the only significant falls on the lee side are associated with the front. Aircraft measurements of liquid water content, ice particle concentrations, and riming of ice particles showed maximum values in the clouds leeward of the crest and closest to the frontal location. The concentration of the frontal precipitation in an 80 km-wide mesoscale convective cloud band is another noteworthy feature of this situation. Such band structures are now widely recognized to be a general feature of frontal zones (Browning and Harrold 1969; see also Figure 4.12).

Statistical analyses of total precipitation in mountain areas have used regression techniques incorporating topographic parameters (Spreen 1947; Linsley 1958; Peck and Brown 1962; Sporns 1964). For western Colorado, Spreen showed by a graphical analysis that mean winter precipitation amounts are highly correlated ($R = 0.94$) with the combined influence of station elevation, the maximum relative relief within an 8 km radius, 'exposure' (the fractional circumference of a circle 32 km in radius not containing a barrier more than 300 m higher than the station) and 'orientation' (the direction to the sector of greatest exposure). By comparison, correlation of precipitation with elevation alone gave a value of only 0.55. The use of modern multiple regression techniques allows spatial patterns of residuals from the regression to be analysed and this may enable additional parameters to be incorporated in the equations in order to improve the statistical explanation (Hutchinson 1968; Bleasdale and Chan 1972). This type of procedure has been used to construct precipitation frequency maps for mountain areas of the United States by Miller (1972) and to analyse elevation, orientation and inland distance effects on precipitation in West Africa (Gregory 1968) and in the Great Basin of the western United States (Houghton 1979).

A similar regression approach can also be used to estimate snow depth. Rhea and Grant (1974) find that 80 per cent of the variance of water content on snow courses in Colorado and Utah can be accounted for in terms of two parameters: the directionally adjusted slope which potential precipitation-bearing air currents must cross for a distance of 200 km upwind; and the number of upwind barriers to the airflow.

Formerly, precipitation maps for mountain areas were obtained by extrapolation from existing station data, empirical altitude–precipitation relationships, and adjustments for windward/leeward location (Steinhauser 1967). There may be serious errors with this approach, especially where the mountains do not form a simple range.

Estimation of *probable maximum precipitation* (PMP) – the maximum that can fall over a particular drainage basin during a specified time interval and season – is a particular problem in mountainous areas (Miller 1982). In general, PMP estimates are based on an assessment of major storms over the basin, or closely similar regions, and the transportation of such storms to the area in question. The moisture in the storm is maximized based, for example, on the surface dewpoint value attained for 12 hours or more. For the storm model it is generally assumed that the atmosphere is saturated with a pseudo-adiabatic lapse rate. Moisture inflow is determined from maximized wind speeds from representative directions. PMP estimates are made for areas of different size, up to that of the basin, and different duration, and the largest value for each duration and area is determined. In mountainous regions, various terrain influences will modify the precipitation characteristics. Miller describes two different approaches to estimating PMP in such situations. One possibility is to estimate PMP for lowlands adjacent to the mountains and then to make adjustments for orography. The second method is termed 'orographic separation'; it involves estimating precipitation generated by convergence using the above-described procedures for the non-mountainous portions of the basin. The orographic precipitation is then calculated using a multilayer flow model similar to Myers (1962) (see below p. 244) for several adjacent cross-mountain profiles. Values are then mapped for different time intervals. The two components – due to convergence and orographic uplift – are then combined. Other, purely statistical, approaches based on precipitation frequency data and probabilities of extreme values are also used if long records are available.

Snowfall and snow cover

Several studies show that, at lowland stations in middle latitudes, a threshold temperature can be used to discriminate type of precipitation on a statistical basis. For lowland Britain, there is an equal probability of precipitation occurring as rain or snow when the screen temperature is 1.5°C, implying a freezing level about 250 m above the surface (Murray 1952; Lamb 1955). In the mountains of central Asia, however, Glazyrin (1970) finds that this threshold temperature increases from about 1°C at 500 m elevation to 4°C at 3500–4000 m. Correspondingly, the range about the threshold value, delimiting the temperatures below (above) which snow (rain) always occurs, increase from ±2.5°C at 500 m, to ±5°C at 3500–4000 m. This apparently reflects a higher frequency of snow occurring in shower form and perhaps also a locally steeper lapse rate. When rain is falling at near-freezing temperatures, the removal of latent heat of fusion from the air by melting snowflakes is sufficient to cool a layer of 200–300 m deep (Stewart 1985). Thus, if the wet-bulb temperature is 0°C at 300–400 m altitude when precipitation begins, snow may fall at altitudes of only 150–200 m within an hour according to Lumb (1983). Wet bulb temperature is often better correlated with precipitation type than dry bulb temperature, because snowfalls at positive temperatures are commonly associated with low relative humidity. In central Europe, snowfalls are rare when wet bulb temperatures exceed 2°C (Steinacker 1983). Moreover, the intensity of such falls is usually low (Rohrer 1989).

The fraction of annual precipitation falling as snow obviously increases with altitude. Even in July–August, 65 per cent falls in solid form (excluding hail) at 3000 m in the eastern Alps, whereas the figure drops to only 12–15 per cent at 2000 m (Lauscher 1976b). For European mountain stations, Lauscher has developed a general relationship between solid precipitation, altitude and mean monthly temperature (Figure 4.16). The annual solid fraction, for a mean temperature of 0°C, increases from about 40 per cent near sea level to 75 per cent at 3000 m, with the most rapid increase taking place in the lowest thousand metres. Thus, for the Sonnblick Observatory (3106 m), the relationship between solid precipitation (S) as a percentage of total, and mean monthly temperature (°C) is:

$$S = 75 - 8T,$$

whereas the global relationship for stations near sea level is:

$$S = 50 - 5T.$$

A study using data from 32 Swiss stations located between 300 and 1800 m (mean station altitude 716 m) by Sevruk (1985) indicates

$$S = 61 - 5T,$$

between 1000 and 1800 m altitude, and

$$S = 41 - 3.5\ T,$$

between 500 and 700 m altitude. Sevruk illustrates graphically the variation of these relationships over the cold season. The mean monthly temperature for which half of the monthly precipitation falls as snow varies between −1.9°C in January and 3.9°C in April; it is 0°C in November and February.

When the snowfall fraction is expressed directly in terms of altitude, there is also a general linear relationship (Conrad 1935). The same is true for days with snowfall (Yoshino 1975: 218) and for snow cover duration in the Tatra (Koncek 1959) and in the French Alps, although the relationship of snowcover duration to altitude in the inner Savoy Alps may become curvilinear with an upward increase (Poggi 1959). Slatyer *et al.* (1984) cite linear relationships for snow cover duration in the Australian Alps (38°S), the Swiss Alps (47°N), and Britain (53°N), which differ only in the constants relating to the latitudinal variation of mean temperature. A snow cover duration ≥ 90 days is found above 930 m in Switzerland, but only above 1700 m in Australia.

Many regional studies of snowfall and snow cover have been performed and it is now beginning to be possible to compare empirical relations developed in different areas. Jackson (1978) has re-evaluated British data using median rather than mean duration since the latter is strongly biased by abnormal values. Given the sea-level median duration of snow cover (D_0) (defined as a day with more than half of the ground covered at 09.00 GMT), the duration at any altitude (D_H) is given by:

$$D_H = D_0 \exp (H/300) \text{ for } H < 400 \text{ m}$$

$$\text{and } D_H = 3.75\ D_0[1 + (H - 400)/310] \text{ for } H > 400 \text{ m.}$$

Hence, the relationship is linear above 400 m. Jackson notes that data for Vancouver, British Columbia, fit a similar profile. In both areas, the steep lapse rates associated with maritime air masses cause substantial increases in snow

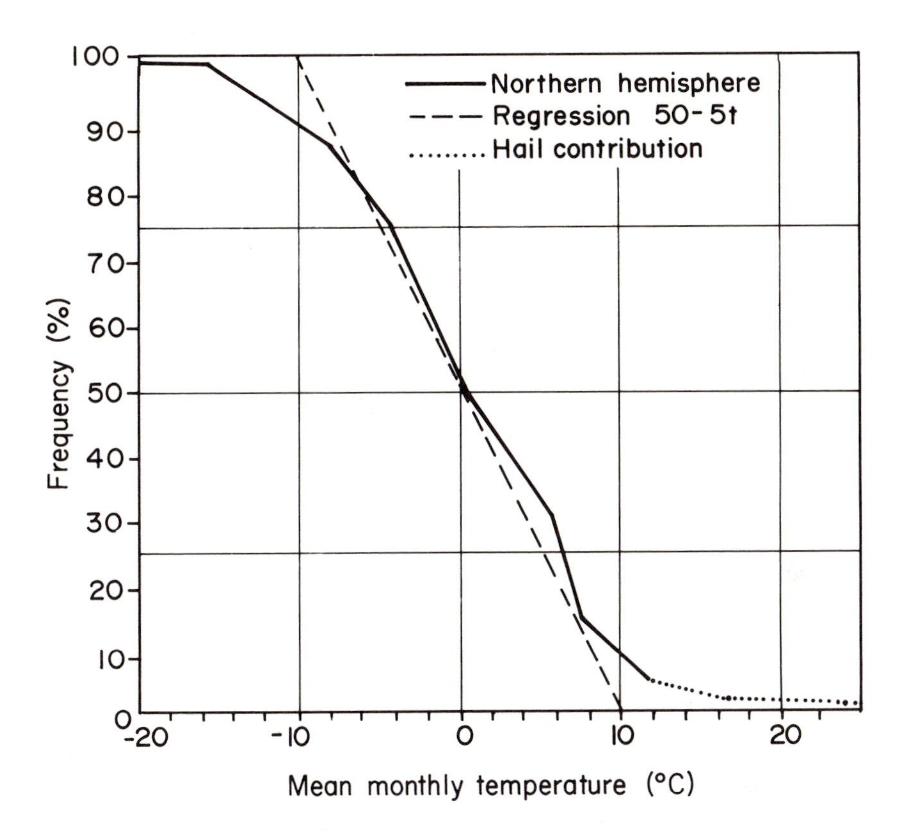

Figure 4.16 Empirical relationship between monthly mean temperature (°C) and annual percentage frequency of solid precipitation in the northern hemisphere compared with the regression (50–5t)
Source: From Lauscher 1976b

cover with altitude. The rate of increase varies latitudinally from 5 days per 100 m in south-west England to 15 days per 100 m in the Scottish Highlands where there are typically 220 days with snow cover at 1200 m (Manley 1971). Mountains in central and eastern Scotland have considerably greater durations of snow cover than mountains in western Britain (Figure 4.17).

The variation of snow depth with altitude is generally more complex, even when local differences due to relief effects and small-scale terrain features are ignored. For example, in Austria the average maximum depth of snow increases (although not linearly) up to about 1000 m, where there is a pronounced decrease, followed by a further increase above this level (Steinhauser 1948). This pattern is attributed to the frequent winter occurrence of an inversion between about 900 and 1100 m, with cold foggy conditions in the valley and clearer skies with more radiation falling on the slopes above. In the Swiss Alps, snow depths (water equivalent) appear to be maximal around 2700 m altitude, and may

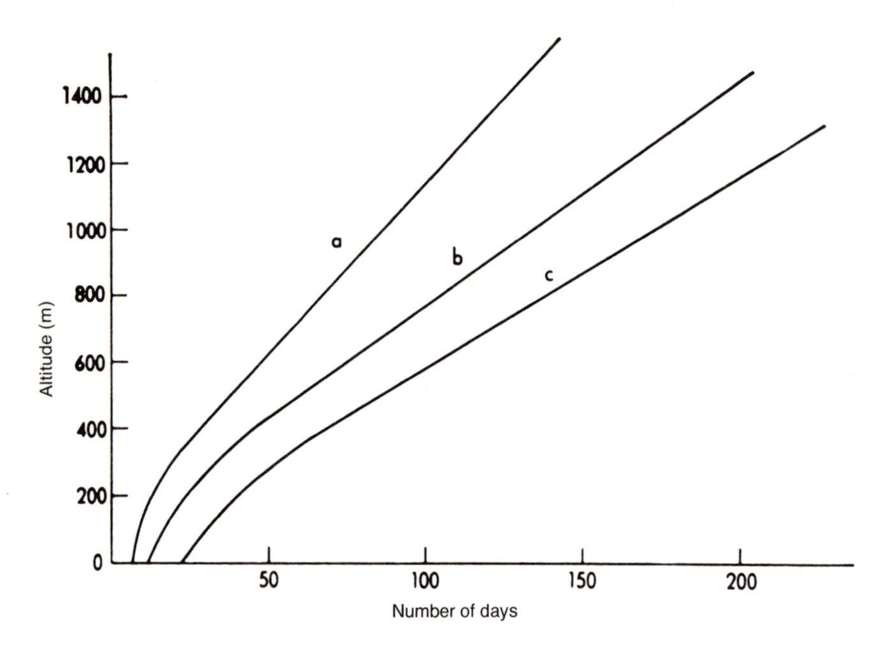

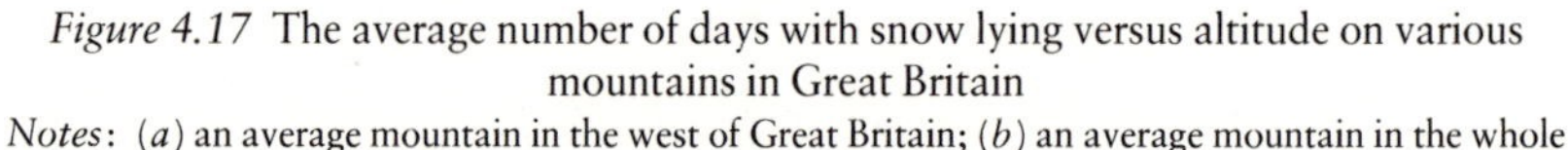

Figure 4.17 The average number of days with snow lying versus altitude on various mountains in Great Britain

Notes: (*a*) an average mountain in the west of Great Britain; (*b*) an average mountain in the whole of Great Britain; (*c*) an average mountain in central Scotland

Source: From Jackson 1978

decrease slightly above, according to Martinec (1987). His estimates are based on depletion curves of basin snowcover, where the usual time axis is replaced by the cumulative snow melt derived from calculated daily values for several years. From satellite observations of the seasonal reduction in basin snow-covered area, an estimate of snow water equivalent at the beginning of the melt season can be obtained. In the San Juan Mountains, Colorado, Caine (1975) finds an elevational influence in the relative variability of maximum snowpack as well as on the accumulation. The latter increases linearly at a rate of 65.5 cm km^{-1} (with a correlation coefficient of 0.66 for 24 snow courses) from a zero-accumulation level at 2400 m. Variability, on the other hand, decreases with increasing elevation. Caine notes that this may in part reflect the longer season of snow accumulation, which at 3500 m is about twice that at 2600 m, with consequently a greater number of storms helping to even out the year-to-year differences. However, since this inverse relationship is concentrated in years with above-normal snowfall, whereas below-normal years show a direct relationship, Caine suggests that there is an interaction between topography and atmospheric factors. In snowy winters, there is a high frequency of cold lows, which cause widespread precipitation with relatively more falling at low elevation, whereas in winters with light accumulation there is a higher frequency of local storms which

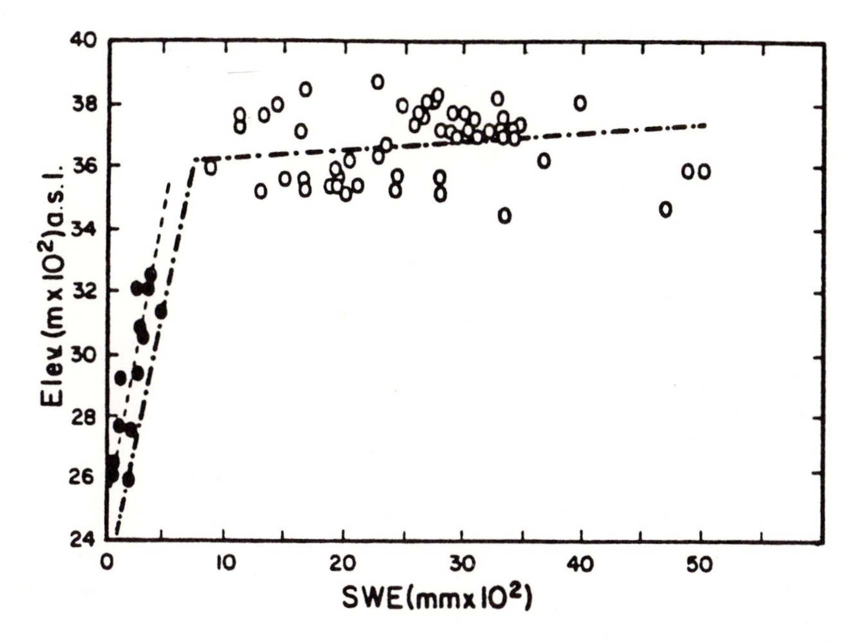

Figure 4.18 The altitudinal gradient of winter precipitation (solid dots) determined from snow course data (––– 1969; – • – • mean) and accumulation measured on glaciers in the Front Range, Colorado in 1970 (open circles), expressed as snow water equivalent (SWE, 10^2 mm). This shows the disjunct relationship between such measurements
Source: From Alford 1985

nevertheless provide sufficient snowfall to keep the pack on snow courses at high elevations closer to average depth.

In high mountain areas, snow accumulation is sometimes recorded only in glacier basins. Figure 4.18 illustrates the contrast between measurements from snow courses below tree line and accumulation on cirque glaciers in the Front Range, Colorado (Alford 1985). The high accumulation on the glaciers is apparently caused by large-scale rotors on the lee side of the Front Range summits that transport snow back into the cirque basins.

Theoretical models of orographic precipitation

Theoretical analysis of precipitation rates over mountains has a long history. Pockels (1901) used the hydrodynamic equations for frictionless two-dimensional flow to calculate vertical velocities and condensation due to adiabatic ascent over an idealized slope, specified by Fourier functions. He concluded that slope angle was more important than absolute height for the precipitation rate on a windward slope. Wagner (1937) noted that Pockels' model ignored the effect of air motion on fallout. Thus snowflakes, in the presence of strong winds, may not show any altitudinal maximum, whereas large drops are little affected by wind.

He cited winter and summer observations from the vicinity of the Sonnblick to support this argument. Another similar theoretical analysis to that of Pockels was carried out by Ono (1925) and tested with Japanese data.

Modelling of orographic precipitation has subsequently followed two broad lines of approach. Some analytical studies have used a combination of the Bernoulli equation, the continuity equation and hydrostatic flow for mountains of arbitrary shape. Other analyses have been based on the perturbation method (see below) and idealized barriers. Most models at present are two-dimensional, for a mountain cross-section, although a few numerical models are now three-dimensional. The treatment of water substance is quite variable, it has received considerable attention since many studies of orographic precipitation are related to cloudseeding assessments. In some models, all of the condensed moisture is precipitated, others use various 'precipitation efficiency' factors; for 'warm' clouds (with temperatures above about −10°C), such as occur over the low coastal ranges of California, the ratio of precipitation to condensate is about 0.3, increasing to about 0.6 for 'cold' clouds in winter cyclones over Colorado mountains (Marwitz 1974). In some recent models, such as those of Young (1974) and Nickerson *et al.* (1978), cloud microphysics is also incorporated.

The essential components of an orographic precipitation model include measures of the vertical displacement of air over the barrier, adiabatic ascent (descent), condensation (evaporation), and precipitation of some fraction of the condensate. It may also be important to include 'blocking' of the low-level airflow by the barrier and lee-wave effects.

For simple barriers, several models use a calculation of the vertical transport of water vapour due to the mean terrain slope. The horizontal wind may be adjusted directly to the mean flow perpendicular to the barrier (Elliot and Shaffer 1962; Myers 1962), or for actual topography the vertical wind component (w_s) relative to the slopes can be determined by:

$$w_s = \frac{u\partial h}{\partial x} + \frac{v\partial h}{\partial y}$$

This is used by Danard (1971). Illustrative maps of terrain-induced vertical motion have been prepared for the Appalachian Mountains (Jarvis and Leonard 1969) and the Canadian Arctic (Fogarasi 1972), indicating values of ±2–10 cm s^{-1} for winds of 10 m s^{-1}. The theoretical decrease with height of terrain-induced vertical motion has an approximately parabolic profile (Berkovsky 1964). The magnitude is less than half of the surface value at 700 mb.

Myers (1962) uses an airflow model based on the work of Scorer (see p. 131) for the mechanical lifting of stable air. The orographic precipitation in *t* hours is expressed:

$$P_t(10^{-3}\ \text{cm}) = \frac{\Delta p}{g}\left(\frac{\bar{V}t}{\Delta x}\right)(\bar{q}_1 - \bar{q}_2),$$

where Δp = depth of the air current to windward of the barrier (mb),
$\bar{V}$ = mean wind speed in layer Δp, perpendicular to the barrier ($km\ h^{-1}$),
Δx = downwind distance (km) over which precipitation falls,
$g = 9.81\ m\ s^{-2}$,
$\bar{q}_1$ and $\bar{q}_2$ = mean specific humidities ($g\ kg^{-1}$) of the air to windward and leeward.

The calculated precipitation with this model is about one-third too large, due to the saturation assumption.

An elaboration of this type of approach by Elliott (1977) expresses the orographic component (P_0), defined as the mountain precipitation minus the upwind lowland amount, by:

$$P_0 = \bar{V} \tan s(-\partial q/\partial z)\Delta C$$

where $\bar{V}$ = mean wind perpendicular to the slope (assumed uniform),
s = slope angle,
ΔC = cloud depth.

Again, the model assumes cloudy air with all of the moisture being condensed. Rhea (1978) has adapted this scheme to develop an operational model for winter precipitation in the Rocky Mountains of Colorado. It is a steady-state multi-layer model incorporating upstream barrier effects, but not allowing for horizontal streamline displacement. After calibration of model parameters against two seasons of snow course and precipitation data, an analysis of 13 winters was performed using twice-daily upper-air soundings as input. Calculated seasonal totals correlate well with observed spring runoff, and a derived map of mean precipitation for the 13 winters compares favourably with one based on point observations and altitudinal–topographic regressions. The model gives the best results for ridges and high plateaux, but overestimates amounts in narrow mountain valleys and underestimates for broad intermontane basins.

More complex models have been developed based on the perturbation approach. The motion in an x,z plane (where x here is along the wind direction), can be expressed as a perturbation superimposed on a steady basic current of velocity, V. The linearized equations for the perturbation vertical velocity were originally developed by Queney (1948) and Holmboe and Klieforth (1957). The following abbreviated presentation is based on Walker (1961) and Wilson (1978). The approximate vertical velocity equation is:

$$\frac{\partial^2 \omega(z, k)}{\partial z^2} + [f(z) - k^2]\omega(z, k) = 0$$

where ω = amplitude of the vertical velocity (sinusoidal with wave number k),
k = wave number of the ground profile (assumed sinusoidal) in the x direction, and

$$f(z) = \frac{\partial(\ln\theta)}{\partial z}\frac{g}{V^2} - \frac{1}{V}\frac{\partial^2 V}{\partial z^2}$$

$\partial(\ln\theta)/\partial z$ is the static stability expressed in terms of potential temperature, θ. This expression for $F(z)$ is used extensively in meso-scale studies (see p. 129). For rainfall situations in British Columbia, $f(z) \simeq 0.09\ \text{km}^{-2}$, in unstable situations $f(z) \to 0$.

The lower boundary condition, assumed to be sinusoidal, has an amplitude:

$$\zeta_0 = f(k)\cos kx,$$
$$= ah\exp(-ak)\cos kx,$$

where h = maximum height of the perturbation in the surface profile,
a = the 'half width' (the distance along x from the mountain crest to the point where $\zeta_0 = h/2$).
ζ = streamline displacement.

The idealized surface profile is shown in Figure 4.19 for $h = 1$ km, $a = 10$ km. Two airflow cases are illustrated from Walker (1961).

Case 1. Static stability $\partial(\ln\theta)/\partial z = 0$, and $\delta V/\delta z = 0$. Thus $f(z) = 0$. This leads to the perturbation vertical velocity:

$$W = -\frac{2axhV(a+z)}{[(a+z)^2 + x^2]^2}$$

and

$$\zeta_z = \frac{ah(a+z)}{[(a+z)^2 + x^2]}$$

for the streamline perturbation at some upper level, z. The numerical example shown in Figure 4.19a is fairly representative of mid-latitude west coast mountains in summer.

Case 2. $k = 0$. This represents the general mountain perturbation, excluding possible lee waves. Here,

$$W = -\frac{2axhV}{(a^2 + x^2)}\left\{a\cos\surd[f(z)]Z - x\sin\surd[f(z)]Z\right\}$$
$$-\frac{ahV}{a^2 + x^2}\sin\surd[f(z)]Z$$

and

$$\zeta_z = \frac{ah}{a^2 + x^2}\left\{a\cos\surd[f(z)]Z - x\sin\surd[f(z)]Z\right\}.$$

The solution for ζ_z is periodic in the vertical with wavelength $= 2\pi/\surd[f(z)]$. In Figure 4.19b, $\surd[f(z)]$ is approximately $0.3\ \text{km}^{-1}$.

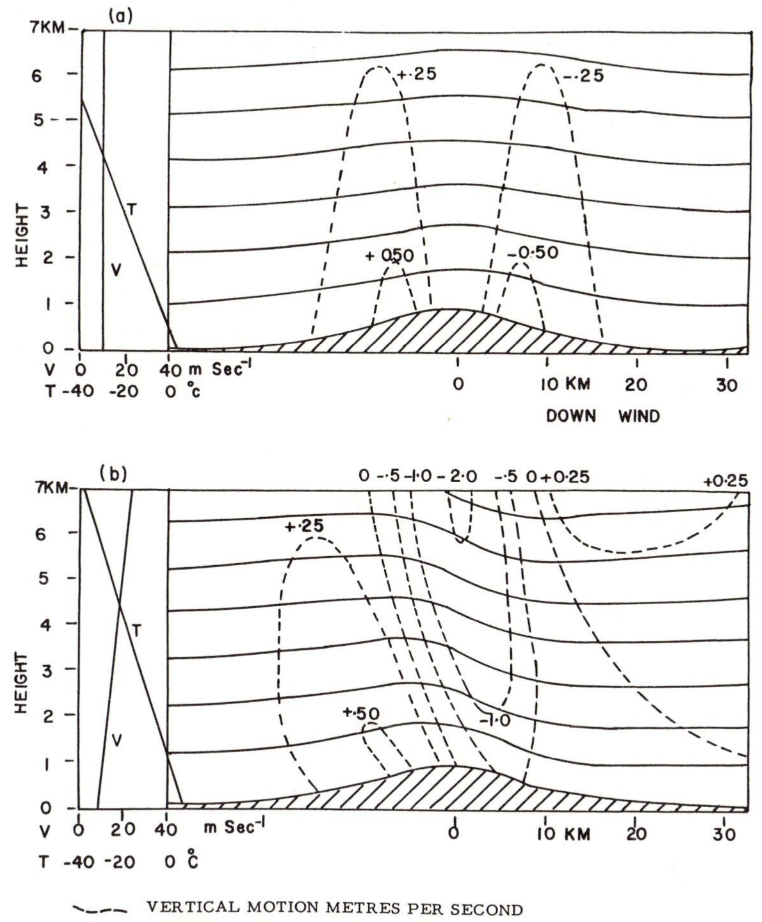

Figure 4.19 Orographically induced vertical motion for a simple barrier with westerly flow: (*a*) $f(z) = 0$; (*b*) $k = 0$

Notes: The solid lines are streamlines; the dashed lines are vertical motion (m s^{-1}). The assumed gradients of wind speed and temperature are shown.

Source: See text and from Walker 1961

There is a good agreement between Walker's (1961) version of this model and climatological profiles across British Columbia, where most of the precipitation falls with westerly or south-westerly tropospheric flow during the cold season. Precipitation amounts are calculated from the estimated condensation rate for saturated airflow (Fulks 1935) crossing an idealized mountain range. The assumed air temperature and wind speed profiles are shown in Figure 4.19b. Two alternative precipitation rates are determined: one assumes instantaneous fallout, with terminal velocities of 5 m s^{-1} for raindrops and 1 m s^{-1} for snowflakes; the other assumes delayed fallout with appropriate droplet coalescence

(50 per cent of the droplets fall out in 10 minutes for a rainfall rate of 2.5 mm h^{-1}). Precipitation is assumed to fall through the evaporation zone both with and without loss. The frequency of airflow directions at 700 mb in 1956 was used to compute mean seasonal profiles. An annual profile, combined with observations of precipitation frequency and amounts, also allowed Walker to construct improved mean precipitation maps over southern British Columbia.

The same basic model has also been used independently by Sarker (1966, 1967) for the Western Ghats of India. The results correctly position the maximum falls near the crestline on days of strong monsoon flow, and on average the model accounts for about 65 per cent of the coastal precipitation. It also suggests that 'spill over' on the leeward slope, due to the winds, extends less than 10–15 km beyond the crest. This result may not be general, however. Walker suggests that most of the leeside evaporation of cloud droplets and precipitation due to streamline descent takes place from the smaller particles, based on calculated and observed precipitation profiles (Figure 4.20). Studies of westerly airstreams crossing the Cascade Mountains in California indicate that solid precipitation can be transported up to 50–70 km downwind (Hobbs *et al.* 1973; Hobbs 1975). This is most likely when crystal aggregation is unimportant, if ice particle concentrations are about 100 l^{-1} and growth is mainly by deposition rather than riming. All of these effects favour particles of low density and small fall speed.

The most recent modelling studies use the primitive equations of motion. A meso-scale numerical model developed by Colton (1976), originally in two-dimensional form, has seven equations as follows: the horizontal momentum equations (to predict the horizontal components of motion), the diagnostic mass continuity equation (to determine vertical motion), the thermodynamic equation (to obtain potential temperature), the continuity equations for water vapour and for liquid water, and the hydrostatic equation. Computations for cases of westerly air flow over the Sierra Nevada Mountains used geostrophic winds and sounding data as input at 13 levels (0 to 11 km) with a fine mesh (4.3 km) resolution, except towards the lateral boundaries. Results for a situation on 21–23 December, 1964, which caused severe flooding in northern California (Figure 4.21) indicate a realistic vertical motion field, correct positioning of maximum precipitation rates, and an appropriate degree of spillover. A different meso-scale boundary-layer model has been used to study the effect of arctic outbreaks over the foothills of the Rocky Mountains in Alberta (Raddatz and Khandekar 1977). They demonstrate the sensitivity of 'upslope' weather to surface heating through its triggering of convective precipitation.

The advent of numerical modelling techniques and aircraft observation of orographic clouds has greatly augmented our understanding of orographic precipitation. Up to now, however, most of the attention has focused on synoptic investigations with an emphasis on possibilities for precipitation augmentation via cloud seeding. The interpretation and generalization of many of these findings in a climatological context is still largely unexplored.

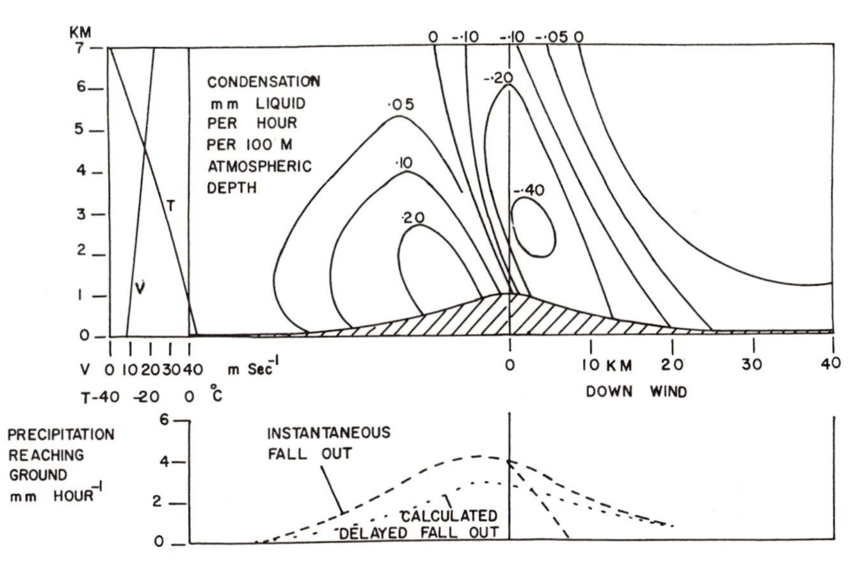

Figure 4.20 Calculated condensation and precipitation (mm h^{-1}) for the airflow case shown in Figure 4.19b
Source: From Walker 1961 and see text

The prediction of orographic precipitation presents special problems because the spatial resolution, even of fine mesh models, is inadequate to account for most orographic features. One procedure is to use the output of a prognostic model as input to a fine resolution diagnostic model. Tucker and Reiter (1988) use a six-layer primitive equation model over the United States, within which is nested a 1000 × 1000 km domain over the southern Rocky Mountains. Terrain influences on the lowest two levels cause modifications to the wind fields and, therefore, affect mass and moisture divergence and precipitation. Realistic simulations of two terrain-related heavy rainstorm and flood events are obtained. Another approach is to use satellite and radar data for short-term forecasts and flood warnings ('nowcasting') (Browning 1980). Both model predictions and 'nowcasting' information can be refined by empirical data on the local-scale precipitation climatology, categorized according to synoptic types. Using a dense local gauge network in north Wales, UK, Nicholass and Harrold (1975) analysed the precipitation over subcatchments of about 60 km^2 (Rc) versus the average amount over the entire 1000 km^2 area (R). The mean ratio $\bar{R}c/\bar{R}$ varied from about 0.5 to 2.0 according to subcatchment, synoptic type and surface wind. Such statistics can be incorporated into forecasts.

Observational problems

It has been assumed up to now that precipitation amounts can be reliably measured. In actuality, this is far from the case. This section examines rainfall

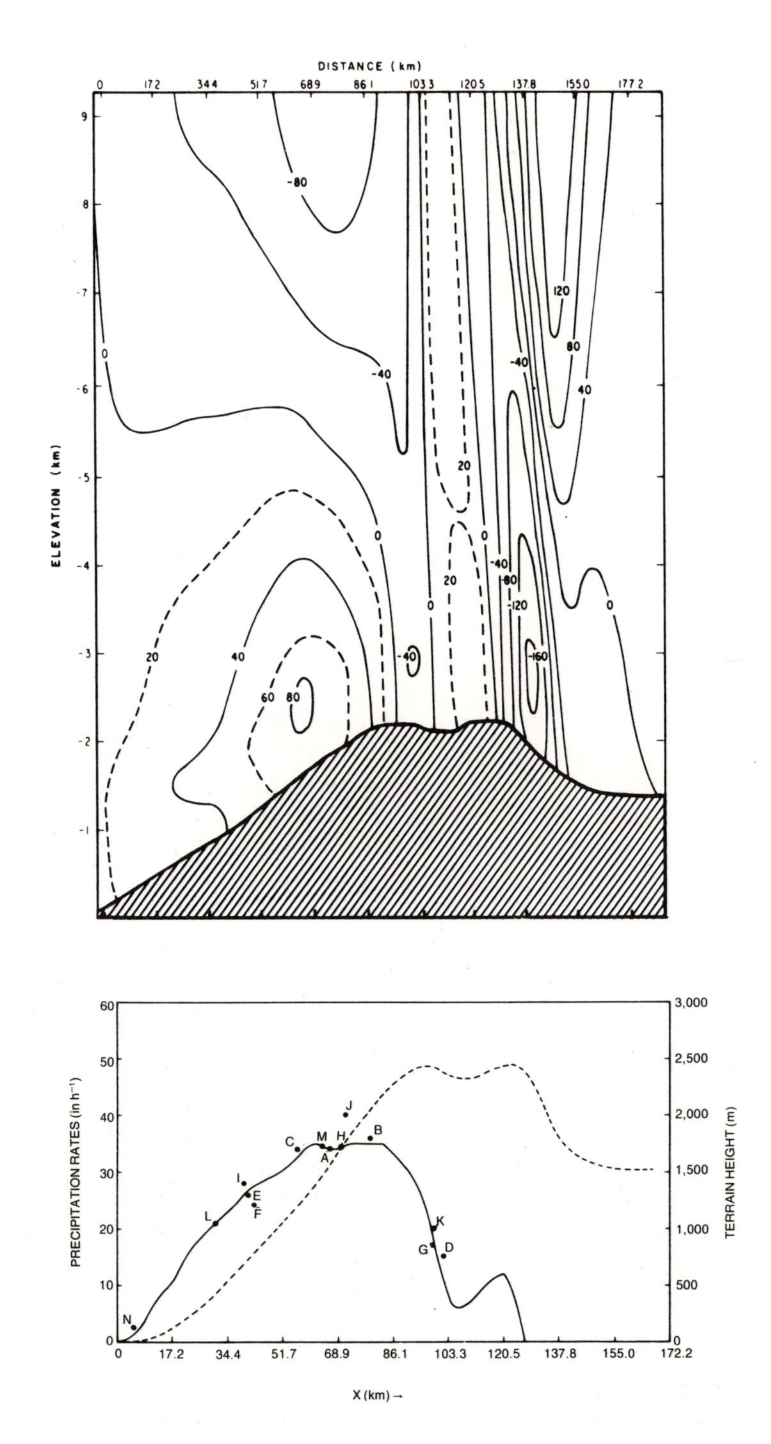
DISTANCE (km)
0
17.2
34.4
51.7
68.9
86.1
103.3
120.5
137.8
155.0
177.2
ELEVATION (km)
-80
120
80
40
0
-40
20
-120
-160
60
a
PRECIPITATION RATES (in h⁻¹)
TERRAIN HEIGHT (m)
3,000
2,500
2,000
1,500
1,000
500
X (km) →
172.2
b

Notes: A Blue Canyon
B Big Bend Ranger Station (Cisco Grove)
C Deer Creek Power House
D Truckee Ranger Station
E Grass Valley
F Nevada City
G Squaw Valley
H Fuller Lake
I North San Juan
J Lake Spaulding
K Donner Memorial State Park
L Dobbins
M Bowman Lake
N Marysville

Source: From Colton 1976

Figure 4.21 Model simulations for a storm with westerly airflow over northern California, 21–23 December 1964; (*a*) Computed steady-state vertical motion field (cm s^{-1}); (*b*) Computed precipitation rate (solid line) and observed values (dots); the dashed line is the terrain cross-section

and snowfall observations; other moisture inputs are discussed in the following section.

The errors involved in precipitation measurement, illustrated in Figure 4.22, have been investigated minutely in lowland situations. Rodda (1967) shows, for example, that rain catch in a standard gauge mounted on the ground with a rim at 25 cm height is systematically 6–8 per cent less than that caught by a ground level (sunken) gauge. This error increases as the gauge top is raised above the ground; a problem first noted over a century ago by Jevons (1861). Systematic errors in precipitation measurement are estimated to be the following (Sevruk 1986a): windfield deformation above the gauge, 2–10 per cent for rainfall and 10–50 per cent for snowfall; losses from wetting of the internal walls of the collector and measuring container, 2–10 per cent; evaporation losses, 0–4 per cent; splash-out/in, 1–2 per cent. The overall systematic error is about 5–15 per cent for rainfall and 20–50 per cent for snowfall. However, correction factors are valid only for particular gauge types and sites creating severe problems of inhomogeneity in time and space (Sevruk 1989).

In mountain areas, gauge catch is strongly affected by local and micro-scale wind effects. The effect of slope aspect on precipitation has been the subject of various investigations with rather differing conclusions. In Yorkshire, England, Reid (1973) maintained two grids, each of six gauges, across the 12°–15° slopes of a west–east valley about 60–m in depth. For a 50-week period, the north-facing slope received 8 per cent less than the south-facing one. The latter was a windward slope for 49 per cent of the time with precipitation and leeward 28 per cent of the time. In contrast, Hovkind (1965) recorded the maximum on the lee of the crest of a conical hill near Santa Barbara, but he considered that these gauges may have over-recorded relative to ones on the windward side. Indeed, such a pattern of over- and under-representation has been found on a larger scale.

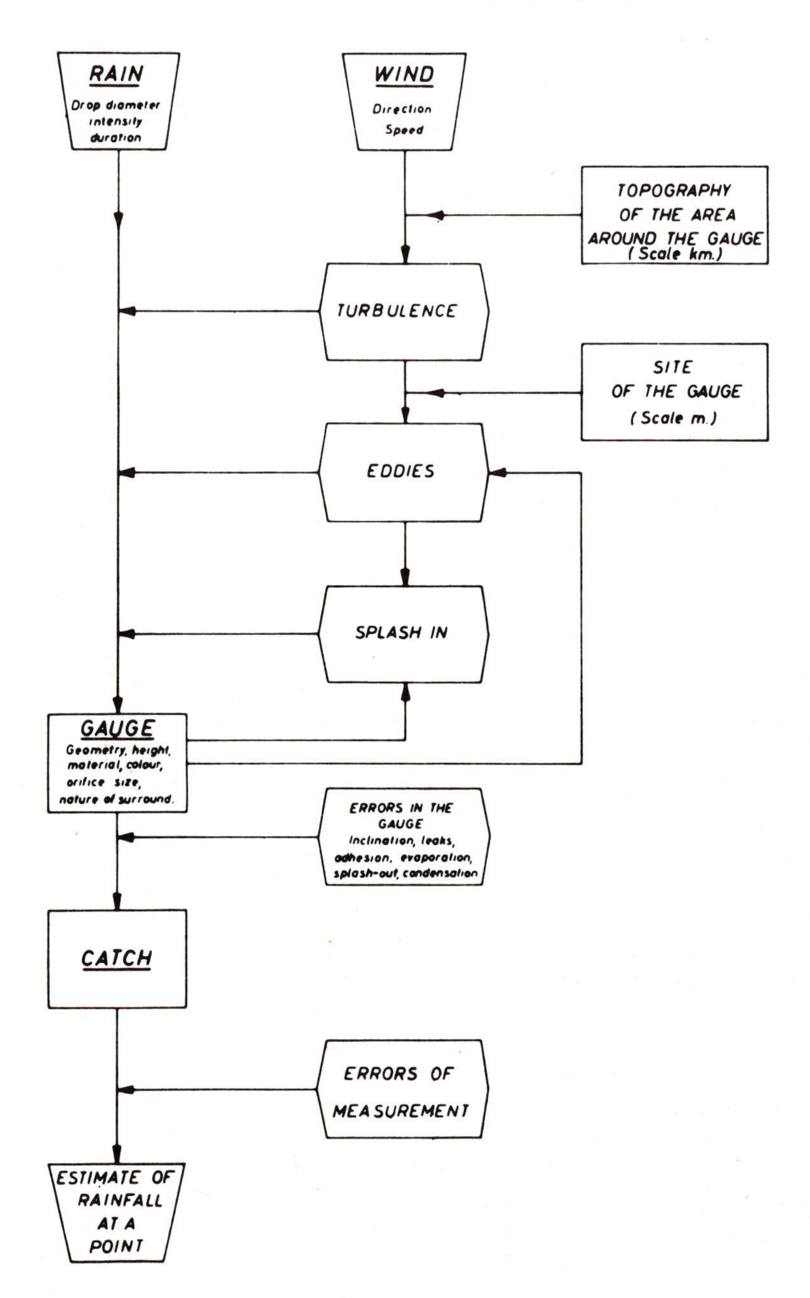

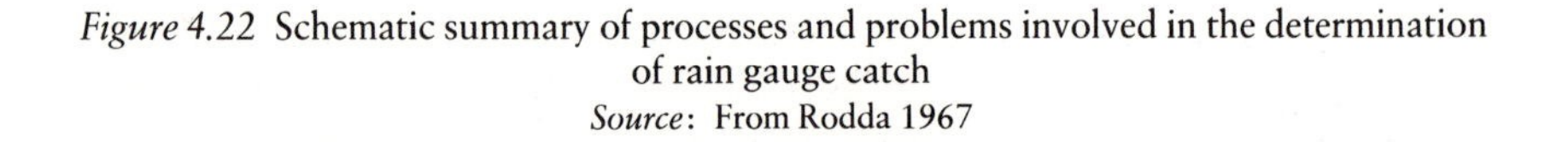
Figure 4.22 Schematic summary of processes and problems involved in the determination of rain gauge catch
Source: From Rodda 1967

Grunow (1960) shows that at Hohenpeissenberg Observatory (975 m) in Bavaria, the catch on the windward slope may be under-estimated 14 per cent and that on the lee slope overestimated 10 per cent. These errors are particularly serious in the case of snowfall where the catch is a function of wind speed over the gauge, and snowflake fall velocity. The latter, of the order of 0.5–1.5 m s^{-1}, depends on crystal shape which is also temperature dependent. Snowfall catch is also affected by the gauge dimensions and by the geometry of the orifice rim, which modify airflow, and by the effects of gauge wetting and losses due to evaporation. Currently, some 50 different types of standard gauge are in use by national meteorological services (Sevruk and Klemm 1989). Storage gauges, which are often used in remote areas are also affected by evaporation losses when the gauge is read infrequently. A film, at least 5 mm thick, of a 'suppressant' such as a glycol can be used to prevent this (Sevruk 1972b).

Considerable attention has been given to the problem of gauge shielding to minimize wind eddies around a gauge. However, neither the solid shield, invented by F.E. Nipher (1878), nor the flexible Alter (1937) type is adequate by itself in windy mountain locations where a significant fraction ($>$ 20 per cent) of the precipitation falls as snow. Shields of the Alter type commonly blow apart in strong winds and the rigid shield often causes a larger underestimate due to 'bridging' of snow over the orifice. While no absolute reference standard exists, Larson (1971) demonstrated that the catch of a gauge with an Alter shield in Wyoming was only 66–75 per cent of a similarly shielded gauge surrounded by snow fencing. The purpose of the snow fence is to minimize local site eddies. Further experiments of this type have been conducted in the same locale (Rechard 1972), using various configurations of 1.2 m vertical-lath fences of 50 per cent density with a gap at the base of between 15–30 cm. One design, the 'blow fence', has the shielded gauge within two concentric circles of fence; the inner one has a 1.5 m radius and slopes at 60°, with the base of the fence positioned 0.6 m above the base of the gauge which is itself 1.5 m above the ground. The object is to keep the area beneath the gauge free of snow accumulation by directing airflow downward. Preliminary results suggest that while this installation catches less than reference gauges in forest clearings, the degree of protection seems to be relatively constant for wind speeds of 4–9 m s^{-1}, compared with other configurations.

An international comparison of national procedures for determining solid precipitation, organized through World Meteorological Organization (Goodison *et al.* 1989), is currently assessing gauge errors relative to a standard reference gauge. An octagonal vertical double-fence shield, surrounding a Soviet Tretyakov (shielded) precipitation gauge mounted at 3 m, is the adopted Intercomparison Reference (DFIR). The vertical 50 per cent density fences form 12 m and 4 m diameter circles; the outer one is 3.5 m and the inner one 3 m high, with gaps between the base of the fences and the ground of 2 m and 1.5 m, respectively. Goodison *et al.* report that the catch of both the Canadian AES–Nipher shielded gauge and the Universal Belfort gauge with a Nipher shield were almost identical

Table 4.4 Ratios of precipitation measured at Valdai, USSR, by Tretyakov gauges with various fence configurations compared with a control gauge surrounded by bushes

Fence type[a]	*Fence heights (m) Outer (inner)*	*Gauge orifice height (m)*	*December*	*April*	*July*	*Year*
1V	2.5 (2.0)	2.0	0.96	0.96	0.99	0.95
1I	2.5 (2.0)	2.0	0.92	0.98	0.99	0.96
DFIR	3.5 (3.0)	3.0	0.91	0.95	0.90	0.93
1V	2.5 (2.0) circle diameters 4 m, 2 m	2.0	0.88	0.89	0.95	0.91
Wyoming (I)		2.0	0.82	0.87	0.93	0.88
—	Unfenced	2.0	0.66	0.87	0.98	0.84

Notes: [a]See text for further description of DFIR and Wyoming type. Type 1 differs in having shorter fences with basal gaps of 1 m and 0.5 m. V denotes vertical fences, I = inclined Outward 15°.
Source: After Golubev 1986

to that of the DFIR in a one month field test in January 1987. Golubev (1986) reported the results of a long-term study of the Tretyakov gauge in an open plot surrounded by double fence structures of different dimensions (installed both vertically and inclined outward 15°), with a control gauge in a plot sheltered by bushes. The results (Table 4.4) show the necessity for the double fence installation, especially for solid precipitation. Moreover, at wind speeds >5 m s^{-1}, some correction for wind speed is required even for fenced gauges.

Correction factors for monthly totals are typically about ×1.06 for shielded gauges in summer, but are of the order of ×2 for winter precipitation, with light winds (3 m s^{-1} at 2 m) (Goodison *et al.* 1989).

A general equation for estimating corrected monthly precipitation has been developed by Sevruk (1986b):

$$P_k = k(P_m + W)$$

where P_m = precipitation total from daily gauges,
k = conversion factor for wind speed over the gauge orifice,
and W = correction for wetting losses.

Sevruk provides detailed tables of these coefficients for Switzerland. k varies with gauge type, the wind speed during precipitation, and a precipitation parameter which depends on the number of days with rainfall in a month and the fraction falling with an intensity ≤ 1.8 mm h^{-1} (or, in winter, the snowfall fraction). For the Hellman gauge in Switzerland, k is between 1.015 and 1.72 for wind speeds at 2 m of 0.5–4.0 m s^{-1}. The average annual correction for stations below 2000 m with daily gauges is about ×1.07 in summer and ×1.11 in winter. Above 2000 m, where storage gauges are used, the corresponding figures are ×1.15 and ×1.35.

A simpler and reasonably accurate approach for snowfall determination in mountain areas is to use snow boards, which are reset on the snow surface daily (or whenever practicable) (Föhn 1977). Flat shoulders are the most suitable sites for such installations. The depth is read against a vertical scale and water equivalent is obtained by concurrent density measurements. Over a 30-year winter record, near the Weissfluhjoch station, a Nipher-shielded gauge averaged only 502 mm compared with 794 mm on snow boards. By determining the water equivalent of new snowfall measured on snow boards, corrections can be determined for individual gauge measurements (Sevruk 1983). Martinec (1985) shows that the ratio may vary considerably from year-to-year. For example, corrections to a Hellman recording gauge ranged from ×1.13 to ×1.45 during 1951–80, with a mean of ×1.26.

Net snow accumulation in alpine watersheds can be investigated more thoroughly using accumulation data collected at snow pits located in a stratified random sampling arrangement. For a study in the Sierra Nevada (Elder *et al.* 1989), the survey points were carefully located using orthographically-corrected aerial photographs, topographic maps and compass survey. The topographic data were available in a 5 m resolution digital elevation model (DEM) grid. Zones of similar snow properties were identified based on elevation, slope and net radiation values calculated using the DEM. In North America, at sites below timber line, considerable use has been made of snow courses to determine net accumulation (as water equivalent) from depth–density surveys on a monthly basis. At some sites, comparable data are now gathered by telemetry from a snow pressure-pillow which weighs the snow pack (Warnick and Penton 1971) and, therefore, gives the water content directly.

An extensive network of more than 500 snow reporting stations has been installed by the Soil Conservation Service in the western United States. The stations (snow 'pillows', accumulation gauges, and air temperature sensors), many in remote mountain locations, report with high frequency to data collection stations in Boise, Idaho, and Ogden, Utah, using meteor-burst relay (Rallison 1981). This SNOTEL (Snow Telemetry) network transmits VHF signals that are reflected by the trails of numerous small meteorites in the ionosphere to activate the remote station.

In steeply sloping terrain the gauge orifice is sometimes constructed to be parallel to the ground and presumably the air flow. Such 'stereo' or 'tailored' gauges have been tested in various locations. Sevruk (1972a, 1974) argues from long-term experiments in Switzerland that they give more representative measurements on steep, open slopes exposed to rain-bearing winds, although on other slopes they may give a smaller catch than a horizontal gauge (Grunow 1960). The subject is reviewed in detail by Peck (1972a). He notes that if the aim is to measure the precipitation on a horizontal plane at or near the surface (the 'meteorological precipitation' which is mapped) then a normal gauge is appropriate. Moreover, he emphasizes that the angle of inclination of falling precipitation particles has no direct effect on the catch of a horizontal orifice (see Figure

4.23). The stereo-top gauge (or a tilted gauge) may be appropriate, however, if the precipitation falling on unit area parallel to a slope (the 'hydrological precipitation') is of interest. Peck also indicates that, in sites protected from strong winds, there is little difference between gauges with horizontal or stereo orifices.

Even when the catch at a given site is adequately registered, the question of site representativeness remains. Below the timber line, gauges may be set in forest clearings. These must be large enough to ensure a 30° horizon from gauge to the tree crowns to avoid oversheltering. However, Soviet research shows that in moderately sheltered locations the catch of snow exceeds that in open terrain by considerable amounts, particularly at moderate wind speeds (Struzer *et al.* 1965); see Table 4.5.

Table 4.5 Excess catch of snow in a sheltered location versus open terrain (per cent)

		2m wind speed ($m s^{-1}$)	
		2.5	*5.0*
	⩾0	32	57
Air temperature (°C)	−20 to −5	53	120
	<−20	80	200

Source: After Struzer *et al.* 1965

Comparisons of winter (October–March) precipitation and water equivalent of the snowpack on 1 April have been made over four to six seasons at 30 sites above 2400 m in Utah (Brown and Peck 1962). The sites were subjectively classified as to their exposure, with the following results:

Site exposure	*Sets of measurements*	*Winter precipitation minus 1 April w.e.(in)*	*Percentage excess/deficit*
Well-protected	77	2.5	+17
Fairly well-protected	9	0.0	0
Moderately windy	6	−2.5	−10
Windy	26	−2.6	−15
Very windy	13	−3.6	−16
Overprotected	11	−0.3	−2

The excess (or deficit) of measured precipitation over the snowpack water equivalent shows good general agreement with the subjective exposure categories in terms of anticipated wind effects. The figures also indicate the magnitude of average seasonal differences that might result from such site differences. The elevation of these stations avoided possible effects of snow melt on the ground during the winter months.

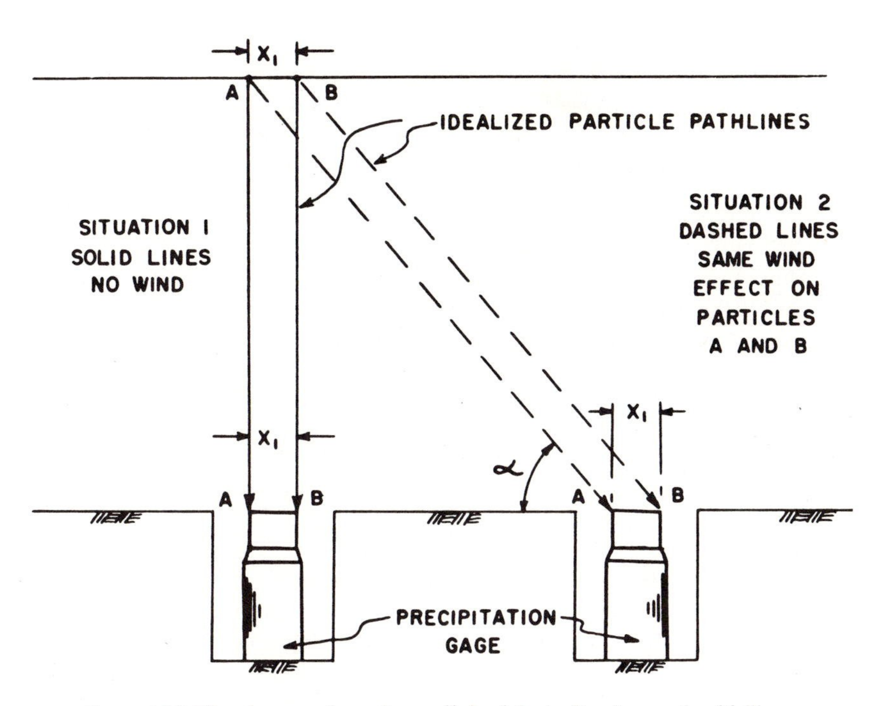

Figure 4.23 The absence of any direct effect of the inclination angle of falling precipitation on gauge catch
Source: From Peck 1972a and see text

Well-protected sites in Brown and Peck's study are those with sheltering on all sides given by objects subtending angles of 20°–30° from the gauge orifice. The objects must be broad enough to exclude eddy effects and the general terrain should afford some wind protection. Overprotected sites, on the other hand, have some objects extending above the 45° angle from the gauge. The other categories denote subjective rankings of shelter provided by nearby objects and the general terrain.

An expression for sheltering of individual gauge sites has been developed by Catterall (1972). Protection (P), on a scale of zero (exposure in all directions) to one (complete protection), is given by:

$$P = \sum_{a=1}^{8} \left(\frac{H_{\text{Amax}} - H}{D_{\text{Amax}}} \right)^2 \Big/ W_a,$$

where H is the elevation of the gauge sites, $(H_{\text{Amax}} - H)$ is relative relief, D_{Amax} is the distance (m) between the gauge and the height maximum along direction A W_{a} is the time when sheltering is effective in terms of the percentage annual frequency of wind direction a. In practice P rarely exceeds 0.5. The distance considered in determining D_{Amax} and H_{Amax} is based (subjectively) on the size of the catchment and the scale of local relief.

The areal representativeness of a gauge is obviously less in complex terrain than in lowlands. Thus, the recommended densities of gauge networks for hydrometeorological applications are considerably larger in mountain regions. The minimum network densities suggested by the World Meteorological Organization (1984 p. 3.13) are as follows:

1 flat terrain in temperate, subtropical and tropical zones, 1 station per 600 to 900 km^2;
2 mountainous regions in these zones, 1 station per 100–250 km^2;
3 arid and polar regions, 1 station per 1500 to 10,000 km^2;
4 mountainous islands with irregular precipitation, 1 station per 25 km^2.

The error in areal estimates of precipitation increases not only in relation to network density, but also the time interval (day, month) for which the estimates are made. However, the gauge locations within a sub-area, such as a mountain basin, also affect the accuracy of basin averages. Sevruk (1989) suggests that the most representative gauges are centrally-located and close to the mean basin altitude.

In view of the many problems associated with point measurements, experiments have been conducted to test the use of radar determinations of precipitation volume over extensive mountain watersheds (Harrold 1966). Anderl *et al.* (1976) found, for two small basins near Hohenpeissenberg, in Bavaria, that results were as good as could be obtained with a gauge network density of 1 per 25 km^2, and much better than with the regular network at 1 per 500 km^2. More recently, over the Dee catchment in North Wales, Collier and Larke (1978) showed that even for snowfall, an accuracy similar to that for areal rainfall is feasible. Their mean accuracy was within 13 per cent. Calibrations based on snow depth at an upland and a lowland site can be used to allow for the effect of snow melt at lower elevations.

The availability of reliable automatic instrumentation and transmission capabilities that provide frequent sampling of mountain weather, in contrast to the higher costs and practical difficulties associated with remote stations that need to be visited at weekly/monthly intervals, has led many countries to implement such new systems. In Switzerland, for example, sixty automatic stations (ANETZ) installed in 1980 provide observations via telephone lines every ten minutes, that are averaged into hourly values. Nevertheless, the question of designing an appropriate network of precipitation gauges remains. For the Auvergne region of France, Benizou (1989) has used an optimization technique to propose an expanded network that takes account of topographic parameters (west–east slope, north–south slope, mean altitude, and relative relief, over a scale of 10–50 km). The procedure analyses the spatial pattern of errors in monthly rainfall predicted by a linear regression of precipitation and relief, weighted by the topographic parameters at an individual station. Results using the existing network and alternative expanded networks are compared over a grid of interpolated precipitation amounts.

A more practical approach in remote mountain areas is to estimate a hydrological budget for entire drainage basins. Flohn (1969, 1970) shows that in subtropical mountain ranges, snowfall accumulation on glaciers and runoff may be a more reliable guide to basin precipitation than precipitation data from valley stations, since mountain and valley wind systems cause the valleys to be much drier than the surrounding ridges. Also, in mapping mountain precipitation, estimates of seasonal altitudinal relationships using data from storage gauges or snow courses can usefully supplement the regular gauge network (Peck and Brown 1962). Basin snow water equivalent in mountain areas is also being estimated experimentally using a combination of visible band and passive microwave satellite data (Rango *et al.* 1989). However, the present low spatial resolution of the microwave data permits the use of this method only for large basins.

OTHER HYDROMETEORS

Hydrometeors include not only the common precipitation forms, discussed above, but also liquid or solid water particles suspended in the atmosphere (clouds, fog), wind-raised particles (blowing snow), and liquid or solid water particles deposited on the surface (dew, fog deposition, hoar frost, rime ice and glaze) (World Meteorological Organization, 1975). While the information on most of these phenomena is limited, all are important components of mountain weather and climate.

Fog

The visible suspension of water droplets in the air as fog, or cloud, is a prominent feature of most mountain areas. It occurs in valleys and basins, as well as over summits and slope, according to the particular physical processes in different weather regimes. Nocturnal radiation and cold air drainage, associated with low-level inversions, cause rather persistent *ground fog* in mountain valleys (see Plate 12). Fog on the summits, in contrast, may simply be a result of the height of the mean condensation level, and therefore cloud base, in a particular airstream. Orographically induced adiabatic cooling also occurs in upslope flow, especially with large-scale advection in the warm sector of mid-latitude cyclones. In both cases, such summit and slope cloud is simply termed *hill fog* in the British Isles. *High fog*, which may affect slopes, also forms by radiative and mixing processes associated with high-level inversions.

For mountains in Japan, Yoshino (1975: 205) shows a pronounced altitudinal maximum of fog occurrence at about 1500 m where fog frequency (visibility <1 km) exceeds 300 days year^{-1}. In contrast, in the European Alps, the frequency is greatest at the highest stations, such as Sonnblick (275 days year^{-1}), while there are zones of minimal frequency around 600 m and 1000 m along the valley slopes (Fliri 1975; Wanner 1979). This distribution is related to the occurrence of both ground-based radiation fog and high fogs associated with upper inversions in winter, as well as orographic cloud in all seasons. In maritime climates in middle latitudes high frequencies may occur at much lower elevations. For example, thick fog (with visibility below 200 m) occurs at 60 per cent of observations at Dun Fell (848 m) in northern England. The single altitudinal-maximum pattern observed in Japan is apparently determined by the dominance of the summer monsoon cloud regime. This pattern resembles the vertical distribution on Mauna Loa, Hawaii, where there is a well-defined fog belt on the windward slope between 1500 and 2500 m (Juvik and Ekern 1978). Here it is related to the level of summer orographic clouds below the trade-wind inversion, which is typically located at about 2000 m over Hilo, on the east coast, but occurs somewhat higher over the mountain slopes (Mendonca and Iwaoka 1969). The summit zone (3400 m) of the mountain, in contrast, has a weak winter maximum of fog frequency associated with high-level cloud in synoptic disturbances in the upper westerlies.

Fog precipitation

On many mountain slopes, especially in the tropics and subtropics, frequent orographic cloud augments the total moisture budget through the interception of fog droplets by the vegetation. This effect is noticeable at the edge of forest stands (Geiger 1966: 348–50), but it occurs generally on forested slopes.

Normal rain gauges miss this moisture deposition and special traps have been devised, using wire mesh or louvres above the gauge, to simulate the filtering effect of vegetation. Studies at Hohenpeissenberg show that, whereas a rain gauge caught 682 mm during May–September 1950, one with a wire mesh fog-trap caught 853 mm (25 per cent more) (Grunow 1952a). On days when only fog precipitation occurred, however, the fog-trap caught only 4.6 mm more. Using the same device on Table Mountain, Capetown, Nagel (1956) estimated that fog-drip may contribute an additional 70 per cent to the measured rainfall, while at Thodung (3100 m) in eastern Nepal, an additional 22 per cent (582 mm) was recorded during the monsoon months, June–September 1963, (Kraus 1967). Similar findings are reported from many forested coastal mountain ranges (Parsons 1960; Vogelman 1973) and other foggy environments. Fog-traps recorded 100–150 per cent of precipitation gauge totals during winters November–May 1959–65 on the Sonnblick, Austria (Grunow and Tollner 1969). Deposition rates as high as 6.9 mm h^{-1} were observed during warm air incursions from the Mediterranean. The fog contribution in the Alps appears to increase with height above the mean cloud base level. Grunow-type collectors used along a traverse in the Dischma Valley, near Davos, Switzerland, for summer 1961–5, showed no additional increment in the valley, 5–10 per cent more than in the regular gauge at 2100 m (the forest limit), and 20 per cent more catch on hill crests at 2550 m (Turner 1985).

Uncertainties exist as to the absolute amounts. A careful study has been carried out on Mauna Loa, Hawaii, by Juvik and Ekern (1978) using louvred-

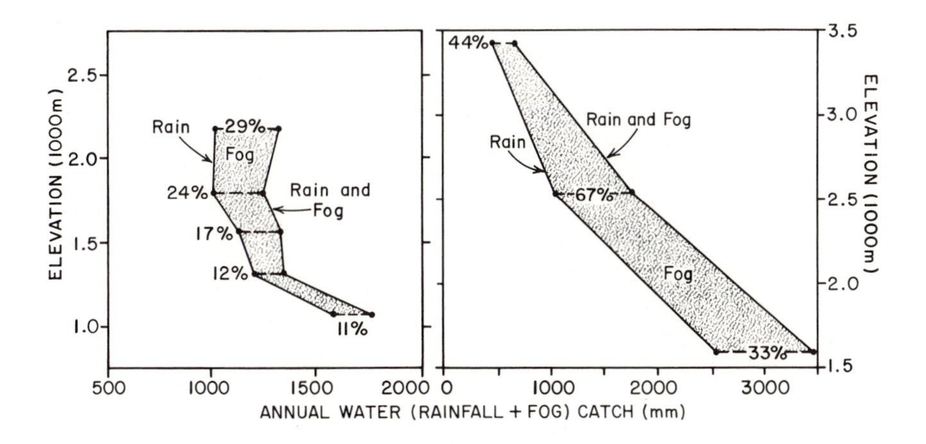

Figure 4.24 Annual rainfall and fog catch versus elevation on Mauna Loa, Hawaii. Left: lee slope. Right: windward slope

Note: The percentage increase in annual total due to fog catch is shown at the elevations of measurements

Source: After Juvik and Ekern 1978

screen fog gauges and calculations based on drop-size distributions, to separate the rain and fog ($<100\ \mu$m diameter) components during precipitation episodes. On the windward slope there is a well-defined fog belt between 1500 and 2500 m, where fog-drip amounts to about 750 mm, augmenting the annual precipitation by 50 per cent or more. Above 2000 m, the relative contribution of the fog-drip adds 65–70 per cent to the rainfall, although amounts are small. On the leeward slope of Mauna Loa, fog-drip contributes about 250 mm (a 25 per cent increase in annual total). Vertical profiles of the moisture, components illustrated in Figure 4.24, are controlled to a large measure by the trade-wind inversion and its seasonal variations. There is a clear maximum of fog moisture on the windward slope in summer, related to the development below the inversion of shallow orographic clouds that bank against the slopes. On the lee slope, fog-drip increases in both relative and absolute amounts above 1300 m elevation. Here upslope flow is associated with the sea breezes, supplemented by a valley–mountain wind on the higher slopes. There is a nocturnal maximum of fog precipitation on the windward slopes and an afternoon maximum on the lee slopes, in both cases closely corresponding to the diurnal occurrence of rainfall.

In the tropics, fog interception has long been regarded as an important determinant of the occurrence of montane *cloud forest* which is characterized by an abundance of mosses and liverworts. The role of fog interception has been investigated in northern South America where a stunted tree canopy (elfin cloud forest) is often present (Cavelier and Goldstein 1989). The water intercepted by a cylindrical plastic screen (40 mesh/cm^2) was scaled to measured forest throughfall per unit horizontal surface area. In northern Colombia (Serrania de Macuira)

Plate 13(a) Soft rime needles formed on a twig with an air temperature of −4°C, with 96 per cent relative humidity in a light south-easterly wind, 1030, 20 January 1953 (courtesy of Deutscher Wetterdienst, Hohenpeissenberg Observatory).

fog interception occurs almost daily, but rainfall is much less frequent. Fog interception increased linearly with altitude from 600 to 750 m and in one nine-day period in July 1984 it was 2.5 times greater on a windward slope than a leeward one. In the coastal Caribbean mountains of Colombia and Venezuela, rainfall increases eastward, whereas fog interception decreases. Thus, in 1985 rainfall totalled 853 mm at Macuira, Colombia (865 m) and fog interception 796 mm, while on Cerro Sta. Anna, Peninsula de Paraguana, Venezuela (815 m), the corresponding figures were 1630 mm and 518 mm, and on Cerro Copey, Margarita Island, Venezuela (987 m), 4461 mm and 480 mm, respectively. In the driest month at each site, fog interception accounted for over 60 per cent of the water income at Macuira and Sta. Anna, but only 9 per cent at Cerro Copey.

Plate 13(b) Hard rime 'feathers' formed on an anemometer support cable in light wind conditions at a temperature of −3°C, 100 per cent relative humidity, with occasional light drizzle and fog, 0900, 28 November 1953 (courtesy of Deutscher Wetterdienst, Hohenpeissenberg Observatory).

Even in the wettest month at Macuira it represented 33 per cent of the water income (Cavelier and Goldstein 1989).

Rime

Rime ice is deposited when supercooled fog droplets encounter standing objects such as trees or buildings and also wires (Plates 13 and 14). Three types of rime are distinguished (Table 4.6). However, rime deposits are frequently heterogeneous due to the more or less simultaneous, or consecutive, development of several forms with slightly varying conditions.

Table 4.6 Characteristics of rime deposits

Type[a]	*Characteristics*	*Mode of formation*	*Surface adhesion*	*Temperature criteria* (°C)
SOFT RIME (Rauhreif)	Fine needles/dendrites, clear crystalline structure, mean density ~ 0.2 g cm^{-3}	Deposition of vapour in solid form from super-cooled fog, or air super-saturated with respect to ice. Fog *not* essential	Easily detached	< −8
HARD RIME (Rauhfrost)	White granular structure with crystalline branches more or less separated by entrapped air. The deposit grows triangularly downwind, mean density ~ 0.5 g cm^{-3}	Fog essential. Rapid freezing of super-cooled fog droplets, leaving interstices	Brittle, easily detached at low temperatures	−2 to −10
CLEAR ICE (Rauheis)	Amorphous compact ice with alternate transparent and opaque (bubbles of air) layers, mean density 0.8 g cm^{-3}	Slow freezing of fog drops with released heat hindering crystallization	Firmly adheres to surface	0 to −3
GLAZE (Glatteis)	Clear ice layer, density of pure ice	Freezing of super-cooled raindrops on impact with surface; or of rain/fog drops contacting a super-cooled surface	Firmly adheres on surface	0 ± 3

Notes: [a]This terminology follows WMO (1975) and Grunow (1952b). Kuroiwa (1965) uses soft rime for what is here listed as hard rime, and hard rime for what is here clear ice.
Source: After Grunow 1952b; Waibel 1955; Kuroiwa 1965; and World Meteorological Organization 1975

Studies in Czechoslovakia (Hrudicka 1937), Poland (Baranowski and Liebersbach 1977), Romania (Tepes 1978) and eastern Germany (Kolbig and Beckert 1968) indicate that the frequency of rime (all forms) increases from about 20 days year^{-1} at lowland stations to 80–100 days year^{-1} at 1000–1500 m, and up to 180 days year^{-1} above 2500 m. Hrudicka suggests an *S*-form curvilinear trend with the major increase between about 750 and 1250 m related to wintertime cloud levels. Frontal situations with altostratus/nimbostratus cloud gave rise to most of the rime events recorded in the Tatra during 1952–3 (Koncek 1960), but in the West Carpathians they occur with cold maritime arctic air advection or due to radiative cooling of polar maritime air (Tepes 1978).

Plate 14 A weather screen on Mount Coburg, in the Canadian Arctic (76°N, 79.3°W), heavily coated with rime ice on the leeward side. Open water areas in the North Water, northern Baffin Bay (visible in the background) provide a ready moisture source during spring; air temperatures were −15°C (courtesy of Dr K. Steffen, Institute of Geography, ETH, Zurich).

Heavy rime accumulations during strong winds can pose a serious problem for structures and power lines (Phillips 1956) by increasing the dead-weight load and increasing the wind resistance. Ice growth on power lines may also lead to galloping excitation, due to wind forces (Sachs 1972). Mountain observatories in cloudy 'maritime' climates, such as Mt. Washington, New Hampshire, and the former Ben Nevis Observatory in Scotland, have experienced severe difficulties in maintaining reliable weather records due to rime accretion on instrument shelters and anemometers (Alexeiev *et al.* 1974). Half of the icing events on Mt. Washington lead to thicknesses in excess of 2.7 cm (18 kg m^{-2} loading), but the frequency of events and accretion rates are much less on nearby Mt. Mansfield (1339 m) (Ryerson 1988). Icing events lasting 3–20 hours with rates of 7 mm h^{-1} are typical in the Colorado Rockies (Hindman 1986). Riming is less severe in the western Cordillera around Tahoe, but it occurs frequently and the water equivalent may amount to 14–66 per cent of the precipitation during comparable periods (Berg 1988).

A 17-year study of rime accretion on Mt. Vitosha (2286 m), Bulgaria, indicates a mean duration of events in winter of about 36 hours (Stanev 1968).

Almost half the cases occur with temperatures between −2° and −6°C with winds of 10–12 m s^{-1}. On the Feldberg (1490 m) in the Black Forest, Germany, the accretion of ice on power lines (measured over two winters) averaged 50 g m^{-1} h^{-1}, with a maximum daily load of 3.2 kg m^{-1} (Waibel 1955). The loading is directly related to wind speed and water content according to Diem (1955), although Sachs (1972) states that the accretion depends on drop size, with the wind effect being debatable, and it is also inversely related to cable diameter. According to observations on Mt. Fuji and Mt. Neseko (1300 m) in Japan, wires are more often cut by the dynamic wind pressure – a 10–15 m s^{-1} wind produced a tension of 30 kg on 20 m length wires – than by ice load (Kuroiwa 1965). Clear ice build-up with winds of 20 m s^{-1} and temperatures of −5°C to −10°C was 2–5 kg m^{-1} over 20 hours. However, Kuroiwa also noted that snow accretion with temperatures near 0°C and winds of only 1–2 m s^{-1} can give loads of up to 5 kg m^{-1} for a 20 cm diameter ribbon of snow. On 19 March, 1969, a 260 m television transmitter mast on Emley Moor (400 m elevation), in the Pennines of northern England, collapsed following several days with glaze conditions (Page 1969). Fog and cloud shrouded the area for the preceding week with temperatures between −3°C and +2.5°C. Apparently, the build-up of ice on the support stays, to a radial thickness of 10 cm or more, was a major contributor to the structure's collapse.

There are few quantitative measurements of the moisture contributions from this source. Studies of the weight of rime accretions on branches of lodgepole pine in the Cascade Range, Washington (about 1900 m elevation), indicate that daily amounts average 0.014 cm w.e. (Berndt and Fowler 1969). The total contribution during periods without snowfall in the winter of 1966–7 amounted to 3.8 to 5 cm compared with a mean annual precipitation of 90 cm of which 85 per cent falls as snow. This contribution is similar to the estimate of 5–12.5 cm for the mountains of south-eastern Australia (Costin and Wimbush 1961). In northern Norway, data on rime provided by H. Köhler from the vicinity of the Haldde Observatory, indicate an order of magnitude increase on summits only 100–200 m above the observatory, due to stronger winds and more frequent hill fog (Landsberg 1962: 186).

Deposition on horizontal surfaces

The deposition of dew (or its frozen form hoarfrost) on the ground or vegetation surfaces has received little attention in mountain areas. In part, this may reflect the predominance of fog droplet or rime deposition in cloudy mountain environments. However, there are also observational problems. In a survey of records for the Austrian Alps, Lauscher (1977) concluded that many inhomogeneities were present in the data on hoarfrost, apparently due to failure to report such occurrences. He suggested that calculations may be preferable. In the presence of a snow cover, incursions of warm humid air commonly cause condensation on the surface. The saturation vapour pressure over melting snow is 6.1 mb and with an air temperature of 5°C, for example, the *maximum* relative humidity of the air corresponding to this limiting vapour pressure is only 70 per cent. (The snow surface temperature in the morning is about 1°C below the screen wet-bulb psychrometer reading according to Lauscher.) Based on more or less reliable observations of hoarfrost for four stations during 1946–75, and calculated condensation on snow, Lauscher estimated that average annual frequencies in the Austrian Alps vary with elevation approximately as follows:

Elevation (km)	*Days with snow cover*	*Hoarfrost on bare ground*	*Hoarfrost on snow*	*Condensation on snow*	*Total days*
3	365	7	81	49	137
2	200–250	16	55	33	104
1	110–145	25	29	17	71
(0)	30–45	35	3	1	39

Apart from making a contribution to the total moisture budget, such ice crystal deposits must also have an effect on the physical properties of the snow surface. These may be worthy of investigation in terms of their possible significance for spectral radiation signatures.

Blowing and drifting snow

Redistribution of snow by the wind is a major feature of the winter environments of the northern continents, as well as polar and alpine regions. Windblown snow is a strikingly visible element of the climatic environment of many mountain areas, yet the literature specifically addressing the problem is sparse. It has been investigated in Antarctica, primarily from the scientific standpoint, and in middle latitudes mainly in light of practical engineering concerns, such as snow control on highways.

Meteorological observations make a distinction between blowing snow, which is raised to a height of 1.8 m or more so as to obscure visibility, and drifting snow near the surface. The critical wind speed ('threshold' velocity) at which snow is picked up from the surface by turbulent eddies obviously depends on the state of the snow cover, including the temperature, size, shape and density of the snow particles and the degree of intergranular bonding (Pugh and Price 1954; Tabler 1975a). For loose unbonded snow, the typical threshold windspeed (at 10 m) is 5 m s^{-1}, whereas for a dense and bonded snow cover winds >25 m s^{-1} are necessary to cause blowing. Blowing snow will also, of course, occur with modest winds whenever snow is falling.

Three modes of snow transport are illustrated in Figure 4.25. *Creep* involves the rolling of dry snow particles along the surface. *Saltation* – where particles bounce along the surface in a layer about 10 cm deep, dislodging other particles as they return to the surface – occurs with winds of 5–10 m s^{-1} over cold loose snow. Creep accounts for perhaps 10 per cent of the saltation load (Berg and Caine 1975). *Suspension* is caused by turbulent diffusion lifting particles tens of metres above the surface, and this is considered to be the major mechanism for snow redistribution in windy environments (Radok 1977; Takeuchi 1980), although Soviet scientists regard saltation as the most significant factor for snow drifting (Mikhel *et al.* 1971; Dyunin and Kotlyakov 1980). Measurements at Weissflujoch Peak (2690 m) in the Alps show that saltation flux in the lowest 1 m is enhanced by stronger winds in this layer over crests, and on slopes that lie across wind direction (Meister 1987). Radok (1968) shows that the transition from saltation to suspension occurs with winds of about 15 m s^{-1}, when 60 per cent of the upcurrents in the surface boundary layer exceed the fall velocity of the snow particles.

Snow particles rapidly become rounded by mechanical abrasion on the surface, since natural concentrations in blowing snow are too low for particle interaction according to Schmidt (1972). Particles have a typical size of 0.1 mm and a terminal velocity of 0.5 m s^{-1}. The mass transport above the 50 cm surface layer is in the range 0.05–0.40 kg $m^{-2}s^{-1}$ for winds of 20–30 m s^{-1} in Antarctica (Mellor 1965), but is an order of magnitude smaller over an Alpine crest (Föhn 1980). Meister (1987) reports that for 75 runs averaging 3 hours in length, with an overall mean wind speed of 11 m s^{-1} at 4.3 m height, the average flux in the lowest 5 m at Weissflujoch was 48 g m^{-1} s^{-1}. Theoretical aspects of blowing snow transport cannot be dealt with here; convenient summaries are provided by Mellor (1965), Radok (1968) and Dyunin and Kotlyakov (1980). Direct measurement of the volume of snow transported is possible by means of an aerodynamically shaped trap coupled to a precipitation gauge (Jairell 1975) or by photoelectric devices which determine the attenuation of a light beam (Schmidt 1977).

In the mountain environment, snow distribution as well as drift profiles (Pedgley 1967) are strongly affected by meso- and micro-scale topography, including the vegetation structure. Hollows 10–100 m across are filled in during the course of each winter season until an equilibrium level of the snow surface is reached where erosion balances deposition. This is illustrated in Figure 4.26 for an alpine site in Colorado. The minimum change of ground slope angle necessary to cause streamline separation and the initiation of drift development is about 10° (Berg and Caine 1975). For 17 high elevation sites in Colorado and Wyoming, with fetches of 600–6000 m, Tabler (1975a) shows that the major controls of the slope of the drift surface, D, are the mean exhaust slope angle of the ground downwind of the break of slope, E, and the mean approach slope angle for 45 m upwind, A:

$$D(\%) = 0.25A + 0.55E_1 + 0.15E_2 + 0.05E_3$$

where E_1 = mean ground slope 0–15 m downwind (as a percentage), E_2 is for 15–30 m and E_3 for 30–45 m downwind. This equation accounted for 87 per cent of the variance in Tabler's data. On the lee side of hill tops and other

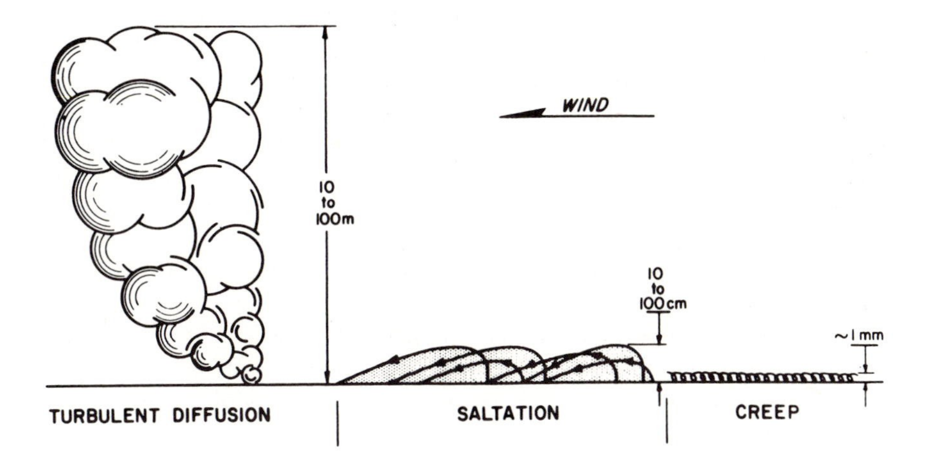

Figure 4.25 Modes of snow transport by the wind
Source: From Mellor 1965

locations where there is a sharp surface discontinuity, cornices may form due to the flow separation. The angle of the approach slope seems to be critical in determining whether a cornice or snowdrift forms (Santeford 1972). Föhn (1980) measured snow deposition on 30° ridge slopes at 2700 m near the Jungfraujoch following three storm periods with winds of 5–13 m s^{-1}. Totals were 61 kg m^{-2} on the windward slope and 236 kg m^{-2} on the lee slope, within ± 200 m of the crest, compared with 125 kg m^{-2} on a level surface. Snow transport contributed 87.5 kg m^{-2} on to the lee slope, primarily through the recurrence of a rotor in the lee of the crest. More extensive measurements during 4 winters show that the topography simply redistributes the snow. Amounts within 20–50 m of an elongated crest line average 1.6 times more on the lee than on the windward slope, but the entire ridge accumulates only the same amount as level terrain (Föhn and Meister 1983). The across-ridge profile of snow depth resembles a damped sine wave whose amplitude depends on the size of the ridge and the crest angle. On steep lee slopes ($>25°$), the first depth maximum is an attached cornice.

In topographic traps, such as Figure 4.26, the rate of infill and shape of the drift are determined by the amount of snow transported into the trap, which in turn is dependent upon its size, the amount of precipitation and the wind velocity. The 'blowpast' increases during the winter as the trap size decreases through infilling.

Deposition in drifts is clearly an important component of alpine water storage. For the Colorado Front Range, water contents of 400 m^3 per linear metre width of the drift are estimated (Martinelli 1973); figures of up to 1000–1200 m^3 per metre are cited for mountain areas such as the Altai in the Asiatic USSR (Mikhel and Rudneva 1971).

The wind transport of snow is now thought to be a major factor in snow sublimation in windy environments. Schmidt (1972) indicates that the sublimation rate at any given time is proportional to the 3/2 power of the diameter of the grains. A 100 μm diameter ice particle ventilated at 1 m s^{-1} loses 20 per cent of its weight in one minute at an air temperature of −20°C, 90 per cent relative humidity and sea level pressure. For the same temperature, humidity, and ventilation velocity, this rate may be increased 50 per cent at 4 km altitude, considering the increased diffusivity due to decreased air pressure (which more than offsets the decreased ventilation due to lower air density) and the increased solar radiation.

In a column of blowing snow, the situation is complicated by the vertical temperature and moisture profiles. Schmidt suggests that, in order to exhaust the vapour created by sublimation, an upward-directed gradient of moisture must exist, but estimates of probable sublimation rates are not provided. Tabler (1975b) expresses the ratio of the residual mass, M, of a snow particle to its original value, M_0, in terms of the distance travelled, D:

$$\frac{M}{M_0} = e^{-2(D/\bar{D})},$$

where $\bar{D}$ = the mean transport distance of the average size particle. Measurements of $\bar{D}$ in Wyoming, based on snow accumulation behind fences and in natural traps compared with precipitation data, indicate a value of about 3000 m

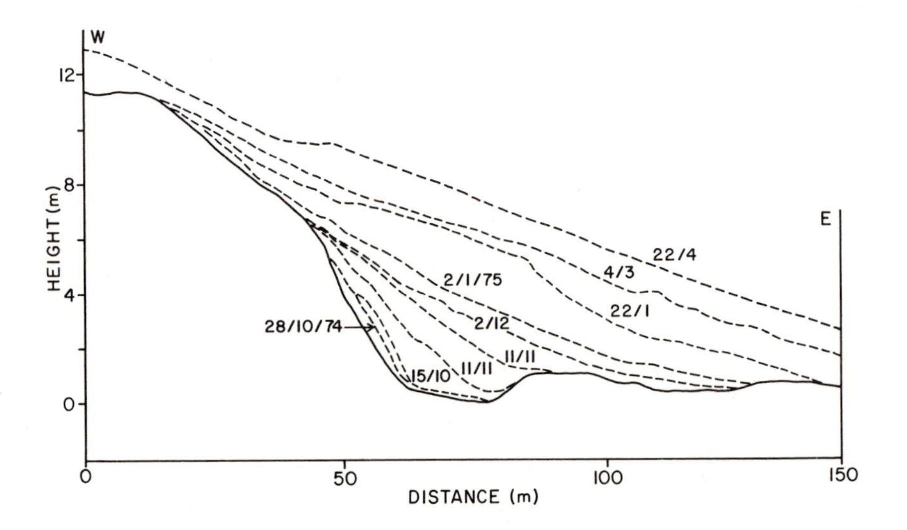

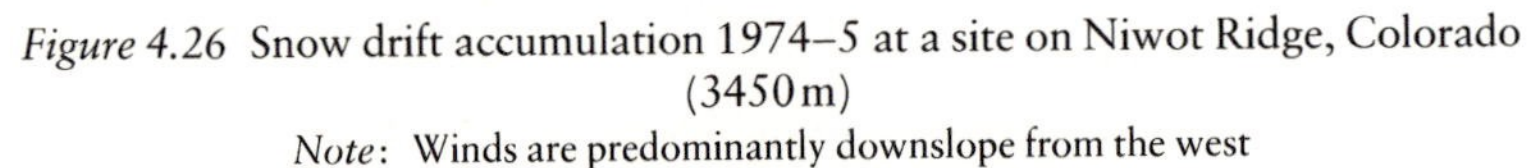
Figure 4.26 Snow drift accumulation 1974–5 at a site on Niwot Ridge, Colorado (3450 m)

Note: Winds are predominantly downslope from the west

Source: After Berg and Caine 1975

in rolling terrain and 850 m at a mountain site. The residual mass decreases to 39 per cent of the initial value for $D/\bar{D} = 0.5$, 11 per cent for $D/\bar{D} = 1$, and 0.4 per cent for $D/\bar{D} = 2$. The percentage of relocated precipitation lost to evaporation/sublimation, in relation to distance between natural traps, increases from about 60 per cent for $D/\bar{D} = 1$, to 80 per cent for $D/\bar{D} = 3$.

Mesoscale models of snow transport and accumulation are beginning to be developed, incorporating sub-models of the terrain-modified windfield, of snow transport and diffusion, and of cloud microphysics for the solid, liquid and vapour phases. Tesche (1988) describes a three-dimensional hydrometeorological model for cold orographic storms that features these three components. Snow accumulation is treated in three steps – transport by advection and diffusion just above the surface, transport to the surface as modified by terrain shape, and the incorporation of deposited snow into the ice crystal matrix. Also, snow erosion is computed in relation to the critical shear stress for re-suspension. The complete model and its individual submodels were validated using various data sets for the Lake Tahoe basin area.

EVAPORATION

Processes and methods of calculation

The transfer of water vapour from a water surface or from bare soil (*evaporation*) depends on both the properties of the ambient air and the energy supply to the surface. A number of meteorological factors are involved: the surface–air difference in vapour pressure; the temperatures of the air and of the evaporating surface (since temperature determines the saturation vapour pressure); the rate of air movement over the evaporating surface; and the energy supply via absorbed radiation, warm air advection, and heat storage beneath the air/surface interface. Lower atmospheric pressure, which leads to a higher evaporation rate, is also a factor but the pressure reduction due to high altitude is more than compensated by the decrease in air temperature.

In the case of bare soil, the availability of soil moisture is an additional factor and the transpiration of water from vegetation also involves consideration of plant physiological characteristics, particularly the stomatal characteristics of the leaves, the root structure, and any special drought-resistant adaptations. The total moisture loss from a vegetated surface is commonly referred to as *evapotranspiration*; where moisture is non-limiting, the term *potential evapotranspiration* is used. This value is more readily estimated from energy balance considerations (Henning and Henning 1981).

Where the surface is snow covered, there may be a direct crystal to vapour exchange, or *sublimation*. This phase change requires more energy than evaporation: the latent heat of vaporization for a water surface is 2.50 MJ kg^{-1}, to which must be added the latent heat of fusion. For snow at 0°C, the energy requirement is 2.83 MJ kg^{-1}.

A brief summary of theoretical approaches to evaporation estimation must suffice in view of the complexity of the problem and the limited number of studies relating specifically to conditions at high altitude. Four main approaches can be distinguished.

Hydrological balance calculations

This is primarily of use for estimating evaporation from water bodies. The water budget equation is written:

$$E = P - \Delta F - \Delta S,$$

where E = evaporation volume from the water body,
P = precipitation on the water body,
ΔF = net volume of outflow (above and below ground),
ΔS = net storage change in the water body.

Methods for estimating water budgets from monthly climatic data, extensively developed by C.W. Thornthwaite, are described in standard climatological texts; Willmott (1977) gives algorithms for the Thornthwaite formulation. The same principle can be applied to a snowpack (Krestovisky, 1962). Here the precipitation component includes condensation; runoff involves surface and groundwater; and the storage changes must include both the snowpack and the soil.

Energy balance method

The basic formulation is as follows:

$$LE = R_n - H - G - \Delta S + LP,$$

where L = latent heat of vaporization (or sublimation),
R_n = net radiation,
H = sensible heat flux to the atmosphere,
G = sensible heat flux to the soil interface,
ΔS = change of heat storage in the snowpack,
P = precipitation and condensation.

Since the sensible heat flux to the atmosphere is not readily determined with any reliability, it is usual to employ the ratio of sensible to latent heat (the Bowen ratio, β).

$$\beta = H/LE = \frac{0.61(T_s - T_a)}{(e_s - e_a)} \cdot \frac{p}{1000},$$

where T = temperature (°C),
e_a = vapour pressure in the air (mb),
e_s = saturation vapour pressure for the temperature of the water,
p = air pressure (mb).

β is close to unity for moist vegetated surfaces and increases to 5–10 for dry surfaces.

Substituting β in the equation above (ignoring the ΔS and LP terms),

$$LE = (R_n - G)/(1 + \beta).$$

However, it is more common to use this approach in some combination method (described below).

Evaporation estimates based on this procedure, in the case of snow covers in the Valdai and at Omsk, gave random errors of 0.44 mm day^{-1} according to Kuz'min (1972: 151).

The aerodynamic method

This method considers the role of turbulent diffusion in the vertical flux of water vapour. The basic expression, for the so-called 'bulk method', is known as Dalton's equation.

$$E = K_w(e_w - e_a)$$

where K_w = the turbulent exchange coefficient for water vapour.

K_w is expressed as a function of the wind speed. For a water surface, evaporation (mm) can be determined from the empirical equation:

$$E = 0.13u_2(e_s - e_2),$$

where u_2 = wind speed (m s^{-1}) at 2 m,
e_2 = vapour pressure at 2 m (Gangopadhyaya *et al.* 1966).

For snow, Kuz'min (1970) gives the following expression.

$$E = (0.18 + 0.098u_{10})(e_s - e_2)$$

where: u_{10} = wind speed (m s^{-1}) at 10 m,
e_s = saturation vapour pressure (mb) at snow surface temperature.

More generally, equations for vertical vapour flux take account of the profiles of wind and vapour pressure. Assuming *neutral* stability and therefore a logarithmic wind profile, and assuming that the coefficient of vapour exchange is equivalent to that for momentum exchange, then

$$E(\text{cm s}^{-1}) = \frac{k^2\rho(q_2 - q_1)(u_2 - u_1)}{\ln(z_2/z_1)^2}$$

where k = von Karman's constant ($\simeq 0.37$)
ρ = air density (g cm^{-3})
q = specific humidity
u = wind speed (cm s^{-1})
z = height.

This is essentially Sverdrup's equation (see Light 1941).

Combination methods

The most widely used methods combine the aerodynamic and energy budget equations. Penman's (1963) equation for evapo-transpiration from a vegetated surface (E_T, mm day^{-1}) is:

$$E_T = \left(\frac{\Delta}{\gamma}\frac{R_n}{L} + E_a\right) \Big/ \left(\frac{\Delta}{\gamma} + 1\right),$$

where: γ = the psychrometric coefficient (0.66 mb K^{-1} at sea level)
Δ = the slope of the saturation vapour pressure curve at mean air temperature (Δ/γ = 1.3 at 10°C, 2.3 at 20°C and 3.9 at 30°C at sea level)
R_n = net radiation over the natural surface
L = latent heat of vaporization

E_a (mm day^{-1}) is an aerodynamic term depending on wind speed and saturation vapour pressure deficit:

$$E_a = (0.263 + 0.138\,u)(e_s - e)$$

where u = wind at 2 m (m s^{-1})
$(e_s - e)$ = saturation deficit of the air at screen level (mb).

A computer solution for the equations is described by Chidley and Pike (1970). At elevated stations, corrections must be made to the psychrometric coefficient (Storr and den Hartog 1975; Stigter 1976, 1978) since:

$$\gamma = c_p p / \varepsilon L,$$

where p = atmospheric pressure (mb),
c_p = specific heat of dry air constant pressure (J kg^{-1} K^{-1}),
ε = the ratio of the molar mass of water vapour to that of dry air (= 0.622).

Tables for calculating Penman's estimate incorporating this altitude pressure effect have been published by McCulloch (1965). Stigter (1978) demonstrates that to modify the aerodynamic term for pressure dependence is unnecessary, except in high-altitude (and therefore cold) windy environments where the ratio of the net radiation/aerodynamic terms in the Penman equation is less than one; in such environments evaporation tends to be low anyway. Altitudinal corrections for γ are appropriate, however, since use of the 'standard' value may cause underestimates of evaporation at mountain stations (Storr and den Hartog 1975; Stigter 1976, 1978). More important, however, are the demonstrated effects of the type of vegetation cover. Various field programmes confirm that evaporation losses from forested basins exceed those from other surfaces due to the changes in surface albedo, interception and roughness (Bosch and Hewlett, 1982; Calder, 1990). It is inappropriate to attempt a detailed discussion of this topic here, especially since the theory is still being examined (Thom and Oliver 1977).

The term γE_a in the Penman equation can under certain conditions be estimated as a constant fraction of the energy term. Priestley and Taylor (1972) define a proportionality factor

$$\alpha = \mathrm{LE} \bigg/ \left(\frac{\Delta}{\Delta + \gamma}\right) (\mathrm{R_n} - \mathrm{G})$$

and show that over-large areas potential evapo-transpiration is given by

$$\mathrm{E_T} = 1.26\left(\frac{\Delta}{\Delta = \gamma}\right) (\mathrm{R_n} - \mathrm{G})$$

The proportionality factor, α decreases linearly as LE/Rn decreases; $\alpha \simeq 0.8$ for $\mathrm{LE/R_n} = 0.5$ (Thompson, 1975).

Measurements

There are several approaches to direct measurement of evaporation and evapo-transpiration. Sensitive instruments are now available for accurate instantaneous measurements of the vertical wind component and vertical differences in vapour content. By determining the instantaneous vertical transports of moisture in each direction the net vertical flux can be calculated. In practice, departures of vertical wind, temperature and humidity from their mean values are measured and the cross-correlation of these is determined computationally; the technique is referred to as 'eddy correlation'. Fast-response, accurate measurements are essential and, therefore, this technique is only suitable as a research procedure (Oke 1978: 323). More widely applicable is the use of a weighing lysimeter. A block of soil with its plant cover is weighed and the change due to evaporation losses is determined from knowledge of the amounts of precipitation and drainage. Simple systems have been designed for field use (LeDrew and Emerick 1974). Lake evaporation is commonly estimated on the basis of water losses from evaporation pans (Gangopadhyaya *et al.* 1966). Appropriate pan coefficients must be determined for each location but, unless additional corrections are made for seasonal climatic factors, only annual totals of lake evaporation can be estimated by this means. In the case of a snow cover, the weighing of plexiglass monoliths sunk in the snow is reported to give relatively good results with daily errors of about 0.1 mm (Kuz'min 1972: 151). Snow pans should be deep enough (30 cm) to eliminate radiation absorption by the bottom; 'false bottoms' are desirable to drain meltwater in spring; and it should be large enough to avoid disturbance to the normal air and vapour circulation in the pack (mesh walls are recommended) (Sabo 1956). Most of these criteria are usually ignored, however. As a result of the large horizontal variability in snow pack properties, both size of

Plate 15 Penitent forms, approximately 5 m high, formed by differential ablation on the Khumbu Glacier (5000 m) in the Nepal Himalayas. The ice fall of the Khumbu Glacier, where the Mount Everest Base Camp is located, is visible in the background (courtesy of Dr K. Steffen, Institute of Geography, Zurich).

lysimeter used (2 m^2, 5 m^2 or larger), and the spatial arrangement of the lysimeter installation (usually an array of many trays), can have a significant effect on the results. Careful sampling is required.

Little consideration has been given to the applicability of theoretical formulations in mountain environments, and few special observational programmes have been carried out. Clearly, the altitudinal effect may increase the solar radiation theoretically available for evaporation, although cloud cover may reverse this tendency. The atmospheric vapour content and air pressure both decrease with height. However, the most important factor on mid-latitude mountains at least is probably wind velocity. Many mountain locations are exposed to the wind regime of the free atmosphere, or to special local wind systems, such that enhanced advection effects occur. On tropical mountains the decrease in temperature and occurrence of cloud and fog belts must be of primary importance.

The irregular terrain and heterogeneous vegetation cover of mountains make areal assessments of evaporation based on point observations, or on theoretical grounds, difficult and of indeterminate accuracy. Analysis of evaporation rates in upland north Wales calculated by the Penman method using automatic weather station data for 1970–3 shows that spatial correlations between sites range from 'strong' (0.76) to unacceptably poor (Ovadia and Pegg 1979). Water budget calculations are commonly unreliable due to the problem of measuring precipitation accurately, while advection effects and other theoretical limitations make an aerodynamic approach inapplicable. An energy budget (or combination method) calculation seems likely to be most reliable, although the use of lysimeters, or snow pans in the case of a snowpack, may give the best results.

Evaporation and water balance

Mountaineers and botanists have long recognized the physiologically desiccating effect of low absolute humidity at high elevations and this, coupled with the commonly observed strong winds, has created a widespread impression of large evaporation losses in mountain areas. The occurrence of 'penitent' forms on many tropical glaciers (Plate 15) has also been interpreted in a like manner. Since the evaporation coefficient in equations of the Dalton type varies inversely with atmospheric pressure, an increase in evaporation with elevation has been argued on physical grounds, although Horton (1934) noted that theoretical and empirical considerations contradict this view. There are many conflicting observational results in the literature on evaporation and an attempt must therefore be made to select those where commonly accepted observational procedures were employed. Slaughter (1970) provides a convenient review.

Remarkably, at first sight, more data have been obtained for evaporation/sublimation losses from snow cover than from vegetated surfaces. That this is so reflects the importance of mountain snow packs for water supplies in many parts of the world. Careful studies of alpine snowpacks in the central Sierra Nevada, California, and the Rocky Mountains of Colorado, indicate that direct evaporation/sublimation has a negligible role compared with melt followed by subsequent evaporation of the meltwaters (Martinelli 1959, 1960; West 1959, 1962; Hutchison 1966). Martinelli, for example, reported that evaporation determined by Sverdrup's equation (p. 274) accounted for only 1–2 per cent of the summer ablation of snowfields at 3500–3800 m in the Rocky Mountains, 39°N. Energy balance measurements for a few days in March 1964 at Fraser Experimental Forest, Colorado (2700 m), gave daytime sublimation amounts of 1.0–1.5 mm and nocturnal condensation of 0.2–0.3 mm (Bergen and Swanson 1964), although work by Köhler (1950; Lauscher 1978) at Haldde Observatory, Norway (70°N, 23°E, 893 m) and studies in southern Finland by Lemmela and Kuusisto (1974) suggest that evaporation and condensation on snow cover are more or less in balance in winter. In a forest opening at the Central Sierra Snow Laboratory (2100 m), California, West (1959, 1962) measured an annual average snowpack evaporation of only 2.5–3.5 cm (2–3 per cent of the ablation) using small polyethylene pans sunk in the snow surface. In clearings and on exposed ridge top sites, amounts two to three times larger than in the forested areas were indicated. Nevertheless, the loss of 16.5 cm of water in four weeks from an ice block exposed on a 3.7 m tower in Wyoming by Peak (1963) seems unlikely to be representative of natural conditions. Similar results have been obtained in the USSR. In the Krestovoi Pass in the central Caucasus, observed daily sublimation averaged 0.11 mm in mid-winter for coarse-grained snow (Kuvaeva 1967). Slight condensation was observed at night. During 5 days in March 1965, the mean daily rate was only 0.21 mm from coarse-grained snow (0.11 mm for 7–10/10 cloud and 0.36 mm for 0–3/10 cloud) and 0.13 mm from fine-grained snow (0.10 mm for 7–10/10 and 0.23 mm for 0–3/10 cloud).

Few studies have specifically examined altitudinal differences in snow evaporation. Church (1934) reported figures two to four times greater on Mt. Rose, Nevada (3292 m) than from open meadow locations near Lake Tahoe (1897 m) and attributed the difference to wind conditions. More detailed measurements at the Weissflujoch, Switzerland, show a similar altitudinal increase during the daytime (de Quervain 1951). Between 13.00–15.00 hours on 9 March, 1950, for example, the snow evaporation was approximately 40 g m^{-2} h^{-1} (0.04 mm h^{-1}) at Davos (1550 m), 60–75 g m^{-2} h^{-1} at the Weissflujoch (2670 m), and 90–110 g m^{-2} h^{-1} at the summit (2850 m). Based on data for 20 days without precipitation in each month, a seasonal trend of evaporation was determined:

	January	*February*	*March*	*April*	*May*	*June*	*Total*
Sunny location	4	5	10	15	13	−5	42 mm
Shade	1	1	4	7	5	−6	12 mm

The figures show an increasing trend in spring, followed by condensation on the snow surface in summer. Again, the seasonal totals are quite modest. These results have been confirmed by other studies of evaporation from snow cover at altitudes of 2000–2500 in the Alps (Lang 1981).

Calculations by Lauscher and Lauscher (1976) using Kuz'min's equation (p. 274) and daily meteorological data for October 1969–September 1974 at the Sonnblick Observatory, Austria (3106 m), show that evaporation there is

Table 4.7 Snow evaporation and condensation (mm) at the Sonnblick (3106 m) October 1969–September 1976

	J	*F*	*M*	*A*	*M*	*J*	*J*	*A*	*S*	*O*	*N*	*D*	*Yr*
Evaporation	5.5	4.2	3.5	2.5	3.6	2.3	1.8	2.1	5.2	17.9	8.9	10.3	67.8
Condensation	−0.2	−0.2	−0.3	−0.1	−0.6	−4.9	−16.6	−20.1	−4.9	−0.4	−0.2	−0.1	−48.6
Balance	5.3	4.0	3.2	2.3	3.0	−2.5	−14.8	−18.1	0.3	17.4	8.7	10.1	19.0

Source: From Lauscher and Lauscher 1976

relatively constant from January through May and reaches a maximum in October (Table 4.7). Conversely, condensation values are large in July–August. The annual net evaporation is only 19 mm, compared with 270 cm of precipitation (90 per cent of which falls as snow), although the October–January net evaporation totals 42 mm (Table 4.7). Lauscher and Lauscher also show that almost two-thirds of days during February–May have zero snow evaporation at each of the three daily observations (0700, 1400 and 1900 or 2100 hours).

In contrast, other work has produced results which indicate that evaporation from mountain snow cover is far from negligible. In the dry, high-radiation environment of the White Mountains, California (38°N, 118°W; 3800 m), springtime sublimation/evaporation exceeds melt by a factor of four for 'fresh' snow and by 1.5 times for old drifts, according to Beaty (1975). In the central Rocky Mountains of Colorado, measurements and hydrological balance calculations led Santeford (1972) to the conclusion that, in the alpine area above the timber line, 80 per cent of the winter snowfall must be removed by the sublimation of blowing and drifting snow, or by sublimation *in situ*. The first factor is dominant in early and mid-winter, but later in the season sublimation/evaporation becomes important and may account for about half of the observed storage change in ridgeline cornices according to Santeford. In (artificial) drift situations on Pole Mountain, Wyoming (2470 m), losses by evaporation/sublimation of 20–30 per cent of the volume at the start of the ablation period are attributed to sensible heat advection from surrounding bare ground (Rechard and Raffelson 1974). Nevertheless, this situation is unlikely to be common in much alpine terrain.

In many mid-latitude continental areas, forest cover extends to 3000–3500 m and sublimation/evaporation of snow intercepted in the crowns then becomes an additional complication (Miller 1955). Again, however, Miller (1962) shows that the maximum daily evaporation rate from snow-covered trees would not exceed about 0.7 mm in the Sierra Nevada or on the Allegheny Plateau. The duration of snow cover on tree canopies may be relatively limited, at least in dry sunny climates. At Fraser Experiment Station in the Colorado Rockies, a south-east facing slope in a valley location was snow-free 31 per cent of the time between 1 December and 31 March and had ⩾2/3 cover during 55 per cent of this time (Hoover and Leaf 1967). It was also suggested that snowfall redistribution from tree crowns into openings would reduce evaporation losses from openings compared with forested sites since melt rates are higher in the former, but other work contradicts this. Measurements on the east slope of the Rocky Mountains, at Pingree Park, Colorado (2740 m), using evaporation pans designed to simulate snowpack conditions, indicate losses of 135 mm from a forest opening and 122 mm within lodgepole pine forest during a five-month winter period (Meiman and Grant 1974). This represents 45 per cent of seasonal snowfall. For 83 observation periods during two winter seasons 1972–3, 1973–4, average evapo-sublimation losses were as follows:

	Morning	*Afternoon*	*Night*
Opening	0.048	0.070	0.024 (mm H_2O) h^{-1}
Forest site	0.039	0.048	0.029

Condensation on the surface was rarely observed. Meiman and Grant (1974) also note that calculated evaporation rates from on-site measurements of temperature, vapour pressure and wind using an aerodynamic formulation accounted for 83 and 76 per cent of the observed variances for the opening and forest sites, respectively.

To summarize, evaporation/sublimation losses from mountain snowpacks seem to be a minor component of snow ablation in many environments. However, carefully conducted measurement programmes are not numerous and in warm/dry and windy conditions important exceptions do occur. Usually in spring/summer warm air masses are also moist and the moisture and temperature gradients produce condensation on the snow surface. While blowing snow is subject to significant sublimation, it is not yet clear to what extent this may occur from snow on ground. The environment of snow-covered mountains in winter is one where there is sufficient energy available to vaporize the snow grains, despite the generally dry air.

Data on evaporation from mountain areas in summer are similarly indeterminate. Measurements in 1904 by F. Adams using sunken pans on the east slope of Mt. Whitney indicated that evaporation rates decrease with elevation to about 3000 m, thereafter becoming more or less constant to 4000 m (Horton 1934). Estimates for water surfaces based on extensive meteorological data and pan measurements in the upper San Joaquin River basin of central California (37°N, 119°W) show a similar altitudinal pattern (Figure 4.27) (Longacre and Blaney 1962). However, the derivation of pan coefficients to estimate evaporation for the winter half-year makes the calculated mean annual values in Figure 4.27 of uncertain reliability. Summer measurements, also using evaporation pans, in the Wasatch Mountains, Utah, show a greater complexity (Peck and Pfankuch 1963). Mean daily wind speed and elevation together are shown to be more important, on a seasonal basis, than elevation or temperature alone.

Aspect must also play an important role. Peck and Pfankuch found higher evaporation rates on south-facing slopes, but attributed this to southerly airflow and increased instability effects rather than to excess radiation. Comparison of diurnal differences between a station on the canyon floor and one on a ridge top suggested that nocturnal drainage winds substantially augment evaporation rates. On lee slopes, föhn winds can be a major factor in enhancing evaporation. Calculations of potential evaporation during winter chinook conditions in Alberta gave 1.2 mm day^{-1} for 19 days in 1975 and 2.0 mm day^{-1} for 20 days in 1976 (Golding 1978).

Evapo-transpiration data in mountain areas are almost non-existent, yet it is

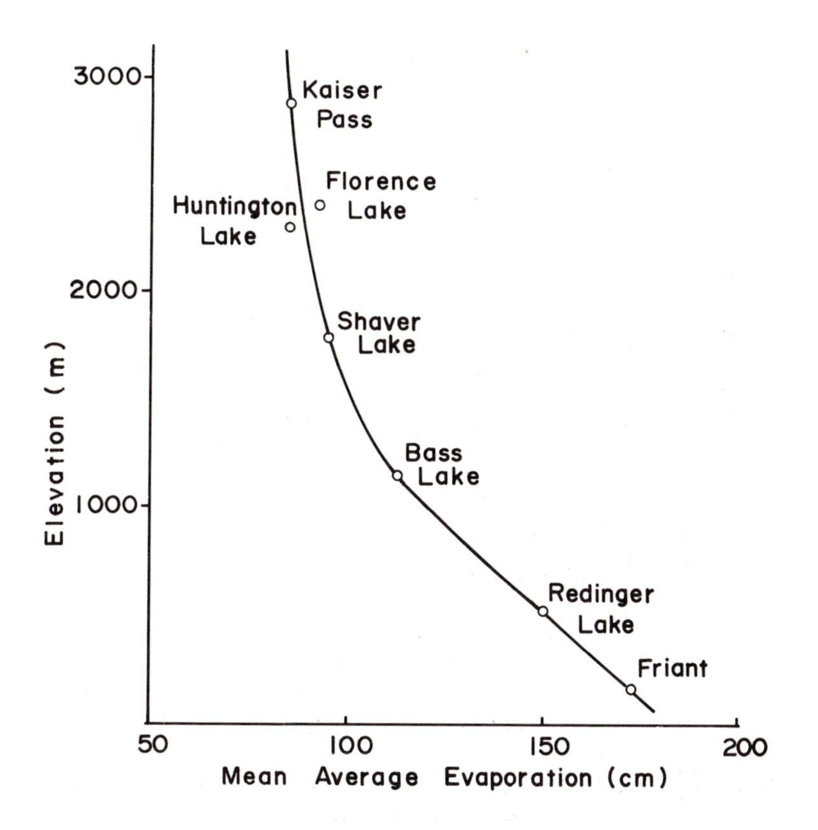

Figure 4.27 Estimated annual evaporation (cm) from water surfaces in central California based on pan measurements and meteorological data
Source: After Longacre and Blaney 1962

these losses rather than potential evaporation which are of most significance for vegetation growth. For an alpine tundra at 3500 m on Niwot Ridge, Colorado, LeDrew (1975) estimated a rate of 1.9 mm day^{-1} in July 1973. This was obtained through an empirical formula based on measurements with a small weighing lysimeter and measured vertical profiles of temperature, humidity, and wind speed. As a result of strong advection at this exposed site, evapo-transpiration continues through the night if soil moisture is available, although rates are low. Climatic estimates by Greenland (1989) of the water budget terms according to Thornthwaite's method indicate that actual evaporation is only about 28 per cent of annual precipitation at Niwot Ridge (3750 m) compared with 47 per cent at Como (3050 m). In summer, calculated actual evaporation slightly exceeds precipitation at both stations due to the implied use of soil moisture. Energy balance data from the Austrian Alps show a 30 per cent decrease in summer evapo-transpiration from a meadow site at 1960 m to alpine grassland at 2580 m (Staudinger and Rott 1981).

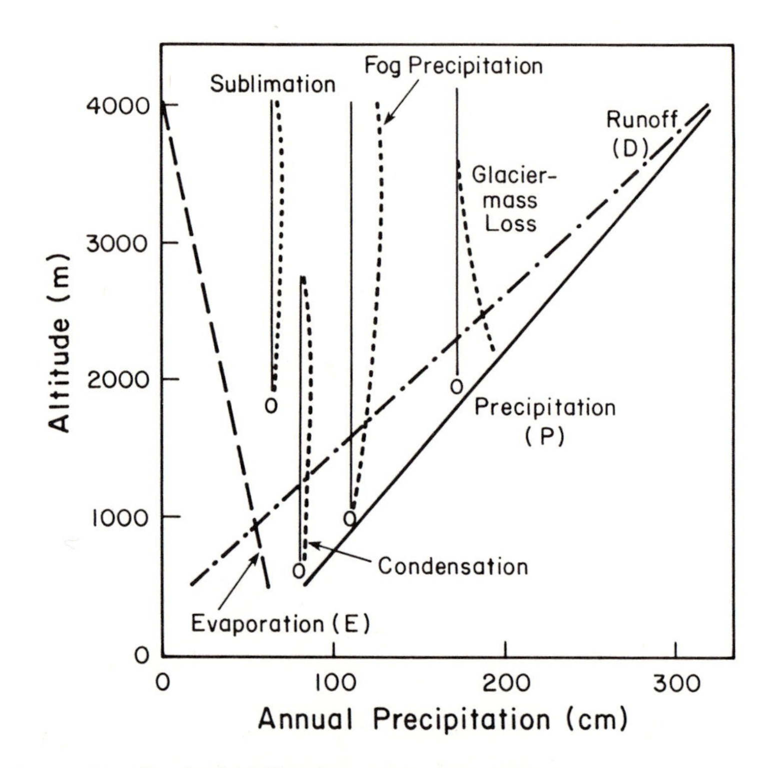

Figure 4.28 Altitudinal profiles of mean annual precipitation (P), evaporation (E) and run-off (D) for the Swiss Alps

Notes: The plots of condensation, sublimation, fog precipitation, and glacier mass loss are offset on the abscissa for clarity

Source: After Baumgartner *et al*. 1983

In the Canadian Arctic, Ohmura (1982) measured the altitudinal variation in evapo-transpiration during the snow-free period using small weighing lysimeters. The rate increased from 1.13 mm/day near sea level on Axel Heiberg Island (79.5°N, 91°W) to 1.65 mm/day at 750 m altitude (averaged over summers 1969 and 1970), in response to increasing wind speeds. However, the shorter snow-free period at the highest ridge site resulted in the seasonal total reaching a maximum of about 100 mm at 600 m compared with 85 mm at 750 m and at sea level. Precipitation, on the other hand, increased linearly with altitude from 60 mm to 110 mm for the snow-free period, so that precipitation exceeds evaporation above 500 m, whereas the opposite holds for this period at the lower elevations.

For the Alps, long-term regional evapo-transpiration averages for 1931–60 have been determined on the basis of water balance calculations (precipitation minus runoff) for 1000 catchments, with checks using the empirical relationships of W. Wundt, C.W. Thornthwaite (see p. 274) and others (Baumgartner *et al.*

1983). Averages for 500 m belts show a small approximately linear decrease with elevation (Figure 4.28). Between 500 m and 2500 m a.s.l. annual evapo-transpiration (E_T) is given by:

$$E_T(\text{mm}) = 680 - 20h,$$

where h is the elevation in hundreds of metres. There is still considerable uncertainty as to the magnitudes of evapo-transpiration determined from water balance calculations, even for basins in the European Alps. Lang (1981) cites a five-fold range in the rate of decrease of mean annual evaporation with altitude from 71 mm km^{-1} to 356 mm km^{-1}, according to difference sources. This range reflects uncertainties in the data rather than regional differences. Measurements on the Hintereisferner basin (Ötztal, Austria) over two years showed an areal evaporation of 180 mm for the 58 per cent ice-covered basin (mean altitude almost 3,000 m); this represents only about 10 per cent of the annual precipitation. Two years of water balance data for the Emerald Lake Basin (2,800–3,400 m) in the Sierra Nevada, California, which is 70 per cent bare rock, show that total evaporation accounts for 19 to 32 per cent of evaporation, the larger amount in the wetter year (Kattelman and Elder, 1991). Snowfall represents 95 per cent of the precipitation and 80 per cent of the evaporative losses are from snowcover. Sublimation averages about 50 mm/month during winter; during non-winter conditions actual evaporation is about 140 mm/month. 75 to 90 per cent of the peak spring snowpack (and subsequent rainfall) goes into streamflow.

As implied above, the effects of wind on snowfall and snow cover make the application of this hydrological method of limited value for evaporation estimates in many alpine locations (Santeford 1972). The implications of decreasing evapo-transpiration, but increasing precipitation with altitude in many mid-latitude mountain areas are that runoff increases with height. This is shown by Baumgartner *et al.* (1983) for the Alps and by Alford (1985) for Colorado (Figure 4.28). Nevertheless, different vertical patterns of runoff can be expected in other latitudes as a result of the profiles of precipitation and evaporation.

REFERENCES

Energy budgets

Aizenshtat, B.A. (1962) 'Nekotorye cherty radiatsonnogo rezhima, teplogo balansa, mikroklimata gornogo perevala' (Some characteristics of the radiation regime, heat balance and microclimate of a mountain pass), *Met. i Gidrol.*, 3, 27–32.

Barry, R.G. (1973) 'A climatological transect along the east slope of the Front Range, Colorado', *Arct. Alp. Res.*, 5, 89–110.

Borzenkova, I.I. (1965) 'K metodike rascheta summarnoy radiatsii dlya uslovii gornogo plato', *Trudy Glav. Geofiz. Obs.*, 179, 98–107.

Borzenkova, I.I. (1967) 'K voprosy o vliyanii mestnikh faktorov na prikhod radiastsii v gornoi mestnosti', *Trudy Glav. Geofiz. Obs.*, 209, 70–7.

Brazel, A.H. and Marcus, M.G. (1979) 'Heat exchange across a snow surface at 5365 meters, Mt. Logan, Yukon', *Arct. Alp. Res.*, 11, 1–16.

Goodin, D.G. and Isard, S.A. (1989) 'Magnitude and sources of variation in albedo within an alpine tundra', *Theoret. Appl. Climatol.*, 40, 50–60.
Greenland, D. (1978) 'Spatial distribution of radiation on the Colorado Front Range', *Climat. Bull.*, Montreal, 24, 1–14.
Hastenrath, S.L. (1978) 'Heat-budget measurements on the Quelccaya ice cap, Peruvian Andes', *J. Glaciol.*, 20 (82), 85–97.
Hennemuth, B. and Köhler, U. (1984) 'Estimation of the energy balance of the Dischma Valley', *Arch. Met. Geophys. Biocl.*, B, 34, 97–119.
Korff, H.C. (1971) 'Messungen zum Wärmehaushalt in den äquatorialen Anden', *Ann. Met.*, N.F. 5, 99–102.
Kraus, H. (1971) 'A contribution to the heat and radiation budget in the Himalayas', *Arch. Met. Geophys. Biokl.*, A, 20, 175–82.
LeDrew, E.F. (1975) 'The energy balance of a mid-latitude alpine site during the growing season, 1973', *Arct. Alp. Res.* 7, 301–14.
LeDrew, E.F. and Weller, G. (1978) 'A comparison of the radiation and energy balance during the growing season for arctic and alpine tundra', *Arct. Alp. Res.* 10, 665–78.
Marcus, M.G. and Brazel, A.J. (1974) 'Solar radiation measurements at 5365 meters, Mt. Logan, Yukon', V.C. Bushnell and M.G. Marcus (eds) *Icefield Ranges Research Project, Scientific Results*, vol. 4, pp. 117–19, New York, American Geographical Society.
Marcus, M.G. and LaBelle, J.C. (1970) 'Summertime observations at the 5360 meter level, Mount Logan, Yukon, 1968–1969', *Arct. Alp. Res.*, 2, 103–14.
Martinec, J. (1989) 'Hour-to-hour snowmelt rates and lysimeter outflow during an entire ablation period', in *Snow Cover and Glacier Variations*, Internat. Assoc. Hydrol. Sci. Publ. no. 183, Wallingford, Oxfordshire, pp. 19–28.
Müller, H. (1985) 'On the radiation budget in the Alps', *J. Climatol.*, 5:445–62.
Pleško, N. and Šinik, N. (1978) 'The energy balance in the mountains of Croatia', *Arbeiten, Zentralanst. Met. Geodynam.*, Vienna, 31, 9, 1–9, 16.
Rott, H. (1979) 'Vergleichende Untersuchungen der Energiebilanz in Hochgebirge', *Arch. Met. Geophys. Biokl.*, A, 28, 211–32.
Sheaffer, J.D. and Reiter, E.R. (1987) 'Measurements of surface energy budgets in the Rocky Mountains of Colorado', *J. Geophys. Res.*, 92(D4), 445–62.
Tappenier, U. and Cernusca, A. (1989) 'Canopy structure and light climate of different alpine plant communities: analysis by means of a model', *Theoret. Appl. Climatol.*, 40, 81–92.
Voloshina, A.P. (1966) *Teplovoy Balans Poverkhnosti Vysokogornykh Lednikov v Letnii Period*, Moscow, Nauka.
Wendler, G. and Ishikawa, N. (1973) 'Heat balance investigations in an arctic mountainous area in northern Alaska', *J. appl. Met.*, 12, 955–62.

Temperature

Aulitsky, H. (1967) 'Lage und Ausmass der "warmen Hangzone" in einen Quertal der Innenalp', *Ann. Met.*, 3, 159–65.
Baumgartner, A. (1960–2) 'Die Lufttemperatur als Standortsfaktor am Grossen Falkenstein', 1, *Forstwiss. Centralblatt*, 79, 362–73; 2, *Forstwiss. Centralblatt*, 80, 107–20; 3, *Forstwiss. Centralblatt*, 81, 17–47.
Buchan, A. (1890) 'The meteorology of Ben Nevis', *Trans. R. Soc. Edin.*, 34, xvii–lxi.
Chickering, J.W., Jr. (1884) 'Thermal belts', *Am. Met. J.*, 1, 213–18.
Cox, H.J. (1923) 'Thermal belts and fruit growing in North Carolina', *Mon. Weather Rev., Suppl.*, 19.
Douguédroit, A. and Saintignon, F.F. de (1970) 'Methode de l'étude de la décroissance

des temperatures en montagne de latitude moyenne: exemple des Alpes françaises du sud', *Rev. Géog. Alp.*, 58, 453–72.

Dreiseitl, E. (1988) 'Slope and free air temperatures in the Inn valley', *Met. Atmos. Phys.*, 39, 25–41.

Dunbar, G.S. (1966) 'Thermal belts in North Carolina', *Geog. Rev.*, 56, 516–26.

Fujita, T., Baralt, G. and Tsuchiya, J. (1968) 'Aerial measurement of radiation temperatures over Mt. Fuji and Tokyo areas and their application to the determination of ground- and water-surface temperatures', *J. appl. Met.*, 7, 801–16.

Furman, R.W. (1978) 'Wildfire zones on a mountain ridge', *Ann. Ass. Am. Geog.*, 68, 89–94.

Geiger, R. (1965) *The Climate Near the Ground*, pp. 417–18, 430–42, 453–4, Cambridge, Mass., Harvard University Press.

Hennessy, J.P., Jr. (1979) 'Comments on "Use of mesoscale climatology in mountainous terrain to improve the spatial representation of mean monthly temperatures"', *Mon. Weather Rev.*, 107, 352–3.

Hess, M., Niedzwiedz, T. and Obrebska-Starkel, B. (1975) 'The methods of constructing climatic maps of various scales for mountainous and upland territories, exemplified by the maps prepared for southern Poland', *Geog. Polonica*, 31, 163–87.

Koch, H.G. (1961) 'Die warme Hangzone. Neue anschauungen zur nachtlichen Kaltluftschichtung in Tälern und an Hängen', *Zeit. Met.*, 15, 151–71.

McCutchan, M.H. (1976) 'Diagnosing and predicting surface temperature in mountainous terrain', *Mon. Weather Rev.*, 104, 1044–51.

Obrebska-Starkel, B. (1970) 'Uber die thermische Temperaturschichtung in Bergtälern', *Acta Climat.*, 9, 33–47.

Pielke, R.A. and Mehring, P. (1977) 'Use of mesoscale climatology in mountainous terrain to improve the spatial representation of mean monthly temperatures', *Mon. Weather Rev.*, 105, 108–12.

Reiter, R. and Sladkovic, P. (1970) 'Control of vertical transport of aerosols between 700 and 3000 metres by lapse rate and fine structure of temperature', *J. Geophys. Res.*, 75, 3065–75.

Saintignon, M.F. de, (1976) 'Décroissance de températures en montagne de latitude moyenne: exemple des Alpes françaises du Nord', *Rev. Géog. Alp.*, 64, 483–94.

Steinhauser, F. (1967) 'Methods of evaluation and drawing of climatic maps in mountainous countries', *Arch. Met. Geophys. Biokl.*, B, 15, 329–58.

Wagner, A. (1930) 'Uber die Feinstruktur des Temperaturgradienten längs Berghängen', *Zeit. Geophys.*, 6, 310–18.

Waldemann, G. (1959) 'Schnee und Bodenfrost als Standortsfaktoren am Grossen Falkenstein', *Forstwiss. Centralblatt.*, 78, 98–108.

Yoshino, M.M. (1975) *Climate in a Small Area*, pp. 429–34. Tokyo, University of Tokyo Press.

Yoshino, M.M. (1984) 'Thermal belt and cold air drainage on the mountain slope and cold air lake in the basin at quiet, clear night', *Geojournal* 8, 235–50.

Cloudiness

Barry, R.G. and Chorley, R.J. (1987) *Atmosphere, Weather and Climate*, 5th edn, London, Methuen.

Fliri, F. (1967) 'Uber die klimatologische Bedeutung der Kondensationshöhe im Gebirge', *Die Erde*, 98, 203–10.

Fujita, T. (1967) 'Mesoscale aspects of orographic influences on flow and precipitation patterns', in E.R. Reiter and J.L. Rasmussen (eds), *Proceedings of the Symposium on*

Mountain Meteorology, Atmos. Sci. Pap. No. 122, pp. 131–46, Fort Collins, Colorado, Colorado State University.

Fujita, T., Baralt, G. and Tsuchiya, K. (1968) 'Aerial measurements of radiation temperatures over Mt. Fuji and Tokyo areas and their application to the determination of ground- and water-surface temperatures', *J. appl. Met.*, 7, 801–16.

Hauer, H. (1950) 'Klima und Wetter der Zugspitze', *Berichte d. Deutschen Wetterdienstes in der US-Zone*, 16.

Hjermstad, L.M. (1975) *Final Comprehensive Operations Report 1970–75 Season, Colorado River Basin Pilot Project*, EG & G (Report AL-1200), Albuquerque, New Mexico.

Hosler, C.L., Davis, L.G. and Booker, D.R. (1963) 'Modification of convective systems by terrain with local relief of several hundred metres', *Zeit. angew. Math. Phys.*, 14, 410–18.

Kuo, J.T. and Orville, H.D. (1973) 'A radar climatology of summertime convective clouds in the Black Hills', *J. appl. Met.*, 12, 359–73.

Küttner, J. and Model, E. (1948) 'Verschlüsselungsschwierigkeiten auf Bergstationen', *Zeit. Met.*, 2, 139–41.

MacCready, P.B. (1955) 'High and low elevations as thermal source regions', *Weather*, 10, 35–40.

Orville, H.D. (1965a) 'A numerical study on the initiation of cumulus clouds over mountainous terrain', *J. Atmos. Sci.*, 22, 684–99.

Orville, H.D. (1965b) 'A photogrammetric study of the initiation of cumulus clouds over mountainous terrain', *J. Atmos. Sci.*, 22, 700–9.

Orville, H.D. (1968) 'Ambient wind effects on the initiation and development of cumulus clouds over mountains', *J. Atmos. Sci.*, 25, 385–403.

Pedgley, D.E. (1971) 'Some weather patterns in Snowdonia', *Weather*, 26, 412–44.

Raymond, D.J. and Wilkening, M.H. (1980) 'Mountain-induced convection under fair weather conditions', *J. Atmos. Sci.*, 37, 2693–706.

Scorer, R.S. (1955) 'The growth of cumulus over mountains', *Arch. Met. Geophys. Biokl.*, A8, 25–34.

Silverman, B.A. (1960) 'The effect of a mountain on convection', in C.E. Anderson (ed.) *Cumulus Dynamics*, pp. 4–27, New York, Pergamon Press.

Tucker, G.B. (1954) 'Mountain cumulus', *Weather*, 9, 198–200.

World Meteorological Organization (1956) *International Cloud Atlas*, (abridged), Geneva, World Meterological Organization.

World Meterological Organization (1975) *Manual on the Observation of Clouds and Other Meteors: International Cloud Atlas*, vol. I, Geneva, World Meteorological Organization, no. 407.

Precipitation

Alford, D. (1985) 'Montain hydrologic systems', *Mountain Res. Devel.*, 5, 349–63.

Alpert, P. (1986) 'Mesoscale indexing of the distribution of orographic precipitation over high mountains', *J. Clim. appl. Met.*, 25, 532–45.

Alter, J.C. (1937) 'Shielded storage precipitation gages', *Mon. Weather Rev.*, 65, 262–5.

Anderl, B., Altmanspacker, W. and Schultz, G.A. (1976) 'Accuracy of reservoir inflow measurements based on radar rainfall measurements', *Water Resources Res.* 12, 217–23.

Andersson, T. (1980) 'Bergeron and the oreigenic (orographic) maxima of precipitation', *Pure Appl. Geophys.*, 119, 558–76.

Armstrong, C.F. and Stidd, C.K. (1967) 'A moisture-balance profile in the Sierra Nevada', *J. Hydrol.*, 5, 252–68.

Atkinson, B.W. and Smithson, P.A. (1976) 'Precipitation', in T.J. Chandler and S. Gregory (eds) *The Climate of the British Isles*, pp. 129–82, London, Longman.

Benizou, P. (1989) 'Taking topography into account for network optimization in mountainous areas', in B. Sevruk (ed.) *Precipitation Measurement*, WMO/IAHS/ETH Workshop on Precipitation Measurement, pp. 307–12, Zurich, Swiss Federal Institute of Technology.

Bergeron, T. (1949) 'Problem of artificial control of rainfall on the globe', *Tellus*, 1, 32–43.

Berkovsky, L. (1964) 'The fall-off with height of terrain-induced vertical velocity', *J. appl. Met.*, 3, 410–14.

Bleasdale, A. and Chan, Y.K. (1972) 'Orographic influences on the distribution of precipitation', in *The Distribution of Precipitation in Mountainous Areas*, vol. II, pp. 322–33, Geneva, World Meterological Organization no. 326.

Bonacina, L.C.W. (1945) 'Orographic rainfall and its place in the hydrology of the globe', *Q. J. R. Met. Soc.*, 71, 41–55.

Brown, M.J. and Peck, E.L. (1962) 'Reliability of precipitation measurements as related to exposure', *J. appl. Met.*, 1, 203–7.

Browning, K.A. (1980) 'Structure, mechanism and prediction of orographically enhanced rain in Britain', in Hide, R. and White, P.W. (eds) *Orographic Effects in Planetary Flows*, GARP Publ. Series no. 23, pp. 85–114, Geneva, WMO–ICSU Joint Scientific Committee, World Meteorological Organization.

Browning, K. and Harrold, T.W. (1969) 'Air motion and precipitation growth in a wave depression', *Q. J. R. Met. Soc.*, 95, 288–309.

Browning, K.A. and Hill, F.F. (1981) 'Orographic rain', *Weather*, 36, 326–9.

Browning, K.A., Hill, F.F. and Pardoe, C.W. (1974) 'Structure and mechanism of precipitation and the effect of orography in a wintertime warm sector', *Q. J. R. Met. Soc.*, 100, 309–30.

Browning, K.A., Pardoe, C.W. and Hill, F.F. (1975) 'The nature of orographic rain at wintertime cold fronts', *Q. J. R. Met. Soc.*, 101, 333–52.

Caine, N. (1975) 'An elevational control of peak snowpack variability', *Water Res. Bull.*, 11, 613–21.

Carruthers, D.J. and Choularton, T.W. (1983) 'A model of the seeder-feeder mechanism of orographic rain including stratification and wind-drift effects', *Q. J. R. Met. Soc.*, 109, 575–88.

Catterall, J.W. (1972) 'An *a priori* model to suggest rain gauge domains', *Area*, 4, 158–63.

Chuan, G.K. and Lockwood, J.G. (1974) 'An assessment of the topographic controls on the distribution of rainfall in the central Pennines', *Met. Mag.*, 103, 275–87.

Collier, C.G. and Larke, P.R. (1978) 'A case study of the measurement of snowfall by radar: an assessment of accuracy', *Q. J. R. Met. Soc.*, 104, 615–21.

Colton, D.E. (1976) 'Numerical simulation of the orographically induced precipitation distribution for use in hydrologic analysis', *J. appl. Met.*, 15, 1241–51.

Conrad, V. (1935) 'Beiträge zur Kenntnis der Schneedeckenverhältnisse', *Gerlands Beitr. Geophys.*, 45, 225–36.

Danard, M.B. (1971) 'A simple method for computing the variation of annual precipitation over mountainous terrain', *Boundary-layer Met.*, 2, 41–55.

Douglas, C.K.M. and Glasspoole, J. (1947) 'Meteorological conditions in heavy orographic rainfall in the British Isles', *Q. J. R. Met. Soc.*, 73, 11–38.

Elder, K., Dozier, J. and Michaelsen, J. (1989) 'Spatial and temporal variation of net snow accumulation in a small alpine watershed, Emerald Lake basin, Sierra Nevada, California, *Ann. Glaciol.*, 13, 56–63.

Elliott, R.D. (1977) 'Methods for estimating areal precipitation in mountainous areas',

Rep. 77–13, Goleta, California, North American Weather Consultants (for National Weather Service, NOAA-77-111506) (NTIS:PB-276 140/IGA).

Elliott, R.D. and Shaffer, R.W. (1962) 'The development of quantitative relationships between orographic precipitation and air-mass parameters for use in forecasting and cloud seeding evaluation', *J. appl. Met.*, 1, 218–28.

Erk, F. (1887) 'Die vertikale Verteilung und die Maximalzone des Neiderschlags am Nordhange der bayrischen Alpen im Zeitraum November 1883 bis November 1885', *Met. Zeit.*, 4, 55–69.

Fliri, F. (1967) 'Uber die klimatologische Bedeutung der Kondensationshöhe im Gebirge', *Die Erde*, 98, 203–10.

Flohn, H. (1969) 'Zum Klima und Wasserhaushalt des Hindukushs und benachtbaren Gebirge', *Erdkunde*, 23, 205–15.

Flohn, H. (1970) 'Comments on water budget investigations, especially in tropical and subtropical mountain regions', in *Symposium on World Water Balance*, pp. 251–62, *Int. Assoc. Sci. Hydrol. Publ*, 93, vol. 2, UNESCO.

Flohn, H. (1974) 'Contribution to a comparative meteorology of mountain areas', in J.D. Ives and R.G. Barry (eds) *Arctic and Alpine Environments*, pp. 55–71, London, Methuen.

Fogarasi, S. (1972) 'Weather systems and precipitation characteristics over the Arctic Archipelago in the summer of 1968', Inland Waters Directorate, Tech. Rep. no. 16, Ottawa, Environment Canada.

Föhn, P.M.B. (1977) 'Representativeness of precipitation measurements in mountainous areas', *Proceedings of the Joint AMS/SGBB/SSG Meeting on Mountain Meteorology and Biometeorology*, pp. 61–77, Geneva, Blanc et Wittwer.

Frère, M., Rijks, J.Q. and Rea, J. (1975) '*Estudio Agroclimatologico de la Zona Andina*', Informe Technico, Rome, Food and Agricultural Organization of the United Nations.

Fulks, J.R. (1935) 'Rate of precipitation from adiabatically ascending air', *Mon. Weather Rev.*, 63, 291–4.

Glazyrin, G.E. (1970) 'Fazovoe sostoyanie osadkov v gorakh v zavisimosti ot prizemnoy temperaturiy vozdukha', *Met. i Gidrol.*, 30–4.

Golubev, V.S. (1986) 'On the problem of standard conditions for precipitation gauge installation', in B. Sevruk (ed.) *Proceedings, International Workshop on the Correction of Precipitation Measurements*, Instruments and Observing Methods, Report no. 24 (WMO/TD no. 104), pp. 57–9, Geneva, World Meteorological Organization.

Goodison, B.E., Sevruk, B. and Klemm, S. (1989) 'WMO solid precipitation measurement intercomparison: objectives, methodology, analysis', in J.W. Delleur (ed.) *Atmospheric Deposition, Internat. Assoc. Hydrol. Sci.*, Publ. no. 179, pp. 59–64, Wallingford, UK, IAHS Press.

Gregory, S. (1968) 'The orographic component in rainfall distribution patterns', in J.A. Sporck (ed.) *Mélanges de Géographie. I. Géographie Physique et Géographie Humaine*, pp. 234–52. Gembloux, Belgium, J. Duculot, S.A.

Grunow, J. (1960) 'Ergebnisse mehrjähriger Messungen von Niederschlägen am Hang und im Gebirge', *Int. Assoc. Sci. Hydrol. Publ.*, 53, 300–16.

Hanson, C.L. (1982) 'Distribution and stochastic generation of annual and monthly precipitation on a mountainous watershed in southwest Idaho', *Water Resour. Bull.* 18: 875–83.

Harrold, T.W. (1966) 'The measurement of rainfall using radar', *Weather*, 21, 247–9, 256–8.

Hastenrath, S.L. (1967) 'Rainfall distribution and regime in central America', *Arch. Met. Geophys. Biokl.* B, 15, 201–41.

Havlik, D. (1968) 'Die Höhenstufe maximaler Niederschlagssummen in den Westalpen', *Freiburger Geogr. Hefte*, 7.

Henz, J.F. (1972) 'An operational technique of forecasting thunderstorms along the lee slope of a mountain range', *J. appl. Met.*, 11, 1284–92.

Hill, F.F. (1983) 'The use of average annual rainfall to derive estimates of orographic enhancement of frontal rain over England and Wales for different wind directions', *J. Climatol.*, 3, 113–29.

Hill, F.F., Browning, K.A., Bader, M.J. (1981) 'Radar and raingauge observations of orographic rain over south Wales', *Q. J. R. Met. Soc.*, 107, 643–70.

Hill, S.A. (1881) 'The meteorology of the North-West Himalaya', *Ind. Met. Mem.*, Calcutta, 1(VI), 377–429.

Hindman, E.E. (1986) 'An atmospheric water balance over a mountain barrier', in Y-G. Xu (ed.) *Proceedings of the International Symposium on the Qinghai-Xizang Plateau and Mountain Meteorology*, Beijing, Science Press, pp. 580–95.

Hjermstad, L.M. (1970) *The Influence of Meteorological Parameters on the Distribution of Precipitation across Central Colorado Mountains*, Atmos. Sci. Pap. no. 163, Fort Collins, Colorado State University.

Hobbs, P.V. (1975) 'The nature of winter clouds and precipitation in the Cascade Mountains and their modification by artificial seeding. Pt. 1. Natural conditions', *J. appl. Met.*, 14, 783–804.

Hobbs, P.V. (1978) 'Organization and structure of clouds and precipitation on the mesoscale and microscale in cyclonic storms', *Rev. Geophys. Space Phys.*, 16, 741–55.

Hobbs, P.V., Easter, R.C. and Fraser, A.B. (1973) 'A theoretical study of the flow of air and fallout of solid precipitation over mountainous terrain. Pt. II: Microphysics', *J. Atmos. Sci.*, 30, 813–23.

Hobbs, P.V., Houze, R.A., Jr. and Matejka, T.J. (1975) 'The dynamical and microphysical structure of an occluded frontal system and its modification by orography', *J. Atmos. Sci.*, 32, 1542–62.

Holmboe, J. and Klieforth, H. (1957) 'Investigations of mountain lee waves and the air flow over the Sierra Nevada', Final Report, Contract AF19-(604)-728, Los Angeles, Meteorology Department, University of California.

Houghton, J.G. (1979) 'A model for orographic precipitation in the north–central Great Basin', *Mon. Weather Rev.*, 107, 1462–75.

Hovkind, E.L. (1965) 'Precipitation distribution round a windy mountain peak', *J. Geophys. Res.* 70, 3271–8.

Hutchinson, P. (1968) 'An analysis of the effect of topography on rainfall in the Taieri catchment, Otago', *Earth Sci. J.*, 2, 51–68.

Jackson, M.C. (1978) 'Snow cover in Great Britain', *Weather*, 33, 298–309.

Jarvis, E.C. and Leonard, R. (1969) 'Vertical velocities induced by smoothed topography for Ontario and their use in areal forecasting' Met. Branch, Tech. Mem. 728, Department of Transport, Toronto.

Jevons, W.S. (1861) 'On the deficiency of rain in an elevated rain gauge as caused by wind', *London, Edinburgh and Dublin Phil. Mag.*, 22, 421–33.

Koncek, M. (1959) 'Schneeverhältnisse der Hohen Tatra', *Ber. dtsch. Wetterdienst*, 54, 132–3.

Lamb, H.H. (1955) 'Two-way relationships between the snow or ice limit and 1000–500 mb thickness in the overlying atmosphere', *Q. J. R. Met. Soc.*, 81, 172–89.

Lang, H. (1985) 'Höhenabhängigkeit der Niederschläge', in B. Sevruk (ed.) *Die Niederschlag in der Schweiz* (Berträge zur Geologie der Schweiz – Hydrologie, no. 31), pp. 149–57, Berne, Kummerly and Frey.

Larson, L.W. (1971) *Shielding Precipitation Gages from Adverse Wind Effects with Snow Fences*, University of Wyoming, Laramie, Water Resources Ser. no. 25

Lauer, W. (1975) 'Klimatische Grundzüge der Höhenstufung tropischer Gebirge', in *Tagungsbericht und wissenschaftliche Abhandlungen, 40 Deutscher Geographentag,*

Innsbruck, pp. 76–90, Innsbruck, F. Steiner.
Lauscher, F. (1976a) 'Weltweite Typen der Höhenabhängigkeit des Niederschlags', *Wetter u. Leben*, 28, 80–90.
Lauscher, F. (1976b) 'Methoden zur Weltklimatologie der Hydrometeore. Der Anteil des festen Niederschlags am Gesamtniederschlag', *Arch. Met. Geophys. Biokl.* B, 24, 129–76.
Lauscher, F. (1978) 'Typen der Höhenabhängigkeit des Niederschlags bei verschiedenen Witterungslagen im Sonnblick Gebiet', *Arbeiten, Zentralanst. für Met. Geodynam.* (Vienna), 32 (95), 1–6.
Linsley, R.K. (1958) 'Correlation of rainfall intensity and topography in northern California', *Trans. Am. Geophys. Un.*, 39, 15–18.
Lumb, P.E. (1983) 'Snow on the hills', *Weather*, 38, 114–15.
Manley, G. (1971) 'The mountain snows of Britain', *Weather*, 26, 192–200.
Martinec, J. (1985) 'Korrektur der Niederschlagsdaten durch Schneemessungen', in B. Sevruk (ed.) *Die Niederschlag in der Schweiz* (Beiträge zur Geologie der Schweiz-Hydrologie no. 31), pp. 77–96, Berne, Kummerly and Frey.
Martinec, J. (1987) 'Importance and effects of seasonal snow cover', in B.E. Goodison, R.G. Barry, and J. Dozier (eds) *Large Scale Effects of Seasonal Snow Cover*, IAHS Publ. no. 166, Int. Assoc. Hydrol. Sci., Wallingford, UK, pp. 107–20.
Marwitz, J.D. (1974) 'An airflow case study over the San Juan Mountains of Colorado', *J. appl. Met.*, 13, 450–8.
Marwitz, J.D. (1980) 'Winter storms over the San Juan Mountains. Part I. Dynamical processes', *J. appl. Met.*, 19, 913–26.
Marwitz, J.D. (1987) 'Deep orographic storms over the Sierra Nevada. Part II. The precipitation process', *J. Atmos. Sci.*, 44, 174–85.
Miller, J.F. (1972) 'Physiographically adjusted precipitation-frequency maps', in *Distribution of Precipitation in Mountainous Areas*, vol. 2, pp. 264–77, Geneva, World Meteorological Organization no. 326.
Miller, J.F. (1982) 'Precipitation evaluation in hydrology', in E.J. Plate (ed.) *Engineering Meteorology*, pp. 371–428, Amsterdam, Elsevier.
Murray, R. (1952) 'Rain and snow in relation to the 1000–700 mb and 1000–500 mb thickness and the freezing level', *Met. Mag.*, 81, 5–8.
Myers, V.A. (1962) 'Airflow on the windward side of a ridge', *J. Geophys. Res.*, 67, 4267–91.
Nicholass, C.A. and Harrold, T.W. (1975) 'The distribution of rainfall over subcatchments of the River Dee as a function of synoptic type', *Met. Mag.*, 104, 208–17.
Nickerson, E.C., Smith, D.R. and Chappell, C.F. (1978) 'Numerical calculation of airflow and cloud during winter storm conditions in the Colorado Rockies and Sierra Nevada mountains', in *Conference on Sierra Nevada Meteorology*, pp. 126–32, Boston American Meteorological Society.
Nipher, F.E. (1878) 'On the determination of the true rainfall in elevated gages', *Proc. Am. Ass. Adv. Sci.*, 27, 103–8.
Nullet, D. and McGranaghan, M. (1988) 'Rainfall enhancement over the Hawaiian Islands', *J. Climate* 1, 837–9.
Ohmura, A. (1991) 'New precipitation and accumulation maps for Greenland', *J. Glaciol.* 37(125), 140–8.
Ono, S. (1925) 'On orographic precipitation', *Phil. Mag.*, 6th Ser., 49, 144–64.
Peck, E.L. (1964) 'The little used third dimension', *32nd Annual Meeting of the Western Snow Conf.*, Nelson, British Columbia, pp. 33–40.
Peck, E.L. (1972a) 'Discussion of problems in measuring precipitation in mountainous areas', in *Distribution of Precipitation in Mountainous Areas*, vol. I, pp. 5–16, Geneva, World Meteorological Organization no. 326.

Peck, E.L. (1972b) 'Relation of orographic precipitation patterns to meteorological parameters', in *Distribution of Precipitation in Mountainous Areas*, vol. II, pp. 234–42, Geneva, World Meteorological Organization no. 326.

Peck, E.L. and Brown, M.J. (1962) 'An approach to the development of isohyetal maps for mountainous areas', *J. Geophys. Res.*, 67, 681–94.

Pedgley, D.E. (1970) 'Heavy rainfalls over Snowdonia', *Weather*, 25, 340–9.

Pedgley, D.E. (1971) 'Some weather patterns in Snowdonia', *Weather*, 26, 412–44.

Pockels, F. (1901) 'The theory of the formation of precipitation on mountain slopes', *Mon. Weather Rev.*, 29, 152–9, 306–7.

Poggi, A. (1959) 'Contribution à la connaissance de la distribution altimétrique de la durée de l'enneigement dans les Alpes françaises du nord', *Ber. dtsch. Wetterd.*, 54, 134–49.

Poulter, R.M. (1936) 'Configuration, air mass and rainfall', *Q. J. R. Met. Soc.*, 62, 49–79.

Queney, P. (1948) 'The problem of airflow over mountains: a summary of theoretical studies', *Bull. Am. Met. Soc.*, 29, 16–26.

Raddatz, R.L. and Khandekar, M.L. (1977) 'Numerical simulation of cold easterly circulations over the Canadian Western Plains using a mesoscale boundary-layer model', *Boundary-Layer Met.*, 11, 307–28.

Rallison, R.E. (1981) 'Automated system for collecting snow and related hydrological data in mountains of the western United States', *Hydrol. Sci. Bull.*, 26, 83–9.

Rangno, A.L. (1979) 'A re-analysis of the Wolf Creek cloud seeding experiment', *J. appl. Met.*, 18, 579–65.

Rango, A., Martinec, J., Chang, A.T.C., Foster, J.L. and van Katwijk, V. (1989) 'Average areal water equivalent of snow in a mountain basin using microwave and visible satellite data', *IEEE Trans. Geosci. Remote Sensing* 27: 740–45.

Rauber, R.M., Grant, L.O., Feng, D.-X, and Snider, J.B. (1986) 'The characteristics and distribution of cloud water over the mountains of northern Colorado during wintertime storms. Part I. Temporal variations', *J. Clim. appl. Met.*, 25, 468–88.

Rechard, P.A. (1972) 'Winter precipitation gauge catch in windy mountainous areas', in *Distribution of Precipitation in Mountainous Areas*, vol. I, pp. 13–26, Geneva, World Meteorological Organization no. 326.

Reid, I. (1973) 'The influence of slope aspect on precipitation receipt', *Weather*, 28, 490–3.

Reinelt, E.R. (1970) 'On the role of orography in the precipitation regime of Alberta', *Albertan Geographer*, 6, 45–58.

Rhea, J.O. (1978) *Orographic Precipitation Model for Hydrometeorological Use*, Atmos. Sci. Pap. No. 287, Fort Collins, Colorado State University.

Rhea, J.O. and Grant, L.O. (1974) 'Topographic influences on snowfall patterns in mountainous terrain', in *Advanced Concepts and Techniques in the Study of Snow and Ice Resources*, pp. 182–92, Washington, D.C., National Academy of Science.

Rodda, J.C. (1967) 'The rainfall measurement problem', in *Geochemistry, Precipitation, Evaporation, Soil-Moisture, Evaporation*, pp. 215–30, IUGG General Assembly of Bern.

Rohrer, M. (1989) 'Determination of the transition air temperature from snow to rain and intensity of precipitation' in B. Sevruk (ed.) *Precipitation Measurement*, WMO/IAHS/ETH Workshop on Precipitation Measurements, pp. 475–82, Zurich, Swiss Federal Institute of Technology.

Salter, M. de C.S. (1918) 'The relation of rainfall to configuration', *British Rainfall*, 1918, 40–56.

Sarker, R.P. (1966) 'A dynamical model of orographic rainfall', *Mon. Weather Rev.*, 94, 555–72.

Sarker, R.P. (1967) 'Some modifications in a dynamical model of orographic rainfall', *Mon. Weather Rev.*, 95, 673–84.

Sawyer, J.S. (1956) 'The physical and dynamical problems of orographic rain', *Weather*, 11, 375–81.
Schermerhorn, V.P. (1967) 'Relations between topography and annual precipitation in western Oregon and Washington', *Water Resources Res.*, 3, 707–11.
Sevruk, B. (1972a) 'Precipitation measurements by means of storage gauges with stereo and horizontal orifices in the Baye de Montreux watershed', in *Distribution of Precipitation in Mountainous Areas*, vol. 1, pp. 86–95, Geneva, World Meteorological Organization no. 326.
Sevruk, B. (1972b) 'Evaporation losses from storage gauges', ibid., pp. 96–102.
Sevruk, B. (1974) 'The use of stereo, horizontal and ground level orifice gages to determine a rainfall-elevation relationship', *Water Resources Res.*, 10, 1138–42.
Sevruk, B. (1983) 'Correction of measured precipitation in the Alps using the water equivalent of new snow, *Nordic Hydrol.*, 14, 49–58.
Sevruk, B. (1985) 'Schneeanteil am Monatsniederschlag', in B. Sevruk (ed.) *Die Niederschlag in der Schweiz* (Beiträge zur Geologie der Schweiz – Hydrologie no. 31), pp. 127–37, Berne, Kummerly and Frey.
Sevruk, B. (1986a) 'Correction of precipitation measurements', in B. Sevruk (ed.) *Proceedings, International Workshop on the Correction of Precipitation Measurements*, Instruments and Observing Methods, Report no. 24 (WMO/TD no. 104), pp. 13–23, Geneva, World Meteorological Organization.
Sevruk, B. (1986b) 'Correction of precipitation measurements: Swiss experience', in B. Sevruk (ed.) *Precipitation Measurement.* Swiss Federal Institute of Technology, Zurich, pp. 187–93.
Sevruk, B. (ed.) (1989) '*Precipitation measurement*', WMO/IAHS/ETH Workshop on Precipitation Measurement, Zurich, Swiss Federal Institute of Technology.
Sevruk, B. and Klemm, S. (1989) 'Types of standard precipitation gauges', in B. Sevruk (ed.), *Precipitation Measurement*, Swiss Federal Institute of Technology, Zurich, pp. 227–32.
Slatyer, R.O., Cochrane, P.M., Galloway, R.W. (1984) 'Duration and extent of snow cover in the Snowy Mountains and a comparison with Switzerland', *Search* 15: 327–31.
Smith, R.B. (1979) 'The influence of mountains on the atmosphere', *Adv. Geophys.*, 21, 87–230.
Smith, R.B. (1982) 'A differential advection model of orographic rain', *Mon. Wea. Rev.*, 110, 306–9.
Smith, R.B. (1985) 'Comment on "Interaction of low-level flow with the Western Ghat mountains and offshore convection in the summer monsoon" (Reply D.R. Durran and R.L. Grossman)', *Mon. Wea. Rev.*, 113, 2176–81.
Smith, R.B. and Lin, Y.-L. (1983) 'Orographic rain on the Western Ghats', in E.R. Riter, B.-Z. Zhu and Y.-F. Qian (eds) *Proceedings of the First Sino-American Workshop on Mountain Meteorology*, Beijing, Science Press, pp. 71–94.
Smithson, P.A. (1970) 'Influence of topography and exposure on airstream rainfall in Scotland', *Weather*, 25, 379–86.
Sporns, U. (1964) 'On the transportation of short duration rainfall intensity data in mountainous regions', *Arch. Met. Geophys. Biokl.*, B, 13, 438–42.
Spreen, W.C. (1947) 'Determination of the effect of topography on precipitation', *Trans. Am. Geophys. Un.*, 28, 285–90.
Steinacker, R. (1983) 'Diagnose und Prognose der Schneefallgrenze', *Wetter u. Leben*, 35, 81–90.
Steinhauser, F. (1948) 'Die Schneehöhen in den Ostalpen und die Bedeutung der winterlichen Temperaturinversion', *Arch. Met. Geophys. Biokl.*, B, 1, 63–74.
Steinhauser, F. (1967) 'Methods of evaluation and drawing of climatic maps in mountain-

ous countries', *Arch. Met. Geophys. Biokl.*, B, 15, 329–58.
Stewart, R.E. (1985) 'Precipitation types in winter storms', *Pure Appl. Geophys.*, 123, 597–609.
Storebö, P.B. (1968) 'Precipitation formation in a mountainous coast region', *Tellus*, 20, 239–50.
Struzer, L.R., Nechayer, I.N. and Bogdanova, E.G. (1965) 'Systematic errors of measurements of atmospheric precipitation', *Soviet Hydrol.*, 4, 500–4.
Sugden, D.E. (1977) 'Reconstruction of the morphology, dynamics, and thermal characteristics of the Laurentide ice sheet at its maximum', *Arct. Alp. Res.*, 9, 21–47.
Tucker, D.F. and Reiter, E.R. (1988) 'Modeling heavy precipitation in complex terrain', *Met. Atmos. Phys.*, 39, 119–31.
Wagner, A. (1937) 'Gibt es im Gebirge eine Höhenzone maximalen Niederschlages?' *Gerlands Beitr. Geophys.*, 50, 150–5.
Walker, E.R. (1961) *A Synoptic Climatology for Parts of the Western Cordillera*, Arctic Meteorology Research Group, Publ. in Met. no. 35, Montreal, McGill University.
Warnick, C.C. and Penton, V.E. (1971) 'New methods of measuring water equivalent of snow pack for automatic recording at remote mountain locations', *J. Hydrol.*, 13, 201–15.
Weischet, W. (1969) 'Klimatologische Regeln zur Vertikalverteilung der Niederschlage in Tropengebirgen', *Die Erde*, 100, 287–306.
Wilson, H.P. (1978) 'On orographic precipitation', in *Climatic Networks: Proceedings of the Workshop and Annual Meeting of the Alberta Climatological Association, Inform. Rep.*, NOR-X-209, pp. 82–21. Fisheries and Environment Canada, Edmonton, Alberta, Northern Forest Research Centre.
World Meteorological Organization (1984) '*Guide to Hydrological Practices. Vol. 1. Data Acquisition and Processing*' (4th edn,), Geneva, World Meteorological Organization, no. 168.
Yoshino, M.M. (1975) *Climate in a Small Area: An Introduction to Local Meteorology*, Tokyo, University of Tokyo Press.
Young, K.C. (1974) 'A numerical simulation of wintertime, orographic precipitation', *J. Atmos. Sci.*, 31, 1735–48; 1749–67.

Other hydrometeors: fog, fog precipitation, rime, deposition on horizontal surfaces

Alexeiev, J.K., Dalyrymple, P.C. and Gerger, H. (1974) *Instrument and Observing Problems in Cold Climates*, Geneva, World Meteorological Organization no. 384.
Baranowski, S. and Liebersbach, J. (1977) 'The intensity of different kinds of rime on the upper tree line in the Sudety Mountains', *J. Glaciol.*, 19, 489–97.
Berg, N.H. (1988) 'Mountain-top riming at sites in California and Nevada, U.S.A', *Arct. Alp. Res.*, 20, 429–47.
Berndt, H.W. and Fowler, B.W. (1969) 'Rime and hoarfrost in upper-slope forests of eastern Washington', *J. Forestry*, 67, 92–5.
Cavelier, J. and Goldstein, G. (1989) 'Mist and fog interception in elfin cloud forests in Colombia and Venezuela', *J. Trop. Ecol.*, 5, 309–22.
Costin, A.B. and Wimbush, D.J. (1961) 'Studies in catchment hydrology in the Australian Alps, IV. Interception by trees of rain, cloud and fog', C.S.I.R.O., Australia Division of Plant Industry Tech. Pap. 16.
Diem, M. (1955) 'Höchstlasten der Nebelfrostablagerungen am Hochspannungsleitungen im Gebirge', *Arch. Met. Geophys. Biokl.*, B,7, 84–95.
Fliri, F. (1975) *Das Klima der Alpen im Raume von Tirol*, Innsbruck, Universitätsverlag Wagner.

Geiger, R. (1966) *The Climate Near the Ground*, Cambridge, Mass., Harvard University Press.

Grunow, J. (1952a) 'Nebelniederschlag: Bedeutung und Erfassung einer Zusatzkomponente des Niederschlags', *Berichte deutsch. Wetterdienst.-US Zone*, 7 (42), 30–4.

Grunow, J. (1952b) 'Kritische Nebelfroststudien', *Arch. Met. Geophys. Biokl.*, B, 4, 389–419.

Grunow, J. and Tollner, H. (1969) 'Nebelniederschlag im Hochgebirge', *Arch. Met. Geophys. Biokl.*, B, 17, 201–28.

Hindman, E.E. (1986) 'Characteristics of supercooled liquid water in clouds at mountaintop sites in the Colorado Rockies', *J. Clim. appl. Met.*, 25, 1271–9.

Hrudicka, B. (1937) 'Zur Nebelfrosttage', *Gerlands Beitr. Geophys.*, 51, 335–42.

Juvik, J.O. and Ekern, P.C. (1978) *A Climatology of Mountain Fog on Mauna Loa, Hawaii Island*, University of Hawaii, Water Resources Res. Cen. Tech. Rep. no. 118.

Kolbig, J. and Beckert, T. (1968) 'Untersuchungen der regionalen Unterschiede im Auftreten von Nebelfrost', *Zeit. Met.*, 20, 148–60.

Koncek, M. (1960) 'Zur Frage der Nebelfrostablagerungen im Gebirge', *Studia Geophys. Geodet.*, 4, 69–84.

Kraus, H. (1967) 'Das Klima von Nepal', *Khumbu Himal*, 1(4), 301–21.

Kuroiwa, D. (1965) 'Icing and snow accretion on electric wires', *CRREL Res. Rep.* 123, Hanover, N.H., U.S. Army.

Landsberg, H. (1962) *Physical Climatology*, p. 186, DuBois, Pa., Gray Printing Co., Inc.

Lauscher, F. (1977) 'Reif und Kondensation auf Schnee und die wahre Zahl der Tage mit Reif', *Wetter u. Leben*, 29, 175–80.

Mendonca, B.G. and Iwaoka, W.T. (1969) 'Tradewind inversion at the slope of Mauna Loa', *J. appl. Met.*, 8, 213–19.

Nagel, J.F. (1956) 'Fog precipitation on Table Mountain', *Q. J. R. Met. Soc.*, 83, 452–60.

Page, J.K. (1969) 'Heavy glaze in Yorkshire-March 1969', *Weather*, 24, 486–95.

Parsons, J. (1960) '"Fog drip" from coastal stratus, with special reference to California', *Weather*, 15, 58–62.

Phillips, P.E. (1956) 'Icing of overhead high-voltage power lines in the Grampians', *Met. Mag.*, 85, 376–8.

Ryerson, C.C. (1988) 'New England mountain icing climatology', CRREL Rep. 88–12, Hanover, N.H., US Army.

Sachs, P. (1972) *Wind Forces in Engineering*, p. 270, Oxford, Pergamon Press.

Stanev, S. (1968) 'Nebelfrostablagerungen an Freileitungen unter Gebirgsbedingungen', *Zeit. Met.*, 20, 161–4.

Tepes, E. (1978) 'Ice depositions in Romanian mountainous regions', *Veröff, Schweiz. Met. Zentralanstalt*, Zurich, 40, 308–12.

Turner, H. (1985) 'Nebelniederschlag', in B. Sevruk (ed.) *Die Niederschlag in der Schweiz*', (Beiträge zur Geologie der Schweiz – Hydrologie, no. 31), pp. 77–96, Bern, Kummerly and Frey.

Vogelman, H.W. (1973) 'Fog precipitation in the cloud forests of eastern Mexico', *BioScience*, 23, 96–100.

Waibel, K. (1955) 'Die meteorologischen Bedingungen für Nebelfrostablagerungen am Hochspannungsleitungen im Gebirge', *Arch. Met. Geophys. Biokl.*, B 7, 74–83.

Wanner, H. (1979) 'Zur Bildung, Verteilung und Vorhersage winterlicher Nebel im Querschnitt Jura-Alpen', *Geographica Bernensia*, G., 7.

World Meteorological Organization (1975) *Manual on the Observation of Clouds and other Meteors. International Cloud Atlas Vol. I*, Geneva, WMO no. 407.

Yoshino, M.M. (1975) *Climate in a Small Area*, Tokyo, University of Tokyo Press.

Blowing and drifting snow

Berg, N. and Caine, N. (1975) 'Prediction of natural snowdrift accumulation in alpine areas', *Final Report to Rocky Mountain Forest and Range Expt. Station* (USFS 16-388-CA), Boulder, Department of Geography, University of Colorado.

Dyunin, A.K. and Kotlyakov, V.M. (1980) 'Redistribution of snow in the mountains under the effect of heavy snow storms', *Cold Regions Sci. Technol.*, 3, 287–94.

Föhn, P.M. (1980) 'Snow transport over mountain crests', *J. Glaciol.*, 26(94), 469–80.

Föhn, P. and Meister, R. (1983) 'Distribution of snowdrifts on ridge slopes: measurements and theoretical approximations', *Ann. Glaciol.*, 4, 52–7.

Jairell, R.L. (1975) 'An improved recording gage for blowing snow', *Water Resources Res.*, 11, 674–80.

Martinelli, M. Jr. (1973) 'Snow-fence experiments in alpine areas', *J. Glaciol.*, 12 (65), 291–303.

Meister, R. (1987) 'Wind systems and snow transport in alpine topography', in B. Salm and H. Gubler (eds) *Avalanche Formation, Movement and Effects*, Assoc. Hydrol. Sci. Publ. 162, Wallingford, UK, pp. 265–7.

Mellor, M. (1965) *Blowing Snow* (Cold Regions Science and Engineering. Part III, Section A3c), Hanover, N.H., US Army, Cold Regions Research Engineering Laboratory.

Mikhel, V.M. and Rudneva, A.V. (1971) 'Description of snow transport and snow deposition in the European USSR', *Soviet Hydrol.*, 10, 342–8.

Mikhel, V.M., Rudneva, A.V. and Lipovskaya, V.I. (1971) *Snowfall and Snow Transport during Snowstorms over the USSR*, Jerusalem, Israel Prog. Sci. Transl.

Pedgley, D.E. (1967) 'The shape of snowdrifts', *Weather*, 22, 42–8.

Pugh, H.L.D. and Price, W.I.J. (1954) 'Snow drifting and the use of snow fences', *Polar Rec.*, 7, (47), 4–23.

Radok, U. (1968) 'Deposition and erosion of snow by the wind', *CRREL Res. Rep. 20*, Hanover, N.H., US Army, Cold Regions Research Engineering Laboratory.

Radok, U. (1977) 'Snow drift', *J. Glaciol.*, 19(81), 123–39.

Santeford, H.S., Jr. (1972) 'Management of windblown alpine snows', Unpubl. Ph.D. thesis, Fort Collins, Colorado, Colorado State University.

Schmidt, R.A., Jr. (1972) 'Sublimation of wind-transported snow – a model', *USDA, Forest Serv. Res. Pap. RM-90*, Fort Collins, Colorado, Rocky Mountain Forest and Range Experimental Station.

Schmidt, R.A. (1977) 'A system that measures blowing snow', *USDA., Forest Serv.Res.Pap. RM-194*, Fort Collins, Colorado, Rocky Mountain Forest and Range Experimental Station.

Tabler, R.D. (1975a) 'Predicting profiles of snowdrifts in topographic catchments', *Proc. 43rd Western Snow Conference*, pp. 87–97.

Tabler, R.D. (1975b) 'Estimating the transport and evaporation of blowing snow', in *Snow Management on the Great Plains*, Publ. 73, pp. 85–104, Lincoln, University of Nebraska, Agriculture Experimental Station.

Takeuchi, M. (1980) 'Vertical profile and horizontal increase of drift-snow transport', *J. Glaciol.*, 26(94), 481–92.

Tesche, T.W. (1988) 'Numerical simulation of snow transport, deposition and redistribution', *Proc. Western Snow Conf. 56th Annual Meeting*, 93–103.

Evaporation

Alford, D. (1985) 'Mountain hydrologic systems', *Mountain Res. Devel.*, 5, 349–63.

Baumgartner, A., Reichel, E. and Weber, G. (1983) '*Der Wasserhaushalt der Alpen*', Munich, Oldenbourg.

Beaty, C.B. (1975) 'Sublimation or melting: observations from the White Mountains, California, and Nevada, U.S.A.' *J. Glaciol.*, 14(71), 275–86.
Bergen, J.D. and Swanson, R.H. (1964) 'Evaporation from a winter snow cover in the Rocky Mountain forest zone', *Proc. 32nd Western Snow Conference*, pp. 52–8.
Bosch, J.M. and Hewlett, J.D. (1982) 'A review of catchment experiments to determine the effect of vegetation change on water yield and evapotranspiration', *J. Hydrol.*, 55, 2–23.
Calder, I.R. (1990) *Evaporation in the Uplands*, Chichester, J. Wiley and Sons.
Chidley, T.R.E. and Pike, J.G. (1970) 'A generalized computer program for the solution of the Penman equation for evapotranspiration', *J. Hydrol.* 10, 75–89.
Church, J.E. (1934) 'Evaporation at high altitudes and latitudes', *Trans. Am. Geophys. Un.*, 15(2), 326–51.
de Quervain, M.R. (1951) 'Zur Verdunstung der Schneedecke', *Arch. Met. Geophys. Biokl.*, B, 3, 47–64.
Gangopadhyaya, M., Harbeck, G.E., Jr., Nordenson, T.J., Omar, M.H. and Uryvaev, V.A. (1966) *Measurement and Estimation of Evaporation and Evapotranspiration*, Geneva, World Meteorological Organization Tech. Note no. 83.
Golding, D.L. (1978) 'Calculated snowpack evaporation during chinooks along the eastern slopes of the Rocky Mountains', *J. appl. Met.*, 17, 1647–51.
Greenland, D. (1989) 'The climate of Niwot Ridge, Front Range, Colorado, USA', *Arct. Alp. Res.* 21, 380–91.
Henning, D. and Henning, D. (1981) 'Potential evapotranspiration in mountain geo-ecosystems of different altitudes and latitudes', *Mountain Res. Devel.* 1, 267–74.
Hoover, M.D. and Leaf, C.E. (1967) 'Process and significance of interception in Colorado subalpine forest', in Sopper, W.E. and Lull, H.W. (eds) *Symposium on Forest Hydrology*, pp. 213–24, Oxford, Pergamon Press.
Horton, R.E. (1934) 'Water losses in high latitudes and at high elevations', *Trans. Am. Geophys. Un.* 15(2), 351–79.
Hutchison, B.A. (1966) 'A comparison of evaporation from snow and soil surfaces', *Bull. Int. Ass. Sci. Hydrol.*, 11, 34–42.
Kattleman, R. and Elder, K. (1991) 'Hydrologic characteristics and balance of an alpine basin in the Sierra Nevada'. *Water Resources Res.*, 27, 1553–62.
Köhler, H. (1950) 'On evaporation from snow surfaces', *Arkiv Geofis.*, 1, 159–85.
Krestoviskiy, O.I. (1962) 'The water balance of small drainage basins during the period of high water', *Soviet Hydrol.*, 1, 362–411.
Kuvaeva, G.M. (1967) 'Nekotorye resultaty nablyudeniy nad ispareniem s poverkhnosti snezhnogo pokrova v vysokogornoy zone tsentralnogo Kavkaza (Some results of observations on evaporation over snow surfaces in the high mountain zone of the Central Caucasus 1)', *Trudy Vysokogornyi Geofiz. Inst.*, 12, 40–6.
Kuz'min, P. (1970) 'Methods for the estimation of evaporation from land applied in the USSR', in *Symposium on World Water Balance*, vol. 1, pp. 225–31, Paris, Publ. no 92, Int. Ass. Sci. Hydrol., UNESCO.
Kuz'min, P. (1972) *Melting of Snow Cover*, Jerusalem, Israel Prog. Sci. Transl.
Lang, H. (1981) 'Is evaporation an important component in high alpine hydrology', *Nordic Hydrol.*, 12, 217–24.
Lauscher, A. and Lauscher, F. (1976) 'Zur Berechnung der Schneeverdunstung auf dem Sonnblick', *72–73 Jahresber. des Sonnblick-Vereines für die Jahre 1974–1975*, 3–10.
Lauscher, F. (1978) 'Eine neue Analyse von Hilding Köhler's Messungen der Schneeverdunstung auf dem Haldde-Observatorium aus dem Winter 1920/21', *Arch. Met. Geophys. Biokl.*, B, 26, 193–8.
LeDrew, E.F. (1975) 'The energy balance of a mid-latitude alpine site during the growing season, 1973', *Arct. Alp. Res.*, 7, 301–14.

LeDrew, E.F. and Emerick, J.C. (1974) 'A mechanical balance-type lysimeter for use in remote environments', *Agric. Met.*, 13, 253–8.

Lemmela, R. and Kuusisto, E. (1974) 'Evaporation from snow cover', *Hydrol. Sci. Bull.*, 19, 541–8.

Light, P. (1941) 'Analysis of high rates of snow-melting', *Trans. Am. Geophys. Un.*, 22(1), 195–205.

Longacre, L.L. and Blaney, H.F. (1962) 'Evaporation at high elevations in California', *Proc. Am. Soc. Civ. Eng. J. Irrig. Drainage Div.*, 3172, 33–54.

McCulloch, J.S.G. (1965) 'Tables for the rapid computation of the Penman estimate of evaporation', *E. Afr. Agric. Forest J.*, 30, 286–95.

Martinelli, M., Jr. (1959) 'Some hydrologic aspects of alpine snowfields under summer conditions', *J. Geophys. Res.*, 64, 451–5.

Martinelli, M. Jr., (1960) 'Moisture exchange between the atmosphere and alpine snow surfaces under summer conditions', *J. Met.*, 17, 227–31.

Meiman, J.R. and Grant, L.O. (1974) *Snow-Air Interactions and Management of Mountain Watershed Snowpack*, Completion Rep. Ser. No. 57, Fort Collins, Colorado, Environ. Res. Center, Colorado State University.

Miller, D.H. (1955) *Snow Cover and Climate in the Sierra Nevada, California*, Berkeley, University of California Press, Publ. in Geog., 11.

Miller, D.H. (1962) 'Snow in the trees – where does it go?' *Proc. 30th Annual Western Snow Conference*, 21–29.

Ohmura, A. (1982) 'Regional water balance on the arctic tundra in summer', *Water Resour. Res.*, 18, 301–5.

Oke, T.R. (1978) *Boundary Layer Climates*, London, Methuen.

Ovadia, D. and Pegg, R.K. (1979) 'An approach to calculating evaporation rates at remote sites', *Nordic Hydrol.*, 10, 41–8.

Peak, G.W. (1963) 'Snow pack evaporation factors', *Proc. 31st Annual Western Snow Conference*, 20–7.

Peck, E.L. and Pfankuch, D.J. (1963) 'Evaporation rates in mountainous terrain', Int. Ass. Sci. Hydrol. Gentbrugge, Belgium, Publ. no. 62, 267–78.

Penman, H.L. (1963) *Vegetation and Hydrology*, Commonwealth Bureau of Soils Tech. Commun. 53, Farnham Royal, Bucks, England, Commonwealth Agricultural Bureaux.

Priestley, C.H.B. and Taylor, R.J. (1972) On the assessment of surface heat flux and evaporation using large-scale parameters. *Mon. Wea. Rev.* 100: 81–92.

Rechard, P.A. and Raffelson, C.N. (1974) 'Evaporation from snowdrifts under oasis conditions', in *Advanced Concepts and Techniques in the Study of Snow and Ice Resources*, pp. 99–107, Washington, D.C., National Academy Science.

Sabo, E.D. (1956) 'Evaporation from snow in the Ergeni District', in *Selected Articles on Snow and Snow Evaporation*, pp. 14–21, Jerusalem, Israel Prog. Sci. Transl.

Santeford, H.S., Jr. (1972) 'Management of windblown alpine snows', Unpubl. Ph.D. thesis, Fort Collins, Colorado, Colorado State University.

Slaughter, C.W. (1970) *Evaporation from Snow and Evaporation Retardation by Monomolecular Films*, Special Rep. 130, Hanover, N.H., Cold Regions Research Engineering Laboratory, US Army.

Staudinger, M. and Rott, H. (1981) 'Evapotranspiration at two mountain sites during the vegetation period', *Nordic Hydrol.* 12, 207–16.

Stigter, C.J. (1976) 'On the non-constant gamma', *J. appl. Met.*, 15, 1326–7.

Stigter, C.J. (1978) 'On the pressure dependence of the wind function in Dalton's and Penman's evaporation equations', *Arch. Met. Geophys. Biokl.*, A, 27, 147–54.

Storr, D. and den Hartog, G. (1975) 'Gamma – the psychrometer non-constant', *J. appl. Met.*, 14, 1397–8.

Thom, A.S. and Oliver, H.R. (1977) 'On Penman's equation for estimating regional

evapotranspiration', *Q. J. R. Met. Soc.*, 103, 345–58.
Thompson, R.J. (1975) Energy budgets for three small plots – substantiation of Priestley and Taylor's large-scale evaporation parameter. *J. appl. Met.*, 14: 1399–401.
West, A.J. (1959) 'Snow evaporation and condensation', *Proc. 27th Annual Meeting, Western Snow Conference*, pp. 66–74.
West, A.J. (1962) 'Snow evaporation from a forested watershed in the central Sierra Nevada', *J. Forest.*, 60, 481–4.
Willmott, C.J. (1977) 'WATBUG: A FORTRAN IV algorithm for calculating the climatic water budget', Water Resources Center, Newark, DE, University of Delaware.

5

CASE STUDIES

The generalized view of the major climatic controls and characteristics provided in the preceding chapters obviously neglects many significant details and local anomalies, although the data are in many cases so sparse that the generalizations can only be preliminary and partial. As a counterbalance, therefore, it is worthwhile examining particular mountain climates via a series of complementary case studies. These include mountain systems in equatorial, tropical, mid-latitude and sub-polar regions, that have been selected to show both latitudinal and regional climatic differences.

THE EQUATORIAL MOUNTAINS OF NEW GUINEA

The central ranges of New Guinea form a backbone running through the length of the island. Except for the Owen Stanley Range in eastern Papua, which is not considered here due to a total absence of information, the mountains are orientated nearly east–west. The general summit level is about 3500–4000 m, but in Irian Jaya (western New Guinea) peaks rise to over 4700 m with small ice bodies on Mt Jaya (Mt. Carstenz) (4°S, 137°E), which have been studied by two Australian expeditions (Hope *et al.* 1976). The only other mountain area that has received climatological study is Mt Wilhelm in the Bismarck Range (5°40′S, 145°01′E) (Hnatiuk *et al.* 1976). The mountains are generally forested from the cultivation level around 2600 m to about 3500 m, with alpine tussock grassland above.

From May through October, New Guinea is under the influence of deep tropical easterly airflow while in December–March there are equatorial westerly winds at 700 mb and below, associated with the Indonesian–Australian summer monsoon circulation. At all seasons, the region is a frequent locus of airstream convergence and large-scale vertical motion. As a result, there is an average cloudiness of 5–6 oktas over Papua New Guinea and only slightly less farther west.

The seasonal circulation patterns exert a strong influence on precipitation regimes in the Highlands, except in the central part of the island where it is

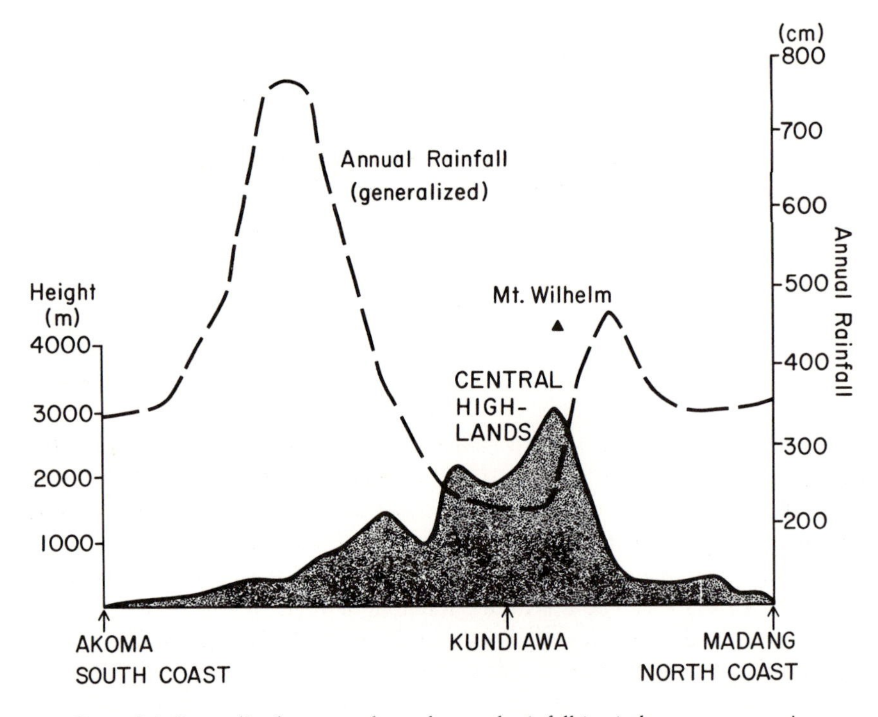

Figure 5.1 Generalized topography and annual rainfall (cm) along a cross-section northward from Akoma on the Gulf of Papua to Kundiawa (145°E, 6°S), then north-eastward to Madang

Source: From Barry 1978; based on a rainfall map produced by the Snowy Mountains Engineering Corporation, 1975

almost equally wet in all months. During the south-east season (May–September/October), the Eastern Highlands of Papua New Guinea (145–146°E) and the central ranges of Irian Jaya experience drier conditions with minimum amounts in June or July. The trade wind inversion, at about 2000 m south of the mountains, is a controlling factor for cloud development, but local convective circulations take on a dominant role at this season. Consequently there is a low inter-station correlation, even for annual precipitation amounts. During the south-east season, cloudiness shows a strong diurnal regime at Mt Wilhelm. Skies are generally clear in the early morning, but convective build-up occurs from 9–10 a.m. and cloud cover is complete by midday, often with afternoon precipitation which may continue during the early part of the night. Sunset and sunrise, of course, vary little from the hours of 6 p.m. and 6 a.m., respectively, in all months.

The 'north-west season' (December–March) is almost continuously wet in all parts of the island and especially at high elevations. However, intensities are consistently modest. Amounts decrease above about 800 m on north and south slopes of the central range (Figure 5.1) (Barry 1978).

Snow and graupel fall quite frequently on Mt Wilhelm above 3800 m, in any month, and sleet or wet snow is the main precipitation form above 4400 m on Mt Carstenz (Allison and Bennett 1976). Occasional heavy falls of dry snow are reported even at the 4250 m level.

The temperature regime in the mountains shows almost no seasonality, but the expected diurnal pattern is present (see p. 18). On Mt Wilhelm, temperatures range between a mean minimum of 6°C and a mean maximum of 10–11°C except in the north-west season when maxima average 8–9°C. Ground frost is not uncommon in the dry season, when skies usually clear before sunrise. Agriculturally-devastating frost events occur periodically in the New Guinea highlands down to 2000–1500 m elevation in association with large-scale subsidence and reduced cloud cover. Such conditions are favoured during El Niño/Southern Oscillation warm events, when the normal area of deep convection over Indonesia–New Guinea shifts eastward over the Pacific (Brookfield and Allen 1989). Allen (1989) shows that frost events in 1965, 1972, and 1982 were also dry years in New Guinea, but the drought-frost/ENSO relationship is by no means definitively established.

A special feature of these equatorial mountains is the light winds. On Mt Carstenz, Allison and Bennett (1976) report an average speed of only 2.1 m s^{-1} for the period 22 December 1971–5 March 1972. Winds in the free air are stronger in July–August with vector mean speeds of 6–7 m s^{-1} at 600 mb. Occasional strong down-valley winds have been reported at Mt Wilhelm during dry spells (Hnatiuk *et al.* 1976) and the highland valleys, generally, are subject to mountain–valley wind systems.

A fuller review of the available material and literature on the climate of the New Guinea mountains is published elsewhere (Barry 1980). Other details may be found in the cited references and McVean (1974).

THE HIMALAYA

The Himalaya are a series of ranges some 300 km in length, between 36°N, 27°E and 27°N, 90°E, varying in width from 80–300 km. Ecologically, they comprise an outer zone of monsoon forest, an inner zone of coniferous forest with heavy winter snows, and a Tibetan zone of arid steppe (Troll 1972). Interest in the meteorology of the Himalaya dates from the latter part of the nineteenth century. Hill (1881) provided a detailed compilation of early records for the north-western Himalaya in Kashmir, although the principal station referred to is Leh (3316 m) which is situated in the upper Indus valley between the Himalaya and Karakoram ranges. Climbing expeditions have provided most of the high-level observations (Somervell and Whipple 1926; Bleeker 1936; Wien 1936; Reiter and Heuberger 1960; Mitsudera and Numata 1967; Raghavan 1979) while, more recently, existing station data have been synthesized to give a more complete and up-to-date picture of conditions (Kraus 1967; Flohn 1968, 1970; Dobremez 1976; Dittmann 1970). Climatic observations have also been made, particularly in the Khumbu Himal, in connection with studies of glaciers (Higuchi 1976).

Despite the sub-tropical latitudes of the Himalayan mountains, their weather is controlled by the Indian monsoon regime from June through September. From October through May, the mountains are just north of the axis of the sub-tropical westerly jet stream in the upper troposphere and, steered by this system, disturbances travel eastward causing gales and blizzards on the peaks. These westerlies decrease markedly in strength from $\geqslant$ 25 m s^{-1} at 9 km to only about 10 m s^{-1} in late May (on average), and are replaced by easterlies a month later (Nedungadi and Srinivasan 1964), as the summer high-level anticyclone develops over the Tibetan Plateau (Flohn 1968). Monthly precipitation totals in fact show a two- to three-fold increase from May to June over Nepal (see Table 5.1), although there is no particular increase in precipitation activity at individual stations that can be distinguished from the pre-monsoonal showers. Nevertheless, an analysis of spatial coherence of rainfall at ten stations in south-eastern Nepal shows that widespread rains occur only during June–September (Ditmann 1970). Although of a convective nature, the 'monsoon rains' in Nepal occur in spells (Ramage 1971). There appears to be a major periodicity of about 10 days over the whole country associated with a general fluctuation in the monsoon regime or oscillations in the Tibetan anticyclone (Yasunari 1976b). A secondary one of about 5 days, which is more pronounced in eastern Nepal is probably due to westward-moving depressions over northern India. The intervening days are relatively dry, apart from local convective showers. Even so, over the highlands of eastern Nepal, July and August typically have about 27 raindays ($\geqslant$ 2.5 mm) with a mean intensity of about 20 mm. Totals and intensities decrease sharply in the higher valleys and at Namche Bazar (3400 m), for example, intensities are only 8 mm rainday^{-1} in summer. Identical intensities occur there in January (Dhar and Narayanan 1965).

Along the outer ranges of the Himalaya, the altitude of maximum precipitation during the monsoon season generally occurs at or a little below 2000 m (see Table 5.1). Reiter and Heuberger (1960) note that this is the level of night and early morning cloud. From their data for September 1954, relative humidity is $\geqslant$ 90 per cent between 2 and 4 km (Figure 5.2). Humidities of 90–100 per cent during the monsoon months are confirmed by observations at Lhajung (4420 m) in the Khumbu Himal (Inoue, 1976). Stations in the central Himalayas (Kosi) exhibit seasonal maxima in the foothills of the Siwalik Range and around 2000–2400 m elevation in the Lesser Himalayan Range; further northward amounts decrease towards the main Himalayan range (Dhar and Rakhecha 1980). In the period of the upper westerly wind regime, however, precipitation may increase above 3000 m (Flohn 1970). The relative importance of 'winter' precipitation increases with latitude and altitude although actual amounts are larger farther west. Precipitation decreases sharply north of the mountain barrier (Gyangtse in Table 5.1) and the summer precipitation over the Tibetan Plateau is convectional from large cumulonimbus cells (Lu 1939; Flohn

Table 5.1 Precipitation in eastern Nepal-Sikkim-southern Tibet, between approximately 86.5°–89°E

Station	*Latitude*	*Elevation (m)*	*Annual total*	*June–May ratio*	*June–September (% annual)*	*December–March (% annual)*
Gyangtse	28°56′N	3996	271	3.2	90	1.5
Khumbu Glacier	27°59′N	5300	450	–	(73)	(16)
Lhajung						
(1974–6	27°53′N	4420	527	–	(84)	(8)
Thanggu	27°55′N	3000	738	1.5	55	26
Namche Bazar	27°50′N	3400	939	3.2	75	11
Lachen	27°43′N	2697	1707	1.7	47	16
Wallungchung						
Gola	27°42′N	3048	1695	2.3	74	9
Chaunrikharka	27°42′N	2700	2284	3.0	86	2.5
Jiri	27°38′N	1895	2387	3.9	81	4
Gangtok	27°20′N	1764	3452	1.3	66	7
Darjeeling	27°03′N	2265	3082	2.6	80	3

Source: After Kraus 1967, Flohn 1968, 1970; Higuchi 1976; Yasunari and Inoue 1978

1968). However, at a temporary station on the northern (lee) side of Dhaulagiri Himal, at 5055 m, major precipitation events in 1974 were associated with large-scale disturbances of the monsoon trough to the south (Shresta *et al.* 1976). The July–August total (200 mm) exceeded amounts at lower stations in windward locations. Here, winter precipitation is minimal and the excess of potential evaporation over summer precipitation produces an arid environment. As a consequence of the greater aridity, the snowline is 400–500 m higher on the north side of the Nangpa Pass (28°07′N, 85°36′E, 5500 m) than on the south side. Reiter and Heuberger (1960) contrast this situation with the Alps where the snow line is 200 m higher on the south side since radiation and temperature are the primary controls.

The autumn transition in the upper troposphere between the easterly winds and the re-establishment of the westerlies is unlike the gradual change in spring. Observations from the 1954 Austrian expedition showed that the westerly jet re-established itself at the end of September although föhn effects on the south side of the range maintained relatively warm conditions during October (Reiter and Heuberger 1960). Persistent strong winds are to be expected on the high Himalaya once this wind shift has taken place.

Precipitation varies considerably on a local scale in relation to location on windward/leeward slopes, and as a result of the local wind circulations. Large annual totals are observed on the southern slopes of the eastern Himalayan foothills, exposed to the southerly monsoon flow. The Khasi hills of Assam, which extend 300 km east–west and rise to about 1800 m, are exceptional in this respect. The annual rainfall at Cherrapunji, 1313 m altitude (25.3°N, 91.7°E) averaged 1,149 cm for 1931–60 (Chagger 1983), with 250 cm or more falling in both June and July; in 1974 Cherrapunji received 2440 cm. However, nearby Mawsyuram (1401 m) averaged 1,221 cm for 1941–69 while Cherrapunji averaged only 1,102 cm. Rao (1976) notes that monsoon season amounts on the adjacent plains are only about 160 cm compared with 800–900 cm around 1300–1400 m; corresponding amounts near the crest of the hills, however, are around 250 cm. Rao points out that the exceptionally wet middle slopes are partly attributable to a funnel-shaped catchment in the hills, opening at the south, that augments the airflow convergence. The overall slope from the plains is about 1:8.

Rainfall decreases northward and altitudinally along the major valleys in the Khumbu Himal (Ageta 1976), for example, but there are local altitudinal increases. Yasunari and Inoue (1978) note that short-term records in the monsoon season of 1976 show totals around 5000–5500 m peaks and ridges were 4–5 times those at Lhajung in the main valley. They attribute this to orographically and thermally induced convection; radiative heating below the snowline may be an important trigger. In summer, convective cloud spreads northward up the main Imja Valley, reaching Lhajung by late afternoon on fine days and by 09.00–10.00 on unstable days (Yasunari 1976a). Much of the monsoon precipitation in the valley occurs nocturnally, associated with these upvalley motions.

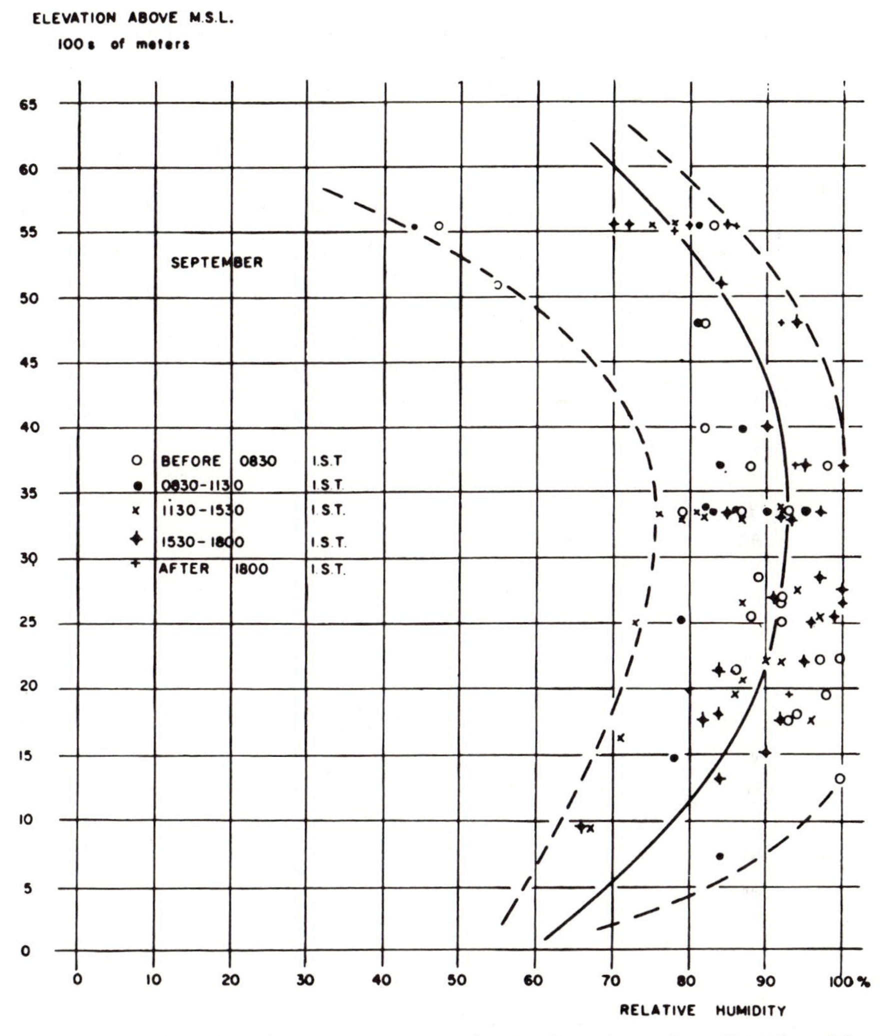

Figure 5.2 Relative humidity measurements in the Himalaya, September 1954. The solid line is an approximate mean
Source: From Reiter and Heuberger 1960

During 21–31 August 1974, 70 per cent of the precipitation (37 mm total) occurred at night at Lhajung (4420 m) compared with only 35 per cent (of a total of 107 mm) at a glacier col at 5360 m, 10 km to the north-west (Ageta 1976). The instability diminishes at night and stratiform clouds fill the valleys in the early morning. Middle and high level clouds may affect the highest peaks, especially in association with westerly disturbances in winter. A schematic view of these cloud systems is illustrated in Figure 5.3.

Studies in Langtang Valley, Nepal (28.1°N, 85.4°E) between 3920 m and

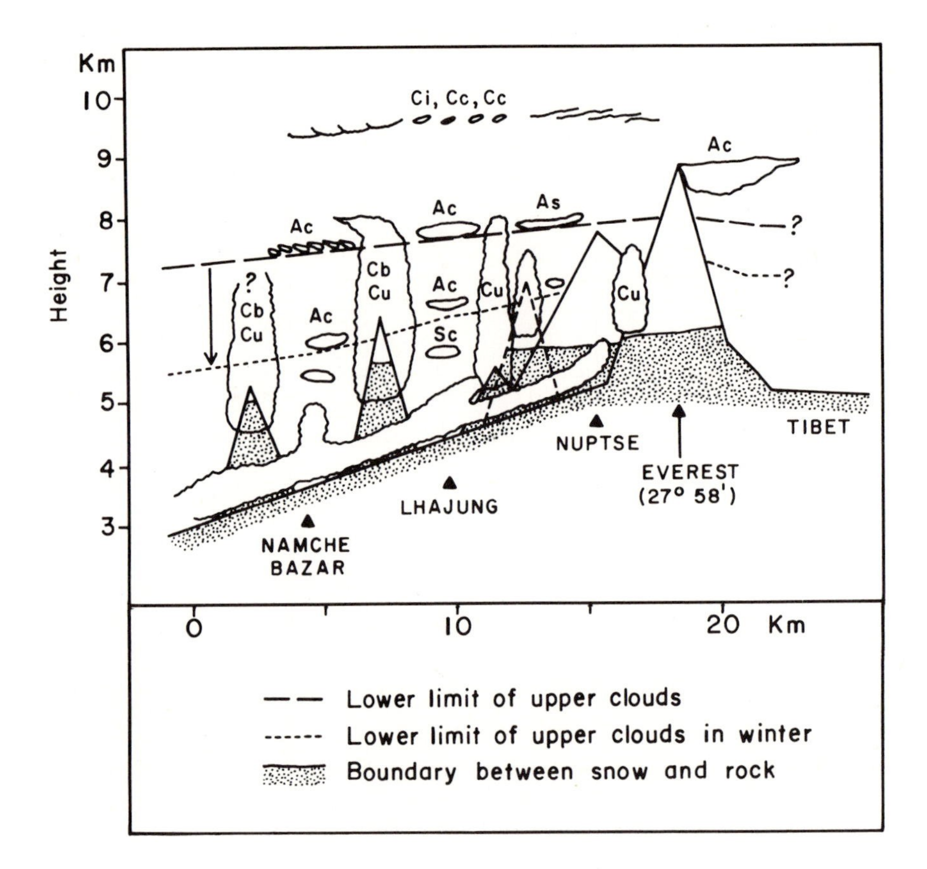

Figure 5.3 Schematic cloud patterns in summer and winter in the Khumbu region, Nepal
Notes: Ac = Altocumulus, As = Altostratus, Ci = Cirrus, Cc = Cirro-cumulus, Cu = Cumulus, Cb = Cumulonimbus
Source: After Yasunari, 1976a

5090 m show that, in winter, snowfall from westerly disturbances increase with height. During the monsoon season it decreases with altitude from Kathmandu (60 km to the south-west) to 3920 m and then increases slightly to 5090 m as a result of convective activity over the ridges. During mid-August to October, however, there are distinct diurnal patterns (Ueno and Yamada 1990). From 1600/1800 hours to 1100/0200 hours (local time) about half of the daily precipitation occurs simultaneously over the valley, in up-valley winds. During the night (midnight to 0500/0600 hours) rain falls in the valley bottom from stratiform cloud, while in the morning hours to mid-afternoon it occurs over ridges and peaks associated with upslope winds. In the north-western Himalaya, the shower regime, which is strongly influenced by local wind systems, largely determines the distribution of summer precipitation since the role of monsoon depressions is much reduced in the far north-west.

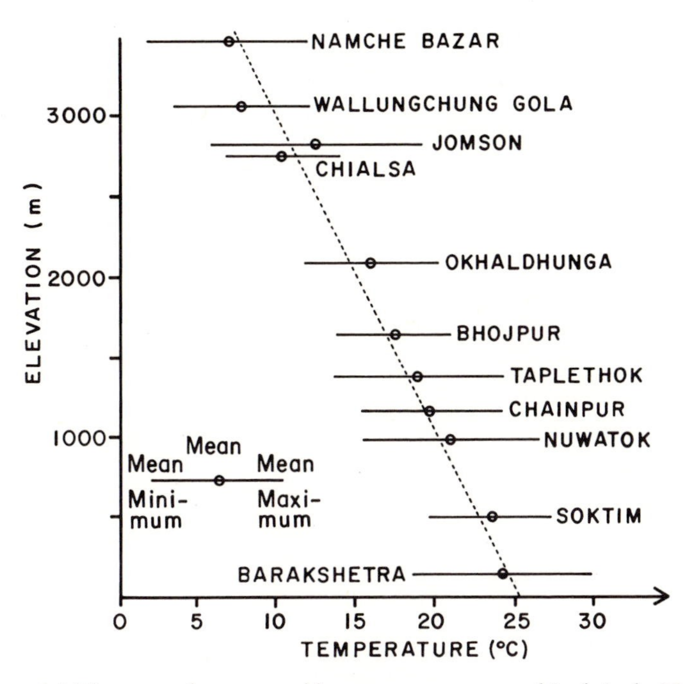

Figure 5.4 The range of mean monthly temperatures versus altitude in the Nepal Himalaya
Source: After Dobremez 1976

Troll (1951) and Schweinfürth (1956) have both shown how the vegetation patterns in the major Himalayan valleys reflect the fact that it is much wetter on slopes and ridges and drier along the valley bottoms. This is attributed to the tendency for subsidence over the valleys in the slope wind circulation. Since many observing stations are located in valleys, such wind systems may result in serious underestimation of areally-averaged precipitation (sometimes referred to as the 'Troll effect'). Flohn (1974) illustrates this for the Hindu Kush and Karakoram ranges. Annual totals at five stations in the valleys of the Karakoram are only 8–16 cm, whereas hydrological and glaciological estimates show that at least 2–3 m must fall annually over the surrounding mountains and glaciers. In the Himalaya, the upvalley wind system is continuous through the night (until about 4 a.m.) during the summer monsoon, whereas a normal diurnal mountain–valley wind regime is observed during the dry months, October–May (Ohata *et al.* 1981). Flohn (1970) shows that this is due to the regional-scale circulation between the southern edge of the Tibetan Plateau and the Indo-Pakistan lowlands. As discussed on p. 57, the Plateau and Himalaya act as a major heat source, one component of which is the release of latent heat in cumulonimbus cells. This maintains the convective process, and thereby the regional-scale thermally driven circulation. Substantial local differences in the seasonal distribution of precipitation amounts can also occur due to the orientation of major drainage

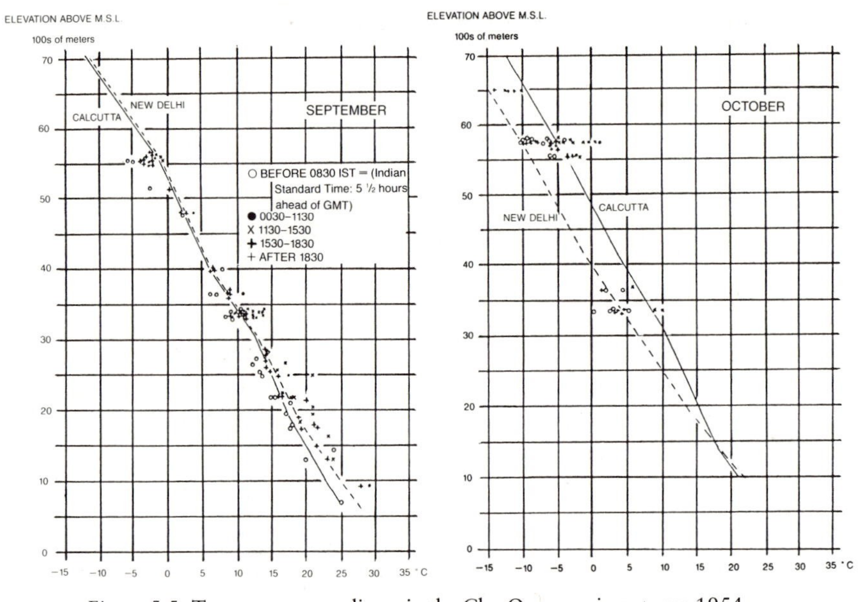

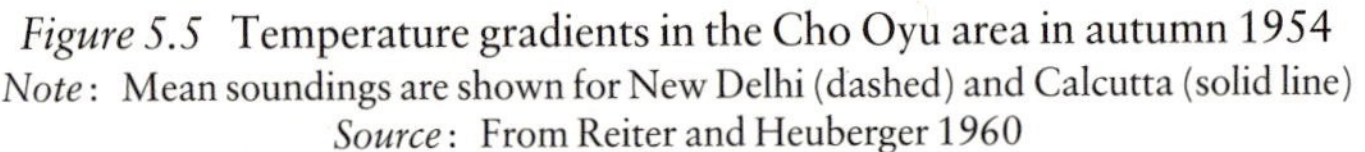
Figure 5.5 Temperature gradients in the Cho Oyu area in autumn 1954
Note: Mean soundings are shown for New Delhi (dashed) and Calcutta (solid line)
Source: From Reiter and Heuberger 1960

basins. Upper-level easterly airflow in summer and westerly airflow in winter may interact with the topography in different ways to affect the low-level cloud motion in the valleys.

To complete this brief overview of climatic conditions in the Himalaya, mean monthly temperatures versus altitude are summarized in Figure 5.4. Interestingly, there is no apparent decrease in seasonal range with altitude, probably due to the widespread significance of the monsoon regime (Dobremez 1976), although the higher stations are sited in valley locations. The mean diurnal range is strongly site-dependent and with major seasonal differences. Mitsudera and Numata (1967) show that large diurnal variations occur in spring with a daily range of 21°C between 1000 and 4000 m in April on Mt. Numbur, eastern Nepal, compared with only 2–4°C range in late June. Inoue (1976) notes that solar radiation intensities reach a maximum in May at Lhajung and decrease with the beginning of the monsoon in June. The mean daily range at Lhajung (4420 m) is about 12°C in December–January compared with 5–6°C in the monsoon months. A temperature profile for the Cho-Oyu area (28°06′N, 86°40′E) in September 1954 shows conditions close to those in the free air over Calcutta and New Delhi (Figure 5.5). In October and early November, when the upper westerlies are re-established, temperatures are lower over the Himalaya and New Delhi than Calcutta, reflecting the normal latitudinal gradient (except for some pronounced föhn cases at 5700 m in October) (Reiter and Heuberger 1960).

SUB-TROPICAL DESERT MOUNTAINS – THE AHAGGAR

Two mountain massifs rise above the Saharan desert to elevations close to 3000 m. They are the Ahaggar (Hoggar) of southern Algeria and the Tibesti of northern Chad. The latter area was the site of scientific research organized by the Free University, Berlin, from 1966 to 1974 (Jäkel 1977). Weather stations were established at Bardai (1020 m) and Trou au Natron (2450 m). Climatological studies are reported by Heckendorff (1972) and Indermühle (1972). The Ahaggar massif, which exceeds 2000 m over an area some 20–30 km wide and about 100 km long, has been the location of considerable research since the 1950s by the Institut de Recherches Sahariennes, University of Algiers. This group established a first-order weather station at Asekreme (2706 m), 23°16′N, 50°38′E, in March 1955 and an autographic station at the Tahat summit (2900 m) in 1959 (Dubieff 1963). Between 1959 and 1962 a network of 120 rain gauges was maintained on and around the mountains (Yacono 1968). A long-term record, since 1925, is provided by the observatory at Tamanrasset (1376 m) to the south-west.

The climatic regime and its synoptic controls have been described in some detail by Yacono (1968) and the following account is based largely on her studies. The lapse rate of temperature averages 0.5°C 100 m^{-1} in January, increasing to 0.8°C 100 m^{-1} during the more unstable conditions of July. More interesting is the increase of days with minimum temperatures below 0°C, from 39 days $year^{-1}$ at Tamanrasset to 114 days $year^{-1}$ at Asekreme. Temperatures below −5°C have been observed at Asekreme in association with outbreaks of polar air in deep cold lows. There is a typical Sudan-Sahelian type of rainfall regime with about 60 per cent of the annual total falling during August–September and another 25 per cent in May–June. The annual mean for 1955–62 was a low 35 mm at Tamanrasset, but increased to 145 mm at Asekreme.

The altitudinal effect is strongly dependent on total amount. Figure 5.6 shows that, in a dry year, altitude has only a limited effect whereas in years with more depression rain there tend to be maximum amounts at about 2500 m. However, this is not simply a result of increased intensity, since there are more days with measurable rainfall, and also with traces of precipitation, at the higher stations.

The aridity of the area is primarily due to low humidity rather than to an absence of disturbances. Relative humidity ranges from a mean of 30 per cent in April and 46 per cent in November at Segueika (2450 m) to between 17 and 29 per cent, respectively, for the same months at Tamanrasset probably due to a tendency for downslope motion in the predominantly north-easterly flow. Yacono (1968) shows that cloud systems associated with westward-moving disturbances regularly affect the Ahaggar. Analysis of satellite imagery and weather maps for 1968 by Winiger (1972) indicated that the 'ITC', extending northward as a 'cloud bridge' from Niger, was over or north of Ahaggar on 42 days with mean monthly cloud cover exceeding 50 per cent in May–July and September–October. Precipitation, however, is invariably light. Rare,

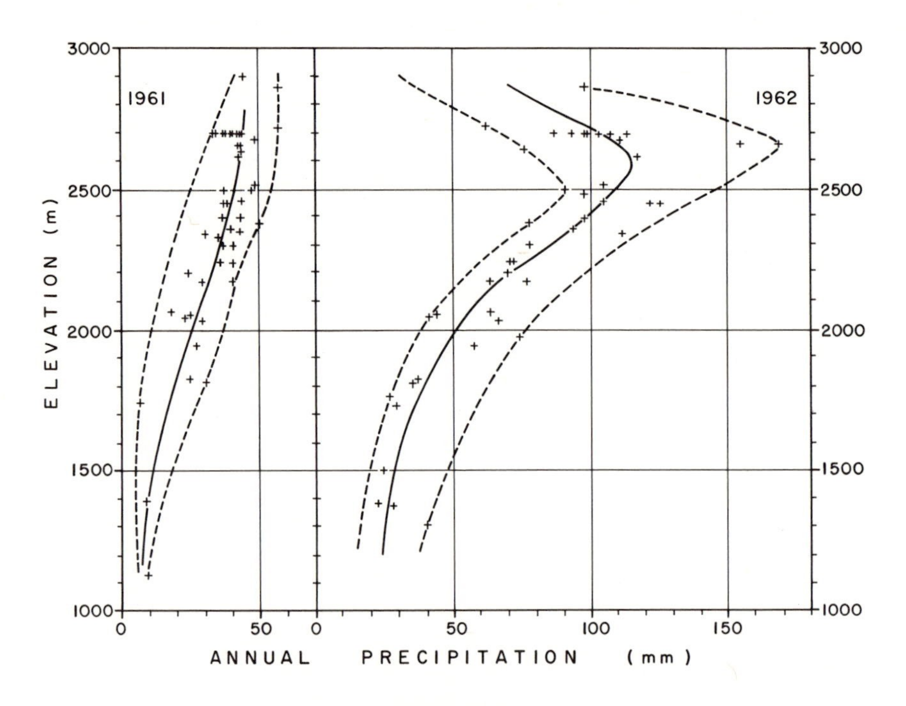

Figure 5.6 Annual rainfall (mm) versus altitude in the Ahaggar in a wet year and dry year
Source: After Yacono 1968

exceptional rainstorms are usually caused by North African lows such as that which gave 111 mm at Asekreme on 25–26 November 1968. Moist, tropical maritime air in a southwesterly flow was drawn northward by a deepening depression over Morocco–Tunisia. In winter and spring, Sudano-Saharan depressions may affect the Ahaggar after recurving northeastward, whereas in autumn they are usually too far west (Figure 5.7). Their movement is dependent on the flow around the mean upper-level high-pressure cell over northern Africa. Yacono (1968: 111) notes that these systems are major sources of Saharan precipitation. Occasional winter precipitation is derived from polar front depressions or polar air advected southward on the western limb of an upper trough. Such incursions may produce light falls of snow at Asekreme averaging about one per year.

The Ahaggar is rather windier than might be expected from its location with respect to the sub-tropical anticyclone and the predominance of subsiding air. Limited data cited by Yacono (1968) indicated a mean annual windspeed of 2 m s^{-1} at Tamanrasset, but 7 m s^{-1} at Asekreme for 10 months of observation in 1960. A southerly storm on 16 December 1960 gave winds of up to 54 m s^{-1} at Asekreme while the maximum at Segueika (2450 m) reached only 15 m s^{-1}.

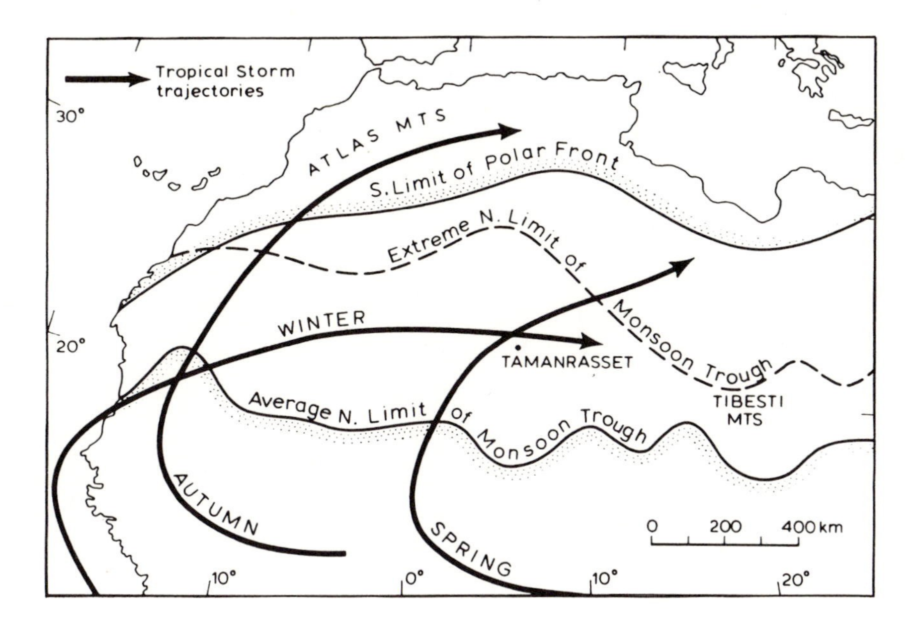

Figure 5.7 Extent of precipitation systems affecting western and central North Africa and typical tracks of Sudano-Saharan depressions
Source: After Dubief 1963; Yacono 1968

THE ROCKY MOUNTAINS IN COLORADO

The Rocky Mountains extend from the Yukon Territory to New Mexico, but we are concerned only with their southern part in Colorado, at about latitude 40°N, where fairly extensive data are available. Here the mountains form a pronounced north–south barrier rising to 4000 m a.s.l. To the east they front the high plains and to the west there are intermontane basins, both with elevations around 1500 m. The location, 1500 km from the Pacific coast, provides a continental climatic setting, although the Rockies create their own distinctive altitudinal climatic belts, to the extent that permafrost patches are present under wind-swept sites above about 3750 m a.s.l. (Ives 1973). Here mean annual air temperatures are around −4°C.

Despite the relative ease of access to the alpine zone above treeline, located about 3500 m, there are no permanent mountain observatories. The Pike's Peak Observatory (4311 m) was operated in the 1870–90s (see Table 1.2, p. 6), (Diaz *et al.* 1982), but recent high-level stations have been of a climatological nature (Judson 1965, 1977; Marr 1961). Four stations established for ecological purposes by J. W. Marr in 1952 along an altitudinal transect on the east slope of the Front Range west of Boulder, and subsequently maintained by the Institute of

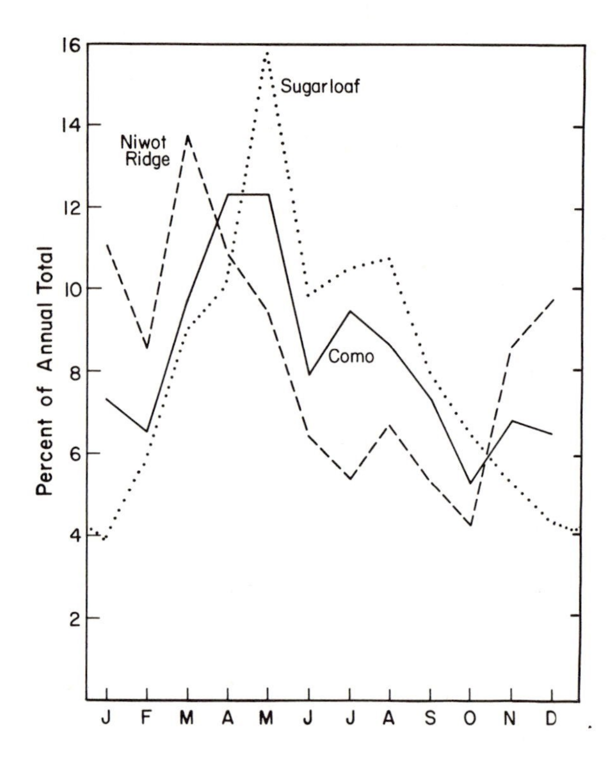

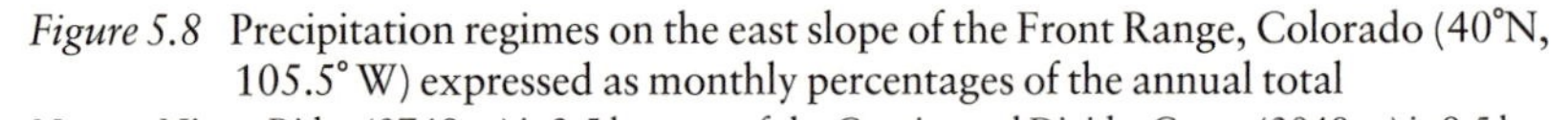

Figure 5.8 Precipitation regimes on the east slope of the Front Range, Colorado (40°N, 105.5° W) expressed as monthly percentages of the annual total

Notes: Niwot Ridge (3749 m) is 2.5 km east of the Continental Divide, Como (3048 m) is 9.5 km east of Sugarloaf 22.5 km east of the Divide. The record length is 18 years at Niwot Ridge and 20 years at Como, adjusted to 1951–85 averages

Source: Based on Greenland 1989; the Sugarloaf record is for 1952–70, Barry 1973

Arctic and Alpine Research at the University of Colorado, provide a valuable picture of year-round conditions in the mountains. These, and other records collected in connection with avalanche studies and cloud-seeding programmes, provide the basis for the present discussion.

Topics of special interest in the Rocky Mountain area are the precipitation characteristics, especially snowfall, and the wind regime. Analysis of the precipitation records from the four-station transect on the east slope of the Front Range demonstrates the importance of the large-scale atmospheric circulation and its interaction with the topography (Barry 1973). In the winter months, westerly air circulation is dominant with Pacific storms affecting particularly the western side of the range and its higher elevations. Nevertheless, some meridional flow patterns cause upslope flow on the east slope. In spring, and also in autumn, meridional troughs and occasionally deep cut-off cold lows draw moist air northward from the Gulf of Mexico. The temperature structure is potentially unstable and forced ascent in upslope flow causes heavy precipitation, falling as snow at higher elevations and even in the foothills, on the east slope. On 14–15 April 1921, for example, a deep storm with this type of circulation established a US record of 193 cm for a 24-hour snowfall at Silver Lake, Boulder County (3170 m). In summer, precipitation is mainly of a convective kind, although sufficient moisture must be transported into the area for the instability to be released by heating and orographic effects. At this season, patterns of equivalent potential temperature show sharp gradients paralleling the Rocky Mountains, with higher values indicating warm moist air to the east of the mountains, which act as a climatic divide (Mitchell 1976).

On an annual basis there is a sharp transition across the Front Range from a Great Plains regime, with a spring precipitation maximum which is apparent up to at least 3050 m at Como station on the east slope, to a west slope pattern, characterized by a winter maximum and autumn minimum, at Niwot Ridge (3750 m), located just 2.5 km east of the Continental Divide (Figure 5.8). The two stations are separated by a distance of only 7 km. At Berthoud Pass (3448 m) on the Continental Divide some 30 km further south, Judson (1977) reports a mean annual total of 93 cm, comparable to the figure for Niwot Ridge, but there April receives the highest total. On the east slope, annual totals increase from less than 60 cm in the foothills (2200 to 2600 m) to about 100 cm (75 per cent occurring as snowfall) at 3750 m on Niwot Ridge where the gauge is screened by a modified Wyoming-type of snow fence (see p. 253).

A larger-scale study of winter storm precipitation across the Continental Divide from Grand Junction to Denver by Hjermstad (1970) shows the effects of orography (Table 5.2). Altitudinal increases on the west slope are most pronounced between 2100 m and 3200 m, where there is, on average, a six-fold increase. Analysis by wind direction at 500 mb shows that the orographic effect is largest for north-westerly winds in excess of 25 m s^{-1}. Table 5.2 shows that totals are 50 per cent larger at the base of the east slope than at the same elevation on the west slope. This is attributable, in part, to cases of flow patterns with an easterly upslope component. Precipitation in these situations may not reach west of the Divide. However, even with westerly upper-level flow, storm systems give almost equal amounts of precipitation, on average, at Grand Junction and Denver. This may partly reflect carry-over effects and partly easterly components in the low-level flow ahead of the travelling storm systems. In this area, simple concepts of sheltering effects due to the mountains must be treated with caution.

The occurrence of extreme precipitation events and flooding is of considerable importance in the Rocky Mountains foothills for flood control planning purposes. In the eastern foothills of the Front Range, Colorado, snowmelt accounts for peak discharges above an elevation of approximately 2300 m, whereas at lower elevations they are caused by extreme rainfall events according to Jarrett (1990b). His analysis of 97 intense rainstorms shows that 6-hour totals decrease sharply from 500 mm at elevations ≤2100 m to barely 50 mm above

2400 m. The elevational limitation on large-magnitude floods is also confirmed by the contrasting types of erosional and depositional features in the river channels above and below a similar altitudinal threshold. The upper elevation of such rainstorm-producing floods decreases northward. In New Mexico, it is around 2400 m, in Wyoming 2000 m, and Idaho–Montana only 1600 m. This reflects the increasing distance from moisture-bearing air from the Gulf of Mexico.

Many of the extreme events are associated with thunderstorms that rarely affect an entire drainage basin. A massive system on 31 July 1976 resulted in a major disaster in Big Thompson Canyon, north-west of Denver. A quasi-stationary thunderstorm complex with very high moisture contents (14 g kg^{-1} at low levels) located over the mountains west of Loveland, Colorado produced rainfalls totals of 25–30 cm between 1830 and 2230 hours local time. A massive flood wall, estimated from geomorphological evidence to have been the largest in the Big Thompson drainage since glacial meltwater flows occurred 8,000–10,000 years ago (Jarrett, 1990a), caused at least 139 deaths, mainly of tourists and campers along the canyon highway and property damage of $35 million (Maddox *et al.* 1978). In other years, such as May 1969, more widespread flooding may occur in the Colorado foothills in association with a deep quasi-stationary cold low. Such pressure systems draw moist air from the south and south-east into an upslope flow. Thunderstorms may also be generated within the low.

Cloud and radiation conditions are also strongly influenced by the mountains, although fewer data are available to illustrate this. In winter, when westerly flow prevails, a crest cloud or föhn wall is commonly observed over the Continental Divide while the east slopes remain generally clear apart from occasional lee wave clouds. Under upslope conditions with an inversion, however, the lower slopes may be below stratiform cloud not reaching above 2800–3000 m, leaving the higher zones in sunlight. Clark and Peterson (1967) found this situation to be more prevalent than the crest cloud pattern in 1964–5, although their relative frequencies must vary from year to year. These regimes are apparent also from the radiation studies of Greenland (1978) (see p. 207). In summer, in contrast, towering cumulus rapidly develops over the mountains on most mornings, and spreads eastward. Radar studies indicate that convective build-up begins by mid-morning over the east slope of the Front Range, with preferred locations near Estes Park, Idaho Springs, and south-west of Pueblo, Colorado (Karr and Wooten 1976). A further investigation using half-hourly GOES imagery for summer 1983–5 by Banta and Schaaf (1987) and Schaaf *et al.* (1988) extends the radar results. Thunderstorm initiation is related both to isolated mountains such as Pike's Peak, lifting and channelling by the San Juan Mountains, and leeside convergence zones (see p. 166) such as along the eastern slopes of the Front Range and the Sangre de Cristo Mountains. The latter zone is well-developed under northwesterly to southwesterly flows. For southeasterly flow, there is lee convergence west of the southern Sangre de Cristo Mountains.

Table 5.2 A profile of winter storm precipitation across the Colorado Rockies for 265 storms during winters 1960/61–1967/68

	West Slope							*East Slope*				
	Grand Junction						*Vail Pass*					*Denver*
Height (m)	1525	1830	2135	2440	2745	3050	3200	3050	2590	2285	1830	1525
Precipitation (cm)	25.9	34.4	22.4	79.4	112.0	153.9	151.3	110.9	21.2	30.9	48.4	39.3
Ratio to Grand Junction	1.0	1.325	0.85	3.05	4.30	5.93	5.83	4.28	0.78	1.20	1.85	1.50

Source: From Hjermstad 1970

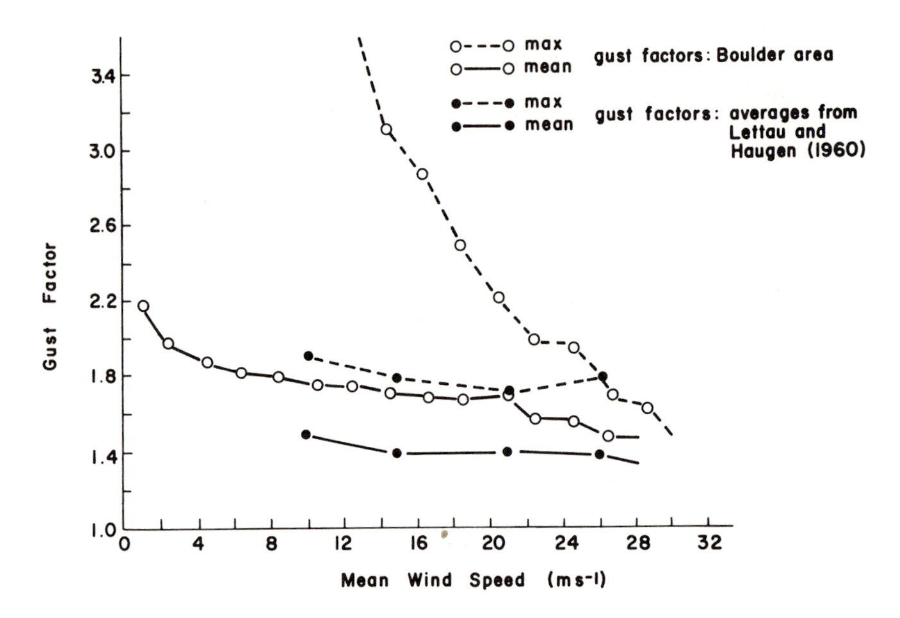

Figure 5.9 Ratios of mean and maximum gust speeds to 5-min mean wind speeds in Boulder, Colorado, compared with averages cited by H.H. Lettau and D.A. Haugen (*Handbook of Geophysics*, 1960)
Source: From Brinkmann 1973

Wind conditions in the Colorado Rockies have received much attention, both in terms of intrinsic interest in the phenomena themselves and in view of their significance for subalpine vegetation and snow transport. The summits are exposed to high mean wind speeds during the winter half-year with prevailing westerly flow; the summer months are much less windy. Records on an 11.6 m tower on Mines Peak (3808 m) on the Continental Divide indicate a mean speed of 15.4 m s^{-1} in January 1968–75 (Judson 1977), while just east of the Divide at Niwot Ridge (3750 m), the averages for October–March 1965–70 on a 2 m mast are 12–14 m s^{-1} (equivalent to 15.6–18.2 m s^{-1} at 11.6 m) (Barry 1973). On a knoll (3608 m) 1 km farther east on Niwot Ridge, short-term measurements on 117 days in winter 1975–6 showed that, at 6 m above the surface, winds exceeded 18 m s^{-1} during 50 per cent of the time and exceeded 27 m s^{-1} for 16 per cent of the time. The average daily maximum recorded at the knoll site was 39 m s^{-1} compared with 24 m s^{-1} (at the same height) at the Niwot Ridge station, and the extreme gust recorded on the knoll was 62 m s^{-1}.

These high mean velocities and gust speeds are a basic result of the height and position of the Continental Divide in relation to the westerly wind belt. Wind speeds in the valleys are only about one quarter of those at summit sites. However, in a narrow zone along the east slope and its foothills gust/mean speed ratios are well above average during downslope windstorm conditions. Figure

5.9 illustrates the results of Brinkman (1973) during 20 windstorms in the Boulder area. The occurrence and mechanisms of these winds are described in Chapter 3 (p. 151). The airstream parameters are occasionally such that downslope windstorm conditions occur at east slope locations up to at least 3050 m elevation.

One major consequence of the strong winter winds is the transport, redistribution and sublimation of snow. These aspects have been treated in Chapter 4 (pp. 267 and 277), but information on the frequency of blowing snow in the Colorado Rockies and its large-scale consequences for alpine snow hydrology seems to be lacking. Within the montane forest belt, winds serve mainly to redistribute snow and deposit it in clearings where it is subject to more rapid melt than in the forest (Hoover and Leaf 1967). This has the potential for lessening the losses by sublimation and evaporation (see p. 279).

THE ALPS

The European Alps were the birthplace of mountain meteorology (see p. 5) and have been the location of so many meteorological studies that it is only feasible to give a general impression here of their variety and content. First, there are the climatological accounts for the major observatories – the Sonnblick in Austria, Hohenpeissenberg and the Zugspitze in Germany, and the Jungfraujoch in Switzerland (see Table 1.1, p. 6). Second, there are several syntheses – for the French Alps (Bénévent 1926; Bezinge 1977), the Tirol (Fliri 1962, 1975, 1977, 1982), the Hohe Tauern (Dobesch 1983), French Switzerland (Bouët 1972), as well as the entire area (Flohn 1954; Fliri 1974, 1984; Schüepp and Schirmer 1977; Kerschner, 1989, and Sevruk 1985). The publications of the biennial *International Tagung für Alpine Meteorologie* (see References, p. 17) are a further major reference source.

A particular focus of research on weather and climate in the Alps has been the application of synoptic–climatological catalogues of pressure or airflow patterns (see Barry and Perry 1973: 151–8, for a review). Fliri's (1962) analysis of the climatic characteristics in the Tirol uses the system developed for the eastern Alps by Lauscher (1958), which was based on the *Grosswetterlagen* scheme of Hess and Brezowsky. Other classifications systems, developed by Schüepp (1959), and more recently for the 500 mb level by Kirchhofer (1976, 1982), have been used in Switzerland. Kirchhofer analyses temperature, precipitation, and sunshine data for Säntis, Davos and three lowland stations for each of 24 upper air circulation patterns.

The Alps are arcuate in form at their western end where they extend some 250 km north–south. Their orientation is almost west–east through Switzerland, where the ranges are less than 100 km in total width, and through Austria where they again broaden to 150 km, but are subdivided by pronounced longitudinal valleys such as the Inn and Drau. The Rhône and upper Rhine valleys in Switzerland are similar, but less extensive. The Alps accentuate the general

Table 5.3 District averages of seasonal precipitation characteristics for the Alps

District	*Mean elevation*	*Winter*			*Spring*			*Summer*			*Autumn*		
	(m)	*p*	*spd*	d_{30}	*p*	*spd*	d_{30}	*p*	*spd*	d_{30}	*p*	*spd*	d_{30}
N. Foreland	670	169	5.2	0.3	208	6.2	0.3	407	9.7	2.0	186	6.6	0.4
N. Edge (W)	953	412	10.8	2.1	388	10.7	2.0	681	13.7	4.8	384	11.1	2.3
N. Edge (E)	768	320	9.6	1.6	336	9.1	1.4	606	12.0	3.6	288	9.3	1.2
Central Inn	809	159	6.8	0.5	171	6.5	0.3	397	9.5	1.8	172	7.6	0.5
Silvretta	1541	261	9.6	1.5	214	7.5	0.4	448	9.8	2.0	236	9.0	0.8
Ötztal Alps	1619	160	7.1	0.7	143	5.7	0.3	301	7.2	0.9	163	7.1	0.4
Vintschgau	1130	95	6.8	0.3	98	6.0	0.2	236	7.9	0.9	151	8.2	0.6
Bolzano	809	91	7.0	0.2	143	7.2	0.1	302	9.4	0.8	164	8.9	0.6
Veltlin	1540	214	11.4	1.5	270	11.2	1.6	414	11.9	2.8	338	14.5	3.2
SE Dolomites	1315	209	11.0	1.6	269	9.3	1.0	412	10.0	1.9	321	13.3	2.6
Trentino	236	167	12.2	1.1	217	10.4	1.0	279	10.9	1.7	287	13.9	2.7

Notes: p = mean seasonal precipitation (mm); *spd* = specific precipitation density (mean precipitation per rain day, mm day^{-1}); d_{30} = number of days with $\geqslant$ 30 mm precipitation
Source: After Fliri 1962

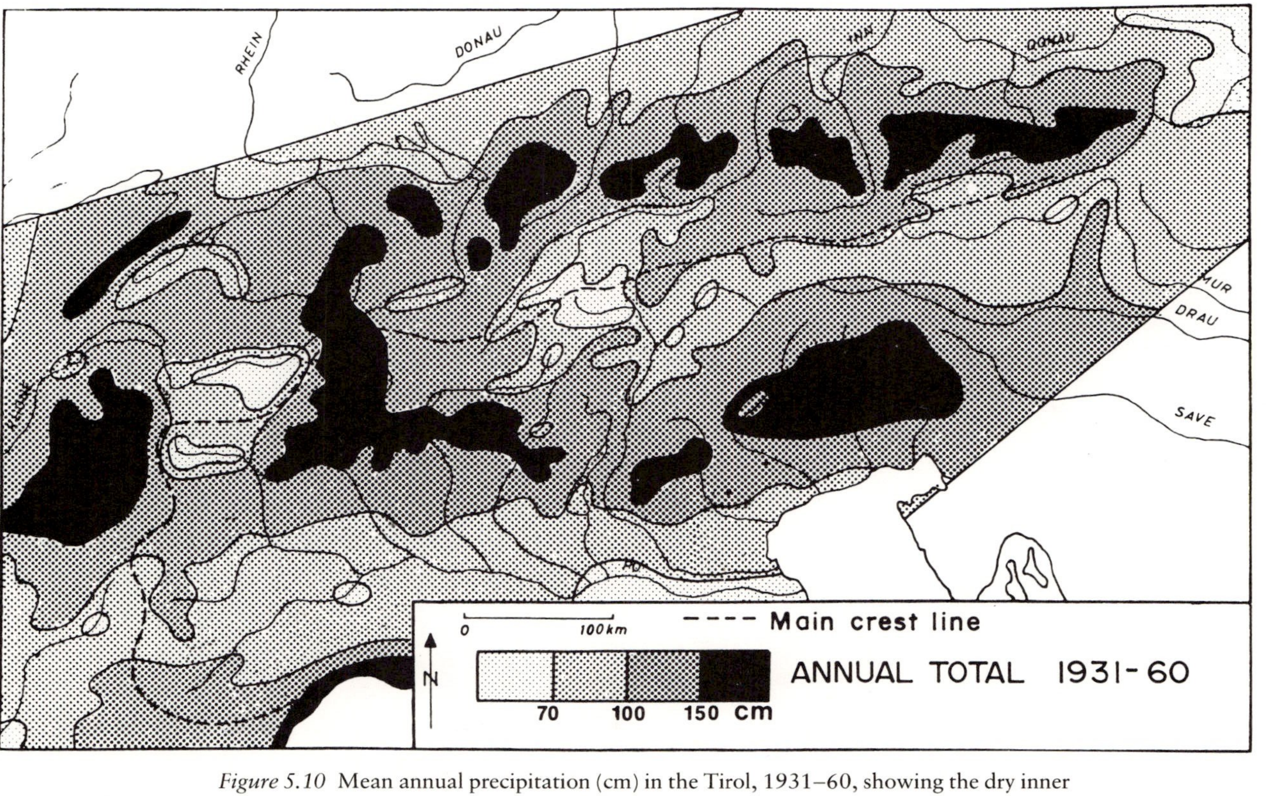

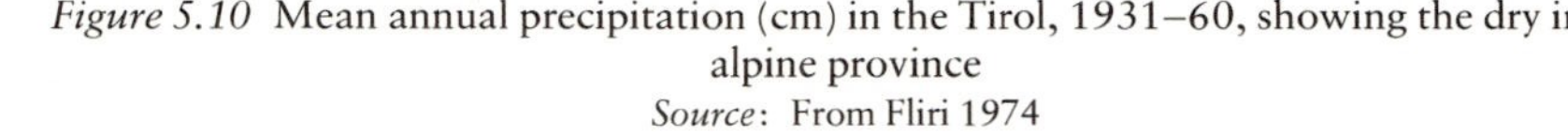
Figure 5.10 Mean annual precipitation (cm) in the Tirol, 1931–60, showing the dry inner alpine province
Source: From Fliri 1974

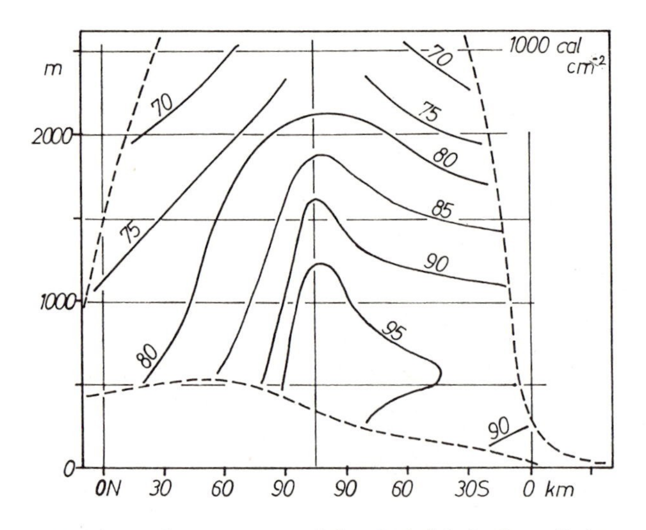

Figure 5.11 North–south cross-section of absorbed global solar radiation over the Alps
Source: From Fliri 1971

climatic gradients between the Mediterranean and central Europe and the contrast between northern and southern slopes is pronounced in the vicinity of the St. Gotthard Pass. In the Tirol, however, there is an 'inner alpine province' with its own distinctive climatic characteristics (Fliri 1975). Table 5.3, showing mean precipitation on raindays ($\geqslant$ 1 mm), illustrates the dryness of this area (see Figure 5.10), which is due to the dual sheltering effect of the ranges to both the north and south. Precipitation amounts also generally decline eastward through the Alps; compare for example the western and eastern parts of the 'northern edge'; also Silvretta and the Öztal Alps in Table 5.3. The inner alpine province is continental in character with a summer precipitation maximum and a winter or spring minimum. In contrast, the northern Alps have a primary maximum in summer and a secondary one in winter in the western section, while the southern slopes of the southern Alps have an autumn maximum and winter minimum, typical of the northern Mediterranean. Precipitation on the northern slopes of the Alps is mainly of cyclonic origin, enhanced by orographic effects, and of moderate intensity, but with summer convective precipitation (Table 5.3). The southern ranges more commonly experience showery precipitation, with thunderstorms and hail, particularly in summer and autumn. The central Alps have less thunderstorm activity (Bouët 1972).

Interactions between orography and weather systems simultaneously lead to heavy precipitation on one side of the Alps and much lesser amounts on the opposite side, as well as creating corresponding patterns of snow accumulation, or melt. Lang and Rohrer (1987) illustrate the atmospheric conditions for several cases with deeper snow cover on the northern slopes of the Alps, or south of the Alps. Southeasterly flows of Mediterranean air have higher humidities and temperatures than the northwesterly flows, for example. However, these differences for individual storm events are largely cancelled out when longer time intervals are considered.

The greater continentality of the inner Alps is also apparent in the greater sunshine duration – 60 per cent of possible on an annual basis – compared with only 45–50 per cent on the northern and southern margins. As a result, the central zone has higher totals of absorbed global solar radiation (Figure 5.11) and a larger annual range of mean daily maximum temperature; 23°C in the inner zone compared with 20°C at the edges at 500 m elevations and 17°C and 15°C, respectively, at 2000 m elevation. The various data and Fliri's (1975: 139) analysis bear on the question of the so-called *Massenerhebung* (mass-elevation) effect discussed earlier (p. 57). Tollner (1949) demonstrated that, in the Alps, stations located on mountains, in high valleys, and in passes, are all colder than the free air on a mean annual basis at 7 a.m. (Figure 5.12) and there are ample observations to confirm this result. Mountain temperatures have a tendency to exceed those in the free air, by a degree or so, only on summer afternoons (see p. 55). The fact that vegetation boundaries extend to higher elevations in the central Alps is related primarily to the reduced snow cover duration, greater radiation income, and longer growing season (see Figure 5.11). The Bernese Oberland, which on topographic grounds alone might be expected to show such a mass-elevation effect, in fact experiences a more 'maritime' climatic regime as a result of its general windward location (H. Turner, personal communication, 1975). The relative dryness of the central Alps, together with the high radiation amounts, implies less energy going into latent heat and more into sensible heat to the atmosphere. However, this transfer is not the sole mechanism for raising atmospheric temperatures over the central Alps since the effect is already present in April when all of the higher elevations are still snow-covered (Fliri 1975).

Despite the problem of precipitation measurements in high mountains, the various high-level stations and observatories in the Alps indicate that amounts increase with elevation to the highest levels (3000–3500 m). Figure 5.13 illustrates this in the vicinity of the Jungfrau massif, and Steinhauser (1938: 96) shows the same pattern on a local scale around the Sonnblick observatory. Baumgartner *et al.* (1983) show the generality of these results based on 1000 Alpine catchments. Havlik (1968) demonstrates convincingly that most of this orographic effect is associated with cyclonic weather situations. For the French Alps, Bénévent (1926: 64) concluded that the elevation of maximum precipitation increases towards the Alps proper since it occurs around 2000 m in the Pre-Alps, 2500 m in the central massifs of the Dauphiné-Savoie, and 2500–3000 m in the mountains farthest into the intra-alpine zone, but no summit observatory data are available in this area. Only about 10 per cent of the altitudinal increase in precipitation amount is attributable to more frequent raindays, according to Havlik (1968); differences in intensity are the primary factors. Nevertheless, variations in precipitation intensity are also attributable to location with respect to the entire mountain area and the orientation of particular ranges in relation to

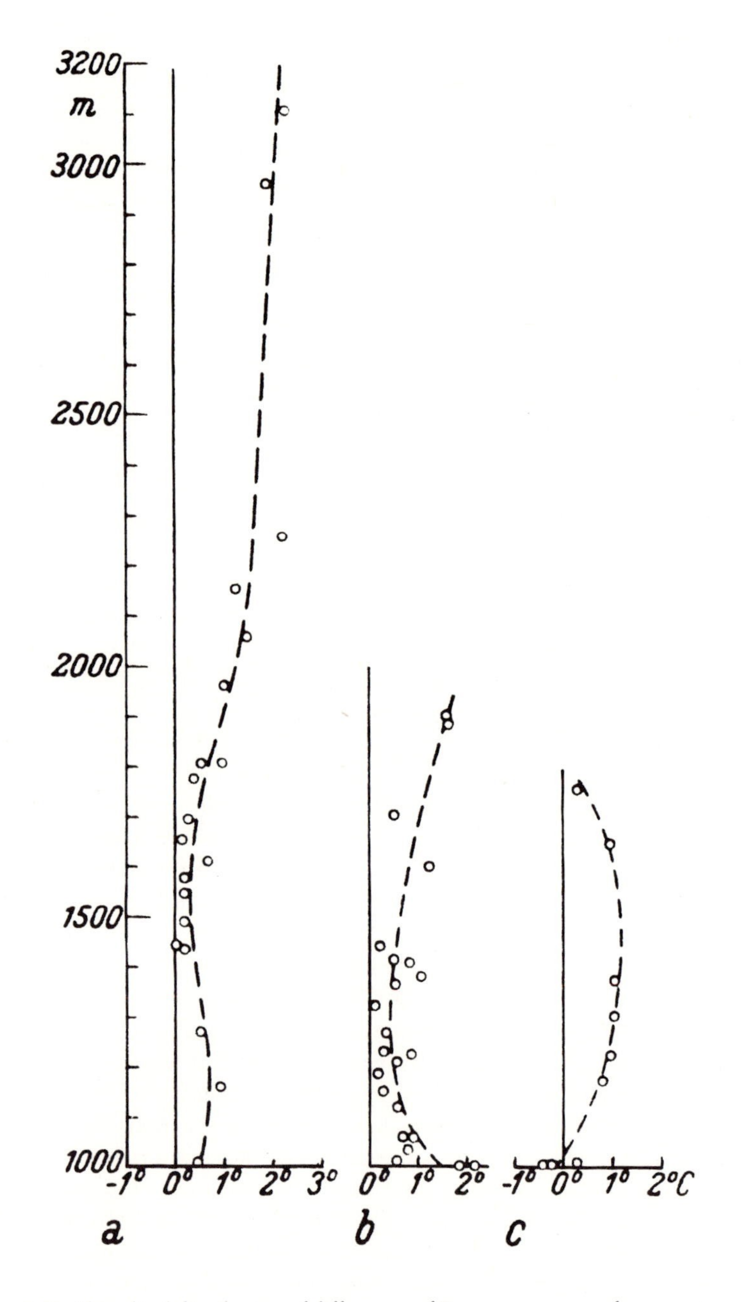

Figure 5.12 Altitudinal distribution of differences of 7 a.m. mean annual temperature at alpine stations and in the free air

Notes: (*a*) Free air minus mountain stations; (*b*) Free air minus valley stations; (*c*) Free air minus pass stations

Source: From Tollner 1949

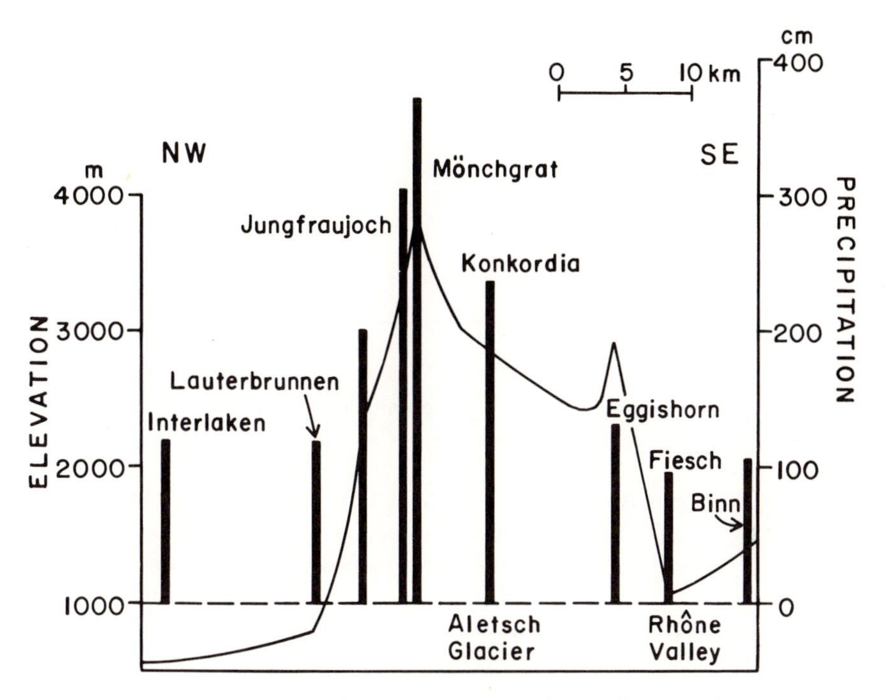

Figure 5.13 Annual precipitation across the Jungfrau massif
Source: After Maurer and Lütschg 1931

the major air currents (Bénévent 1926).

Snow cover in Switzerland has been examined in detail by Witmer *et al.* (1986) using the networks of the Swiss Meteorological Institute and the Federal Institute for Snow and Avalanche Research (ELSLF) for October–June, 1950/51–1979/80 (or sub-periods). The mean gradient of total new snowfall below 1100 m altitude for November–April (1970/71–1979/80) varies from 8 cm/100 m in the Valais (south-western Alps) to 73 cm/100 m in the Bernese Alps (north slope of the Alps). Above 900 m, the mean gradients range from 26 cm/100 m in the south-eastern Engadine (Grisons) to 107 cm/100 m in the north-east (Glarner Alps). Using data for 1960/61–1979/80, Witmer *et al.* show that snow depths above 900 m in April have a range of gradients (for median depth values) from 9 cm/100 m in the Valais to 26 cm/100 m in the Glarner Alps with values > 20 cm/100 m through the north slope of the Alps and in the southern Tessin (Gotthard area). There are also large differences between north- and south-facing slopes in winter and spring. Depth ratios compared with a horizontal surface are approximately 1.2/0.7 for 20° N/S-facing slopes at 1800 m in early February, increasing to 2.0/0.3, respectively, in late March. The altitudinal variation of snow cover duration for 1960/61–1970/80 increases almost linearly for depths ≥ 30 cm, from 5 days at 400 m to 160 days at 2000 m, whereas it is curvilinear for depths ≥ 1 cm, ranging from 25–50 days around 400 m to 175 days at

Table 5.4 Mean potential temperature (°C) during 12 cases each of north föhn and south föhn, 1942–5

	Elevation (m)					
	Lugano 276	*Airolo* 1170	*Gotthard* 2096	*Göschenen* 1107	*Altdorf* 456	*Zurich* 493
S Föhn	8.3	11.2	12.8	14.3	15.7	12.4
N Föhn	13.8	10.5	9.5	6.2	4.8	5.1

Source: After Frey 1953

2000 m and above. The problem of mapping snow depths is overcome by the preparation of maps for all of Switzerland or the Alps at fixed levels (1000, 1500, and 2000 m).

A general picture of altitudinal variations in moisture balance for Austria is provided by Liang (1982). Potential evaporation is calculated from Penman's method (p. 274); actual evaporation is determined using an empirical relationship proposed by Turc. Potential evaporation (PE) decreases rapidly and then more slowly with altitude while actual evaporation (AE) decreases nearly linearly. AE increases from 90 per cent of PE between 500 and 1500 m to over 95 per cent above 2000 m. AE accounts for about 45 per cent of the annual precipitation between 500–1500 m, but only 25 per cent around 20000m.

A major climatic characteristic of the alpine valleys and foothill locations is the occurrence of föhn winds. Numerous general and local studies have been made of their effects. Tables 5.4 and 5.5 illustrate profiles of mean potential temperature for 12 cases each of north and south föhn during the peak phases with valley föhn (Frey 1953). It is notable that the warming is already apparent over the higher elevations (at St. Gotthard) implying subsidence effects. North föhn generally occurs with a northerly flow of modified cold polar air behind a meridional trough. South föhn is typically associated with warm sector air. At Sierre, in the central part of the Valais of south-west Switzerland, the south föhn occurs on about 33 days per year, with a maximum in April (Bouët 1972). The winds at summit level (4000 m) are usually southwesterly, but valley chanelling causes them to be southerly in the lower Valais and northeasterly at Sierre where they have a mean speed of 6 m s^{-1}. In a third of the cases the maximum wind may exceed 15 m s^{-1}. At Innsbruck the south föhn has an average annual frequency of 60 days, with 40 per cent of these in the spring. Other months each have 3–5 days with south föhn (Fliri 1975). However, the greatest departures of daily temperature from normal (+8°C) at Innsbruck are experienced during autumn and winter cases.

The south föhn does not always penetrate to ground level in the valleys and northern Foreland since shallow cold air lifts it above the surface. Occasionally, when the föhn wall extends across the crest of the Alps and winds are strong at

Table 5.5 Characteristics of south föhn in the eastern Alps and Foreland

	Elevation (m)					
	Lake Garda 90	*Innsbruck* 575	*Partenkirchen* 715	*Hohenpeissenberg* 994	*Munich* 528	*Regensburg* 343
θ_{14} (°C)	8.3	15.5	17.4	16.1	13.4	9.7
$R.H._{14}$ (%)	69	39	33	40	49	56
Mean cloud (%)	73	40	30	43	44	49

Notes: θ_{14} = Potential temperature (at 500 m) at 14 hours; $R.H._{14}$ = Relative humidity at 14 hours
Source: After von Ficker and de Rudder 1943

summit level, the air is unable to follow the terrain and only reaches the ground where the valleys open out onto the lowlands. This is known as 'dimmer-föhn', meaning blocked or damned-up föhn (Frey 1953). Apart from the much-discussed but poorly understood symptoms of 'föhn sickness', föhn occurrences in spring cause rapid snowmelt, with a risk of avalanche activity and also floods. Föhn gales may result in damage to forest through windfalls, as well as enhanced fire risk due to the low humidity. On 7–8 November 1982 a violent south föhn in the northern Alpine foreland of Switzerland brought peak gusts of 40 m s^{-1} at Jungfraujoch and 54 ms^{-1} at Gütsch near Altdorf (2282 m), associated with a south–north pressure gradient of 18 mb/100 km – the highest on record (Frey 1984). Temperatures rose to 20°C at Zurich compared with readings of 2°C in the Po lowlands of Italy. This '100-year' event is an example of the persistent type of south föhn that occurs ahead of a deep, slow-moving upper trough, with strong southerly flow aloft (Schüepp 1990). Short-lived south föhn events are associated with upper westerlies, when upper-level divergence generates surface pressure falls north of the Alps.

A local wind of importance in the northern and western Alps is the *bise*. Cold northerly flows of polar air in winter and spring are typically blocked and channelled by the Jura and Alps (Wanner and Furger, 1990). It occurs, on average, on 94 days per year at Lausanne. The speed of the bise averages 3–4 m s^{-1} at Lausanne with gale-force maxima in about 10 per cent of cases. In the vicinity of Geneva, and also in the sheltered Valais, these winds give dry fine weather, but further east the orientation of the ranges in the Pre-Alps and Alps produces heavy cloud, due to the forced uplift, with precipitation in about one third cases (Bouët 1972). On the southern slopes of the Alps, of course, these situations usually give rise to north föhn, provided the flow is forced across the mountains.

Fog and low stratus is a significant feature of the winter climate of the Bernese Plateau. A detailed analysis by Wanner (1979) shows that fog occurs at least locally on 50 per cent of mornings between September and March. There is clear

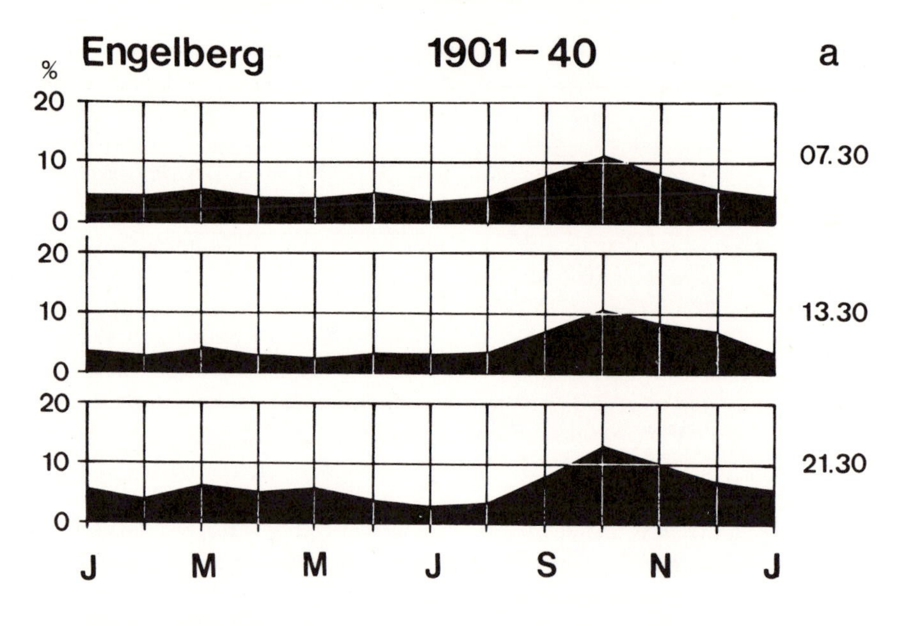

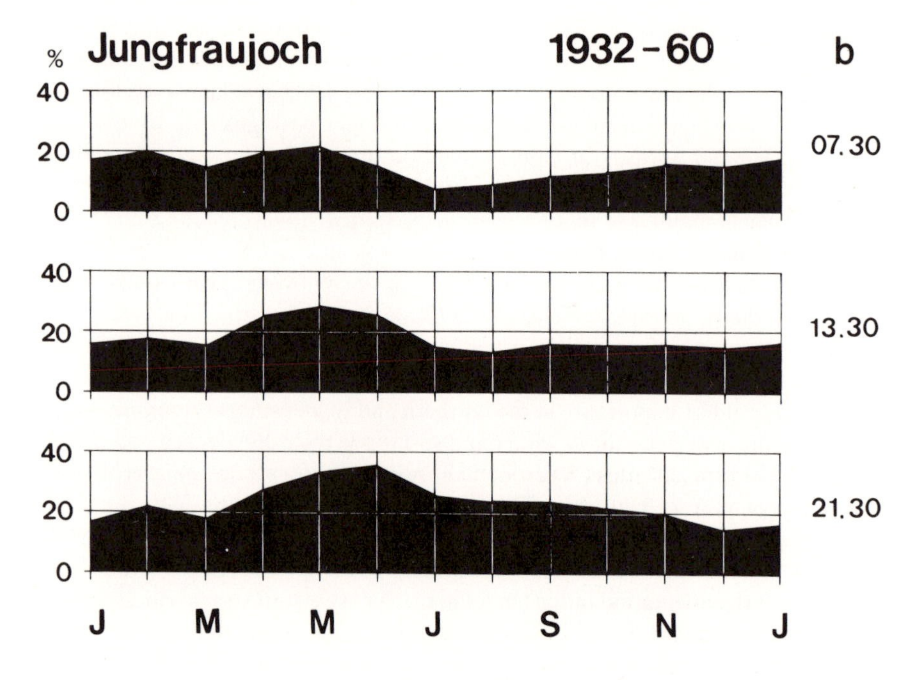

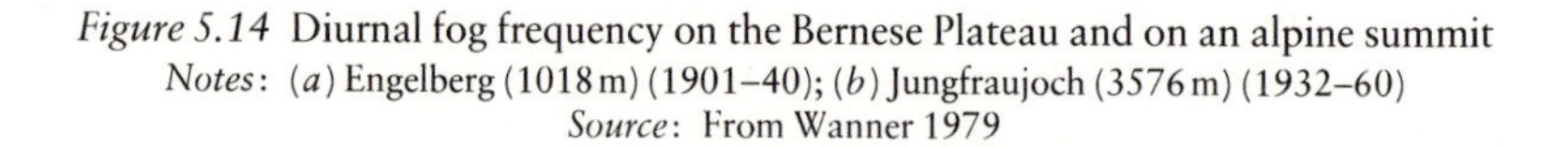

Figure 5.14 Diurnal fog frequency on the Bernese Plateau and on an alpine summit
Notes: (*a*) Engelberg (1018 m) (1901–40); (*b*) Jungfraujoch (3576 m) (1932–60)
Source: From Wanner 1979

distinction between seasonal fog regimes on the plateau, where there is a pronounced maximum in October–November and a summer minimum, and at summit stations which have a high frequency year-round, but with a spring maximum (Figure 5.14). In many alpine valleys, and in the Valais, however, the frequency of early morning fog in winter is less than 10 per cent. An inversion top is frequently located about 800 a.s.l. above Payerne corresponding to the typical fog limit reported by Wanner. It is also noteworthy that, at Montreux, stratus occurs on an average of 80 mornings between September and March, with its base between 700–900 m on 87 per cent of occasions (Bouët 1972). Low level inversions and radiation fog are most common during weak anticyclonic situations, while cold northeasterly flow with anticyclonic patterns (bise) tends to give rise to high-level inversions and shallow high fog or stratus. Upslope fog, in contrast, usually occurs with cyclonic south-west airflows.

The final phenomenon to be discussed is common to all mountain systems with east–west valleys. It is the well known *adret* (sunny) *ubac* (shaded) contrast between slopes, which has been most thoroughly investigated in the Alps. The differences are largest in winter when the solar angle is low and the period of radiation income is minimal. Bénévent (1926: 105) cites typical mean temperature differences of 0.5–1.0°C, with up to 3°C difference at 13.00 in appropriate locations. Bouët (1972) notes that snow-cover duration at 1500 m is 160 days on the shady slope, but 30 days less on the sunny slope. This difference finds expression in the pattern of land use and also in the preferred location of alpine settlement.

THE MARITIME MOUNTAINS OF GREAT BRITAIN

The mountains of north-western Europe are notable for their extremely maritime character. This is manifested in the limited seasonal variation of cloudiness, precipitation and temperature, in the high humidities, in the steep lapse rates, and in the low tree line (Manley 1945; Green 1955; Taylor 1976). The most detailed observational material in the British Isles was provided by the Ben Nevis Observatories (McConnell 1988). The records for 1883–1904 at Ben Nevis (1343 m), 6 km inland from the observatory at Fort William (13 m), are fully tabulated by Buchan (1890) and Buchan and Omond (1902, 1905, 1910), although Buchan's death in 1907 halted the planned thorough analysis of the data. (Some of the data were also summarized by von Hann (1912)). The maintenance of good observational records on the summit of Ben Nevis was a major problem due to frequent riming of instruments (see, for example, Curran *et al.* 1977), summit cloud, and winter snowfall. (In addition, the cost was borne by private sources and the unwillingness of the government to provide assistance led to the eventual closure.) The thermometer screens were mounted on ladders and their height adjusted in relation to the snow surface which could exceed 3 metres. The site was adjacent to a steep 550 m drop on the north side of the mountain and the effect of this location on air motion caused frequent 'pumping' of the barometer.

Currently, there are few meteorological stations in Britain above 400 m and no actual mountain stations (Taylor 1976). Manley (1936, 1942, 1943, 1980) has summarized some records for the northern Pennines, but data for the mountains of Scotland, Wales and the Lake District are very fragmentary. Automatic stations can help to fill this gap, at least for the major weather elements (Barton 1984).

The western highlands of Britain have a high proportion of cloudy days. On Ben Nevis, for example, the mean cloudiness exceeds 80 per cent in all months and in winter months only about 10 per cent of the possible sunshine hours are recorded. The summit is clear of fog only 21 per cent of the time from November to March, with the majority of clear weather occurring in spring (45 per cent of the time). Buchanan (1902) showed that, during 1885–97, there were 185 spells, averaging 4.0 days in length, with *no* clear observations.

Glasspoole (1953) used data of R.C. Mossman (in Buchan and Omond, 1910: 444) to show that summit cloud around Ben Nevis during winter 1901–2, averaged almost 50 per cent at 1000 m, 67 per cent at the summit and only 20 per cent at 700 m, whereas in the northern Pennines of England (Great Dun Fell, 847 m) and the Southern Uplands of Scotland (Lowther Hill, 725 m) the frequency was also approximately 60 per cent.

As a result of cloud cover, solar radiation amounts in Wales and the Pennines decrease with elevation below 500 m a.s.l. by roughly 2.5–3 MJ m^{-2} day^{-1} km^{-1} (Harding 1979b). However, this effect operates on a regional scale. Receipts are fairly uniform within a given upland area, regardless of altitudinal differences. Above 500 m, the higher summits in summer may have shallow cloud, or no cloud, above them. Harding notes that a 35 per cent increase in solar radiation from Glenmore Lodge (341 m) to the Cairngorm Summit (1245 m) was observed between 25 May–22 July 1977 (for 10.00–14.00 hours).

Annual precipitation is high in north-west Britain due to the northeastward movement of frontal depressions from the Atlantic, and the location of the mountains causes large altitudinal increases. In the western highlands of Scotland, measurable precipitation falls more than 1500 hours per year, although annual rates are only about 2.0 mm h^{-1} (Atkinson and Smithson 1976). Using data from more than 6500 stations over Britain, Bleasdale and Chan (1972) determine that average annual precipitation, R (mm) can be expressed:

$$R = 714 + 2.42\ H,$$

where H = height (m). The line of zero anomaly from this regression closely approximates the main east–west water divide in Scotland, northern England and Wales. Positive residuals $\geqslant$ 600 mm occur in western Scotland, the Lake District and Wales with negative residuals $\geqslant$ 600 mm over the Cairngorms, where there are strong lee (sheltering) effects with the prevailing air flow. An annual total of 4084 mm was recorded on Ben Nevis for 1883–1904, although this is only 120 mm in excess of the 'predicted' amount. The orographic effect is exhibited more in terms of increased intensity than duration. For example, the annual mean at Cwm Dyli, Snowdonia (101 m) is 3500 mm with a mean intensity of 2.7 mm h^{-1} whereas at Holyhead (9 m) on the coast, the corresponding figures are 1000 mm and 1.4 mm h^{-1} (Atkinson and Smithson 1976). The total is increased 3.5 times with less than a doubling of the average hours with precipitation. 'Exceptional' daily falls, defined as totals of 125 mm (5 in) which account for at least 15 per cent of the annual average at a given station, are most common in the western mountain districts of Britain in the winter half-year according to Bleasdale (1963), illustrating the importance of orographic intensification of precipitation in deep frontal depressions. Moreover, the Ben Nevis records (Buchan 1890) show that most thunderstorms occur in the winter months, in associations with such systems, in contrast to the summer maximum over the European mountains.

Since surface temperatures are frequently slightly below freezing point on the mountains in winter, the occurrence of rainfall especially with warm front situations, quite commonly gives rise to freezing precipitation and glazed ice. On Ben Nevis, there were 198 cases of such 'silver thaw' (Mossman 1902) during 1885–90, with an average duration of 4.4 h. These cases occurred largely in the winter months with air temperatures between 0° and −2.5°C. In rare instances, ice build-up from such events, perhaps associated with heavy riming (freezing fog/cloud droplets), can set up severe stresses on structures (see p. 265 and Plate 14).

General snowfall conditions in relation to altitude in the British mountains are discussed on p. 240. In connection with a survey of the possibilities for winter sports development at Fort William, Thom (1974) has analysed the snowfall data for 1895–1904 at Ben Nevis. Figure 5.15 shows the percentage of the total precipitation in each month occurring as snow, and the extremes. In April the percentage does not fall below about 60 per cent. At 750 m elevation, the ratio of rain to snow approaches 50:50 from January through April. Thom also shows that, on average, the snowfall fraction on Ben Nevis increases by 9 per cent per 1°C decrease in mean monthly temperature. However, since the scatter is considerable, this relationship may be of limited value in assessing the snowfall fraction in relation to temperature for a particular month.

As already noted, lapse rates in Britain are steep due to the maritime air mass characteristics. Data from 14 pairs of lowland and upland stations show a mean gradient of about 8.5°C km^{-1} for annual mean maximum temperature (Harding 1978). As recognized earlier by Manley (1942), there is a clear winter minimum (6–7°C km^{-1}) and spring maximum (8–10°C km^{-1}) and this pattern seems to be general throughout north-western Europe. The spring maximum apparently reflects a general increase in instability, rather than synoptic airflow patterns (Harding 1978). The gradients of mean temperature and maximum temperature in Britain are closely similar in winter, when the short day-length and large cloud amounts lessen the diurnal variation of lapse rate. The diurnal temperature range on British mountains appears to be remarkably small. The Ben Nevis records for 1883–7 show average departures of hourly temperatures from the daily mean of

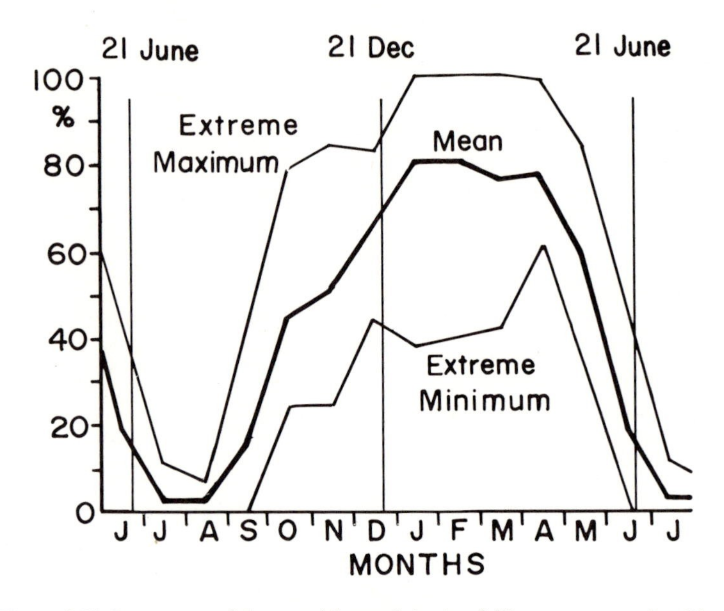

Figure 5.15 Percentage of the monthly precipitation falling as snow on Ben Nevis (1343 m), 1895–1904
Note: Average (solid) and extremes (fine lines)
Source: After Thom 1974

± 1°C in summer and only ± 0.2°C in winter (Buchan 1890).

The gradient of maximum temperature, which is controlled mainly by radiational heating (Harding 1979a), approximates the dry adiabatic lapse rate in sunny convective conditions, although higher rates may occur due to the existence of a shallow superadiabatic layer at the lowland station. Omond (1910) reported 205 cases (0.2 per cent of observations) with lapse rates > 10.5°C km^{-1} at Ben Nevis between August 1890 and July 1903. They occurred predominantly on sunny afternoons or with dry southeasterly airflow during April–June. The gradient of minimum temperature is subject to local topographic effects, especially valley inversions, making general discussion of regional characteristics impossible.

The length of the frost-free season generally decreases with altitude, although, as shown by S.J. Harrison (cited by Taylor 1976: 279), the use of screen temperatures greatly overestimates conditions at the ground surface. The difference between frost-free seasons, determined from readings of screen and grass minimum thermometers, for example, amounted to 18 weeks (35 versus 17) at sea-level and 25 weeks (33 versus 8) at 450 m a.s.l. in western Wales in 1969.

An unexpected feature of British mountain climate is the occasional observation of very dry air. On Ben Nevis, 2.5 per cent of days (1884–1903) had at least one hourly reading of relative humidity below 20 per cent, based on recalculations using a more accurate hygrometric equation (Green 1967). Most of the cases occurred in autumn and winter, apparently related to subsidence during spells with blocking anticyclones.

Wind data for the British uplands are extremely sparse. The annual mean on Ben Nevis averages 6.5 m s^{-1}, with a range of 9.0 to 4.5 m s^{-1} between the means of January and July. In view of the fact that part of the motivation for the observatory was its location on the Atlantic storm track, these values are lower than might have been anticipated, although Thom (1974) shows good agreement between observed summit winds, estimated geostrophic values, and observed 900 mb soundings at Stornoway in the Hebrides. Winds on Ben Nevis exceeded 22 m s^{-1} about 9 per cent of the time during November through to March. The recently installed automatic station in the Cairngorms (1245 m) indicates that a high degree of gustiness may be expected, despite the smooth dome-shape of that summit (Curran *et al.* 1977).

THE SUB-POLAR ST ELIAS MOUNTAINS – ALASKA/YUKON

Few mountain areas in high latitudes have received any study of their climatic characteristics. Consequently, the intensive investigations organized jointly by the American Geographical Society and the Arctic Institute of North America in the St. Elias Mountains are of special significance. These mountains border the Pacific Ocean, rising sharply to 2600 m some 60–180 km from the coastline, with the mass of Mt Logan exceeding 6000 m. The results of the many diverse studies carried out under the Icefield Ranges Research Project and the related High Mountain Environment Project (Marcus 1974a) are detailed in scientific reports (Bushnell and Ragle 1969–72; Bushnell and Marcus 1974) and other papers. A brief synopsis based on these studies is presented here.

The interactions of the large-scale circulation with the mountain range have been examined by Taylor-Barge (1969). She gives special attention to the role of the St. Elias Mountains in accentuating the normal climatic gradient between coastal maritime conditions and the interior continental regime in summer. At this season the mean circulation is weak westerly at the surface, to southwesterly at 700 mb, alternating between ridge situations and cyclones in the Gulf of Alaska.

Temperature soundings at Yakutat on the coast and Whitehorse, 275 km inland (see Figure 5.16) reveal that lapse rate profiles become closely similar at the 3 km level. The temperature regime at Seward Glacier south-west of the main divide, is similar to that at Yakutat, whereas stations on the Kaskawulsh Glacier (*c.* 1768 m) and at Kluane Lake on the eastern side have a 10–11°C diurnal range in the summer months, resembling Whitehorse (Table 5.6). Stations on the Divide at *c.* 2560 m have intermediate temperature characteristics. However, since daily temperature trends are similar at all stations, Marcus (1965) suggests that the range is not a total barrier to the transference of air mass characteristics.

The degree of climatic similarity between the stations varies according to the parameter. Summer cloud conditions at the Divide resemble those at Seward Glacier, with a predominance of fog and stratus, and similar overall cloud amounts. The Kaskawulsh Glacier stations are more like Kluane Lake, on the east side, in having a summer predominance of cumuliform clouds. In view of this, it is not unexpected that afternoon convective showers occur on the eastern side of the range. These are not experienced at Divide, or stations further west, where most precipitation is of cyclonic origin. Precipitation probabilities and daily amounts decrease eastward in summer (Table 5.6), but observing problems and local variability prevent general inferences from being drawn.

From the available data, Taylor-Barge (1969) concludes that in summer the range acts as a transition zone rather than a sharp climatic divide. An average dividing line between coastal maritime and continental interior regimes can be located on the eastern side of the mountains between Divide station and the Kaskawulsh Glacier, but it shifts location according to synoptic conditions. Thus, an unusual north-easterly flow pattern during 17–19 July 1965 produced cool rainy weather on the east side of the range with precipitation occurring up to the Divide stations, while föhn conditions associated with subsidence in the airflow affected the Pacific slope. This is more or less a mirror image of the normal summer pattern of weather across the range. The barrier seems effective up to about the 3 km level in the free atmosphere, but this is only a few hundred metres above the Divide stations. On the other hand, upper fronts affect the higher parts of the range. As described in Chapter 3, p. 115, frontal systems may be deflected north or southward by the barrier, or become aligned parallel to it, with blocking of the air at lower levels. Thus, for example, the weather on the windward Pacific slope frequently fails to clear up after a cold front passage.

In winter, the climate of the Pacific slope is dominated by Aleutian–Gulf of Alaska low-pressure systems bringing onshore flow, cloudiness and large amounts of precipitation. The circulation of the interior, in contrast, is determined by the Yukon–Mackenzie high-pressure system. Few low-pressure systems cross the Divide, so that the range may act as a climatic divide to a greater degree than in summer (Taylor-Barge, 1969). Whereas the mean low-level flow is southerly, that at 500 mb is northwesterly, parallelling the valleys in the ranges.

Local effects will also be much less pronounced in winter due to the continuous snow cover. In summer, downglacier winds occur 70 per cent of the time over the Kaskawulsh Glacier, with a depth of 50–500 m or more above the surface (Marcus 1974a). Indeed, the lower limit of geostrophic flow may lie above the local summits (Benjey 1969), suggesting large-scale cold air drainage from the extensive high surfaces in the vicinity. On Mt Logan (5360 m) itself, wind directions are similar to those over Yakutat at 500 mb, but speeds averaged only 3 m s^{-1} during July 1968. In other years, however, prolonged high winds have been observed in association with Pacific storms (Marcus and LaBelle 1970).

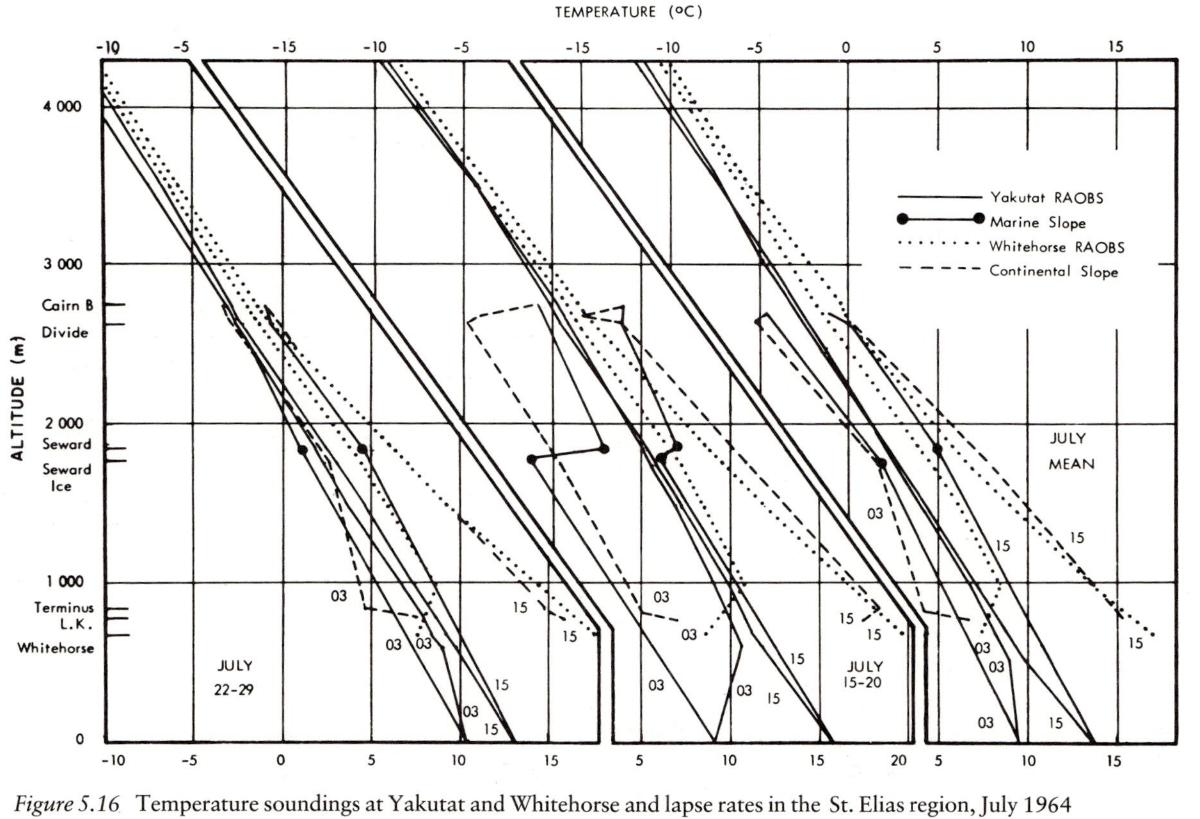

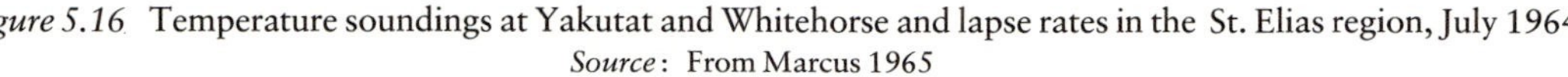

Figure 5.16 Temperature soundings at Yakutat and Whitehorse and lapse rates in the St. Elias region, July 1964

Source: From Marcus 1965

Table 5.6 Summer climatic data for the St. Elias Mountains

Station	Elevation (m)	July 1963–5				June–August 1963–5	
		Mean daily temperature (°C)	Mean daily range (°C)	Vapour pressure (mb)	Solar radiation ($W m^{-2}$)	Daily precipitation probability	Mean daily precipitation ($mm day^{-1}$)
Yakutat	14	11.4	7.1	12.0	–	0.73	8.28
Seward Glacier[c]	*c.* 1860	2.6	8.2	6.6	(252)[f]	0.66	(1.36)
Mt. Logan[a]	5360	−18.2	11.3	1.1	360	–	–
Divide	*c.* 2650	−1.8	8.8	4.7	272[d]	0.66	1.21
Chittistone[b]	1779	5.2	6.6	7.6	198	–	–
Kaskawulsh[e]	*c.* 1768	4.7	7.2	6.4	249[f]	0.48	(0.15)[f]
Kluane Lake	786	11.9	10.8	10.1	212[b]	0.35	0.90
Whitehorse	698	13.8	12.3	8.9	–	0.43	0.92

Notes: [a]1968–70; [b]1967–69; [c]1964 and July 1965 only; [d]1969 only; [e]1964–66; [f]1965 only.
Sources: After Taylor-Barge, 1969; Marcus, 1974a, b.

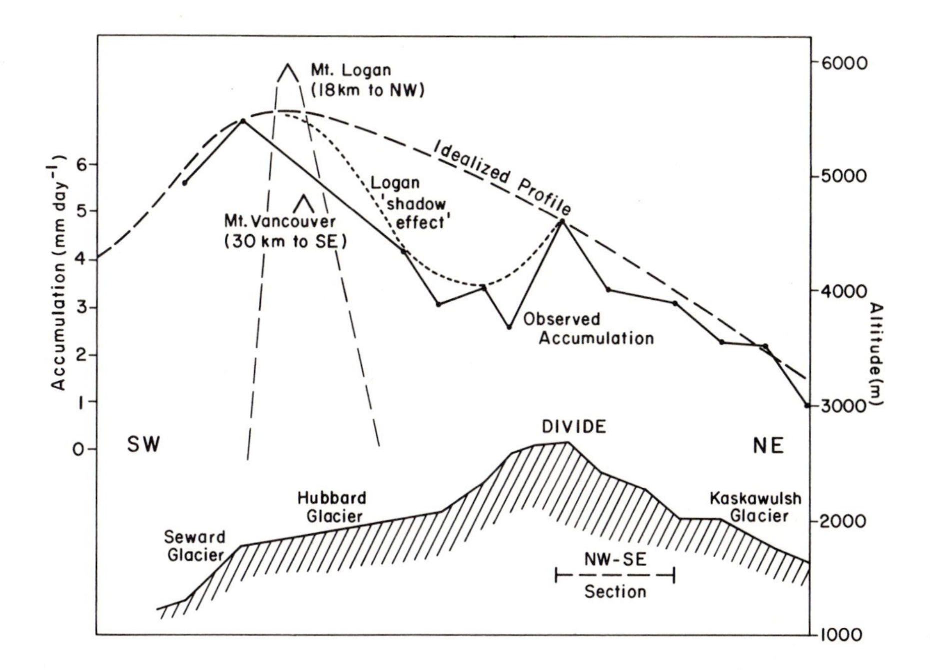

Figure 5.17 Observed and idealized profiles of accumulation rate (mm day^{-1}) across the St. Elias Range based on snow-pit data. The 'shadow' effect of Mt. Logan is indicated
Source: After Taylor-Barge 1969

Only a 'tentative' profile of annual precipitation based on snow-pit data can be given (Figure 5.17). This shows an altitudinal maximum around 1800 m, in line with estimates for the Cordillera in British Columbia (Walker 1961). The graph indicates a pronounced shadow effect in the lee of Mt Logan and a subsidiary maximum at the Divide where, even at the relatively low elevation of 2650 m, only snowfall has been observed. During the 1960s, accumulation of the higher elevations was often unrelated to precipitation anomalies on the coast of Yakutat (Marcus and Ragle 1970), suggesting the importance of upper-level disturbances and upper fronts for precipitation on the mountains. Such altitudinal differences are supported by observations of 2 m snowpack temperatures on Mt Logan in 1965 (Alford and Keeler 1969). These temperatures approximated the dry adiabatic lapse rate, apparently in response to prevailing katabatic drainage, except between 3000 and 3500 m where an *inverted* temperature lapse of 1.3°C 100 m^{-1} occurred. In contrast, in 1696–70, an average lapse rate of 0.61–0.66°C 100 m^{-1} was measured at 2 m in the snow and there was no local inversion zone (Marcus 1974b). The weather regime of 1964–5 was dominated by southwesterly flow across the range from 850–500 mb, which would cause katabatic winds on the lee of Mt Logan. The pressure pattern in 1969–70 was unusual in the presence of a ridge over the area giving northwesterly flow over the eastern slopes. Orographic lifting would account for the observed lapse rate and high accumulation about 4200 m.

OTHER MOUNTAIN AREAS

It is clearly impossible to treat regionally all mountain areas of the world. However, as a basis for further study, some references are provided that will assist the reader in identifying sometimes obscure sources of climate information for other mountain areas. These are listed in arbitrary order below:

Andes, northern: Bradley *et al.* (1991), Flohn (1968), Herrmann (1970), Monasterio (1980), Sarmiento (1986), Troll (1968).
Andes, equatorial: Erikson (1984), Hastenrath (1971, 1981), Kistemann and Lauer (1990).
Andes, southern: Fliri (1968), Lauscher (1977).
Mexico: Lauer and Klaus (1975).
East Africa: Davies *et al.* (1977), Hastenrath (1984), Hurni and Stähli (1982), Kenworthy (1984), Thompson (1966) Winiger (1981).
Rocky Mountains: Dirks and Martner (1982), Finklin (1983, 1986).
Tibetan Plateau: Kang and Xie (1989), Lo and Yanai (1984), Ohata and Kang (1990), Reiter *et al.* (1987), Xu *et al.* (1986), Ye and Gao (1979).

REFERENCES

The equatorial mountains of New Guinea

Allen, B. (1989) 'Frost and drought through time and space, Part 1: The climatological record' (Frost and drought in the highlands of Papua New Guinea, eds B. Allen and H. Brookfield), *Mountain Res. Dev.* 9, 252–78.
Allison, I. and Bennett, J. (1976) 'Climate and microclimate', in Hope, G.S., Peterson, J.A., Allison, I. and Radok, U. (eds) *The Equatorial Glaciers of New Guinea* pp. 61–80, Rotterdam, A.A. Balkema.
Barry, R.G. (1978), 'Aspects of the precipitation characteristics of the New Guinea mountains', *J. trop. Geog.*, 47, 13–30.
Barry, R.G. (1980) 'Mountain climates of New Guinea', in P. Van Royen (ed) *The Alpine Flora of New Guinea. vol. 1. General Part.* pp. 75–110, Vaduz, J. Cramer.
Brookfield, H. and Allen, B. (1989) 'High-altitude occupation and environment' (Frost and drought in the Highlands of Papua New Guinea, eds B. Allen and H. Brookfield), *Mountain Res. Dev.* 9, 201–9.
Hnatiuk, R.J., Smith, J.M.B. and McVean, D.N. (1976) *Mt Wilhelm Studies II. The Climate of Mt. Wilhelm*, Research School for Pacific Studies, Australian National University, Canberra, Publ. BG/4.
Hope, G.S., Peterson, J.A., Allison, I. and Radok, U. (eds) (1976) *The Equatorial Glaciers of New Guinea*, Rotterdam, A.A. Balkema.
McVean, D.M. (1974) 'Mountain climates of the Southwest Pacific', in J.R. Flenley (ed.)

Altitudinal Zonation in Malesia, pp. 47–57, University of Hull, Department of Geography, Miscell. Ser. 16.

Himalaya

Ageta, A. (1976) 'Characteristics of precipitation during monsoon season in Khumbu Himal', *Seppyo*, 38, Special Issue, 84–8.

Bleeker, W. (1936) 'Meteorologisches zu den 3 holländischen Karakorum-Expeditionen', *Proc. R. Acad. Amsterdam*, 39, 746–56, 839–45 and 962–70.

Chaggar, T.S. (1983) 'Highest mean annual point rainfall in the world', *Weather* 38, 220–1.

Dhar, O.N. and Narayanan, J. (1965) 'A study of precipitation distribution in the neighbourhood of Mount Everest, *Ind. J. Met. Geophys.*, 16, 229–40.

Dhar, O.N. and Rakhecha, P.R. (1980) 'The effect of elevation on monsoon rainfall distribution in the central Himalayas', in J,. Lighthill and R.P. Pearce (eds) *Monsoon Dynamics*, pp. 253–60, Cambridge University Press.

Dittmann, E. (1970) 'Statistische Untersuchungen zur Struktur der Niederschläge in Nepal', *Khumbu Himal.* 7 (2), 47–60.

Dobremez, J.F. (1976) 'Climatologie', in *Le Népal. Écologie et Biogéographie*, pp. 31–91, Paris, C.N.R.S.

Flohn, H. (1968) *Contributions to a Meteorology of the Tibetan Highlands*, Atmos. Sci. Pap. No. 130, Ft. Collins, Colorado State University.

Flohn, H. (1970) 'Beiträge zur Meteorologie des Himalaya', *Khumbu Himal.* 7 (2), 25–45.

Flohn, H. (1974) 'A comparative meteorology of mountain areas', in J.D. Ives and R.G. Barry (eds) *Arctic and Alpine Environments*, pp. 55–71, London, Methuen.

Higuchi, K. (ed.) (1976) 'Glaciers and climate of the Nepal Himalayas', *Seppyo*, 38, Special Issue.

Hill, S.A. (1881) 'The meteorology of the North-West Himalaya', *Ind. Met. Mem.*, I(VI), 377–429.

Inoue, H. (1976) 'Climate of Khumbu Himal', *Seppyo*, 38, Special Issue, 66–73.

Kraus, H. (1967) 'Das Klima von Nepal', *Khumbu Himal.*, 1(4) 301–21.

Lu, A. (1939) 'A brief survey of the climate of Lhasa', *Q.J.R. Met. Soc.*, 65, 297–302.

Mitsudera, M. and Numata, M. (1967) 'Meteorology of eastern Nepal', *J. Coll. Arts Sci, Chiba Univ.* 5, 75–83.

Nedungadi, T.M.K. and Srinivasan, T.R. (1964) 'Monsoon onset and Everest expeditions', *Ind. J. Met. Geophys.*, 15, 137–48.

Ohata, T., Higuchi, K. and Ikegami, K. (1981) 'Mountain-valley wind system in the Khumbu Himal, east Nepal', *J. Met. Soc. Japan*, 53, 753–62.

Raghavan, K. (1979) 'Summer weather and climate of the Himalaya', *Weather*, 34, 448–54.

Ramage, C.S. (1971) *Monsoon Meteorology*, New York, Academic Press.

Rao, Y.P. (1976) *Southwest Monsoon.* (Meteorol. Monogr., Synoptic Meteorology), Indian Meteorological Department, Delhi, 367 pp.

Reiter, E.R. and Heuberger, H. (1960) 'A synoptic example of the retreat of the Indian summer monsoon', *Geog. Ann..*, 42, 17–35.

Schweinfürth, U. (1956) 'Uber klimatische Trockentäler in Himalaya', *Erdkunde*, 10, 297–302.

Seko, K. (1987) 'Seasonal variation of altitude dependence of precipitation in Langtang Valley, Nepal Himalayas', *Bull. Glacier Res.* (Tokyo), 5, 41–8.

Shresta, M.L., Fujii, Y. and Nakawo, M. (1976) 'Climate of Hidden Valley, Mukut Himal, during the monsoon in 1974', *Seppyo*, 38, Special Issue, 105–8.

Somervell, T.H. and Whipple, F.J.W. (1926) 'The meteorological results of the Mount Everest expedition', *Q.J.R. Met. Soc.*, 52, 131–42.

Troll, C. (1951) 'Die Lokalwinde der Tropengebirge und ihr Einfluss auf Niederschlag und Vegetation', *Bonner Geog. Abh.*, 9, 124–82.

Troll, C. (1972) 'The three-dimensional zonation of the Himalayan system', in C. Troll (ed.) *Geoecology of the High-Mountain Regions of Eurasia*, pp. 264–75, Wiesbaden, F. Steiner.

Ueno, K. and Yamada, T. (1990) 'Diurnal variations of precipitation in Langtang Valley, Nepal Himalayas', *Bull. Glacier Res.* (Tokyo), 8, 93–101.

Yasunari, T. (1976a) 'Seasonal weather variations in Khumbu Himal', *Seppyo*, 38, Special Issue, 74–83.

Yasunari, T. (1976b) 'Spectral analysis of monsoonal precipitation in the Nepal Himalayas', *Seppyo*, 38, Special Issue, 59–65.

Yasunari, T., and Inoue, J. (1978) 'Characteristics of monsoonal precipitation around peaks and ridges in Shorong and Khumbu Himal', *Seppyo*, 40, Special Issue, 26–32.

Wien, K. (1936) 'Die Wetterverhältnisse am Nanga-Parbat während der Katastrophe auf der deutschen Himalaja – Expedition 1934', *Met. Zeit.*, 53, 26–32. (Bermerkungen, M. Rodewald, *Met. Zeit.*, 53, 182–6).

Ahaggar

Dubieff, J. (1963) 'Résultats tirées d'enregistrements automatiques récents dans des stations élevées du Massif Central Saharien', *Geofis. Met.*, 11, 119–25.

Heckendorff, W.D. (1972) 'Zum Klima des Tibestigebirges', *Berlin. Geog. Abh.*, 16, 165–6.

Indermühle, D. (1972) 'Mikroklimatiche Untersuchung in Tibesti-Gebirge (Sahara)', *Hochgebirgsforschung*, 2, 121–42.

Jäkel, D. (1977) 'The work of the field station at Bardai in the Tibesti Mountains', *Geog. J.*, 143 61–72.

Winiger, M. (1972) 'Die Bewölkungsverhältnisse der zentralsharischen Gebirge aus Wettersatellitenbildern', *Hochgebirgsforschung*, 2, 87–120.

Yacono, D. (1968) 'L'Ahaggar, Essai sur le Climat de Montagne au Sahara', *Travaux de l'Institut de Recherches Sahariennes*, 27, Université d'Alger.

The Rocky Mountains in Colorado

Banta, R.M. and Schaaf, C.L.B. (1987) 'Thunderstorm genesis zones in the Colorado Rocky Mountains as determined by traceback of geosynchronous satellite images', *Mon. Wea. Rev.*, 115, 463–76.

Barry, R.G. (1973) 'A climatological transect on the east slope of the Front Range, Colorado', *Arct. Alp. Res.*, 5, 89–110.

Brinkmann, W.A.R. (1973) *A Climatological Study of Strong Downslope Winds in the Boulder Area*, Inst. Arct. Alp. Res., Occas. Pap. no. 7, Boulder, University of Colorado.

Clark, J.M. and Peterson, E.B. (1967) 'Insolation in relation to cloud characteristics in the Colorado Front Range', in H.E. Wright, Jr. and W.H. Osburn (eds) *Arctic and Alpine Environments*, p. 3–11, Bloomington, Indiana University Press.

Diaz, H.F., Barry, R.G. and Kiladis, G. (1982) 'Climatic characteristics of Pike's Peak, Colorado (1874–1888) and comparisons with other Colorado stations', *Mountain Res. Devel.*, 2, 359–71.

Greenland, D. (1978) 'Spatial distribution of radiation on the Colorado Front Range', *Climat. Bull.*, Montreal, 24, 1–14.

Greenland, D. (1989) 'The climate of Niwot Ridge, Front Range, Colorado, U.S.A.', *Arct. Alp. Res.*, 21, 380–91.
Hjermstad, L.M. (1970) *The Influence of Meteorological Parameters on the Distribution of Precipitation across the Central Colorado Mountains*, Atmos. Sci. Pap. no. 163, Fort Collins, Colorado State University.
Hoover, M. and Leaf, C. (1967) 'Process and significance of interception in Colorado subalpine forest', in W.E. Sopper and H.W. Lull (eds) *Forest Hydrology*, pp. 213–23, Oxford, Pergamon press.
Ives, J.D. (1973) 'Permafrost and its relationship to other environmental parameters in a midlatitude, high-altitude setting, Front Range, Colorado Rocky Mountains', in *Permafrost: The North American Contribution to the Second International Conference*, pp. 121–5, Washington, DC, National Academy of Science.
Jarrett, R.D. (1990a) 'Paleohydrologic techniques used to define the spatial occurrence of floods', *Geomorphology* 3, 181–95.
Jarrett, R.D. (1990b) 'Hydrologic and hydraulic research in mountain rivers', *Water Resour. Bull.* 26, 419–29.
Judson, A. (1965) *The Weather and Climate of a High Mountain Pass in the Colorado Rockies*, Fort Collins, US Department of Agriculture, Forest Service, Res. Pap. RM-16.
Judson, A. (1977) *Climatological Data from the Berthoud Pass Area of Colorado*, Fort Collins, US Department of Agriculture, Forest Service, *General Tech. Rep.* RM-42.
Karr, T.W. and Wooten, R.L. (1976) 'Summer radar echo distribution around Limon, Colorado', *Mon. Wea. Rev.*, 104, 728–34.
Maddox, R.A., Hoxit, L.R., Chappell, C.F. and Caracena, F. (1978) 'Comparison of meteorological aspects of the Big Thompson and Rapid City flash floods', *Mon. Wea. Rev.* 106, 375–89.
Marr, J.W. (1961) *Ecosystems of the East Slope of the Front Range in Colorado*, University of Colorado Studies, Ser. in Biol. 8, Boulder, University of Colorado.
Mitchell, V.L. (1976) 'The regionalization of climate in the western United States;', *J. appl. Met.*, 15, 920–7.
Schaaf, C.L.B., Wurman, J. and Banta, R.M. (1988) 'Thunderstorm-producing terrain features', *Bull. Amer. Met. Soc.*, 769, 272–7.

The Alps

Baumgartner, A., Reichel, E. and Weber, G. (1983) '*Der Wasserhaushalt der Alpen*', Munich, Oldenbourg.
Barry, R.G. and Perry, A.H. (1973) *Synoptic Climatology: Methods and Applications*, London, Methuen.
Bénévent, E. (1926) *Le Climat des Alpes Françaises*, Paris, Mémorial de l'Office Nationale Méteorologique de France, no. 14.
Bezinge, A. (1974) 'Images du climat zur les Alpes', *Bull. de la Murithienne*, 91, 27–48.
Bouët, M. (1972) *Climat et Météorologie de la Suisse Romande*, Lausanne, Payot.
Dobesch, H. (ed.) (1983) '*Die klimatologischen Untersuchungen in den Hohen Tauern von 1974–1980*',(Veröff. Òsterreich, MaB-Programs, Band 6), Innsbruck, Universitäts-verlag Wagner.
Ficker, H. von and de Rudder, B. (1943) *Föhn und Föhnwirkung*, Leipzig, Akad. Verlag, Becker u. Erlerkom. Ges.
Fliri, F. (1962) *Wetterlagenkunde von Tirol*, Innsbruck, Universitäts-Verlag Wagner.
Fliri, F. (1971) 'Neue klimatologische Querprofile der Alpen–ein Energiehaushalt', *Ann. Met.*, N.F. 5, 93–7.
Fliri, F. (1974) *Niederschlag und Lufttemperatur im Alpenraum*, Innsbruck, Universitäts-Verlag Wagner.

Fliri, F. (1975) *Das Klima der Alpen im Raume von Tirol*, Innsbruck, Universitäts-Verlag Wagner.

Fliri, F. (1977) 'Das physiogeographische Regionen des Alpenraumes', in F. Wolkinger (ed) *Natur und Mensch im Alpenraum*, pp. 13–26, Graz, Austria, Ludwig Boltzmann-Institute.

Fliri, F. (1982) *Tirol-Atlas, D. Klima* Innsbruck, Universitätsverlag Wagner. 23 plates.

Fliri, F. (1984) '*Synoptische Klimatographie der Alpen zwischen Mont Blanc und Hohe Tauern*', (with Schüepp, M. '*Alpine Witterungslagen und europäische Luftdruckverteilung*') Innsbruck, Universitätsverlag Wagner.

Flohn, H. (1954) *Witterung und Klima in Mitteleuropa*, Stuttgart, Forsch. dt. Landeskunde, 78.

Frey, K. (1953) 'Die Entwicklung des Sud- und des Nordföhns', *Arch. Met. Geophys. Biokl.*, A, 5, 432–77.

Frey, K. (1984) 'Der "Jahrhundertföhn" von 8 XI.81', *Met. Rdsch.*, 37, 209–20.

Havlik, D. (1968) 'Die Höhenstüfe maximaler Niederschlagssummen in den Westalpen', *Freiburger Geogr. Hefte*, 7, Freiburg.

Kerschner, H. (1989) 'Beiträge zur synoptischen Klimatologie der Alpen zwischen Innsbruck und dem Alpenostrand', *Innsbrucker Geographische Studien* 17, 253 pp.

Kirchhofer, W. (1976) 'Stationsbezogene Wetterlagenklassifikation', *Veröff. Schweiz. Met. Zentralanst.*, 34.

Kirchhofer, W. (ed. in chief) (1982) '*Klimaatlas der Schweiz*', Part 1, Zurich: Schweiz. Met. Anst. (Part 2, 1984; Part 3, 1987).

Lang, H. and Rohrer, M. (1987) 'Temporal and spatial variations of snow cover in the Swiss Alps', in B.E. Goodison, R.G. Barry and J. Dozier (eds) *Large-Scale Effects of Seasonal Snow Cover*, Publ, no. 166, Internat. Assoc. Hydrol. Sciences, pp. 79–92, Wallingford, UK, IAHS Press.

Lauscher, F. (1958) 'Studien zue Wetterlagenklimatologie der Ostalpenlander', *Wetter u. Leben*, 10, 79–83.

Liang, G. (1982) 'Net radiation, potential and actual evapotranspiration in Austria', *Arch. Met. Geophys. Biokl.*, B31, 379–90.

Maurer, J. and Lütschg, O. (1931) 'Zur Meteorologie und Hydrologie des Jungfraugebietes', in *Jungfraujoch Hochalpine Forschungstation*, pp. 33–45, Zurich.

Schüepp, M. (1959) 'Die Klassifikation der Wetterlagen im Alpengebiet', *Geofis. pura appl.*, 44, 242–8.

Schüepp, M. (1990) 'Der Einfluss des Appenines auf die Föhnströmung in den Alpen', *CIMA '88. 20° Congresso Internaz. di Meteorologia Alpina*, Servizio Meteorologico Italiano.

Schüepp, M. and Schirmer, H. (1977) 'Climates of central Europe', in Wallen, C.C. (ed.) *Climates of Central and Southern Europe*, pp. 3–73, Amsterdam, Elsevier.

Sevruk, B. (ed.) (1985) 'Der Niederschlag in der Schweiz', *Beiträge, Geologie der Schweiz-Hydrologie*, 31, Bern, Kümmerly and Frey.

Steinhauser, F. (1938) *Die Meteorologie des Sonnblicks, I. Teil*, Vienna, J. Springer.

Tollner, H. (1949) 'Der Einfluss grosser Massenerhebungen auf die Lufttemperatur und die Ursachen der Hebung der Vegetationsgrenzen in den inneren Ostalpen', *Arch. Met. Geophys. Biokl.*, B, 1, 347–72.

Wanner, H. (1979) 'Zur Bildung, Verteilung und Vorhersage winterlicher Nebel im Querschnitt Jura-Alpen', *Geogr. Bernensia*, G, 7.

Wanner, H. and Furger, M. (1990) 'The Bise – climatology of a regional wind north of the Alps'. *Met. Atmos. Phys.*, 43, 105–15.

Witmer, U., Filliger, P., Kunz, S. and Küng, P. (1986) 'Erfassung, Bearbeitung and Kartieren von Schneedaten in der Schweiz', *Geogr. Bernensia* G25.

The maritime mountains of Great Britain

Atkinson, B.W. and Smithson, P.A. (1976) 'Precipitation', in T.J. Chandler and S. Gregory (eds) *The Climate of the British Isles*, pp. 129–82, London, Longman.

Barton, J.S. (1984) 'Observing mountain weather using an automatic station', *Weather*, 39, 140–5.

Bleasdale, A. (1963) 'The distribution of exceptionally heavy falls of rain in the United Kingdom, 1863 to 1960', *J. Instn. Wat. Engrs.*, 17, 45–55.

Bleasdale, A. and Chan, Y.K. (1972) 'Orographic influences on the distribution of precipitation', in *Distribution of Precipitation in Mountainous Areas*, vol. II, pp. 161–70, Geneva, World Meteorological Organization no. 326.

Buchan, A. (1890) 'The meteorology of Ben Nevis', *Trans. R. Soc. Edinb.*, 34.

Buchan, A. and Omond, R.T. (1902) 'The meteorology of the Ben Nevis observations. Pt. II. Containing the observations for the years 1888, 1889, 1890, 1891 and 1892, with appendices', *Trans. R. Soc. Edinb.*, 42.

Buchan, A. and Omond, R.T. (1905) ibid. 'Pt. III. Containing the observations for the years 1893, 1894, 1895, 1896 and 1897, with appendix', *Trans. R. Soc. Edinb.*, 43.

Buchan, A. and Omond, R.T. (1910) ibid. 'Pt. IV. Containing the observations for the years 1898, 1899, 1900, 1901 and 1902' and 'Pt. V. containing the observations for the years 1903 and 1904, with appendix', *Trans. R. Soc. Edinb.*, 44.

Buchanan, J.Y. (1902) 'Abstract of paper on the meteorology of Ben Nevis in clear and in foggy weather', *Trans. R. Soc. Edinb.*, 42, 465–78.

Curran, J.C., Peckham, G.E., Smith, D., Thom, A.S., McCulloch, J.S. and Strangeways, I.C. (1977) 'Cairngorm summit automatic weather station', *Weather*, 32, 61–3.

Glasspoole, J. (1953) 'Frequency of clouds at mountain summits', *Met. Mag.*, 82, 156–7.

Green, F.H.W. (1955) 'Climatological work in the Nature Conservancy', *Weather*, 10, 233–6.

Green, F.H.W. (1967) 'Air humidity on Ben Nevis', *Weather*, 22, 174–84.

Hann, J. von (1912) 'The meteorology of the Ben Nevis Observatories', *Q.J.R. Met. Soc.*, 38, 51–62.

Harding, R.J. (1978) 'The variation of the altitudinal gradient of temperature within the British Isles', *Geog. Ann.*, A. 60, 43–9.

Harding, R.J. (1979a) 'Altitudinal gradients of temperatures in the northern Pennines', *Weather*, 34, 190–201.

Harding, R.J. (1979b) 'Radiation in the British Uplands', *J. appl. Ecol.*, 16, 161–70.

McConnell, D. (1988) 'The Ben Nevis Observatory log books', *Weather*, 43, 356–62 and 396–401.

Manley, G. (1936) 'The climate of the northern Pennines', *Q.J.R. Met. Soc.*, 62, 103–15.

Manley, G. (1942) 'Meteorological observations on Dun Fell, a mountain station in northern England', *Q.J.R. Met. Soc.*, 68, 151–65.

Manley, G. (1943) 'Further climatological averages for the northern Pennines, with a note on topographical effects', *Q.J.R. Met Soc.*, 69, 251–61.

Manley, G. (1945) 'The effective rate of altitudinal change in temperate Atlantic climates', *Geog. Rev.*, 35, 408–17.

Manley, G. (1980) 'The northern Pennines revisited: Moor House, 1932–78', *Met. Mag.*, 109, 281–92.

Mossman, R.C. (1902) 'Abstract of paper on silver thaw at the Ben Nevis Observatory', *Trans. R. Soc. Edinb.*, 42, 525–7.

Omond, R.T. (1910) 'Large differences of temperature between the Ben Nevis and Fort William Observatories', *Trans. R. Soc. Edinb.*, 44, 702–5.

Taylor, J.A. (1976) 'Upland climates', in T.J. Chandler and S. Gregory (eds), *The Climate of the British Isles*, pp. 264–87, London, Longman.

Thom, A.S. (1974) 'Meteorological report', in B.K. Parnell (ed.) *Anonach Moor: a Planning Report on the Prospect of Winter Sport Development at Fort William*, pp. 85–109, Department of Planning, Glasgow School of Art.

St Elias mountains

Alford, D. and Keeler, C. (1969) 'Stratigraphic studies of the winter snow layer, Mt Logan, St. Elias Range', *Arctic*, 21, 245–54.

Benjey, W.G. (1969) *Upper Air Wind Patterns in the St. Elias Mountains, Summer 1965*, pp. 1–50, Montreal, Arctic Inst. of N. Amer., Res. Pap. no. 54.

Brazel, A.J. and Marcus, M.G. (1979) 'Heat exchange across a snow surface at 5365 metres, Mount Logan, Yukon', *Arct. Alp. Res.*, 11, 1–16.

Bushnell, V. and Marcus, M.G. (eds) (1974) *Icefield Ranges Research Project. Scientific Results*, vol. 4, New York, Amer. Geogr. Soc. and Montreal, Arctic Inst. N. Amer.

Bushnell, V.C. and Ragle, R.H. (eds) (1969–72) *Icefield Ranges Research Project. Scientific Results*, New York, Amer. Geogr. Soc. and Montreal, Arctic Inst. of N. Am.

Marcus, M.G. (1965) 'Summer temperature relationships along a transect in the St. Elias Mountains, Alaska and Yukon territory', in *Man and the Earth*, no, 3, pp. 15–30, Boulder, University of Colorado Press.

Marcus, M..G. (1974a) 'Investigations in alpine climatology: The St. Elias Mountains, 1963–1971', in V.C. Bushnell and M.G. Marcus (eds) *Icefield Ranges Research Project, Scientific Results*, vol. 4, New York, Amer. Geogr. Soc. and Montreal, Arctic Inst. N. Amer., pp. 13–26.

Marcus, M.G. (1974b) 'A note on snow accumulation and climatic trends in the Icefield Ranges, 1969–1970', ibid, pp. 219–23.

Marcus, M.G. and LaBelle, J.C. (1970) 'Summer climatic observations at the 5360 meter level, Mt Logan, 1958–1969', *Arct. Alp. Res.*, 2, 103–14.

Marcus, M.G. and Ragle, R.H. (1970) 'Snow accumulation in the Icefield Ranges, St. Elias Mountains'. *Arct. Alp. Res.*, 2, 277–92.

Taylor-Barge, B. (1969) *The Summer Climate of the St. Elias Mountain Region.* Montreal, Arctic Inst. N. Amerc., Res. Pap. no. 53.

Walker, E.R. (1961) *A Synoptic Climatology for Parts of the Western Cordillera*, McGill University, Montreal, Arctic Met. Res. Group, Publ. in Met. no. 35.

Other mountain areas

Bradley, R., Yuretich, R. and Weingarten, B. (1991) 'Studies of modern climate', in: R. Yuretich, ed., *Late Quaternary Climatic Fluctuations of the Venezuelan Andes*, Contrib. no. 65, Dept. of Geology and Geography, Univ. of Massachusetts, Amherst, MA, pp. 45–62.

Davies, T.D., Brimbecombe, P. and Vincent C. (1977) 'The daily cycle of weather on Mount Kenya', *Weather* 32, 406–17.

Dirks, R.A. and Martner, B.E. (1982) 'The climate of Yellowstone and Grand Teton National Parks', Washington, DC, US Dept. of the Interior, National Park Service, Occas. Pap. no. 6.

Erikson, W. (1984) 'Eco-climatological aspects of the Bolivian puna with special reference to frost frequency and moisture conditions', in W. Lauer (ed.) *Natural Environment and Man in Tropical Mountain Ecosystems*, pp. 197–209, Stuttgart, F. Steiner Verlag.

Finklin, A.I. (1983) '*Weather and Climate of the Selway-Bitterroot Wilderness*', Moscow, Idaho, Univ. Press of Idaho.

Finklin, A.I. (1986) '*A Climatic Handbook for Glacier National Park – with data for*

Waterton Lakes National Park', Ogden, Utah, US Dept. of Agriculture, Forest Service, Intermountain Research Station, Tech. Rep. INT-204.

Fliri, F. (1968) 'Beiträge zur Hydrologie und Glaziologie der Cordillera Blanca/Peru', *Alpenkundliche Studien, Veröff. Univ. Innsbruck*, 1, pp. 25–52, Universität Innsbruck.

Flohn, H. (1968) 'Ein Klimaprofil durch die Sierra Nevada de Mérida (Venezuela)', *Wetter u. Leben* 20, 181–91.

Hastenrath, S. (1971) 'Beobachtungen zur klimatologischen Höhenstufung der Cordillera Real (Boliven)', *Erdkunde* 25, 102–8.

Hastenrath, S. (1981) '*The Glaciation of the Ecuadorian Andes*', Rotterdam, A.A. Balkema.

Hastenrath, S. (1984) 'The Glaciers of Equatorial East Africa', Dordrecht, D. Reidel Publ.

Herrmann, R. (1970) 'Vertically differentiated water balance in tropical high mountains – with special reference to the Sierra Nevada de Santa Marta/Columbia', in *World Water Balance*, Internat. Assoc. Sci. Hydrol., Reading, pp. 266–73.

Hurni, H. and Stähli, P. (1982) 'Contributions to the climate', in H. Hurni *Klima und Dynamik der Höhenstufung von der letzten Kaltzeit bis zur Gegenwart* (Hochgebirge von Semien-Äthiopien, vol. II), pp. 37–82, *Geogr. Bernensia G13*.

Kang, X-Ch. and Xie, Y-Q. (1989) 'The character of weather and climate in the west Kunlun Mountains area in summer, 1987', *Bull. Glacier Res.* (Tokyo), 7, 77–81.

Kenworthy, J.M. (1984) 'Climatic survey in the Kenya Highlands with particular reference to the needs of farmers', in W. Lauer (ed.) *Natural Environment and Man in Tropical Mountain Ecosystems*', pp. 23–39, Stuttgart, F. Steiner Verlag.

Kistemann, T. and Lauer, W. (1990) Lokale Windsysteme in der Charazani-Talung (Bolivien)', *Erdkunde* 44, 46–59.

Lauer, W. and Klaus, D. (1975) 'Geoecological investigations on the timberline of Pico de Orizaba, Mexico', *Arct. Alp. Res.*, 7, 315–30.

Lauscher, F. (1977) 'Ergebnisse der Beobachtungen an den nordchilenischen Hochgebirgsstationen Collahuasi und Chuquicamata', *74–75 Jahresbericht, Sonnblick-Vereines, Jahre 1976–1977*, pp. 43–66.

Lo, H. and Yanai, M. (1984) 'The large-scale circulation and heat sources over the Tibetan Plateau and surrounding areas during the early summer of 1979. Part II. Heat and moisture budgets', *Mon. Wea. Rev.*, 112, 966–89.

Monasterio, M. (ed.) (1980) *Estudios Ecologicos en los Paramos Andinos*, Mérida, Edic. Universidad de los Andes, Venezuela.

Ohata, T. and Kang, X. (1990) 'Full year of surface meteorological data at northwestern Tibetan Plateau using an automatic meteorological station', *Bull. Glacier Res.* (Tokyo), 8, 73–86.

Reiter, E.R., Sheaffer, J.D., Bossert, J.E., Smith, E.A., Stone, G., McBeth, R. and Zheng, Q.L. (1987) 'Tibet revisited – TIPMEX-86', *Bull. Amer. Met. Soc.*, 68, 607–15.

Sarmiento, G. (1986) 'Los principales gradientes ecoclimaticos en los Andes tropicales', in *Ecologia de Tierras Altas, Annales del IV Congreso Latinoamericano de Botanico Vol. 1* (Columbia), pp. 47–64.

Thompson, B.W. (1966) 'Mean annual rainfall of Mount Kenya', *Weather* 21, 48–9.

Troll, C. (1968) 'The Cordilleras of the tropical Americas. Aspects of climatic phytogeographical and agrarian ecology;', in *Colloq. Geogr.* 9 (Bonn), 15–56.

Winiger, M. (1981) 'Zur thermish-hygrischen Gliederungs des Mount Kenya'. *Erdkunde* 35, 248–63.

Xu, Y-G, (ed.) (1986) *Proceedings of the International Symposium of the Qinghai-Xizang Plateau and Mountain Meteorology*, Beijing, Science Press, and Boston, MA, Amer. Met. Soc.

Yeh, D.-Z. and Gao, Y-X. (1979) *Meteorology of Qinghai-Xizang (Tibet) Plateau* (in Chinese) Beijing, Science Press.

6

MOUNTAIN BIOCLIMATOLOGY

HUMAN BIOCLIMATOLOGY

The high altitude environment is one of severe stress for man. Air pressure is reduced from its sea level value by 30 per cent at 3000 m and almost 50 per cent at 5000 m (see Table 2.2, p. 27) and, on average, air temperatures decrease from sea level to the same elevations by about 18°C and 30°C, respectively. Visitors to high mountain areas generally notice the oxygen deficiency as a slight breathlessness at about 2800–3000 m, especially when undergoing any exertion, although Tromp (1978) reports physiological effects and therapeutic applications of high altitude above 1500 m. The consequences for permanent residents of high altitudes are quite different, however, since they acquire long-term acclimatization. In fact, low temperatures, snow cover and, therefore, limited food resources rather than oxygen deficiency are the dominant controls of human occupancy in the high mountains (Grover 1974). We will consider first some basic physiological aspects relating to mountain environments in terms of short-term visitors.

Physiological factors and responses

Oxygen deficiency

With increased altitude, there is a reduction in the capacity of the body to take in oxygen which is distributed throughout the body by haemoglobin in the red blood cells. The proportion of oxygen in air remains a constant 21 per cent by volume, but its partial pressure decreases in relation to the total pressure. The partial pressure of oxygen inspired into the lungs (P_IO_2) is determined by the ambient air pressure, reduced by the saturation water vapour pressure at body temperature (37°C), multipled by the oxygen fraction (0.21). Thus, at sea level, $P_IO_2 = 0.21\ (1013 - 63) = 200$ mb; whereas at 5000 m, $P_IO_2 = 0.21\ (540 - 63) = 100$ mb. The human limit of indefinite tolerance to oxygen deficiency, or *hypoxia*, occurs where P_IO_2 falls below one half of its sea level value, but the effects of a hypoxic environment are physiologically significant above 3000 m, where P_IO_2 is 133 mb (Grover 1974). For a given altitude, say 4000 m (neglecting the small numerical difference between geopotential and geometric height), the corresponding Standard Atmosphere value of pressure is approximately 630 mb at latitudes equatorward of 30°, but at 60°N latitude ranges between 593 mb in January and 616 mb in July (see Table 2.2, p. 27). Thus, at an equivalent height, the effect of reduced air pressure is substantially greater in middle and high latitudes, particularly in winter. The passage of a cyclonic system could further lower the pressure by some 20–30 mb, equivalent at the given altitude to a height difference of ≈ 250–375 m. Such latitudinal and seasonal effects may be important for visitors to high mountains, as well as in terms of high-altitude settlements and work enterprises in mid-latitudes.

The effects of extreme low pressure were graphically demonstrated during balloon ascents up to 8800 m by James Glaisher in the 1860s. On one ascent, he lost the use of his limbs and became unconscious, but fortuitously survived. The immediate response to oxygen deficiency is an increase in air volume inspired, or *hyperventilation.* This maintains the quantity of oxygen in the lungs but not in the blood, which is compensated for initially by an increase in the heart rate. The tendency to hyperventilate results in excessive elimination of carbon dioxide from the lungs, which inhibits respiration. This can be overcome temporarily by taking several breaths into and out of a container, such as a paper bag, to return some of the expired carbon dioxide. After a week or so at high altitude, the volume of blood plasma decreases, thereby increasing the concentration of red blood cells and haemoglobin and allowing more oxygen to be transported by a given volume of blood to the body tissues. Even so, the maximum amount of oxygen that the body can consume, which determines the 'aerobic working capacity', declines 10 per cent per kilometre above about 1500 m (Buskirk 1969). Thus, persons at high altitude tire more rapidly, even if acclimatized. Prolonged high altitude exposure leads to an increase in the total volume of red blood cells (polycythemia) which tends to increase oxygen transport, but the greater viscosity may impede blood flow.

The first laboratory for high-altitude research was established at 4560 m on Mt Rosa, Italy, in 1901 and, about the same time, several studies of mountain sickness were published. Hypoxia can lead to mountain sickness, symptoms of which include headache, dizziness, nausea, loss of appetite, and insomnia. Cheyne-Stokes breathing at night, which is an irregular rhythm fluctuating between deep breathing (giving hyperventilation) and then a cessation of breathing (apnoea) for perhaps 10 seconds, is also a common response to hypoxia. Chest pains, coughing and muscular weakness indicate severe mountain sickness. Although such severe symptoms rarely persist longer than a few days, full recovery may require a month of more for some individuals (Heath and Williams 1977: 105). To avoid mountain sickness, a stop of about a week should be spent around 3000 m and at each 1000 m interval above, before proceeding higher. Dietary precautions include a low-fat, high-carbohydrate diet.

A more serious condition is a pulmonary oedema which occasionally results from a rapid move to high altitude and over-exertion. Here, fluid accumulation in the lungs impairs the oxygen transfer into the blood. The symptoms – excessive fatigue, shortness of breath, and a cough – resemble pneumonia. Active young males at moderate altitudes between 3000–4000 m, who have not acclimatized, seem to be particularly at risk. Ward (1975: 284) notes that deaths have occurred, despite oxygen treatment, even in healthy individuals.

The human energy budget and cold effects

The body core, at a normal temperature of about 37°C, transfers heat by conduction to the muscles and skin layers, and by blood circulation to the extremities which are usually some 8°C cooler. The body surface then loses heat by radiation, conduction, convection and evaporation.

The energy budget of an upright person at high altitude has been studied in the White Mountains of California in July by Terjung (1970). Two important results were demonstrated. First, it was shown that the environmental radiant temperature (determined from $T_c = 0.5\ (T_{sky} + T_s)$, where T_{sky} represents an average radiant sky temperature and T_s an average terrain surface temperature) is subject to much greater extremes than the air temperature. Figure 6.1 compares the altitudinal and diurnal variation of these two parameters. Second, the net radiation on an erect figure was found to have a diurnal variation inverse to that of the ground surface, with minimal values around midday (and also in summer). Computed values of net radiation on the body were largest in spring as a result of the high surface albedo associated with snow cover.

The energy budget of a mountain climber in winter clothing, calculated for light winds and weakly positive air temperatures at 3100 m in the Alps, shows that 60 per cent of the energy loss is by sensible heat; radiation, moisture loss and breathing make up the balance in nearly equal amounts (Hammer *et al.* 1989). Nevertheless, in cold, dry conditions, heat loss by conduction and evaporation from the lungs can account for 20 per cent of the total body heat loss (Steadman 1971). Mitchell (1974) notes that the evaporative heat transfer coefficient is proportional to $(1/p)^{0.4}$ and thus it increases with altitude as the air pressure decreases. The dehydration effects of altitude were recognized and described by H.B. de Saussure in 1787. Wind augments the convective heat loss from the skin and this is measured by the *windchill* index of cooling power or by the windchill equivalent temperature. The latter, which is convenient to use here, denotes the effect of a 5 mile h^{-1} (2.2 m s^{-1}) wind on the skin. Table 6.1 illustrates the windchill equivalent temperatures proposed by Siple and Passel (1945). Steadman (1971) developed indices for clothed persons and Dixon (1991) provides a nomograph to relate these to categories of comfort.

Obviously, windchill is a highly significant factor in mountain environments (Smithson and Baldwin 1979). Greenland (1977) shows that at 3750 m on Niwot Ridge, Colorado, the windchill threshold associated with the freezing of exposed flesh (−31°C windchill equivalent temperature = 1400 kcal m^{-2} h^{-1} or

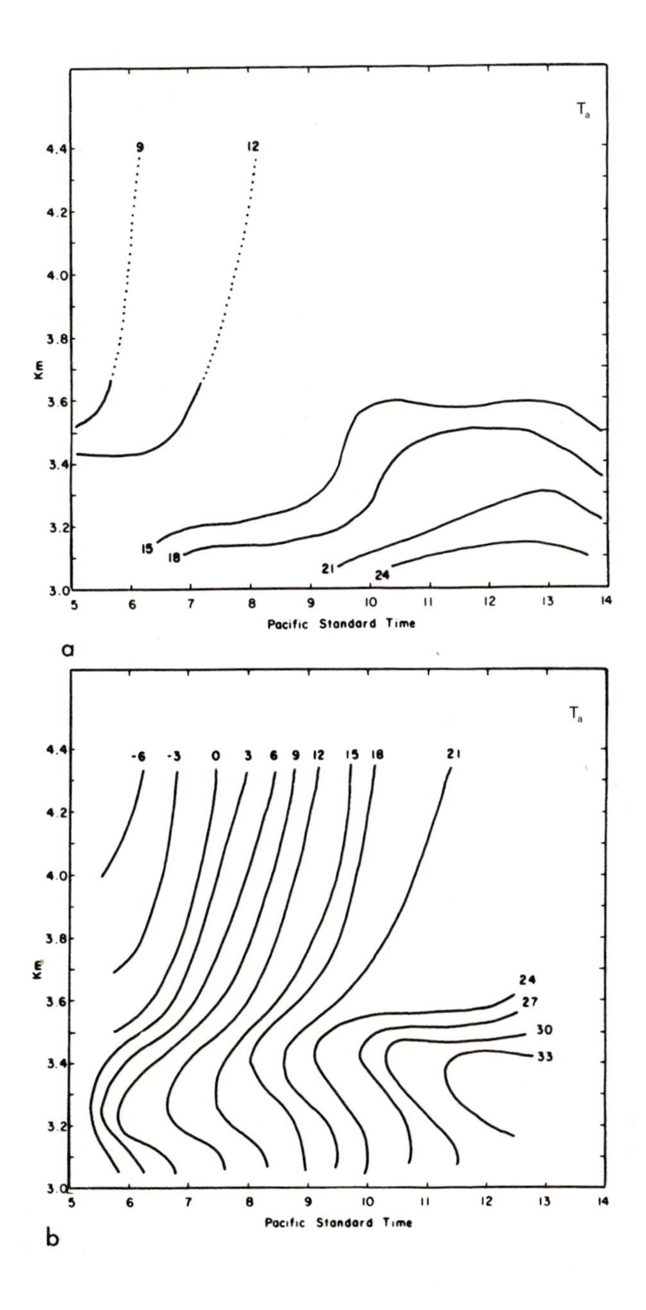

Figure 6.1 Altitudinal and diurnal variations of (*a*) air temperature (°C) and (*b*) environmental radiant temperature (°C) in the White Mountains, California, mid-July
Source: From Terjung 1970

Table 6.1 Windchill equivalent temperatures

Air temperature (°C)	*Wind speed ($m s^{-1}$)*		
	2.5	*10*	*20*
0	0	−12	−18
−10	−11	−26	−33
−20	−21	−40	−48
−40	−41	−68	−78

1628 Wm^{-2} windchill index) occurs more than 50 per cent of the time, December through February). In upland Britain, Baldwin and Smithson (1979) find a linear relation between mean annual windchill (W) and elevation (h). Expressed in Wm^{-2} (where h is in metres),

$$W = 0.456\ h + 394$$

Monthly averages in January are 1308 Wm^{-2} on Ben Nevis (1343 m) and 814 Wm^{-2} at Moor House (561 m), with corresponding hourly extremes of 3925 $W\ m^{-2}$ and 1907 $W\ m^{-2}$, respectively. In windy wet conditions, which are common at low elevations in maritime mountain areas, the cooling effects may be even greater (see below under *clothing*).

Cold-sensitive receptors are located primarily in the skin, and also in abdominal viscera, the spinal cord, and hypothalamus (Webster 1974a). The last of these monitors cold signals and initiates body responses (Van Wie 1974). Cold stress induces several body reactions in the attempt to maintain body temperature. A common and noticeable response is shivering, which serves to increase muscular heat production four- or five-fold (Carlson 1964), although much of this is lost by convection from the skin. Another prompt reaction to cold exposure is the constriction of blood vessels (vasoconstriction) especially in the extremities. This is illustrated in the schematic temperature metabolism curve for homeotherms in general, shown in Figure 6.2. However, at air temperatures below 0°C, blood vessels in the extremities must dilate in order to prevent freezing; this 'cold-induced vasodilation' [Webster (1974a: 59–64) gives a general account of it in homeotherms] increases the sensible (conductive and convective) heat loss. Increases in evaporative loss at very low temperatures (see Figure 6.2) are a result of the increase in ventilation rate associated with the increased metabolic rate.

When the body cannot maintain its core temperature, the condition of *hypothermia* results. Unconsciousness occurs when the core temperature falls to about 30°C and the heart generally ceases at 26°C (Ward 1975: 309).

Recurring, or prolonged, exposure to cold, damp conditions commonly results in swelling and irritation of parts of the skin on the hands and feet, forming chilblains or in more severe cases, immersion foot. If the skin temperature locally drops below freezing, the tissues freeze ('frostbite') (Smith 1970). Following an

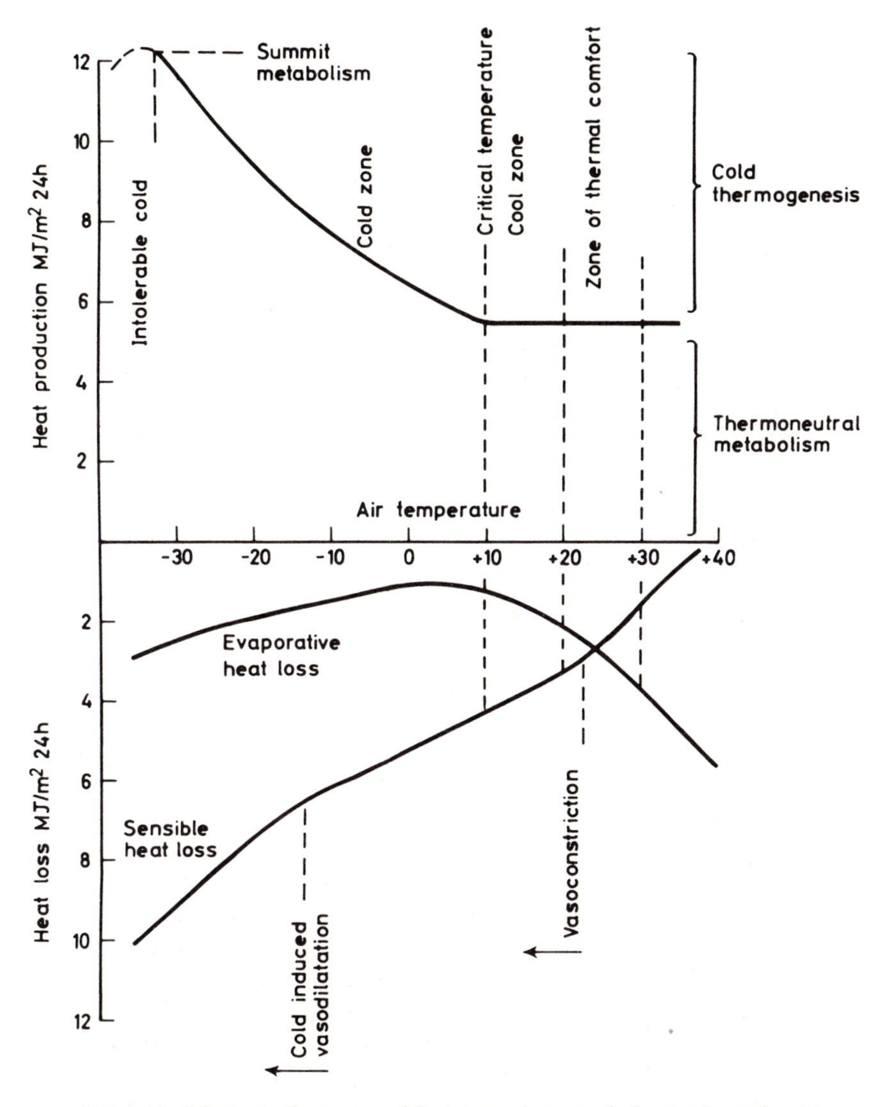

Figure 6.2 Schematic illustration of the heat exchanges of a homeotherm for air temperatures between −40° and +40°C
Source: From Webster 1974a

initial localized burning sensation, the affected area goes numb. If untreated, the area may turn black and develop gangrene. Various medical studies indicate that *rapid* re-warming minimizes tissue loss. Ward (1975) recommends that generalized warming, for example by intake of hot liquids, be combined with local warming to the injury. The latter could take the form of 20-minute applications of a container of water at 44°C, or body contact with another individual. He emphasizes also that surgery is seldom necessary, although complete recovery of an affected part may take from six to twelve months.

Combined influence

High altitudes, especially in winter in middle and high latitudes, combine the problems of cold, dry air and hypoxic environment. Exposure of newborn infants to such stresses may be a special hazard although little is known about this. Ward (1975: 296–9) states that cold injury appears to be more common, for equivalent temperatures, at high altitude. For example, fatigue due to hypoxia, and weight loss if a high-altitude stay is prolonged, decrease the heat that can be produced by activity; greater ventilation augments heat loss from the lungs, and hypoxia may impair judgement with regard to critical behavioural adjustments (such as clothing worn, etc.). However, the cold, dry atmosphere at high altitudes does facilitate the removal of body heat generated by physical exercise and this allows a higher work rate (Lahiri 1974).

Another combined effect at high altitude results from the presence of a snow cover together with the augmented ultraviolet component of solar radiation. The high surface albedo causes intense glare, which can lead to snow blindness, as well as severe facial sunburn, unless protected.

Clothing

The thermal insulation available to a clothed individual comprises three components due to skin tissue, the air, and clothing. The scientific study of cold-weather clothes was given a considerable boost by military needs during the Second World War. Polar expeditions and the growth of high-altitude climbing and winter sports have also contributed to this interest.

The insulation provided by clothing is commonly expressed in units of 'clo'. One clo maintains a seated person at a comfortable temperature indefinitely in an environment of 21°C, air movement of 10 cm s^{-1}, and a relative humidity below 50 per cent (assuming a basal metabolic rate of 58 W m^{-2} (50 kcal $m^{-2}h^{-1}$). Polar or alpine expedition clothing may have a clo value of 3 or more, while the wool garments (excluding poncho or hat) of the Quechua Indians on the Alotplano have a clo value of 1.2–1.4 (Hanna 1976). Calculated requirements at 3100 m for climatic conditions in the Alps at 1300 m range from 2.5 clo in summer to 4.5 clo in winter for moderate activity (climbing 260 m altitude/hour with 1 km hr^{-1} wind; 140 W energy) to 2.0 clo in summer to 2.5 clo in winter for vigorous activity (climbing 520 m/hour with 2 km hr^{-1} wind; 273 W energy) (Hammer *et al.* 1989). For comparison, the clo value of skin tissue may range from 0.3 when exercising, to 0.55 at rest, and that of the air from 0.1 in strong winds to 1.0 when the air is still (Pugh 1966). Weather conditions have a major effect on the insulative properties of clothing. Pugh notes that the insulation of a wind-proof parka may be reduced 30–40 per cent with winds of 12 m s^{-1}.

Clothing with a normal insulation of 1.5 clo may have a value of only 0.2–0.5 when wet.

Apart from the thickness of clothing it is also worth considering its distribution in relation to body heat production and loss. When man is active, the skin and muscles generate 73 per cent of the body's heat, the chest and abdomen 22 per cent, and the brain 3 per cent (van Wie 1974). However, when resting, the brain produces 16 per cent of the total and the chest and abdomen 56 per cent. Clearly, the head and trunk especially should be well protected when an individual is inactive (or injured).

Weather and human comfort

There is a long literature, especially in European countries, on physiological reactions to föhn conditions in and around the Alps (Schmidt 1930). One of the few well-documented investigations of such weather-sensitive reactions was carried out by Richner (1979) in and around Zurich, Switzerland, during south-föhn events. Typically, warm southerly flow crossing the Alps overrides a pool of cold air on the north side creating a sharp temperature discontinuity and wind shear at the interface. Gravity wave propagation sets up pressure fluctuations of about 1 mb with periods of 4–30 minutes. Based on correlation analysis of such events (detected with microbarographs), and of physiological responses for groups of individuals who believed in or disbelieved such effects, Richner (1979) showed an increase in specific ailments and a corresponding decrease in an index of 'well-being' as the amplitude of the pressure fluctuations increased. At the same time, no relationships were found with shorter period (< 4 min) acoustic infra-sound waves. Headaches (the most frequent ailment) also showed no weather relationships. The survey showed the effects to be stronger for women than men, and stronger for workers in air-conditioned buildings, where the pressure fluctuations are also found to be present. The link between pressure fluctuations and human physiology is still hypothetical, but may involve an instability in the normal physiological control system that enables humans to adjust to temporal and altitudinal variations in ambient pressure.

Adaptations

It is appropriate to begin with some definitions. *Adaption* – morphological, physiological and behavioural – denotes an adjustment that is beneficial to well-being and to survival in a particular environment (Webster 1974b; Little 1976). *Acclimatization* is a phenotypic adjustment to a complex of climatic conditions that may or may not be reversible. (*Acclimation* is restricted to changes induced by *one* environmental factor in a controlled experiment (Hart 1957).) While the effects of low pressure and low temperature are not wholly separable in high altitude environments, it is convenient to discuss them in turn.

High altitude

The special capabilities of native populations, such as the Himalayan Sherpas and the Andean Quechua Indians, to live and work effectively at altitudes in excess of 4000 m has been a topic of physiological interest since the 1920s. The available evidence indicates that these peoples develop special features by acclimatization during a lifetime spent at high altitude (Baker 1969; Hock 1970; Baker and Little 1976; Reeves *et al.* 1981). Mountain peoples tend to have a high breathing rate, and high concentrations of red blood cells and haemoglobin. In the Andes, they also have a large chest and lung volume, although this may be a racial trait. According to Grover (1974), their physiology resembles that of lowland athletes; the native likewise having a slow heartbeat and a high aerobic working capacity. Lahiri (1974) emphasizes that Andean Quechua ventilate *less* than acclimatized lowlanders and that this hyposensitivity persists for at least several years if they move to sea level. He cites data for natives in the Andes and Himalaya, which indicate that hypoxic insensitivity develops in childhood, but studies at Leadville (3100 m), Colorado, suggest that this trait increases with duration of high altitude residence, especially beyond 10–12 years. There appears to be no evidence of this occurring in high altitude animals, however. Baker (1976) notes that the Quechua of Nuñoa in southern Peru have almost no cardiovascular diseases, although respiratory diseases are common. Also, it is now well-established that the high-altitude environment of the Altiplano is a cause of reduced fertility in Nuñoa women and of slow rates of physical maturation in the children (Little and Baker 1976). Nevertheless, the work capacity of Andean natives is identical to that of lowland natives and substantially higher than that of European newcomers, even after a year's acclimatization (Baker 1976; Little 1981). The question of a genetic adaptation is still not fully proven and Heath and Williams (1977: 227) favour the view that the Quechua have natural acclimatization with partial adaptation (voluminous chests).

Even permanent residents of high mountains can sometimes lose their ventilatory acclimatization to the hypoxic environment. Severe arterial hyoxia develops, causing low oxygen pressure in the lungs, and stimulating excessive polycythemia. This state, known as chronic mountain sickness or Monge's disease (*soroche* in the Andes), can lead to congestive symptoms and heart failure if the individual remains at high altitude. About 80 cases have been documented, three-quarters of them in males, with a 12 per cent mortality rate (Ward 1975: 270). Many recorded cases are from the Andes, and Heath and Williams (1977: 150) consider that the disease may be a clinical syndrome rather than a pathological entity. Other cases are recorded, however, from Leadville, Colorado (R.F. Grover, personal communication, 1980).

Sub-freezing temperatures

Acclimatization to cold has been most studied in polar regions and experimental situations. It is a controversial subject, but a detailed review by Webster (1974b) argues that certain adaptations are observed. He specifies the different criteria that must be met to establish some type of cold acclimatization. The three major ones are (1) a reduced heat loss by increased tissue insulation; (2) a reduced susceptibility of the extremities to pain or cold injury; and (3) a reduced cutaneous sensory threshold. There seems to be no firm evidence that the thermoneutral metabolic rate (TMR) can be modified by acclimatization. This rate is raised approximately 10 per cent above the basal metabolic rate by food intake and activity (Webster 1974a). While it has been claimed that the Eskimo show metabolic acclimatization, their high metabolic rates (e.g. Folk 1966) are due to their large food intake.

Australian aborigines and Kalahari bushmen can sleep in the open even when night temperatures cause the body heat loss to exceed the TMR. Heat production is increased by shivering, while heat losses are reduced as a result of the high tissue insulation and generalized vasoconstriction. Nevertheless, they can tolerate some degree of hypothermia as shown by the nocturnal decrease of rectal temperature (Webster 1974b). This response may be called hypothermic acclimatization, although Webster prefers the term metabolic habituation. (Habituation is a gradual (quantitative) decrease in physiological response due to a repeated stimulus (Webster 1974b).) Tromp (1978) suggests that exposure to (simulated) high altitude, above 2000 m, can improve thermoregulatory efficiency in asthmatic and rheumatic patients.

Much more common, and of greater interest from the view point of mountain environments, is the tolerance of severe local cooling of hands of feet. This has been identified in Eskimos, Arctic Indians, polar explorers, groups of fishermen, and Quechua Indians of highland Peru. Research with the last group (Little 1976) suggests the maintenance of a high heat loss by enhanced blood flow to the extremities (and perhaps a 5 per cent increase in basal metabolism). Gloves and footwear are not worn and Hanna (1976) considers that adaptation must occur during childhood. Toe temperatures as low as 20°C were recorded in children, who often spend long hours watching and herding the animals. Webster (1974b) refers to this response as vasomotor habituation since the usual vasoconstriction response is diminished.

The Quechua also display behavioural adaptations to their environment. Air temperatures in the coldest month fall several degrees below zero at 4000 m and indoor temperatures in the unheated stone houses may be only 4°C. Families usually sleep in groups of two or more to combat the cold. However, Hanna (1976) notes that some cold stress appears to affect children in the early evening before they go to bed.

WEATHER HAZARDS

Many features of the mountain climatic environment present a hazard to the unwary or inexperienced. Apart from the obvious problems caused by fog due to orographic cloud, strong winds and low temperatures, the general rapidity of weather changes in mountains is a factor not always taken into account. Snow and hail showers can occur on most mid-latitude mountains in the summer months to the surprise of casual tourists from the lowlands. Here, however, we consider two specific weather-related hazards characteristic of different seasons.

Lightning

Lightning is both an individual hazard to hikers and climbers in summer and a major cause of forest fires. Approximately 70 per cent of all forest fires in the western United States are attributed to lightning, and in a 20-year period in the thirteen western states and Alaska, there were 132,000 lightning-caused fires (Fuquay 1962). Based on studies in Montana (Fuquay 1980), the US Forest Service has developed a guide to identify and forecast lightning activity level (LAL) in connection with fire hazard. This is shown in abbreviated form in Table 6.2.

The unit of area considered is a square approximately 80 km on a side, corresponding to the largest area over which lightning activity can be effectively monitored from a fire lookout point. Storms become more intense up to LAL 5, although the area covered by the storm does not increase at the same rate; even with LAL 5, measureable precipitation usually affects less than half of a forecast area. The relationship between lightning and precipitation is not well documented, but Table 6.2 indicates that lightning activity is related to cloud development, as measured by the maximum height of radar echoes. In western

Table 6.2 US Forest Service guide to lightning activity level (LAL)

LAL	*Cloud conditions*	*Average cloud-ground lightning rate (min^{-1})*	*C-G lightning density ($(6500\,km^2)^{-1}$)*	*Maximum radar echo height (m a.s.l.)*
1	No thunderstorms	–	–	–
2	Few towering Cu	$\leqslant 1$	20	<8500
3	Scattered Cu, occasional Cb	Max. 1–2	40	7900–9700
4	1-3/10 Cu, Cu congestus	Max. 2–3	80	9100–11000
5	Extensive Cu congestus, moderate-heavy rain with Cb	Steady flashes at some place during storm; max. >3	160	>11000
6	Scattered towering Cu, high bases; virga common	$\leqslant 0.5$	–	–

Source: After Fuquay 1980

Montana, cloud tops need to be above about 6500 m for thunderstorms to occur.

Lightning density (or risk) increases geometrically with LAL value, LAL 6 is a special category for dry storms which create a high risk of lightning-caused fires. Cloud-to-ground (C–G) flashes occur in an approximately 1:4 ratio to total lightning activity.

In the Rocky Mountains near the Montana–Wyoming border, there are about 44 days with thunderstorms in July–August. This gives rise to 6 C–G flashes per 10 km^2 near Helena decreasing northward to 2 per 10 km^2 at Missoula and only 0.25 per 10 km^2 in Glacier National Park (Fuquay 1962). The average frequency of LAL classes on thunderstorm days in western Montana during summers 1965–7 was 35 per cent each for classes 3 and 4, 18 per cent LAL 5, 10 per cent LAL 2, and 2 per cent LAL 6 (Fuquay 1980). Cloud-to-ground strikes in the western United States increase four-fold between 500 m and 1700 m, but then show little change up to 3000 m altitude, based on two summers' data analysed over 2200 km^2 grid blocks (Reap 1986).

Radar studies in Colorado show that on the large scale there are preferred areas for convective cells to develop (Karr and Wooten 1976; and see p. 314). In the morning, such areas are over the east slopes of the Front Range, with subsequent growth taking place over the mountains. In the late afternoon, development takes place over ridges that extend eastward from the Rocky Mountains into the plains.

On a local scale, lightning strikes are strongly related to the terrain profile. Exposed ridges and summits are widely recognized to be dangerous localities due to their build-up of charge and the usually preferred path of lightning to the nearest high point. Protruding objects sometimes release streamers of current, referred to as coronal discharge (St. Elmo's fire), and in the case of a standing person, this may be dramatically illustrated by the hair literally standing on end! If thunderstorms are nearby, this is a clear hazard warning. Any slope convexity is more prone to a direct lightning strike than a level or concave surface; but even in such sites, ground currents also pose other less well-known dangers (Peterson 1962). When a projection is struck by lightning, the current seeks the path of least electrical impedance. On rock, this path is generally over the surface and downward. Short gaps and crevices tend to be jumped so that it is dangerous to shelter in narrow, vertical gullies or small holes or beneath rock overhangs. These can act as 'spark gaps', which the body bridges for the current path. Large isolated rocks should also be avoided. The recommended procedure for anyone caught out in a thunderstorm is to crouch down in an open slope concavity. One should be at least the body height away from any adjacent cliff or large rock (Peterson 1962). The zone of protection from direct strikes at the base of a 20 m high rock wall is between 2 m and 20 m from the cliff. Insulation between the body and the ground can be obtained by sitting on a rucksack, coil of rope, folded dry clothing or a small *loose* rock.

First aid for lightning injury involves treatment for burns and electric shock.

External chest massage may be necessary in cases of heart fibrillation, and mouth-to-mouth artificial respiration if breathing has ceased.

Snow avalanches

For anyone adequately clothed against windchill effects, the most serious wintertime hazard in mountain areas is the snow avalanche. Most alpine countries operate some form of warning service, based on weather forecast information and snow survey data, for mountain travellers on roads or cross-country ski trails. In the European Alps protective structures such as fences and snowsheds are a prominent feature of the mountain landscape. Even so, property damage and deaths result each year in mountainous areas from avalanche occurrences. Austria, Japan and Switzerland each report an average yearly death-toll of between 25 and 36 victims. Property damage in the United States in 1967–71 was $250,000 per year (Williams 1975) and these figures are increasing with more winter sports activity and the development of mountain areas.

Avalanches form when the snowpack resting on a slope undergoes failure. New dry snow can cling to 40° slopes, whereas wet slushy snow may slide even on 15° slopes. The critical angle of response depends on the temperature and density of the snow, which determine its texture and wetness. Overloading due to newly fallen or wind-blown snow deposition, or to the weight of a skier, may cause the critical angle on a particular slope to be exceeded. Alternatively, physical processes in the snowpack may lead to changes in its structure and cohesion. Failure can occur near the surface, in which case a small mass of snow slips downslope leaving an inverted v-shaped scar. This gives rise to a *loose-snow avalanche*. Most of these are relatively minor 'sluffs', but even so they may be hazardous to an individual skier or mountaineer. The second type of failure involves the fracture of a slab perhaps 1 m or more thick on slopes of 20–45°. *Slab avalanches* may involve dry or wet snow, but both are associated with shear stresses in the snow exceeding the shear strength in some underlying layer. Dry snow falling at $\geqslant 30$ m s^{-1} over a long path may generate an airborne *powder avalanche* ahead of the sliding snow. This powder avalanche travels at high velocities as an atmospheric turbidity current, and causes damage by wind blast up to 100 m beyond the limit of the run-out zone. Details and illustrations of these processes may be found in Mellor (1968), Perla and Martinelli (1975), and Perla (1980).

An avalanche track comprises an upper starting zone, the main track itself which is often clearly delimited by a swath of grassy or shrubby vegetation running down below tree line into the montane forest, and a lower run-out zone which may have a more or less well-marked debris fan at the foot of the slope. Defence structures include snow fences and nets in the starting zone, to help maintain the stability of the snowpack, and deflecting structures (walls and snow sheds) as well as energy-absorbing structures such as concrete or earth mounds, in the run-out zone (Frutiger 1977). In areas of high risk, walls of buildings must be designed to withstand perpendicular pressures of 3 t m^{-2} (Aulitsky 1978). Avalanche mapping and land-use zoning is now being applied in many mountain areas to minimize these problems, but the hazard remains for skiers and other back-country travellers.

Interest here is primarily with the weather phenomena which favour avalanche situations and this aspect is less well understood than the snow mechanics. For the San Juan Mountains of south-west Colorado, Bovis (1977) identified different key meteorological variables for dry- and wet-snow avalanches by discriminant analysis. Dry-snow avalanches are particularly associated with high snowfall amounts over the preceding four days and with wind redistribution of the snow 12–24 hours before the event. Wet-snow slides, which mainly occur in spring, are associated with the antecedent air temperature values. Another situation which is common in dry continental interiors, such as the Rocky Mountains, but less usual in the maritime coastal ranges of North America, is the formation of depth hoar by temperature-gradient metamorphism in the snow. The existence of steep temperature gradients in a snowpack during clear cold weather causes upward transfer of water vapour producing larger and weaker grains. New snow deposition on such an unstable base may cause collapse in the depth hoar layer triggering a slab avalanche. In the San Juan Mountains, Colorado, the large diurnal range of winter radiation and temperature regimes creates a distinctive 'radiation snow climate', where temperature-gradient metamorphism is a factor in most major observed avalanches (LaChapelle and Armstrong 1976).

Avalanche forecasting is based on information on existing snow pack characteristics and on weather conditions – particularly air temperature in the starting zones, windspeeds over the ridge crests, and precipitation amounts and rates. In Colorado, orographic precipitation is now predicted using the numerical model developed by J.O. Rhea, described on p. 245, adapted for 12 hour intervals (Judson 1976, 1977). Critical total water equivalents for snow accumulation for avalanche occurrence are of the order of 2.0–2.5 cm (Perla and Martinelli 1975). Snow loading on slopes is also associated with wind transport, although this factor is mainly important in association with precipitation events. For lee slopes, winds of 5 m s^{-1} are required according to Perla and Martinelli (1975), although this figure seems low compared with theoretical values of the threshold speed for blowing snow transport by turbulent diffusion (p. 267). The development of cornices, where air motion decelerates immediately in the lee of a ridge crest, is an important feature of the snow cover, since cornice collapse will often trigger an avalanche if the mass of snow falls onto the slopes below.

AIR POLLUTION IN MOUNTAIN REGIONS

It is ironic that earlier in this century sanatoria were located in alpine areas because of the purity of mountain air. Today, many mountain valleys face problems of air pollution associated with the development of residential and tourist settlements, industrial establishments and power stations. Emissions from such sources are concentrated by valley inversions at night and more persistently in winter months. Major disasters, as a result of intense valley air-pollution episodes in the Meuse valley, Belgium, in December 1930 that caused 63 deaths, and at Donora, Pennsylvania, in late October 1948 (21 deaths) were among some of the catalysts of research on pollutant transport. Until recently, most studies of effluent dispersion had been carried out in level open country where simple diffusion theories are readily applicable, but measurements of pollution transport from sources in mountain valleys show the necessity for more complex models in order to make reliable predictions relating to proposed new developments.

There is a lack of basic long-term data on the altitudinal variation of aerosols and trace gases in mountain regions. A monitoring programme in southern Bavaria provides illustrative results, however (Reiter *et al.* 1987; Reiter 1988). The measurements show that NO_x levels at Wank Peak (1780 m) are 3–5 ppbv compared with 5–15 ppbv at Garmisch (740 m). SO_2 levels in winter at the mountain site (3–5 ppbv) are also about half those in the valley (5–10 ppbv), but during April–September SO_2 levels are similar (~ 3 ppbv) at both stations, apparently as a result of comparable long-distance transport and convective mixing. An observed positive correlation between SO_2 amounts and concentrations of cloud nuclei, as well as higher levels of sulfate ion during fog conditions at Wank Peak, leads Reiter to point out the possibility that the greater acidity of cloud droplets, compared with rainfall, may accentuate tree damage in zones of fog deposition.

Controls of atmospheric diffusion in complex terrain

The dispersion of pollutants in mountain areas is influenced by the thermal structure of the atmosphere, as well as by the terrain-induced air motion. The following factors are involved to a greater or lesser degree in most locations (Barr *et al.* 1977; Greenland 1979):

1. The deformation and channelling of the streamlines over and around obstacles.
2. The separation of the airflow from the surface over a break of slope or an obstacle (Scorer 1978: 107).
3. Lee waves and rotors and internal gravity waves at an inversion interface.
4. Slope winds and mountain/valley wind systems.
5. The pattern of inversion break-up.
6. Enhanced convection due to differential slope heating.

Four situations where rugged terrain may exacerbate pollution problems from sources in valleys or basins are recognized by Hanna and Strimaitis (1990):

1. plumes impinging on valley sides and hills in stable conditions;
2. cold air pools in valleys and basins where vertical diffusion is minimized;

3 nocturnal drainage flows concentrating pollutants in valleys and basins;
4 valley channelling of gradient winds.

The characteristics of topographically and thermally induced airflows, discussed in Chapter 3, are clearly relevant to the problem of pollutant dispersal, but existing theories seldom prove adequate to make predictions in complex terrain. Special observation programmes are beginning to provide new insights into airflow and pollutant behaviour in particular locations and, although it is too early to generalize such results, it is nevertheless worth illustrating some of the findings.

Studies of katabatic flow over a simple slope in the Geysers area, California (Nappo *et al.* 1989), show that pollutants released at the ground can spread through the depth of the flow, especially over shallow slopes. When releases are elevated above the drainage layer they can be entrained into the layer and so diffused to the ground. Surface concentrations can be high near the source, if a stack releases within the drainage flow, as a result of subsidence in the flow.

Observations with tracers (perfluorocarbon, sulphur hexafluoride) in the Brush Creek Valley, Colorado (see Chapter 3, p. 177 also) show how pollutants are transported and dispersed in valley wind systems. Tracers released at the floor and at 200 m from a balloon, showed well-defined plumes when entrained into the downvalley flow. The external winds appeared to have little effect (Gudiksen and Shearer 1989). Within minutes of sunrise on the valley wall, upslope flows cause ventilation out of the valley to begin (Orgill 1989; Whiteman 1989). This may continue for 2 to 3 hours.

It is usual for the base of inversions to rise during the day. The role of surface heating and entrainment of air from the stable layer above the inversion base into the convective boundary layer was first described by Ball (1960) and subsequently modelled by Carson (1973) and Tennekes (1973). The height of the inversion base is calculated as a function of the sensible heat flux at the surface. Models of the rise of inversions have been incorporated into 'box models' of pollution in mountain valleys (Tennekes 1976; Howard and Fox 1978; Greenland 1979). Such models ignore the details of plume behaviour and examine the total changes in the valley air. Since the walls of the valley prescribe the sides of the box and an elevated inversion is the lid, these simple models have found some success. Greenland (1979) reports that the Tennekes model of inversion rise agreed well with observations made with an acoustic sounder during December 1975–March 1976 at Vail, in the Gore Valley, Colorado. Within this period, observations using a tethered-balloon system demonstrated some unusual features of inversion break-up on 10 December, 1975. Whiteman and McKee (1977) describe how the top of a nocturnal inversion layer descended into the valley, lowering from about 450 m at 8 a.m. to 100 m at 11.30 a.m. They hypothesize that slope heating forms a thin superadiabatic layer from which convective plumes penetrate into the cold stable air above. Entrainment of this valley air is initiated by the plumes along the valley slopes and, as the cold air is removed up the slopes, mass continuity causes the top of the inversion layer to descend. Up-valley winds above the inversion deepen into the valley and assist in eroding the top of the cold air until the inversion finally dissipates (cf. Davidson and Rao 1963). The mechanism of inversion break-up will have important effects for pollutant dispersion (Whiteman and McKee 1978). Assuming that an elevated source has concentrated pollutants in the stable air over the valley floor, growth of the convective boundary layer leads to *fumigation*, or downward transport, of these pollutants to the valley floor through convective mixing of the boundary layer and the stable air above. This sequence is probable in wide shallow valleys where slope flows are less effective in removing air from the centre of the valley. If the convective boundary layer grows slowly, the pollutants sink as the core of the stable layer descends, producing high concentrations at ground level. This pattern is likely over snow-covered terrain. Once the pollutants are entrained into the boundary layer, in this second case, they are advected up the valley sides and dispersed into the air above.

In a study of Huntingdon Canyon, Utah, tracer releases (oil fog and sulphur hexafluoride gas) were used to assess the nature of atmospheric transport and diffusion (Start *et al.* 1974, 1975). The canyon is steep-walled and 400–500 m deep. Releases were made from the 183 m stack of the power plant in the canyon entrance, during lapse to neutral stability conditions, and from sites on the canyon floor or walls during inversions. It was found that in every case of stack release, samples from the canyon floor and walls are 1.5–3 times more dilute than concentrations from the plume centreline and that the plume concentrations are less than those expected theoretically over flat terrain (Pasquill 1961; Scorer 1978, Chapter 2). Under inversion conditions, turbulent mixing produces nearly uniform concentrations along the canyon floor and these concentrations are closer to those expected over flat terrain. Start *et al.* (1975) also examined the characteristics of plume impaction on the canyon walls. They conclude that filament-like plumes are unlikely to occur in steep-walled canyons. The evidence for enhanced mechanical turbulence provided by these observations suggests three special airflow phenomena. These are: turbulence generated around the adjacent summits, with roll eddies which transport momentum down into the canyon; helical circulations triggered by the interaction between katabatic slope winds and air drainage from side canyons; and small-scale wake turbulence associated with airflow around and over protrusions of the canyon walls (see Figure 6.3) (cf. Davidson 1963). The occurrence of roll eddies with cross-canyon flows has also been reported by Clements and Barr (1978) in Los Alamos Canyon.

Other investigations have been carried out by Start *et al.* (1974) at the Garfield copper smelter near the south shore of the Great Salt Lake, Utah, and adjacent to the rugged Oquirrh Mountains. The ground rises 600–1000 m, within 3 km south of the smelter. Gas tracer studies again show dilutions 2–4 times greater than would be estimated for flat terrain in the elevated plume centrelines. Lateral plume spreading is also about twice as great as over flat

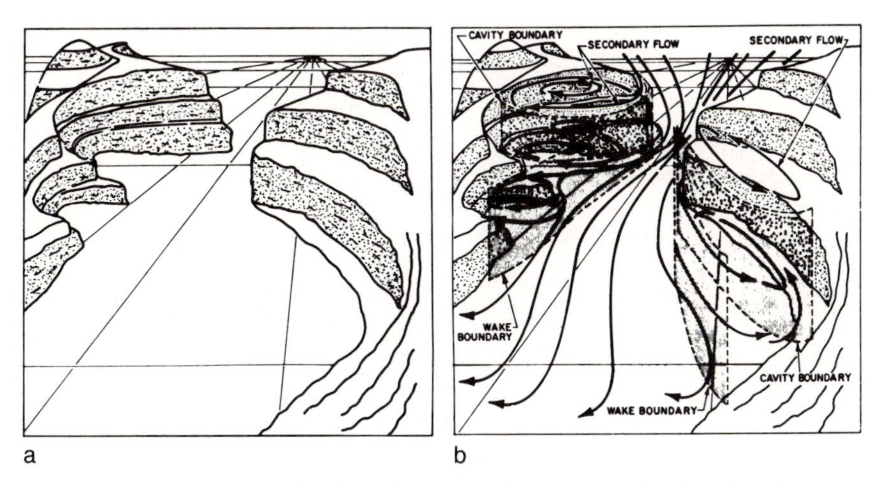

Figure 6.3 Schematic model of turbulent wake effects associated with airflow along a canyon
Source: From Start *et al*. 1975

terrain. Under lapse conditions, plumes are deflected aloft over the ridges and ground-level concentrations are much smaller than those measured aloft. With a strong stable layer just above the ridge tops, a pollution plume may flow in a shallow layer over the mountains. Rapid vertical mixing leads to nearly uniform vertical concentrations downwind, but ground-level concentrations may be up to twice those in the air layer due to ground reflection effects (Zone 4 in Figure 6.4).

Whether a pollutant plume will impact on the front surface of terrain downwind, as sketched in Figure 6.4, or whether it crosses the hilltop, depends on the height of the 'dividing streamline', where the flow crosses the barrier, rather than being deflected around it or blocked (Chapter 2, p. 67). Experiments with a laboratory model indicate that if the source is below the dividing streamline height, then the plume does impact on the mountain front. If the source is above this critical height, the plume crosses the barrier and surface concentrations vary inversely with the height of the source (Snyder and Hunt 1984).

Regional-scale pollutant dispersal and deposition is now being analysed using two- and three-dimensional mesoscale models (see Chapter 3, p. 189). Mass conserving models, such as that described by Sherman (1978) have been used to model the effects of varying atmospheric conditions on pollutant distribution (Lange 1978).

Mesoscale primitive equation models in one and two dimensions have been available for a number of years (see Pielke 1984, 1985). Microscale processes that affect the dispersion of pollutants are now being incorporated into such models. A two-stage approach to modelling pollutant transport can be employed, where estimates of wind components and their departures obtained from a

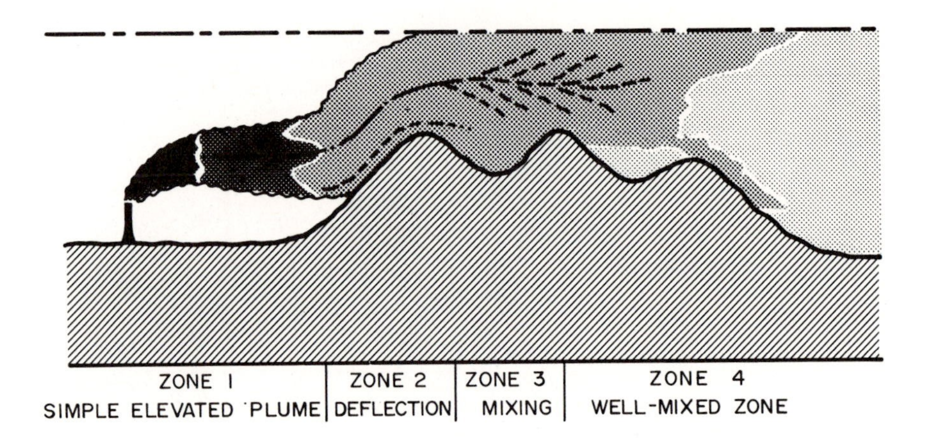

Figure 6.4 The dilution of an airborne plume flowing across nearby elevated terrain. Four zones of plume behaviour and predicted vertical mass distribution are shown; effluent concentrations are indicated by the shading intensity

Notes: Zone 1 'Simple' elevated plume without buoyant rise, becoming bent-over, Near-Gaussian vertical distribution; Zone 2 Deflection zone with the plume nearly parallel to the ground surface. Near-Gaussian vertical distribution; Zone 3 Transitional or mixing zone affected by topographically induced turbulence. Quasi-Gaussian vertical distribution; Zone 4 Well-mixed zone. Quasi-uniform vertical distribution.

Source: After Start *et al.* 1974

mesoscale model are combined with microscale turbulence statistics and used as input to a Lagrangian (i.e. trajectory following) model of particle dispersion (Pielke *et al.* 1987). However, sufficient field data to validate such models fully is seldom available. A specific regional study for the Swiss Jura is described by Beniston (1987). A mesoscale model is developed for an area 20 × 20 km, with a horizontal resolution of 500 m and vertical resolution of 250 m. This is coupled to a microscale model for the city of Bienne (4 × 4 km) with corresponding resolutions of 100 m and 10 m. Plume trajectories are calculated and Beniston shows the effects of channelling by the Jura on synoptic flow directions, as well as the influence of cloud cover and the urban heat island.

The modelling of pollutant dispersal in the boundary layer over complex terrain follows one of three general approaches. Gaussian plume models, mass conservation (box) models, and three-dimensional numerical models based on the diffusion equations (Deardorff 1978; Greenland 1979). The plume model is applicable for the determination of pollutant concentration downwind from a source (at the ground, or elevated). Its application depends on the determination of low-level stability which is a function of cloud cover, solar radiation and wind velocity. The mass conservation model assumes that surface emissions over an area are fully mixed in a specified air volume or box (Venkataram 1978). Concentration changes in the box, typically prescribed by the valley sides and an inversion lid, are determined in terms of an emission rate, mean wind speed, a mixing depth, and the air volume flowing through the box. This approach has been applied to simple mountain valley cases (Greenland 1979) and in complex regional models.

In a broad-scale study for Switzerland, Furger *et al.* (1989) utilize information on the height, depth, intensity and duration of inversions, and 850 mb winds for selected weather situations to map typical patterns of low-level airflow during conditions of anticyclonic circulation, bise, föhn, cold front passages and westerly airflow. Low-level air trajectories were determined by equivalent potential temperature analyses.

A regional air quality model (LIRAQ) using the mass conservation equations integrated vertically to the base of the boundary layer inversion has been developed particularly for use in the San Francisco Bay area by MacCracken *et al.* (1978). Two versions treat the transport and dispersion of photochemical species (non-reactive and reactive). The model requires topographic data, source emission rates, initial and lateral boundary pollution concentrations, and meteorological information (wind, inversion base height, diffusion coefficients and radiative flux). The motion and inversion information are supplied via the MASCON model of Dickerson (1978). Sensitivity studies show that emissions, meteorological factors and vertical boundary conditions are significant for regional scale pollutant concentrations, while initial and horizontal boundary conditions, as well as sub-grid scale effects, are very important for local air quality prediction (Duewer *et al.* 1978).

An important consideration in any pollutant modelling is the availability of emissions data for settlements, industries and traffic in mountain valleys. Additionally, information on winds and mixing depths in the valley must be obtained from observations or modelled, based on comparable studies elsewhere.

Existing regulatory models of the Environmental Protection Agency (EPA) in the United States typically consider 'worst-case' scenarios using simple dispersion formulae such as the Gaussian plume. Hanna and Strimaitis (1990) provide a useful summary of EPA-approved models and the results of field and laboratory experiments to evaluate them. These have been performed for hills (Cinder Cone Butte, Idaho), ridges (Hogback, New Mexico) and narrow valleys (Brush Creek, Colorado, and the Geysers area, California). The dividing streamline concept (p. 67) has been tested in these experiments and good agreement shown with theory. The various investigations suggest that the modelling of pollutant diffusion is satisfactory (within $\pm$ 30 per cent of mean and maximum observed concentrations) *only* for simple terrain situations (Hanna and Strimaitis 1990). Moreover, the accuracy is heavily dependent on the availability of suitable input data (e.g. 12 months of hourly observations of wind speed and direction, mixing depth, and cloudiness). Typically, only the first impact of a plume is modelled and separation zones in the lee of a hill are not treated.

REFERENCES

Human bioclimatology

Baker, P.T. (1969) 'Human adaptation to high altitude', *Science*, 163 (3872), 1149–56.

Baker, P.T. (1976) 'Work performance of highland natives', in P.T. Baker and M.A. Little (eds) *Man in the Andes: A Multidisciplinary Study of the High-Altitude Quechua*, vol. 1. pp. 300–14, Stroudsburg, Penn., Dowden, Hutchinson & Ross, Inc.

Baker, P.T. and Little, M.A. (1976) *Man in the Andes: A Multidisciplinary Study of the High-Altitude Quechua*, vol. 1, Stroudsberg, Penn., Dowden, Hutchinson & Ross, Inc.

Baldwin, H. and Smithson, P.A. (1979) 'Wind chill in upland Britain', *Weather*, 34, 294–306.

Buskirk, E.R. (1969) 'Decrease in physical work capacity at high altitude', in A.H. Hegnauer (ed.) *Medical Climatology*, pp. 204–22, Baltimore, Waverly Press.

Carlson, L.D. (1964) 'Reactions of man to cold', in E. Licht (ed.) *Medical Climatology*, pp. 196–228, Baltimore, Waverly Press.

Dixon, J.C. (1991) 'Wind chill – it's sensational'. *Weather*, 46, 141–4.

Folk, G.E., Jr. (1966) *Introduction to Environmental Physiology*, pp. 77–136, Philadelphia, Zudedn, Lea & Febiger.

Greenland, D. (1977) 'Living on the 700 millibar surface: The Mountain Research Station of the Institute of Arctic and Alpine Research', *Weatherwise*, 30, 233–8.

Grover, R.F. (1974) 'Man living at high altitudes', in J.D. Ives and R.G. Barry (eds) *Arctic and Alpine Environments*, pp. 817–30, London, Methuen.

Hammer, N., Koch, E. and Rudel, E. (1989) 'Klimagerechte Bekleidung im Hochgebirge', *Wetter u. Leben* 41, 235–41.

Hanna, J.M. (1976) 'Natural exposure to cold', in P.T. Baker and M.A. Little (eds) *Man in the Andes: A Multidisciplinary Study of the High-Altitude Quechua*, vol. 1, pp. 315–31, Stroudsburg, Penn., Dowden, Hutchinson & Ross, Inc.

Hart, J.S. (1957) 'Climate and temperature-induced changes in the energetics of homeotherms', *Rev. Can. Biol.* 16, 133.

Heath, D. and Williams, D.R. (1977) *Man at High Altitude*, Edinburgh, Churchill Livingstone.

Hock, R.J. (1970) 'The physiology of high altitude', *Sci. Am.*, 222, 52–62.

Lahiri, S. (1974) 'Physiological responses and adaptations to high altitude', in D. Robertshaw (ed.) *Environmental Physiology*, vol. 7, pp. 271–311, London, Butterworths.

Little, M.A. (1976) 'Physiological responses to cold', in P.T. Baker and M.A. Little (eds) *Man in the Andes: A Multidisciplinary Study of the High-Altitude Quechua*, vol. 1, pp. 332–62, Stroudsburg, Penn., Dowden, Hutchinson & Ross, Inc.

Little, M.A. (1981) 'Human population of the Andes: The human science base for research planning', *Mountain Res. Dev.*, 1, 145–70.

Little, M.A. and Baker, P.T. (1976) 'Environmental adaptations and perspectives', in P.T. Baker and M.A. Little (eds) *Man in the Andes: A Multidisciplinary Study of the High-Altitude Quechua*, vol. 1, pp. 405–28, Stroudsburg, Penn., Dowden, Hutchinson & Ross, Inc.

Mitchell, D. (1974) 'Physical basis of thermoregulation', in D. Robertshaw (ed.) *Environmental Physiology*, vol. 7, pp. 1–32, London, Butterworths.

Pugh, L.G.C. (1966) 'Clothing insulation and accidental hypothermia in youth', *Nature*, 209, 1281–6.

Reeves, J.T., Grover, R.F., Weil, J.V. and Hackett, P. (1981) 'Pulmonary ventilation in peoples living at high altitude', in *Geoecological and Ecological studies of Qinghai-Xizang Plateau*, vol. 2, 1377–83. Beijing: Science Press.

Richner, H. (1979) 'Possible influences of rapid fluctuations in atmospheric pressure on human comfort', in S.W. Tromp and J.J. Bouma (eds) *Biometeorological Survey, Vol. 1*, Part A, pp. 96–102, London, Heyden.

Schmidt, W. (1930) '*Föhnerscheinungen und Föhngebiete*, Innsbruck, Lindauer Verlag.

Siple, P.A. and Passel, C.F. (1945) 'Measurements of dry atmospheric cooling in subfreezing temperatures', *Proc. Am. Phil. Soc.*, 89, 177–99.

Smith, A.U. (1970) 'Frostbite, hypothermia and resuscitation after freezing', in A.U. Smith (ed.) *Current Trends in Cryobiology*, pp. 181–208, New York, Plenum Press.

Smithson, P.A. and Baldwin, H. (1979) 'The cooling power of wind and its influence on human comfort in upland areas of Britain', *Arch. Met. Geophys. Biokl.* B, 27, 361–80.

Steadman, R.G. (1971) 'Indices of windchill of clothed persons', *J. appl. Met.*, 10, 674–83.

Terjung, W.H. (1970) 'The energy budget of man at high altitudes', *Int. J. Biomet.*, 14, 13–43.

Tromp, W.S. (1978) 'Biological effects of high altitude climate and its therapeutic applications', *Veröff. Schweiz. Met. Zent.*, 40, 100–3.

Van Wie, C.C. (1974) 'Physiological responses to cold environments', in J.D. Ives and R.G. Barry (eds) *Arctic and Alpine Environments*, pp. 805–16, London, Methuen.

Ward, M. (1975) *Mountain Medicine: a Clinical Study of Cold and High Altitudes*, St. Albans (New York, 1976).

Webster, A.J.F. (1974a) 'Physiological effects of cold exposure', in D. Robertshaw (ed.) *Environmental Physiology*, vol. 7, pp. 33–69, London, Butterworths.

Webster, A.J.F. (1974b) 'Adaptation to cold', in D. Robertshaw (ed) *Environmental Physiology*, vol. 7, pp. 71–106, London, Butterworths.

Weather hazards: lightning

Fuquay, D.M. (1962) 'Mountain thunderstorms and forest fires', *Weatherwise*, 15, 149–52.

Fuquay, D.M. (1980) 'Forecasting lightning activity level and associated weather', *USDA. For. Serv. Res. Pap.* INT-244, International Forest and Range Experimental Station, Ogden, Utah.

Karr, T.W. and Wooten, R.L. (1976) 'Summer radar echo distribution around Limon, Colorado', *Mon. Weather Rev.*, 106, 728–34.

Peterson, A.E. (1962) 'Lightning hazards to mountaineers', *Am. Alp. J.*, 13 (36), 143–54.

Reap, R.M. (1986) 'Evaluation of cloud-to-ground lightning data from the western United States for the 1983–84 summer seasons', *J. Clim. appl. Met.*, 25, 785–99.

Snow avalanches

Aulitsky, H. (1978) 'State in the avalanche-zoning methods today', *Mitt. Forstl. Bundes-Versuchsanstalt*, 125, 129–43.

Bovis, M.J. (1977) 'Statistical forecasting of snow avalanches, San Juan Mountains, southern Colorado, U.S.A'., *J. Glaciol.* 18 (78), 87–99.

Frutiger, H. (1977) 'Avalanche damage and avalanche protection in Switzerland', in *Avalanches, Glaciol. Data Rep. GD-5*, pp. 17–32, Boulder, Colorado, World Data Center-A for Glaciology.

Judson, A. (1976) 'Colorado's avalanche warning program', *Weatherwise*, 29, 268–77.

Judson, A. (1977) 'The avalanche warning program in Colorado', *Proc. 45th Annual Meeting Western Snow Conf.*, 19–27.

LaChapelle, E. and Armstrong, R.L. (1976) 'Nature and causes of avalanches in the San

Juan Mountains', in R.L. Armstrong and J.D. Ives (eds) *Avalanche Release and Snow Characteristics, San Juan Mountains, Colorado*, Inst. Arct. Alp. Res. Occ. Pap. no. 19, pp. 23–40, Boulder, University of Colorado.

Mellor, M. (1968) *Avalanches*, Cold Reg. Sci. and Eng., Part. III-A3d, Hanover, N.H., US Army, Cold Reg. Res. and Eng. Lab.

Perla, R.I. (1980) 'Avalanche release, motion and impact', in S.C. Colbeck (ed.) *Dynamics of Snow and Ice Masses*, pp. 397–462, New York, Academic Press.

Perla, R.I. and Martinelli, M., Jr. (1975) *Avalanche Handbook*, Fort Collins, Colorado, US Department of Agriculture Forestry Service.

Williams, K. (1975) *The Snowy Torrents. Avalanche Accidents in the United States 1967–71*, Rocky Mt. For. Range Exp. Sta., Tech. Rep. RM-8, Fort Collins, Colorado, US Department of Agriculture, Forest Service.

Air pollution in mountain regions

Ball, F.K. (1960) 'Control of inversion height by surface heating', *Q.J.R. Met. Soc.*, 86, 483–4.

Barr, S., Lunne, R.E., Clements, W.E. and Church, H.W. (1977) *Workshop on Research Needs for Atmospheric Transport and Diffusion in Complex Terrain*, Albuquerque, New Mexico.

Beniston, M. (1987) 'A numerical study of atmospheric pollution over complex terrain in Switzerland', *Boundary-Layer Met.*, 41, 75–96.

Carson, D.J. (1973) 'The development of a dry inversion-capped convectively unstable boundary layer', Q.J.R. Met. Soc., 99, 450–7.

Clements, W.E. and Barr, S. (1978) 'Atmospheric transport at a site dominated by complex terrain', in *Fourth Symposium on Turbulence, Diffusion and Air Pollution*, pp. 430–5, Boston, American Meteorological Society.

Davidson, B. (1963) 'Some turbulence and wind variability observations in the lee of mountain ridges', *J. appl. Met.*, 2, 463–2.

Davidson, B. and Rao, P.K. (1963) 'Experimental studies of the valley-plain wind', *Int. J. Air Water Poll.*, 7, 907–23.

Deardoff, J.W. (1978) 'Different approaches toward predicting pollutant dispersion in the boundary layer, and their advantages and disadvantages', *WMO Symposium on Boundary Layer Physics applied to specific Problems of Air Pollution*, WMO no. 510, pp. 1–8, Geneva, World Meteorological Organization.

Dickerson, M.H. (1987) 'MASCON, a mass-consistent atmospheric flux model for regions with complex terrain.' *J. appl. Met.*, 17, 241-53.

Duewer, W.H., MacCracken, M.C. and Walton, J.J. (1978) 'The Livermore Regional Air Quality Model: II. Verification and sample application in the San Francisco Bay area', *J. appl. Met.*, 17, 273–311.

Furger, M., Wanner, H., Engel, J., Troxler, F.X. and Valsangiacomo, A. (1989) Zur Durchluftung der Täler und Vorlandsenken der Schweiz. *Geogr. Bernensia*, P20. Bern, University of Bern.

Greenland, D.E. (1979) *Modelling air pollution potential for mountain resorts*, Inst. Arct. Alp. Res., Occas. Pap. no. 32, Boulder, University of Colorado.

Gudiksen, P.H. and Shearer, D.L. (1989) 'The dispersion of atmospheric tracers in nocturnal drainage flows', *J. appl. Met.*, 28, 602–8.

Hanna, S.R. and Strimaitis, D.G. (1990) 'Rugged terrain effects on diffusion', in W. Blumen (ed.) *Atmospheric Processes over Complex Terrain, Met. Mongr.* 23 (45), pp. 109–43, Boston, Amer. Met. Soc.

Howard, E.A. and Fox, D.G. (1978) 'Modeling mountain valley airsheds, in *Fourth*

Symposium on Turbulence, Diffusion and Air Pollution, pp. 182–8, Boston, American Meteorological Society.

Lange, R. (1978) 'ADPIC – a three-dimensional particle-in-cell model for the dispersal of atmospheric pollutants and its comparison to regional tracer studies', *J. appl. Met.*, 17, 320–9.

MacCracken, M.C., Wuebbles, J.D., Walton, J.J., Duewer, W.H. and Grant, K.E. (1978) 'The Livermore Regional Air Quality Model I: Concept and Development', *J. appl. Met.* 17, 254–72.

Nappo, C.J., Shankar Rao, K. and Herwethe, J.A. (1989) 'Pollutant transport and diffusion in katabatic flow', *J. appl. Met.*, 28, 617–25.

Orgill, M.M. (1989) 'Early morning ventilation of a gaseous tracer from a mountain valley', *J. appl. Met.*, 28, 636–51.

Pasquill, F. (1961) 'The estimation of the dispersion of windborne material'. *Met. Mag.*, 90, 33–49.

Pielke, R.A. (1984) *Mesoscale Meteorological Modeling*, Orlando, FL, Academic Press.

Pielke, R.A. (1985) 'The use of mesoscale numerical models to assess wind distribution and boundary-layer structure in complex terrain', *Boundary-Layer Met.*, 31, 217–31.

Pielke, R.A., Arritt, R.W., Segal, M., Moran, M.D. and McNider, R.T. (1987) 'Mesoscale numerical modeling of pollutant transport in complex terrain', *Boundary-Layer Met.*, 41, 59–74.

Reiter, R. (1988) 'Messreihen atmosphärische Spurenstoffe an alpinen Station Südbayerns zwischen 0.7 und 3 km Höhe', *Wetter u. Leben*, 40, 1–30.

Reiter, R., Sladkovic, R. and Kanter, H.-J. (1987) 'Concentrations of trace gases in the lower troposphere simultaneously recorded at neighbouring alpine stations. Part III. Sulfur dioxide, nitrogen dioxide and nitric oxide', *Met. Atmos. Phys.*, 37, 114–28.

Scorer, R.S. (1978) *Environmental Aerodynamics*, Chichester, Ellis Horwood.

Sherman, C.A. (1978) 'A mass-consistent model for wind fields over complex terrain', *J. appl. Met.*, 17, 312–19.

Snyder, W.H. and Hunt, J.C.R. (1984) 'Turbulent diffusion from a point source in stratified and neutral flows around a three-dimensional hill'. *Atmos. Environ.*, 18, 1069–2002.

Start, G.E. Dickson, C.R. and Wendell, L.L. (1974) *Effluent Dilutions over Mountainous Terrain*, NOAA Tech. Mem. ERL ARL-51, Idaho Falls, Idaho.

Start, G.E., Dickson, C.R. and Wendell, L.L. (1975) 'Diffusion in a canyon within rough mountainous terrain', *J. appl. Met.*, 14, 333–46.

Tennekes, H. (1973) 'A model for the dynamics of the inversion above a convective boundary layer', *J. Atmos. Sci.*, 30, 558–66.

Tennekes, H. (1976) 'Observations on the dynamics and statistics of simple box models with a variable inversion lid', in *Proceedings of the Third Symposium on Atmospheric Turbulence, Diffusion and Air Quality*, pp. 397–402, Boston, American Meteorological Society.

Venkataram, A. (1978) 'An examination of box models for air quality simulation', *Atmos. Environment* 12, 1243–9.

Whiteman, C.D. (1989) 'Morning transition tracer experiments in a deep narrow valley', *J. appl. Met.*, 28, 626–35.

Whiteman, C.D. and McKee, T.B. (1977) 'Observations of vertical atmospheric structure in a deep mountain valley', *Arch. Met. Geophys, Biokl.*, A, 26, 39–50.

Whiteman, C.D. and McKee, T.B. (1978) 'Air pollution implications of inversion descent in mountain valleys', *Atmos. Env.*, 2, 2151–8.

7

CHANGES IN MOUNTAIN CLIMATES

EVIDENCE

Throughout the description and analysis of climatic regimes in mountain areas, the question of climatic change has so far been ignored. There is considerable evidence, however, of important climatic fluctuations on time scales of human significance and, since the mountain environment is, in many respects, marginal for human activities, it is necessary to understand and take account of such fluctuations (Barry, 1990).

It is of interest to know what changes mountain regions have experienced. Mountain observations can provide an indication of changes in the free atmosphere for a much longer period than sounding data, but up to now such records have received limited attention. Measurements on the Zugspitze and the Sonnblick in the Alps provide such long-term series, but even at these stations problems due to changes of instrumentation or observing practice introduce inhomogeneities into the data (Hauer 1950: 81). Nevertheless, the unique 100-year record on the Sonnblick, Austria (Auer *et al.* 1990) shows the well-known early twentieth-century warming between 1900 and 1950 of about 2°C in spring and summer and 2.5°C between 1910 and 1960 in October (Figure 7.1). In winter, however, no clear trend is apparent although the 1970s were warmer than the early part of the century. This contrast with the general northern hemisphere trend (Folland *et al.* 1990), where the warming is most pronounced in winter, may reflect the free-air lapse rate conditions at the mountain summit.

The documented increases in greenhouse gases (carbon dioxide, methane, chlorofluorocarbons and nitrous oxide) since 1958 represent a global increase in heating of just over 1 Wm^{-2} and the accumulated increases since the beginning of the Industrial Revolution (*c.* AD 1800) have contributed in excess of 2 $W\ m^{-2}$ i.e. 1 per cent of the average absorbed solar radiation of 240 $W\ m^{-2}$ (Hansen *et al.* 1990). Some offsetting cooling may have occurred during the last 20 years as a result of the observed decrease in stratospheric ozone; increasing levels of tropospheric aerosols, predominantly sulphur dioxide, also represent a net cooling effect through the increase in planetary albedo, particularly over the oceans. These changes in atmospheric composition and projected trends are

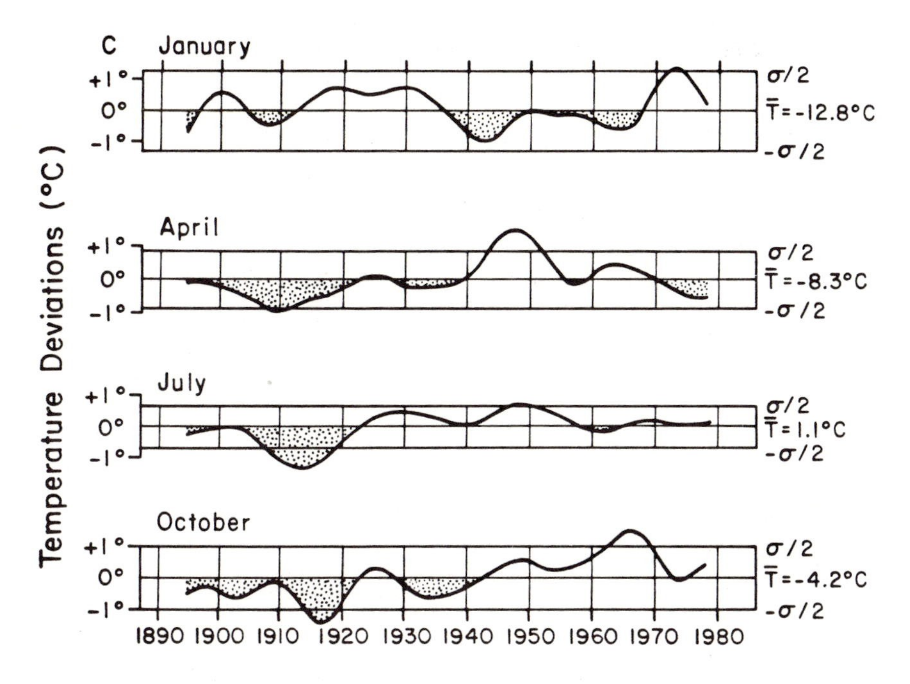

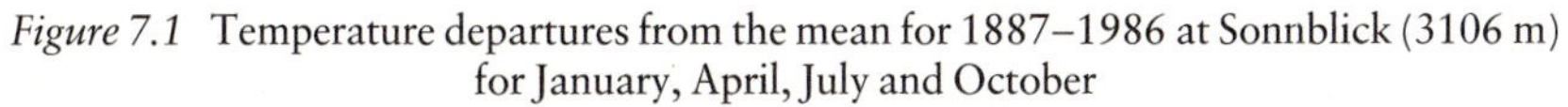

Figure 7.1 Temperature departures from the mean for 1887–1986 at Sonnblick (3106 m) for January, April, July and October

Notes: Departures are smoothed by a 20-year Gaussian moving average. Monthly mean values (°C) and the standard deviation (σ)/2 are indicated

Source: After Auer *et al.* 1990

fairly well established. However, the issue of whether or not the increased concentrations of greenhouse gases can account for a significant part of the observed 20th century global warming of between 0.3°C and 0.6°C is unresolved (Schlesinger 1991). Some of the warming may represent a cessation of Little Ice Age conditions, for example, while the twentieth century fluctuations probably have natural and anthropogenic components (Folland *et al.* 1990).

The records from mountain observatories also enable us to compare the degree of parallelism and relative amplitudes of trends in the mountains and on the adjacent lowlands. Analyses illustrating trends of sunshine and snowfall in lowland and mountain regions of Austria (Steinhauser 1970, 1973) are among the most complete such records available. Annual sunshine totals at Sonnblick (3106 m) and Villach (2140 m) show fluctuations similar in timing and amplitudes to those at four lowland stations (Figure 7.2). This agreement holds also on a seasonal basis, except in winter as a result of inversions associated with lowland fog and stratus. Variations in snow cover duration in Austria since 1900 show inter-regional differences with stations in western Austria and the southern Alps displaying patterns that are different from those in the northern Alpine Foreland and northeastern Austria (Figure 7.3). Temperature trends at Sonnblick

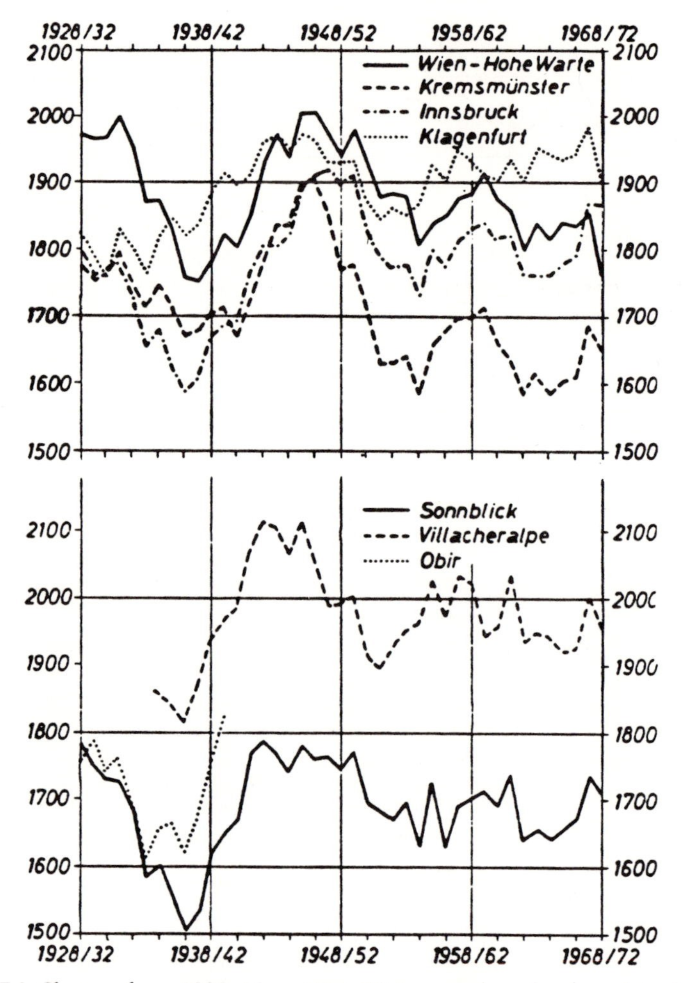

Figure 7.2 Changes from 1928–32 to 1968–72 in annual totals of sunshine hours at four stations in the Austrian lowlands (above) and at three mountain stations (below), plotted as five-year moving averages
Source: From Steinhauser 1973

for 1887–1979, compared with Hohenpeissenberg (994 m) and lowland stations in the surrounding region suggest that differences between them reflect spatial gradients rather than altitudinal contrasts (Lauscher 1980).

Temporal changes of climate variables in the Alps show some interesting contrasts with altitudinal effects. At Sonnblick (3106 m) the average number of days with snow cover during May–September decreased from 82 days for 1910–25 to only 53 days for 1955–70 (Böhm 1986). During the same interval, mean summer temperatures rose about 0.5°C. However, if we compare the difference in days with snow cover over an altitude difference of 100 m (corresponding to a

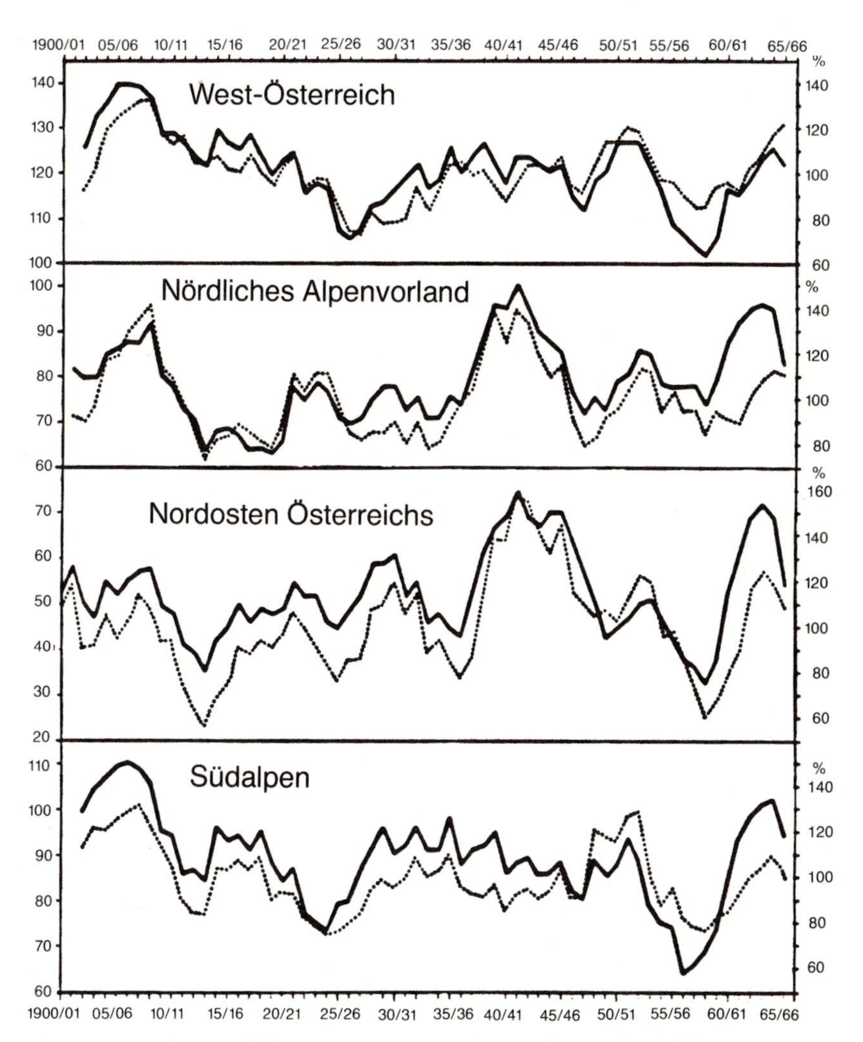

Figure 7.3 Secular trends from 1900–1 to 1965–6 in the number of days with snow cover (solid line, left) and total new snowfall as a percentage of the 1900–1 and 1959–60 mean value (dotted line, right) for four regions in Austria: western Austria, northern Alpine Foreland, north-east Austria, and southern Alps
Source: From Steinhauser 1970

mean lapse rate of 0.5°C/100 m), a decrease of only 10–11 days, rather than 29 days, would be expected in the eastern Alps. Evidently, non-linearities associated with snow melt effects and summer snowfalls must be involved. Snow cover duration at high and low elevations is often poorly correlated (Pfister, 1985).

As a consequence of the limited observational data, climatic proxy sources have to be used. One such line of evidence is provided by the occurrence of tree stumps and logs above present timberlines. In southern California and Nevada, dead tree remains are found up to 150 m above the present upper timberlines, indicating more favourable growing season conditions in the past (LaMarche and Mooney 1972; LaMarche 1973). Wood material can be dated by ^{14}C techniques and by cross-dating of the tree ring chronology with other trees. A detailed account of dendroclimatic methods of reconstructing past climates is given in Fritts (1976).

Remnants of bristlecone pine (*Pinus longaeva* Bailey) on Mt. Washington in eastern Nevada show a 100 m downward shift of the upper tree line zones from their positions between about 4000 and 2000 years ago (LaMarche and Mooney 1972). This can be interpreted as reflecting either cooler summers, more moist summers, or both. Summer moisture plays a significant role in reducing the preconditioning of these trees to winter desiccation stress. For the White Mountains, California, Lamarche (1973) demonstrates three times of timber line lowering: 3500 BP, 2500 BP and 850–450 BP. By comparing changes in different locations, he argues that the first and last change represent primarily cooling, whereas the one around 2500 BP was related to drier conditions.

In the North Island of New Zealand there is evidence for an upward migration of *Nothofagus menziesii* (beech) into the sub-alpine scrub and alpine grassland belts during the last 100 years or so, suggesting warmer and sunnier conditions recently (Burrows and Greenland 1979). However, these authors also note that in other parts of North Island, trees near their upper altitudinal limit are predominantly old and show signs of imminent decline, with little indication that regeneration is occurring. In South Island, the recent rise in snow line has allowed some migration of trees into gullies where seedlings were formerly unable to establish themselves (Wardle 1973). Yet at the same time, a less reliable snowcover encourages frost-heaving of soils and vegetation and this instability may be responsible in part for an observed lowering of the upper limit of continuous alpine grassland (Burrows and Greenland 1979). Clearly, botanical evidence for climatic fluctuations must be interpreted with care since independent or interrelated biotic and pedologic factors may also affect tree growth and regeneration.

The most widely available evidence of climatic change in mountains is provided by alpine glaciers. Snow accumulation and ablation represent more direct responses to climatic parameters than does tree growth, but there are nevertheless considerable problems in developing a precise climatic interpretation of changes observed in glaciers (Paterson 1981). Changes in glacier mass balance are caused by the *net* effect of changes in winter accumulation (involving snow fall and wind drifting) on the one hand, and summer ablation by sublimation and melt (involving the length and warmth of the thaw season, radiation conditions, and wind, humidity and temperature conditions above the glacier) on the other. A glacier's mass is therefore determined by the local micro- and topo-climatic conditions and there is no necessary simple relationship between these and the large-scale climate. Moreover, the glacier's terminal position is what is most commonly recorded and the relationship between mass balance changes and glacier extent is itself complex, involving the bedrock profile, ice thickness and iceflow properties (Nye 1965; Paterson 1981). In glaciers where there is no summer melt, the long-term accumulation can be determined from ice cores. For Quelccaya Ice Cap at 5670 m in the Peruvian Andes, the mean annual temperature is −3°C and there appears to be no melt or evaporation. Thompson *et al.* (1985) identify extended dry regimes during AD 570–610, 1250–1310, and 1720–1860; the periods AD 1500–1720 and to a lesser degree 1870–1984 were wet. At present the wet season, which accounts for about 80 per cent of the annual total, occurs between November and April with convective activity from an easterly direction.

Glaciers in the European Alps have been a subject of scientific interest since the late eighteenth century and in many localities documentary records and landscape paintings provide a means of determining glacier extent and thickness during earlier times. Figure 7.4 for example, illustrates the fluctuations of the Lower Grindelwald Glacier since 1590. While care must be taken in interpreting such a record, as discussed above, in this case it is possible to compare the evidence with temperature observations at Basel from 1755 onward. Messerli *et al.* (1978) show that the post-1860 retreat is correlated with increasing temperatures in spring and autumn. Shorter-term fluctuations in summer temperatures are also involved, especially post-1935. Data on winter accumulation are required to complete the interpretations, but these are unfortunately not available. Since the 1890s, numerous Swiss glaciers have been monitored. The percentage advancing, stationary, or retreating shows distinct differences over the last nine decades (Figure 7.5). Advances in the 1890s are attributed to cool cloudy summers, which diminish ablation, whereas the advances during 1912–20 also represent the effect of a number of winters with heavy accumulation (Hoinkes 1968). The maximal recession during the 1945–54 decade was associated with higher temperatures, increased sunshine duration and reduced precipitation at Alpine stations according to Rudloff (1962). The re-advances during the late 1960s–early 1980s are a result of slightly cooler summers, but these advances which affected small glaciers, have already largely ceased in the Alps. Similar sharp decreases in the rate of recession of glaciers in the Canadian Rocky Mountains during the 1970s–early 1980s also match the records of cooling at neighbouring valley stations (Luckman 1990). It is important to note, however, that summary statistics of glacier terminus fluctuations, such as Figure 7.5, are biased by the larger number of small glaciers. In terms of glacier mass, which is dominated by large valley glaciers, the twentieth century downward trend is essentially uninterrupted except in some maritime coastal ranges where accumulation has increased (Haeberli *et al.* 1989).

The various records for the Alps have enabled LeRoy Ladurie (1971) to characterize the last millenium as follows: a brief cooling with glacier advances took place between about AD 1200–1300, followed by a pronounced cool interval and glacier maximum between about AD 1550 and 1850. This period, known as the Little Ice Age appears to have been a global phenomenon, although

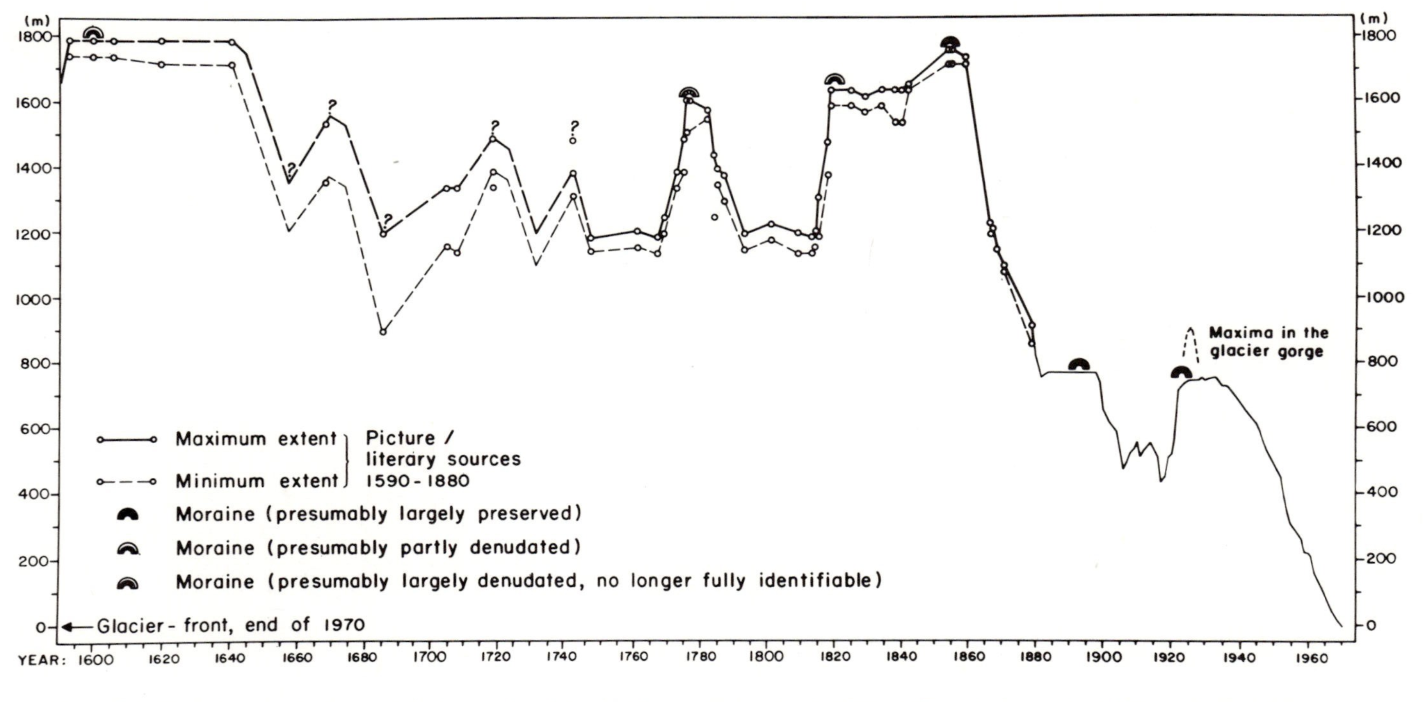

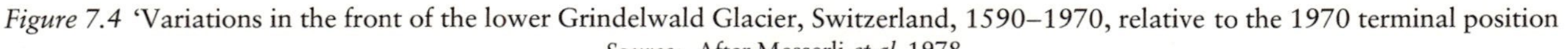

Figure 7.4 'Variations in the front of the lower Grindelwald Glacier, Switzerland, 1590–1970, relative to the 1970 terminal position

Source: After Messerli *et al.* 1978

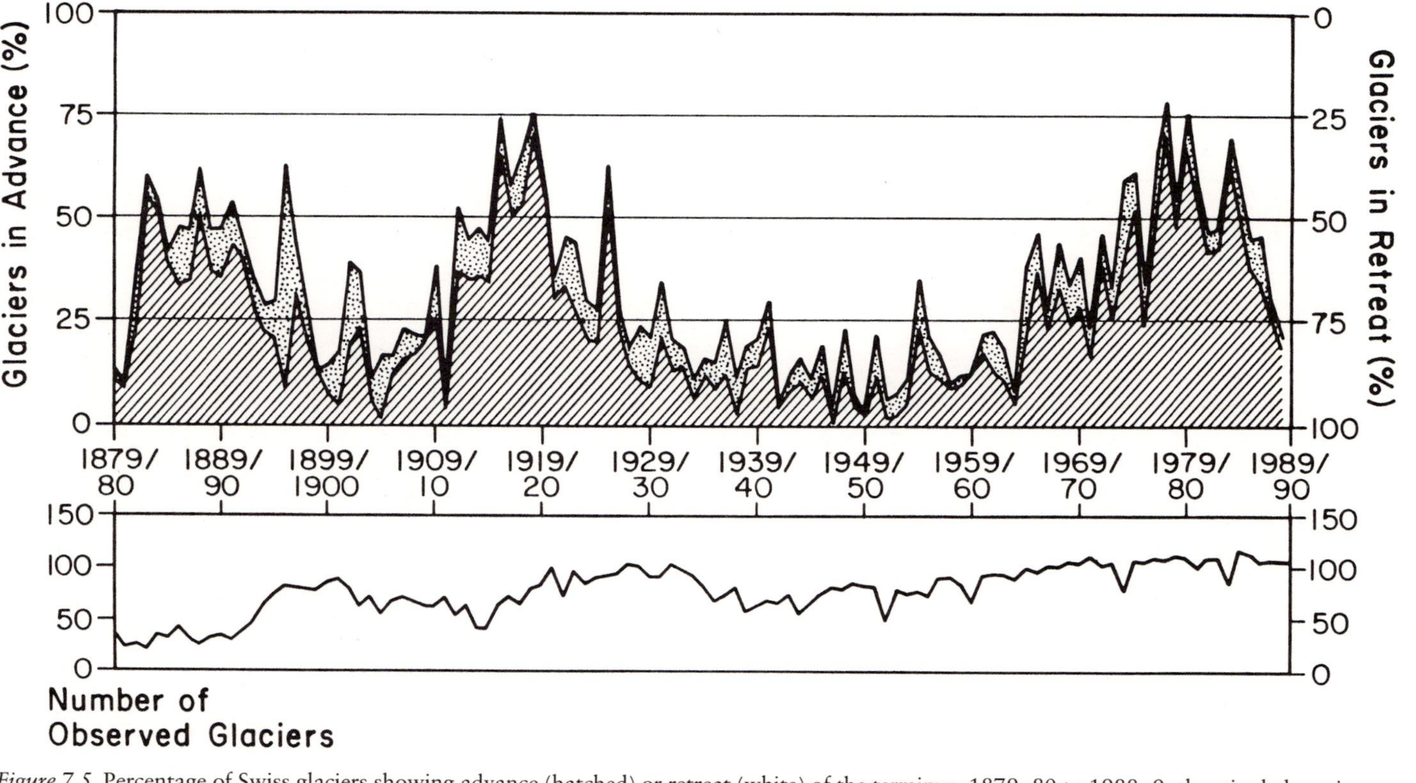

Figure 7.5 Percentage of Swiss glaciers showing advance (hatched) or retreat (white) of the terminus, 1879–80 to 1988–9; the stippled portion denotes stationary glaciers

Note: The number of observed glaciers in each year is shown below

Source: Based on data provided by the World Glacier Monitoring Service, Zurich

it was perhaps not entirely synchronous. In Norway, floods, landslides, avalanches and glacier encroachment on farms also reached a peak between AD 1680 and 1750, according to Grove (1972). However, in the Southern Alps of New Zealand, glacier maxima have been recognized in the thirteenth, fifteenth, seventeenth, mid-eighteenth and late-nineteenth centuries (Burrows and Greenland 1979). The glacier fluctuations since the seventeenth century suggest quasi-periodic climatic fluctuations of 30–60 years duration.

The glacier recession in New Zealand, which began about 1900 and still continues, is the most substantial for at least 1800 years. It appears to be associated with a temperature rise of 1°C, implying that the earlier fluctuations were of similar or lesser magnitude (Burrows and Greenland 1979). In the Himalayan area, Mayewski and Jeschke (1979; Mayewski *et al.* 1980) also find that a general glacial retreat has taken place since AD 1850, but with local differences according to regional and topo-climatic influences. The twentieth-century warming is a global characteristic, it was interrupted from the 1940s–1970s when some cooling occurred in the northern hemisphere, but has resumed in the 1980s to reach higher levels than have been previously recorded. The linear trend between 1890 and 1989 amounts to 0.50°C/100 years (Folland *et al.* 1990). Calculated changes in the dimensions of the small equatorial glaciers of Mt Carstenz (Mt Jaya) at 4°S in western New Guinea (Irian Jaya) have been used in conjunction with a glacier mass balance model by Allison and Kruss (1977) to estimate recent climatic changes there. The calculations indicate a temperature rise of 0.6°C 100 y^{-1} since the glacial maximum of the mid-nineteenth century.

SIGNIFICANCE

The interdisciplinary studies reported above remind us of the temporal dimension of climate in mountains. Scenarios of climatic conditions for the predicted warmer earth resulting from the continued increase in greenhouse gases are numerous. Experiments with GCMs commonly compare the standard concentration of carbon dioxide (nominally 300–325 ppm) with a doubled amount. Taking account of the increases in other greenhouse gases, an equivalent doubling is likely to be attained about AD 2030–2050. Most GCM experiments suggest an associated 3°C ± 1.5°C increase in mean global surface air temperature, with an approximate two- to three-fold amplification in polar latitudes (Schlesinger and Mitchell 1987). The corresponding values at about 3 km (700 mb) in middle latitudes imply that in summer, the lower tropospheric lapse rate is more or less unchanged, whereas in winter it is steeper due to greater warming at the surface. It is, however, uncertain what the temperature changes are likely to be at the ground surface in mountain regions. Moreover, the anticipated changes in moisture balance and snow cover are even less certain. Comparison of GCM simulations, station data and field measurements in July 1988 in the mountains of Ladakh and Kashmir by Brazel and Marcus (1991) illustrate the discrepencies and problems of interpreting GCM results in mountain areas. These arise from the coarse resolution and topographic smoothing in the model simulations, as well as regional and local heating effects over arid surfaces that are not captured by the models. Possible approaches to resolve this problem using mesoscale models in a GCM are noted in Chapter 3 (p. 192).

Calculations of the effect of specified changes on Alpine glaciers by Kuhn (1989), based on a scenario of a 3°C summer warming and an increase in annual accumulation of 100 kg m^{-1} for AD 2050, show an expected rise in altitude of the mean equilibrium line by 190 m (to 3000 m). This would translate into a reduction of almost 50 per cent in ice-covered area in Austria. Such changes in mountain climate regimes may be of considerable practical significance. In countries like New Zealand, Norway, and Switzerland where hydro-electric power is the primary energy source, changes in snowfall and glacier run-off could have serous economic implications. Shortening of the snow-cover season or drastic glacier retreat could also have long-term consequences for the skiing 'industry' and tourism in alpine countries. In the Swiss Canton of Grison, for example, Maisch (1987) estimates that snowline has risen 77 m ± 51 m between 1850 and the present day. Direct information on snowcover conditions at Weissflujoch (2540 m), Switzerland, since 1940 indicates that the disappearance of spring snow cover has got later as a result of cool spring weather and late season snowfalls (Föhn 1989). There is also no long-term trend in snow depth on January 1 since 1935 at Weissflujoch, nor at Davos (1540 m) since 1892. However, Föhn notes that the last 5 to 7 years suggest a diminution in January 1 snowdepth, as a result of a delay in onset of winter snowfalls. Such a trend if it persists could be serious for the ski-industry due to the loss of business over Christmas–New Year. The delay in snow ablation in spring could impact Alpine agriculture by curtailing the early growing season. It would also delay the peak runoff in Alpine river basins, although this could have beneficial effects. Changes in other climatic elements could also have practical implications. Warmer sunnier conditions, as well as favouring tree growth at higher altitudes, could reduce conventional heating costs for buildings and perhaps make solar installations more cost-effective.

As a general outcome of the increasing international concern about climatic variability and change, various climatic 'bench mark' monitoring programmes are now being established. The much-cited record of carbon dioxide trend obtained at Mauna Loa Observatory, Hawaii, is one illustration of the value of permanent mountain stations (Keeling *et al.* 1976). Monitoring of other trace gases and aerosol levels is also needed, however, on both a local and global basis. Likewise, for energy assessments, better networks of solar radiation and wind measurements are essential in many mountainous countries. Pairs of mountain and lowland reference stations, where all climate variables are monitored, need to be established (Cehak 1982). This has become an even more pressing concern in the context of international programmes on Global Change. For example, ultraviolet radiation measurements (0.29–0.33 μ m) at the Jungfraujoch

Observatory, Switzerland, show a 1 per cent increase during 1981–9, apparently in response to the observed 3–4 per cent depletion of stratospheric ozone since 1969 (Blumenthaler and Ambach 1990). In many respects, the future for mountain weather and climate studies seems bright, but a resurgence of mountain meteorology will hopefully be accomplishd via long-term commitments instead of short-term crash programmes. It would also be regrettable if new systems such as automatic stations and remote sensing techniques were allowed to replace rather than complement conventional station observations in remote mountain locations. The histories of the Ben Nevis Observatories in Scotland and the Pike's Peak Observatory in Colorado demonstrate the impossibility of re-establishing mountain facilities once they are closed. Accordingly, it is appropriate to close this book with an expression of appreciation for the commitment of the mountain weather observers in many countries, who have so diligently maintained records under the trials of severe environments, isolation, and sometimes bureaucratic disinterest.

REFERENCES

Changes in mountain climates

Allison, I. and Kruss, P. (1977) 'Estimation of recent climatic change in Irian Jaya by numerical modeling of its tropical glaciers', *Arct. Alp. Res.*, 9, 49–60.

Auer, I., Böhm, R. and Mohnl, H. (1990) 'Die troposphärische Erwarmungsphase die 20. Jahrhunderts im Spiegel der 100-jährigen Messreihe des alpinen Gipfelobservatoriums auf dem Sonnblick', *CIMA '88 Proc. 20th Internat. Tagung f. Alpine Meteorologie* Rome, Italian Met. Service.

Barry, R.G. (1990) 'Changes in mountain climate and glacio-hydrological responses', *Mountain Res. Devel.*, 10, 161–70.

Blumenthaler, M. and Ambach, W. (1990) 'Indication of increasing solar ultraviolet-B radiation flux in alpine regions', *Science*, 248, 206–8.

Böhm, R. (1986) '*Der Sonnblick. Die 100 jährige Geschichte des Observatoriums und seiner Forschungstätigkeit*', Vienna, Osterreichischer Bundesverlag.

Brazel, A.J. and Marcus, M.G. (1991) 'July temperatures in Kashmir and Ladakh, India: Comparisons of observations and general circulation model simulations', *Mountain Res. Devel.*, 11, 75–86.

Burrows, C.J. and Greenland, D.E. (1979) 'An analysis of the evidence for climatic change in the last thousand years: evidence from diverse natural phenomena and from instrumental records', *J.R. Soc. New Zealand*, 9, 321–73.

Cehak, K. (1982) 'Alpenklimatologie – ein Statusbericht', *Wetter u. Leben* 34, 227–40.

Föhn, P.M.B. (1989) 'Climatic change, snow cover and avalanches', in *Landscape Ecological Impact of Climatic Change on Alpine Regions with emphasis on the Alps*, pp. 27–40, Netherlands, Agricultural University of Wagenigen.

Folland, C.K., Karl, T.R. and Vinnikov, K.Ya. (1990) 'Observed climate variations and change', in *Scientific Assessment of Climate Change*. Intergovernmental Panel on Climate Change, WMO/UNEP, pp. 195–242. Cambridge University Press.

Fritts, H.C. (1976) *Tree Rings and Climate*, London, Academic Press.

Grove, J.M. (1972) 'The incidence of landslides, avalanches and floods in western Norway during the Little Ice Age', *Arct. Alp. Res.*, 4, 131–8.

Haeberli, W., Muller, P., Alean, P. and Bösch, H. (1989) 'Glacier changes following the Little Ice Age – a survey of the international data base and its perspectives', in J. Oerlemans (ed.) *Glacier Fluctuations and Climate*, pp. 77–10, Dordrecht, Reidel.

Hansen, J., Rossow, W. and Fung, I. (1990) The missing data on global climate change, *Issues in Science and Technology*, 7 (1), 62–9.

Hauer, H. (1950) 'Klima and Wetter der Zugspitze', *Berichte d. Deutschen Wetterdienstes in der US-Zone*, 16.

Hoinkes, H. (1968) 'Glacier variation and weather', *J. Glaciol.*, 7, 3–21; 129–30.

Keeling, C.D., Bacastow, R.B., Bainbridge, A.E., Ekdahl, C.A., Guenther, P.R., Waterman, L.S. and Chin, J.F.S. (1976) 'Atmospheric carbon dioxide variations at Mauna Loa Observatory, Hawaii', *Tellus*, 28, 538–51.

Kuhn, M. (1989) 'The effects of long-term warming on alpine snow and ice', in *Landscape Ecological Impact of Climatic Change on Alpine Regions with emphasis on the Alps*, pp. 10–20, Netherlands, Agricultural University of Wagenigen.

LaMarche, V.C. (1973) 'Holocene climatic variations inferred from treeline fluctuations in the White Mountains, California', *Quat. Res.*, 3, 632–60.

LaMarche, V.C. and Mooney, H.A. (1972) 'Recent climatic change and development of the Bristlecone Pine (*P. longaeva* Bailey) krummholz zone, Mt. Washington, Nevada', *Arct. Alp. Res.*, 4, 61–72.

Lauscher, F. (1980) 'Die Schwankungen der Temperatur auf dem Sonnblick seit 1887 im Vergleich zu globalen Temperaturschwankungen', in *16 International Tagung für Alpine Meteorologie* (Aix-les Bains), Soc. Météorol. de France, Boulogne-Billancourt, pp. 315–19.

LeRoy Ladurie, E. (1971) *Times of Feast, Times of Famine: A History of Climate Since the Year 1000* (transl. B. Bray), Garden City, New York, Doubleday & Co.

Luckman, B.H. (1990) Mountain areas and global change: a view from the Canadian Rockies. *Mountain Res. Devel.*, 10, 183–95.

Maisch, M. (1987) 'Die Gletscher an "1850" und "Heute" im Bündnerland und in den angrenzden Gebieten: Untersuchungen zur Höhenlage, Veränderungen und räumlich Struktur von Schneegrenzen', *Geogr. Helvet.*, 42, 127–45.

Mayewski, P.A. and Jeschke, P.A. (1979) 'Himalayan and Trans-Himalayan glacier fluctuations since AD 1812',. *Arct. Alp. Res.*, 11, 267–87.

Mayewski, P.A., Pregent, G.P., Jeschke, P.A. and Ahmad, N. (1980) 'Himalayan and Trans-Himalayan glacier fluctuations and the South Asian monsoon record', *Arct. Alp. Res.*, 12 (2), 171–82.

Messerli, B., Messerli, P., Pfister, C. and Zumbühl, H.J. (1978) 'Fluctuations of climate and glaciers in the Bernese Oberland, Switzerland, and their geoecological significance, 1600 to 1975', *Arct. Alp. Res.*, 10, 247–60.

Nye, J.F. (1965) 'A numerical method of inferring the budget history of a glacier from its advance and retreat', *J. Glaciol.*, 5, 589–607.

Paterson, W.S.B. (1981) *The Physics of Glaciers*, 2nd edn., Oxford, Pergamon Press.

Pfister, C. (1985) Snow cover, snowlines and glaciers in central Europe since the 16th century, in M.J. Tooley and G.M. Sheail, eds., *The Climate Scene*, pp. 155–74, London, G. Allen and Unwin.

Rudloff, H. (1962) 'Die Klimaschwankungen in den Hochalpen seit Beginn der Instrumenten-Beobachtungen', *Arch. Met. Geophys. Biokl.*, B, 13, 303–51.

Schlesinger, M.E. (ed.) (1991) *Greenhouse-gas-induced Climatic Change: A Critical Appraisal of Simulations and Observations*, New York, Academic Press.

Schlesinger, M.E. and Mitchell, J.F.B. (1987) 'Climate model simulations of the equilibrium climatic response to increased carbon dioxide', *Rev. Geophys.*, 25, 760–98.

Steinhauser, F. (1938) *Die Meteorologie des Sonnblicks, Teil I*, Vienna, J. Springer.

Steinhauser, F. (1970) 'Die säkularen Änderungen der Schneedeckenverhältnisse in Osterreich', 66–7, *Jahresbericht des Sonnblick-Vereines, 1968–1969*, Vienna, 1–19.

Steinhauser, F. (1973) 'Die Änderungen der Sonnenscheindauer in Osterreich in neurer Zeit, 68–69, *Jahresbericht des Sonnblick-Vereines, 1972–1973*, Vienna, 41–53.

Thompson, L.G., Mosley-Thompson, E., Bolzan, J.F. and Koci, B.R. (1985) 'A 1500-year record of tropical precipitation in ice cores from the Quelccaya Ice Cap, Peru', *Science*, 229, 971–3.

Wardle, P. (1973) 'Variations of the glaciers of Westland National Park and the Hooker Range, New Zealand', *N.Z.J. Bot.*, 11, 349–88.

APPENDIX

SYSTÈME INTERNATIONAL (SI) UNITS

Quantity	*Dimensions*	*SI*	*cgs metric*	*British*
Length	L	m	10^2 cm	3.2808 ft
Area	L^2	m^2	$10^4 cm^2$	10.764 ft^2
Volume	L^3	m^3	$10^6 cm^3$	35.314 ft^3
Mass	M	kg	10^3 g	2.205 lb
Density	ML^{-3}	$kg\,m^{-3}$	$10^{-3} g\,cm^{-3}$	
Time	T	s	s	
Velocity	LT^{-1}	$m\,s^{-1}$	$10^2 cm\,s^{-1}$	2.24 mph
Acceleration	LT^{-2}	$m\,s^{-2}$	$10^2 cm\,s^{-2}$	
Force	MLT^{-2}	newton (N or $kg\,m\,s^{-2}$)	10^5 dynes ($10^5 g\,cm\,s^{-2}$)	
Pressure	$ML^{-1}T^{-2}$	pascal ($N\,m^{-2}$)	10^{-2} mb	
Energy, Work	ML^2T^{-2}	joule (J or $kg\,m^2 s^{-2}$)	10^7 ergs ($10^7 g\,cm^2 s^{-2}$)	
Power	ML^2T^{-3}	watt (W or $kg\,m^2 s^{-3}$)	$10^7 ergs\,s^{-1}$	1.34×10^{-3} hp
Temperature	θ	kelvin (K)	°C	1.8°F
Heat energy	ML^2T^{-2} (or H)	joule (J)	0.2388 cal	9.47×10^{-4} BTU
Heat/ radiation flux	HT^{-1}	watt (W or $J\,s^{-1}$)	0.2388 $cal\,s^{-1}$	3.412 $BTU\,h^{-1}$
Heat flux density	$HL^{-2}T^{-1}$	$W\,m^{-2}$	$2.388 \times 10^{-5} cal\,cm^{-2} s^{-1}$	

ENERGY CONVERSION FACTORS

4.1868 J= 1 calorie
$\mathrm{J\ cm^{-2}}$= 0.2388 $\mathrm{cal\ cm^{-2}}$
watt= $\mathrm{J\ s^{-1}}$
kW hr= 3.60×10^{6}J
$\mathrm{W\ m^{-2}}$= 1.433×10^{-3} $\mathrm{cal\ cm^{-2}\ min^{-1}}$
697.8 $\mathrm{W\ m^{-2}}$ = $\mathrm{cal\ cm^{-2}\ min^{-1}}$

For time sums:
Day: 1 $\mathrm{W\ m^{-2}}$ = 8.64 $\mathrm{J\ cm^{-2}\ day^{-1}}$ = $8.64 \times 10^{4}$$\mathrm{J\ m^{-2}\ day^{-1}}$ =
2.064 $\mathrm{cal\ cm^{-2}\ day^{-1}}$
Month: 1 $\mathrm{W\ m^{-2}}$ = 2.592 $\mathrm{M\ J\ m^{-2}}$ $\mathrm{(30\ day)^{-1}}$ = 61.91 $\mathrm{cal\ cm^{-2}}$ $\mathrm{(30\ day)^{-1}}$
Year: 1 $\mathrm{W\ m^{-2}}$ = 31.536 $\mathrm{M\ J\ m^{-2}\ year^{-1}}$ = 753.4 $\mathrm{cal\ cm^{-2}\ year^{-1}}$
Gravitational acceleration (g) = 9.81 $\mathrm{m\ s^{-2}}$
Latent heat of vaporization (288K) = 2.47×10^{6} $\mathrm{J\ kg^{-1}}$
Latent heat of fusion (273k) = 3.34×10^{5} $\mathrm{J\ kg^{-1}}$
Dry adiabatic lapse rate (Γ) = 9.8 $\mathrm{K\ km^{-1}}$

General Index

Author Index

Names of co-authors that appear only in the references are not included.